ORDERING AND DISORDERING IN ALLOYS

Proceedings of the European Workshop on Ordering and Disordering held in Grenoble, France, 10th-12th July, 1991.

Members of the International Scientific Committee

H. Bakker, Natuurkundig Lab., University of Amsterdam, The Netherlands
M. C. Cadeville, IPCMS, Strasbourg, France
R. W. Cahn, University of Cambridge, UK
M. T. Clavaguera-Mora, Universitat Autonoma de Barcelona, Spain
P. Desré, Institut National Polytechnique de Grenoble, France
A. M. Ektessabi, Kyoto University, Japan
H. Fecht, Universität Augsburg, Germany
B. Fultz, California Institute of Technology, USA
G. Inden, Max-Planck Institut, Dusseldorf, Germany
C. C. Koch, North Carolina State University, USA
G. Martin, CENS Gif-sur-Yvette, France
H. Mori, Osaka University, Japan
A. R. Yavari, CNRS et l'Institut National Polytechnique de Grenoble, France

Members of the Organizing Committee

D. Abert, INPG, Grenoble
M. D. Baro, Universitat Autonoma de Barcelona, Spain
S. Gialanella, University of Cambridge, UK
A. Motta, Centre d'Etude Nucléaire de Grenoble, France
A. R. Yavari *Chairman*, CNRS et l'Inst. National Polytechnique de Grenoble

Executive Assistant

V. Repellin, Grenoble, France

ORDERING AND DISORDERING IN ALLOYS

Edited by

A.R. YAVARI

Amorphous and Microcrystalline Materials Research Group,
Institut National Polytechnique de Grenoble,
Saint Martin d'Hères, France.

ELSEVIER APPLIED SCIENCE
LONDON and NEW YORK

ELSEVIER SCIENCE PUBLISHERS LTD
Crown House, Linton Road, Barking, Essex IG11 8JU, England

Sole Distributor in the USA and Canada
ELSEVIER SCIENCE PUBLISHING CO., INC.
655 Avenue of the Americas, New York, NY 10010, USA

WITH 34 TABLES AND 277 ILLUSTRATIONS

© 1992 ELSEVIER SCIENCE PUBLISHERS LTD
(except pp.223-228, 372-384)

British Library Cataloguing in Publication Data

Ordering and disordering in alloys.
I. Yavari, A. R.
669

ISBN 1-85166-762-8

Library of Congress CIP data applied for

PREFACE

This conference was initially programmed in the framework of the European Science/Twinning project entitled 'Ordering Kinetics in Alloys (1989–92)' with the aim of promoting scientific debate on the mechanisms and kinetics of disordering and ordering in alloys with superlattices. Special attention was reserved for metastable states obtained by heavy deformation, radiation damage or rapid quenching.

In this perspective disordering and ordering phenomena are distinct from those of the equilibrium order–disorder transitions. Nevertheless we welcomed contributions on equilibrium phase transitions which help develop a better understanding of metastable states.

Some authors concentrated on short-range (SRO) ordering and disordering as distinct from long-range-ordering (LRO) while still others considered conditions of destruction of LRO down to the onset of amorphisation. These contributions resulted in a broader and more interdisciplinary debate and the presentation of the latest experimental findings and theoretical models.

This proceedings volume is organized in two parts. Part I concentrates on disordering and ordering in $L1_2$ and other cubic superlattice alloys while Part II covers the other mentioned areas.

We wish to acknowledge substantial financial support from the European Community and partial support from the Centre National de la Recherche Scientifique.

A. R. YAVARI

CONTENTS

Part II

PART I

MECHANISMS AND KINETICS IN ORDERING AND DISORDERING

ROBERT W. CAHN
Department of Materials Science and Metallurgy,
Cambridge University
Pembroke Street, Cambridge CB2 3QZ, England

ABSTRACT

This paper offers a bird's-eye view of the whole subject-matter of the Workshop. — The relation of the kinetics of the establishment of long-range order to the nature of thermal treatment and to the ordering energy is briefly reviewed, and some current issues highlighted. Some matters relating to the measurement of order and to the role of antiphase boundaries are discussed. — Disordering by means of intense cold-work, with special reference to ball-milling, is examined. Thermal disordering, by annealing above T_c, is compared with the aforementioned processes; there is evidence that thermal disordering is preferentially catalysed at antiphase domain boundaries. It is not known whether this also true for the alternative processes. The question is finally addressed whether an alloy partially disordered by mechanical deformation is in a different state from one partially ordered, to the same value of the LRO parameter S, by annealing an initially thermally disordered alloy.

INTRODUCTION

The limited objective of this brief introductory essay is to set the subject-matter of the Workshop in perspective. Ordering has often been examined at conferences, but *disordering* is a relatively unfamiliar topic. This is probably the first conference to combine the two topics, and that decision deserves justification. For a start, terms need to be defined.

The ambiguous term "order" can be used in several different ways. In its most general meaning, the term denotes the presence of crystallinity as opposed to an amorphous arrangement of atoms. In its more specific application, the term *long-range order* (LRO) distinguishes a crystalline solid solution in which the atomic species, two or more, reside on distinct sublattices. Such LRO can be perfect or partial, and some years ago, much debate raged as to the nature and status of states of imperfect LRO. In between these two usages, *short-range order* (SRO) distinguishes a solid in which atoms of one species have more nearest neighbours of the other species than would be found if environments were entirely random. At one time, it was common among x-ray crystallographers to describe imperfect LRO as a state combining generalised LRO with localised SRO; this did not prove very illuminating. Nevertheless, recent findings have made it urgent to analyse critically the distinct states of partially ordered alloys created by different procedures, and this has not yet been attempted.

To confuse usage further, SRO (as opposed to LRO) does not necessarily presuppose crystallinity; the recent literature includes a number of studies of SRO in metallic glasses with two metallic components.

"*Disordering*" most commonly implies the destruction, in whole or in part, of a state of LRO, but it is also used by other investigators to denote the destruction of crystallinity without concomitant melting; in that sense, disordering is another term for amorphisation. (Disordering as a term to denote the reduction of SRO is a very uncommon usage, but as we shall see in this Workshop, such reduction may be an unexpected accompaniment of experiments aimed at destroying LRO.) — It is important nowadays in describing an experiment to distinguish between "disordering" in the sense of destroying LRO and in its other sense of destroying crystallinity or amorphising, especially because in recent years it has become clear that many intermetallic compounds can only be disordered in the second sense when they have previously been disordered in the first sense.

Ordering, in the sense of inducing LRO, can only be achieved thermally, in the sense of giving the material an appropriate heat-treatment below the critical (T_c) or transition temperature (T_t). This applies to *reversibly ordered* solid solutions, for which $T_t < T_m$, where T_m is the melting temperature. Those alloys for which $T_t > T_m$ are *permanently ordered* phases, and in their case the only thermal treatment needed to achieve LRO is to freeze the melt. (Physical purists insist that "T_c" be reserved for ordering transitions of second order such as that in CuZn [this conventional double usage of "order" does not exactly help clarity!] while "T_t" is to be used for ordering transitions of the first order such as that in Cu_3Au, which are the most common kind. Apoplexy can be avoided by keeping to the symbol T_t.)

Variation of the state of SRO in those alloys which do not possess LRO or in which LRO is kinetically hindered by a low T_t has not been examined a great deal, except in a few favoured alloys of which the most important has been Cu-15at.% Al. For many years, the metallurgical literature was full of papers in which the degree of SRO of solid solutions was measured without any specified heat-treatment having been applied beforehand, and this is still the situation in many studies of SRO in metallic glasses. This is a nonsense; SRO is not a characteristic of a particular solid solution alone, but of the solid solution and its thermal treatment in combination.

Disordering, unlike ordering, can be achieved either *thermally*, by heating a reversibly ordered alloy above T_t (or above T_c), or by *mechanically* displacing atoms from their ordered sites; that, in turn, can be achieved either by intense mechanical deformation, or by using ions, electrons or photons as missiles. All these mechanisms find their place in our Workshop.

ORDERING KINETICS

Until recently, almost all studies of long-range ordering kinetics have been done with reversibly ordered alloys, initially put into the disordered (or SRO) state by quenching from above T_t. The literature includes studies of ordering kinetics mainly in Cu_3Au, Ni_3Fe, Ni_3Mn, FeCo and Ni_4Mo, and the physical properties which have been used as measures, direct or indirect, of the degree of LRO include x-ray and neutron diffraction, electrical resistivity (both at temperature and after repeated quenching to room temperature), elastic modulus, Mössbauer spectrometry, differential scanning calorimetry and ferromagnetic property measurements. Dilatometry, in spite of the fact that it is a particularly simple and direct method, has not been favoured.

Such kinetic experiments have been subdivided into two categories — d-o and o-o experiments. d-o experiments are those in which the alloy initially is disordered — i.e., has zero LRO — and the rate of isothermal increase of the Bragg order parameter S (or some quantity linked to S) at a temperature below T_t is measured. O-o experiments are those in which the alloy is initially partially ordered ($0 < S_o < 1$) and then annealed isothermally at a temperature at which the equilibrium value of S is either greater or less than S_o. Most physicists doing these experiments have considered o-o experiments easier to analyse.

In several alloys, it has been clearly established that ordering is fastest at a temperature 30-100 K below T_t; the TTT (time-temperature-transformation) isotherms have a "nose", in metallurgical parlance, at this favoured temperature. This is commonly observed in metallurgical transformations generally, and is simply explained as representing a favourable balance between the increase of the driving force as ΔT, the depression of the annealing temperature below T_t, increases, and the diminution of diffusivity which accompanies the increase of ΔT. Separate from this is the phenomenon of *critical slowing down*, which is the name given to an extreme slowing of ordering or disordering, by orders of magnitude, just below or above T_t; this phenomenon can only be interpreted in terms of the theory of critical phenomena, and it has been little investigated experimentally. It has been clearly demonstrated in Ni_3Mn and Cu_3Au.

Small amounts of ternary (doping) additions can have a major effect on ordering rates; examples of this are Al in Ni_3Fe (which has an accelerating effect) and V in FeCo (which slows down ordering). The reasons for these effects have not been examined. There have also been experiments on broad ternary series of alloys, such as $Ni_3(Fe,Al)$, which are exemplified in this Workshop.

Special interest attaches to the few experiments which have been done on "difficult" alloys with a tendency to form LRO but such low values of T_t that ordering is kinetically impossible. Au_3Cu, Cu_3Zn, Cu_3Al and NiFe are examples of such alloys. Some notable experiments were done here in Grenoble 30 years ago by Paulève and Dautreppe, in Louis Néel's famous laboratory, on the acceleration of ordering in such an alloy (NiFe) by neutron irradiation during annealing; this was at a time when neutrons were plentiful in Grenoble. In this way, LRO was achieved in NiFe, which was estimated to have a T_t of 321°C. The T_t values for the Cu alloys are even lower. — Ordered NiFe has also been observed in meteorites which are estimated to have cooled at about 1K/Myr, which is rather difficult to match on earth.

A number of theories for the kinetics of establishment of LRO were published during the period 1955-182 and were briefly surveyed by the present author in 1987. The formal theories all refer to B2 compounds (in which all atoms have unlike nearest neighbours); rigorous treatment of $L1_2$ (Cu_3Au-type) phases is much more difficult because some atoms have like neighbours. I attempted to relate the relative ordering kinetics of different $L1_2$ alloys, in a non-rigorous way, to the known diffusivities in these phases. It appears that the kinetics are more closely related to the ordering energy (which constitutes the driving force for ordering) than to diffusivities, and a given number of diffusive jumps creates more order in an alloy like Ni_3Al which has a high ordering energy than in an alloy like Cu_3Au whose ordering energy is low. The relation between ordering and diffusion mechanisms is treated in one of the papers at this Workshop.

High concentrations of excess vacancies created by quenching from a high temperature in the disordered domain, or by cold deformation, accelerate ordering kinetics in a transient fashion, and this has been studied in particular in Cu-Pt alloys. The matter is alluded to in this Workshop.

Ordering mechanisms have received a good deal of theorists' attention in recent years. A distinction is made between continuous, or homogeneous, ordering and heterogeneous ordering, and spinodal mechanisms can play a major part. Heterogeneous ordering is nucleated at identifiable locations, notably grain boundaries, and is often associated with grain boundary migration. Such heterogeneous mechanisms have been experimentally studied in detail in several alloys, including CuPt, Ni_2V and FeCo.

When a disordered alloy is ordered below T_t, a population of antiphase domains is formed. In alloys which order without change of symmetry, such as B2 and Ll_2 types, antiphase domain boundaries (APDBs) merely have displacements such as $1/2<111>$ and associated local disorder; in other alloys in which the symmetry changes on ordering, such as Ni_4Mo, the domain structure is far more complicated, because unique axes (e.g., tetragonal, orthorhombic) can have different orientations between adjacent domains; therefore domains are either of the displacement type or of the misorientation type. This second ki nd of domain has been extensively studied by HREM, by Amelinckx's group. — What difference these two modes of domain morphology make to ordering kinetics does not appear to be known. — When an alloy is permanently ordered (i.e., freezes in ordered form) then generally no, or very few and coarse, antiphase domains are observed. This is what would be expected, since antiphase domains are nucleated during ordering in the solid state. Domains can also form during the process of primary recrystallization below T_t after plastic deformation, but again this differs from one alloy to another and the process is imperfectly understood.

More generally, ordering kinetics and recrystallization kinetics have an involved interrelationship, which has been thoroughly analysed for the case of FeCo and Ni_4Mo.

THERMAL AND MECHANICAL DISORDERING

There have been only a few studies of *disordering* kinetics associated with changes of temperature (i.e., thermal disordering); the subject is ripe for more attention. d-o kinetics have been systematically studied in Cu_3Au, the classical "workhorse" for order-disorder studies since the 1920s. —What has been increasingly examined during the past few years is the mechanism by which regions of disorder are nucleated and it is now clear that antiphase domain boundaries (of the displacement type) in the ordered phase furnish heterogeneous nucleation sites for disordering when the temperature is raised above T_t. Such APDBs are sometimes observed to thicken progressively as disordering proceeds. This is not surprising since such boundaries are sites of localised disorder even below T_t. The nucleation of disorder at APDBs has been studied both experimentally and theoretically, the latter via the concepts of the theory of critical phenomena. That theory has also been used to interpret the known experimental fact that ordered alloys may disorder preferentially at free surfaces; that type of observation has however also been interpreted in terms of CVM theory, of the kind which is used in one of the papers at this Workshop.— Several of those who have studied heterogeneous disordering at APDBs are participating in the Workshop.

In addition to the process of thermal disordering by heating an ordered alloy above T_t , it is also possible to disorder reversibly ordered alloys by quenching from above T_t. With some alloys possessing a fairly high T_t, such as $(Co_{78}Fe_{22})_3V$ with $T_t = 910°C$, fairly drastic quenching is required, while others, notably with Ti_3Al with T_t well over 1000°C, have not hitherto been quench-disordered at all. It would be exceedingly interesting to be able to assess the mechanical properties of disordred Ti_3Al! — Recently, it has been found possible by several groups to disorder a permanently ordered phase, Ni_3Al, by drastic quenching from the melt (after many unsuccessful earlier attempts to achieve this), and this work will be presented at the Workshop.

Disordering by mechanical deformation, which nowadays generally means milling in planetary or vibratory ball mills, is a new topic: research on this arose primarily from a strong interest in solid-state amorphization together with the recognition, some years ago, that amorphization by electron irradiation required destruction of atomic order as a prerequisite; it was realized that intense plastic deformation can achieve this just as effectively as various forms of irradiation. This recognition, which owes much to Koch who is participating in the Workshop, has almost crowded out irradiation as the preferred mode of mechanical disordering. (As mentioned before, I like to regard irradiation, including ion implantation, as a form of mechanical damage by minute projectiles). Koch and Jang studied this process in Ni_3Al, the granddaddy of all permanently ordered phases, and also reported remarkable measurements of a very steep increase of microhardness as disordering proceeded, peaking for $S = 0.5$. Other studies on Ni_3Al, reported at this Workshop, demonstrate that the lattice parameter goes on increasing for a long time after all detectable signs of LRO have disappeared. Clearly, milling can attack SRO after it has finished with LRO.

The mechanism by which plastic deformation destroys LRO, especially in its early stages, has not received much attention; only recently has the mechanical disordering of Ni_3Al been closely studied by TEM. Other studies suggest that the simple Bragg LRO parameter, usually denoted by S, may be too blunt an instrument to characterize different states of partial LRO in an appropriately subtle manner. Thus, studies of rolled specimens of an $L1_2$ alloy, $(Co,Fe)_3V$, annealed below T_t , show that the recovery behaviour is drastically different according as the alloy was ordered or disordered before it was rolled; again, the alloy has different mechanical properties according as a given value of S is achieved by mechanical plus thermal treatment or by thermal treatment alone. It seems that the old concern with a critical analysis of states of partial LRO needs to be revived, and that more sophisticated order parameters, such as the Warren parameters, need to be pressed into service. Probably, it will be necessary to combine x-ray diffraction with TEM to make worthwhile progress.

Another recent discovery is that an alloy which was disordered by mechanical milling cannot be compeletely thermally reordered unless the temperature is high enough to recrystallize the deformed alloy. This fact has not yet been properly analyzed.

Most research on mechanical disordering has been done with permanently ordered phases containing no antiphase domains. Many such phases of practical interest, such as Zr_3Al, Ti_3Al, NiAl, remain to be examined, and the rate and even feasibility of mechanical disordering should be examined as a function of ordering energy. Further, it would be interesting to apply various forms of mechanical disordering to phases such as Cu_3Au or (again) Ti_3Al which contain a population of APDBs, to see whether disorder is nucleated at APDBs in mechanical disordering as it is in thermal disordering.

FINAL REMARKS

The above outline has, I hope, shown that disordering presents as many fascinating scientific problems as does ordering, and to that some degree the two mirror-image processes can be expected to shed light on each other. I hope and expect that this Workshop will prove to be such a source of light.

A full bibliography of the topics touched upon here would require several times as much space as this short paper and is in any case unnecessary, since the research papers at this Workshop will furnish many such citations. The very short bibliography appended to this paper merely lists a very few papers   which may help the reader to focus on some of the unsolved problems offered by this fascinating double field of research. Papers covering subject-matter treated elsewhere in the Workshop are generally not cited.

BIBLIOGRAPHY

Burns, F.P. and Quimby, S.L., Ordering processes in Cu_3Au. Phys. Rev., 1958, **97**, 1567-1575.

Cahn, R.W., Ordering kinetics and diffusion in some $L1_2$ alloys. In: Phase Transitions in Condensed Systems—Experiments and Theory, ed. G. S. Cargill III, F.Spaepen and K.-N.. Tu, Mat. Res. Soc. Symp. Proc. Vol. 57 (1987), pp. 385-404.

Cahn, R.W. and Johnson, W.L., Review: the nucleation of disorder. J. Mater. Res. 1986, 1, 724-732.

Cahn, R.W., Takeyama, M., Horton, J.A. and Liu, C.T., Recovery and recrystallization of the deformed, orderable alloy $(Co_{78}Fe_{22})_3$. J. Mater. Res. 1991, 6, 57-70.

Corey, C.L. and Potter, D.I., Recovery processes and ordering in Ni_3Al. J. Appl. Phys., 1967, 38, 3894-3900.

Collins, M.R. and Teh, H.C., Neutron scattering observations of critical slowing down of an Ising system. Phys. Rev. Lett., 1973, 30, 781-784.

Horton, J.A., Baker, I. and Yoo, M.H., Slip-plane disordering in stoichiometric Ni_3Al. Phil. Mag. A. 1991, 63, 319-335.

Irani, R.S. and Cahn, R.W., The mechanism of crystallographic ordering in CuPt. J. Mater. Sci., 1973, 8, 1453-1472.

Jang, J.S.C. and Koch, C.C., Amorphization and disordering of the Ni_3Al ordered intermetallic by mechanical milling. J. Mater. Res. 1990, 5, 498-510.

Koch, C.C., Mechanical milling and alloying. In: Processing of Metals and Alloys, ed. R.W. Cahn, VCH Velagsgesellschaft, Weinheim, 1991, pp. 193-287.

Mitsui, K., Mishima, Y. and Suzuki, T., Composition dependence of ordering kinetics in Cu_4Pt. Phil. Mag. A, 1986, 54, 501-516. The role of excess vacancies in two-stage ordering in ordered alloys. Ibid, 1989, 123-140.

Morris, D.G., Brown, G.T., Piller, R.C. and Smallman, R.E., Ordering and domain growth in Ni_3Fe. Acta Metall., 1976, 23, 21-28.

Paulève, J., Dautreppe, D., Laugier, J. and Néel, L., Irradiation-assisted ordering of NiFe. Compt. Rend. Acad. Sci. (Paris), 1962, 254, 965-968.

Russell, K.C., Irradiation and alloy crystallinity. In: Proceedings of the International Seminar on Solute-Defect Interaction, Kingston, Ont., August 1985, Pergamon Press, Oxford.

Soffa, W.A. and Laughlin, D.E., Decomposition and ordering processes involving thermodynamically first-order order→disorder transformations. Acta Metall., 1989, 3019-3028.

Tanner, L.E. and Leamy, H.J., The microstructure of order-disorder transitions. In: Order-Disorder Transformations in Alloys, ed. H. Warlimont, Springer-Verlag, Berlin, 1974, pp. 180-239.

Weisberg, L.R. and Quimby, S.C., Ordering kinetics of Cu_3Au. Phys. Rev., 1958, 110, 338-348.

Yamauchi, H. and de Fontaine., D., Kinetics of order-disorder. In: Order-Disorder Transformations in Alloys, ed. H. Warlimont, Springer-Verlag, Berlin, 1974, pp. 148-179.

MAGNETIC PROPERTIES OF DISORDERED $L1_2$-Ni_3Al + Fe

A.R. YAVARI

LTPCM-CNRS UA29, Institut National Polytechnique de Grenoble, BP 75, DomaineUniversitaire 38402, St Martin d'Hères, France.

P. CRESPO, E. PULIDO and A. HERNANDO
Instituto de Magnetismo Aplicado 28230 Las Rozas , Madrid, Spain

G. FILLION and P. LETHUILLIER
Laboratoire Louis Néel, CNRS, B.P. 166X 38042 Grenoble cedex, France

M.D. BARO and S. SURINACH
Dept de Fisica, Universitat Autonoma de Barcelona 08193 Bellaterra, Spain

ABSTRACT

Magnetic properties of Ni_3Al-Fe alloys disordered by mechanical grinding have been measured both at low and high temperatures. While the equilibrium high temperature disordered state is paramagnetic, our metastable disordered state is ferromagnetic. Magnetic moment and Curie temperatures of the disordered state have been determined for several compositions. They are lower than those of the ordered state and vary strongly with concentration and especially with Fe content. The weaker ferromagnetism of the disordered state is attributable to a lower number of Fe-Ni nearest neighbors in the disordered fcc state.

INTRODUCTION

While the study of the disordered state of $L1_2$-Ni_3Al-Fe and Ni_3Al alloys obtained by liquid quenching is recent [1,2], the disordered state of Ni_3Al was obtained by mechanical deformation several decades ago [3]. Once the metastable disordered state is obtained, its transport properties and the mechanism and kinetics of its reordering can be studied by various techniques including x-ray and neutron diffraction, calorimetry, dilatometry, TEM observation of antiphase boundaries, resistivity measurements etc. Intensities of superlattice lines in Ni_3Al are quite weak both for x-ray and neutron diffraction and resistivity measurements are difficult on powder samples. We have therefore introduced magnetic measurements as a new technique for the study of ordering of the metastable Ni_3Al disordered fcc state [4]. Previously the disordered state had been available only at high temperatures at equilibrium when it is paramagnetic [5]. Here we will see that magnetic measurements are fully complementary with calorimetric measurements. Furthermore we find that the interesting intrinsic magnetic properties of the disordered fcc state are of theoretical interest.

EXPERIMENTAL TECHNIQUE

The principal alloys were prepared in a cold-crucible with special equipment to avoid aluminium loss or oxidation. They were then melt-spun into ribbon form, annealed to 1000 K to obtain an ordered, brittle state and subsequently disordered by mechanical grinding using a Fritsch Pulverisette-7 planetary mill with vials modified to allow vacuum of 10^{-5} torr before being filled by 1.5 bars of pure argon gas. X-ray diffraction showed that the disordered state was obtained after 4 to 8 hours of grinding depending on the experimental conditions and alloy composition (see figure 1). For Ni_3Al, magnetic measurements indicated significant iron contamination after 24 hours of milling. Most of the calorimetric measurements were obtained on the DSC-7 of the Barcelona group. Low temperature magnetic properties were measured using a SQUID in Madrid and high-T magnetic measurements were performed on the furnace-equiped magnetometer of the Laboratoire Louis Néel in Grenoble.

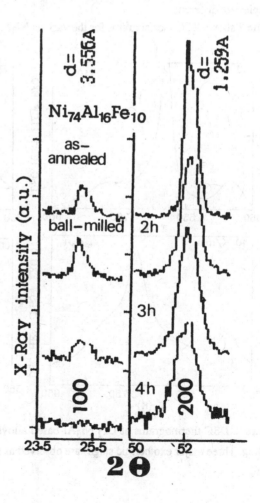

Figure 1. X-ray diffraction patterns showing disappearance of the 100 superlattice peak of $Ni_{74}Al_{16}Fe_{10}$ after 4 hours of ball-milling [4].

RESULTS AND DISCUSSIONS

As it was previously reported [4], the extent of disorder in $L1_2$-type alloys is more easily reproducible when induced by mechanical grinding than when quenched-in by rapid solidification [1]. Furthermore the kinetics of reordering are different in the disordered states obtained by the two techniques. This is understandable since :

a) excess point and line defect densities are expected to attain higher levels during grinding.

b) chemical short-range order (CSRO) in the disordered fcc-phase is difficult to suppress by rapid quenching while it can be destroyed by low-temperature plastic deformation.

c) interface (grain boundary surface) densities are much higher when heavily deformed samples become nanocrystalline [6, 7] while rapidly quenched Ni_3Al-Fe alloys have grain sizes of the order of 0.5 to 1 μm [1].

Consequently the calorimetric signals during the reordering of the two types of disordered samples are different.

Figure 2 shows DSC thermograms for the various Ni_3Al-Fe alloys disordered by grinding.

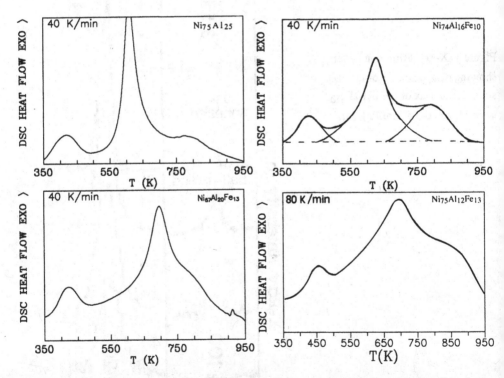

Figure 2. DSC thermograms of $L1_2$ Ni_3Al-based alloys with Fe addition, disordered by ball-milling. Three major exothermic stages are observed as seen in deconvoluted thermogram (top-right).

Three major exothermic events occur during ordering which occurs progressively above room temperature for all conpositions. Some microstructural explanations for these exothermic events have already been discussed [8].

From another point of view it can be said that the mechanism and kinetics of return to the equilibrium ordered $L1_2$ phase should include development of CSRO in the disordered fcc-γ phase, development of long-range-order (LRO or nucleation and growth of the $L1_2$ γ'-phase) and grain growth from the nanocrystalline state to a microcrystalline state. In view of the required diffusion distances, when the fully disordered alloy is reheated, these processes are expected to occur in the order, CSRO-γ, LRO-γ' and grain-growth γ' and it is tempting to attribute the three major consecutive stages of heat release to CSRO, LRO and grain-growth respectively. However it is seen that the third stage gains in relative importance with decreasing Al content and we are not aware of any proof that reduced Al content would increase grain boundary energy of the $L1_2$ phase or allow easier grain refinement by ball-milling. This point therefore requires further investigations.

The three processes influence magnetic properties to varying degrees. A preliminary comparison will be given here and further discussed elsewhere [9]. In this paper we will present results mainly concerning new findings on the magnetic properties of the disordered state.

Figure 3 shows the magnetisation M at H = 2 kOe versus temperature for $Ni_{74} Al_{16} Fe_{10}$ disordered by 4 hours of ball milling, during heating and subsequent cooling [4]. While the heating trace is irreproducible due to the $\gamma \rightarrow \gamma'$ transition as will be further discussed, the cooling curve is superimposed by subsequent heating and cooling cycles and indicates a Curie temperature $T_c \approx 570$ K for the ordered state [4].

While T_c of the ordered phase can be easily deduced from these results, that of the disordered phase is masked by the simultaneous decrease of M with temperature and its recovery because of the ordering reaction beginning just above room temperature. T_c of the disordered phase is therefore more difficult to measure while its magnetisation near room temperature is seen clearly to be about 70 % of that of the ordered state.

Figure 4 shows magnetisation of similarly disordered $Ni_{75} Al_{12} Fe_{13}$ during temperature cycling. Both M and T_c are higher due to the higher Fe content. For the ordered phase $T_c \approx 725 \pm 5$ K and the magnetisation of the disordered state at room temperature is about 70 % of that of the ordered state.

To determine T_c of the disordered state, low-T magnetisation was measured using the SQUID in Madrid. It turned out that the absolute M values measured at T $\approx$ 300 K were somewhat different than to those of the furnace-equiped diffractometer in Grenoble at the same selected applied field of 2 kOe.The reproductivity of the magnetisation scale (emu/g) for different samples of the same composition with each apparatus is better than 5 % for each sample. This includes variations in contamination effects, those due to changes in sample's shape and

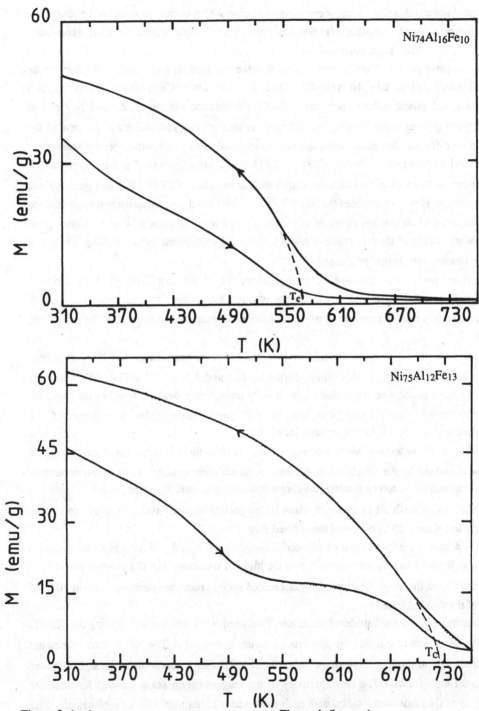

Figure 3. (top)
Magnetisation of disordered $Ni_{74}Al_{16}Fe_{10}$
versus temperature during heating and cooling
in furnace-equiped magnetometer

Figure 4. (bottom)
Magnetisation of disordered $Ni_{75}Al_{12}Fe_{13}$
versus temperature during heating and
cooling in furnace-equiped magnetometer

position in the superconducting coil that generates the applied field, errors in weight measurements as well as those due to temperature control and recording . We chose the applied field H = 2 kOe because of previous magnetisation data for ordered Ni_3Al-Fe [10] but the samples are not fully saturated at this H value and variations in H can give significant variations in the measured M. Whatever the reasons, differences in the absolute scale of M measured on the SQUID and on the magnetometer do not intervene significantly with the variations of M with T which is shown in figure 5 for the disordered state of $Ni_{74} Al_{16} Fe_{10}$ where a slight shift in the vertical scale has allowed the two sets of values to meet smoothly. The extrapolated value appears to be reliable to within ± 15 K thus yielding $T_c \approx 415$ K for this alloy in the γ-fcc disordered state.

From the lower -T data of figure 6, T_c of disordered $Ni_{75} Al_{12} Fe_{13}$ can be estimated to be about 550 K. It is seen that upon heating, M of the disordered phase first begins to decrease as expected with T/T_c for a ferromagnetic phase. We have verified that at room temperature and below, this variation is reversible. A slope change occurs (as best seen in figures 5 and 6) at T ≈ 350 K and 480 K for $Ni_{74}Al_{16}Fe_{10}$ and $Ni_{75}Al_{12}Fe_{13}$ respectively indicating the onset of a process increasing magnetisation which could be the development of CSRO in the fcc-γ phase.

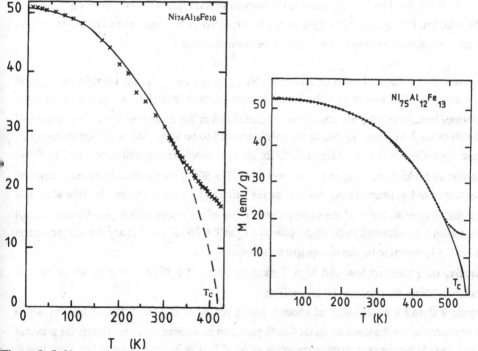

Figure 5. (left)
Magnetisation of disordered
$Ni_{74}Al_{12}Fe_{13}$ measured at low
temperatures in the SQUID (values for
T > 300 K from magnetometer)

Figure 6. (right)
Magnetisation of disordered
$Ni_{75}Al_{12}Fe_{13}$ measured at
lower T in the SQUID
(values for T > 300 K from
magnetometer)

This corresponds to the onset of the first DSC peak of their corresponding thermograms of figure 2 when adjusted for the lower heating rate ≈ 5 K/min, of the magnetometer. The subsequent decrease of the magentisation of the disordered phase at $T > T_c$ (γ) correponds to the magnetisation $M(\gamma)$ of the ordered ($L1_2$) fraction which judging by the principal (second) DSC peak, should be increasing rapidly at these temperatures. However $M(\gamma)$ of the growing ordered phase is also decreasing with increasing $T/T_c(\gamma)$ and the shape of the heating curve in the range $T_c(\gamma) < T < T_c(\gamma)$ is strongly dependent on the heating rate.

Comparison of the DSC and M versus T data in this range requires isothermal measurements which will be presented elsewhere [9].

The third DSC peak, sometimes appearing as the shoulder of the second peak, occurs out of the T-range of the furnace-equiped magnetometer. We therefore preannealed a Ni_{74} Al_{16} Fe_{10} sample as those of figures 3 and 5 by heating up to 950 K. We found that the T_c is hardly different from that of figure 2 while the magnetisation at 310 K increases by about 5 % more than the increase obtained by heating to 760 K. This additional increase in magnetisation is far below the relative contribution of the third (deconvoluted) DSC peak (figure 1) to the total heat-release. If this third peak is attributed to the recrystallisation (grain growth) from a grain-size of about 20 nm [7], its' thermal effect would be expected to be much more important than its impact on the magnetisation, as observed experimentally here.

Figure 7 shows magnetisation of disordered $Ni_{71.6}$ $Al_{20.8}$ $Fe_{7.6}$ during a similar heating and cooling cycle. Consistent with the lower Fe content, both M and T_c are lower than for the previous two alloys. Room temperature magnetisation of the disordered fcc-state is about half of that of the $L1_2$ state. T_c can be roughly estimated to be about 350 K for the ordered $L1_2$ state. From the low-T data of figure 8, T_c for the disordered state is seen to be near 320 K. No significant background magnetic signal remains at $T > 500$ K for the disordered state, however a stronger tail appears during cooling of the ordered $L1_2$ state. On an absolute scale this residual magnetisation is of the same order as those of the alloys with higher Fe content. All these alloys are expected to be single-phased $L1_2$ at $T \leq 760$ K [11, 12] and the magnetisation tail at $T > T_c$ seems to be due to supraparamagnetism.

Finally, we measured low and high T magnetisation of a Ni_3Al alloy in which Fe was introduced from the steel mill by ball milling.

Figure 9 shows a background of about 5 emu/g for NiAl after 65 hours of milling which corresponds to the presence of about 4 at % pure iron as a contaminant and confirms previous x-ray data [6] but the near room temperature value of T_c also indicates presence of some iron in solution [13].

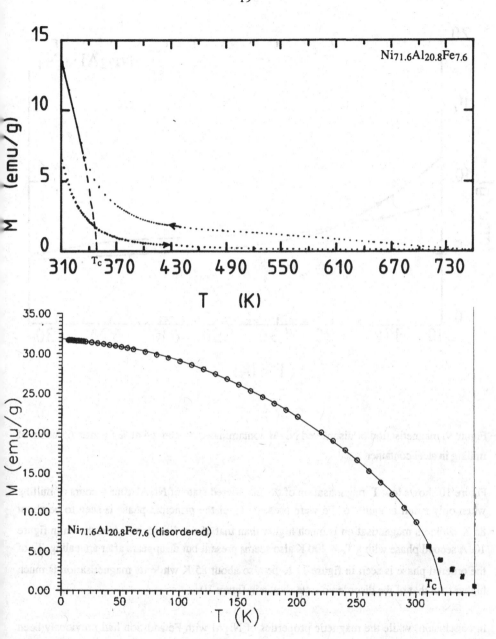

Figure 7. (top)
Magnetisation of disordered $Ni_{71.6}Al_{20.8}Fe_{7.6}$ during heating and cooling in furnace equiped-magnetometer

Figure 8. (bottom)
Magnetisation of disordered $Ni_{71.6}Al_{20.8}Fe_{7.6}$ measured at lower T in the SQUID (values for T > 300 K from magnetometer)

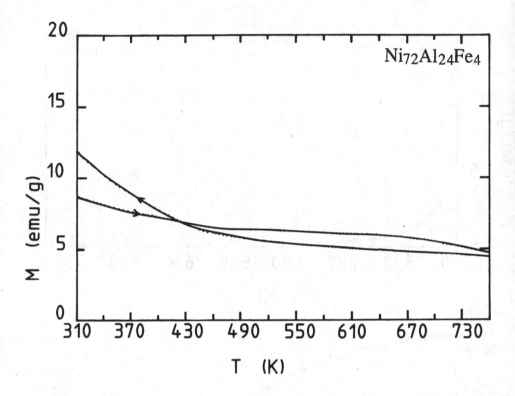

Figure 9. magnetisation of disordered Ni_3Al contaminated by about 4 at % Fe after 65 hours of milling in steel container

Figure 10 shows low T magnetisation of the disordered state of Ni_3Al after 4 hours of milling when only small amounts of Fe were present. T_c of the principal phase is seen to be about 85 K while its magnetisation is much higher than that of the disordered state shown in figure 10. A second phase with a $T_c \approx 200$ K also seems present but disappears after annealing. T_c of the ordered phase is seen in figure 11 to be also about 85 K while its magnetisation is much higher than that of the disordered state shown in figure 10.

In conclusion, while the magnetic properties of Ni_3Al with Fe addition had previously been measured in their $L1_2$ state [13] and attributed to the giant magnetic moment of the Fe-atoms, the present results open the way for a theoretical discussion of the magnetic properties of their disordered fcc-state. We have measured both magnetisation and T_c for some of these alloys as summerized in Table I. In general they are significantly lower than those of their $L1_2$ structures. This has been attributed to a lower number of Fe-Ni magnetic nearest neighbors in

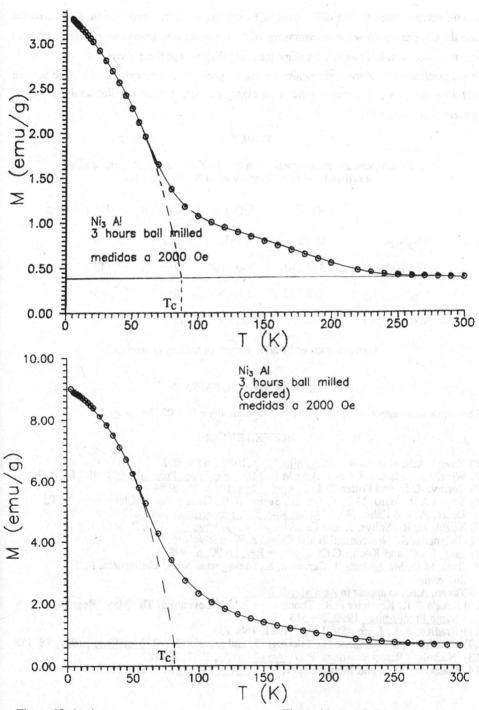

Figure 10. (top)
Low-T magnetisation of disordered Ni₃Al
after 4 hours of ball-milling

Figure 11. (bottom)
Low-T magnetisation of Ni₃Al after 4
hours of ball-milling and subsequent
annealing to reestablish L1₂ ordering

the disordered state [4]. The differences in T_c and magnetisation between the disordered fcc and the $L1_2$ phases allow the determination of the magnetic interaction energy differences that together with chemical interactions shape the phase diagrams of these alloys [14].

In conjunction with calorimetric measurements, magnetic measurements can therefore be used to follow the $\gamma \rightarrow \gamma'$ ordering kinetics with high precision. A more detailed analysis will be presented elsewhere [9].

TABLE I

Curie temperature and magnetisation at 310 K (*) or 0 K (**) and 2 kOe, of ordered and disordered states of Ni_3Al +Fe alloys

	$T_c(\gamma)$,K	$T_c(\gamma')$,K	$M(\gamma)$,emu/g	$M(\gamma')$,emu/g
$Al_{75}Ni_{25}^+$	85 ±5	≈ 85±5	3.4**	9**
$Ni_{71.6}Al_{20.8}Fe_{7.6}$	320 ±10	350 ±10	7*	13.5*
$Ni_{74}Al_{16}Fe_{10}$	415 ±15	570 ±5	30*	45*
$Ni_{75}Al_{12}Fe_{13}$	550±5	725 ±5	45*	60*

+ contains trace of Fe after 4 hours of milling in steel mill

ACKNOWLEDGEMENT

This work was supported by the European Community's SCI-0262-C project.

REFERENCES

1) Yavari, A.R. and Bochu, B., Philo. Mag.,1989 , A59, 697
2) West, J.S. , Manos, J.T and .Aziz,M.J , Mater Res. Soc. Proceedings 1990,213, 859
3) Correy, C.L. and Potter, D.I. J. Appl. Phys.,1967, 38, 3894
4) Yavari,A.R., Baro, M.D, Fillion G., Surinach, S., Gialanella, S., Clavaguera, M.T., Desré, P. and Cahn, R.W, Mater Res. Soc. Proceedings 1990,213, 859
5) Kozubski, R., Soltys, J. and Cadeville, M.C. J. Phys C., 1990, 2, 3451
6) Gialanella, S., Newcomb,S.B.and Cahn, R.W, this volume
7) Jang, J.S.C..and Koch, C.C, J. Mater Res..,1990, 5, 498
8) Baro, M.D, Malagelada, J., Surinach, S., Clavaguera, N.and Clavaguera, M.T, this volume
9) Yavari, A.R., to appear in Acta Metall. Mater
10) Sekula S.T., Kerchner H.R., Thompson J.R. and Leventouri Th., Mater Res. Soc. Symp.Proceedings 1990, 39, 513
11) Bradley, A.J., J. Iron Steel Inst. 1951, 168, 233
12) Masahashi, N., Kanbzoe H., Takasugi, T. and Izumi, O. , Z. Metallkde , 1987, 78, 788
13) Nichols, J.R.and Rawlings, R.D., Acta Metall , 1977, 25, 187
14) Cadeville, M.C. and Moran-Lopez, J. L., Physics Reports, 1987, 153, 331

KINETIC DISORDERING OF INTERMETALLIC COMPOUNDS THROUGH FIRST- AND SECOND-ORDER TRANSITIONS BY RAPID SOLIDIFICATION

JEFFREY A. WEST AND MICHAEL J. AZIZ
Division of Applied Sciences
Harvard University
Cambridge, MA 02138, USA

ABSTRACT

During rapid solidification of intermetallic compounds, the atoms may not have time to find the lowest-energy sites in the crystal, resulting in the growth of a solid with partially or completely suppressed chemical order. A kinetic model has been developed for "disorder trapping" during rapid solidification, which predicts the long range order parameter, composition and temperature at the interface of a chemically ordered phase as functions of interface velocity and liquid composition. The model predicts that as the solidification velocity increases, the long range order parameter decreases. Beyond a critical velocity the order parameter is zero. The predicted transition to solidification of a disordered solid is discontinuous for thermodynamically first-order cases, and continuous for thermodynamically second-order cases.

Ni_2TiAl ($L2_1$) and Ni_3Al ($L1_2$), which in equilibrium are ordered to their melting points, have been kinetically disordered by rapid solidification following pulsed laser melting. The order-disorder transitions are second-order and first-order, respectively. The interface velocity and order parameter can be followed with nanosecond resolution by monitoring the reflectivity and lateral resistance of a thin film sample during and immediately after solidification. X-ray diffraction and Transmission Electron Microscopy (TEM) indicate that metastable fcc Ni_3Al was retained by quenching. In rapidly solidified bulk Ni_2TiAl TEM reveals a fine array of anti-phase boundaries, indicating that the material formed from the melt with the unstable B2 structure and subsequently transformed to the equilibrium Heusler structure during cooling to room temperature. In rapidly solidified Ni_2TiAl thin films, electron diffraction indicates that the unstable bcc structure was formed directly from the melt and retained upon cooling.

INTRODUCTION

Intermetallic compounds are attractive prospects for use as high-temperature, high-strength, corrosion-resistant materials. Their lack of room-temperature ductility, however, can prevent their widespread use by limiting their formability. Many of these materials might become more practical if they could be put into a metastable chemically disordered phase ("disorder trapping") that was less difficult to machine or forge at room temperature to near net shape. Subsequently, the part could be annealed to bring out the chemical order and the associated high-temperature properties. Furthermore, some line compounds might become useful if they could be formed off stoichiometry ("solute trapping"). In addition to the known improvement in ductility that sometimes accompanies deviations from stoichiometry, such materials may become useful intermediate states in the production of practical alloys. Phases of nonequilibrium composition might serve as ideal starting materials for the use of subsequent thermomechanical processing to produce an extremely fine dispersion of precipitates that resist coarsening and re-dissolution at high temperatures, thus pinning grain boundaries and providing superior high-temperature strength. In addition to their potential technological relevance, kinetically disordered materials may serve as ideal starting points for studies of ordering kinetics, or for studies of the effect of ordering on properties.

These possibilities have been investigated by studying theoretically and experimentally the chemical long-range order (LRO) of rapidly solidified intermetallic compounds, and by studying theoretically the expected deviations from stoichiometry that could be induced by rapid solidification. This paper summarizes this work. Boettinger and Aziz developed a model for "disorder trapping" [1]. Boettinger, Bendersky, Cline, West, and Aziz formed the unstable B2 phase in bulk specimens of Ni_2TiAl [2]. West, Manos, and Aziz formed the metastable fcc phase of Ni_3Al [3] and the unstable bcc phase of Ni_2TiAl [4] in thin film samples.

MODEL

The Boettinger-Aziz model for solute and disorder trapping by rapid solidification predicts the deviation from stoichiometry and the deviation from the equilibrium LRO parameter, as functions of the melt composition and the velocity of the crystal/melt interface during rapid solidification. The model predicts that:

(1) Compounds that are normally ordered to the melting point will solidify with trapped anti-site disorder at high solidification velocities. With increasing interface velocity the LRO parameter, η, will decrease. For transitions that are thermodynamically second order, η decreases continuously, reaching zero at a critical velocity as shown in Fig. 1(a). For order-disorder transitions that are thermodynamically first order, η first decreases from the equilibrium value and then makes a discontinuous jump to zero at a critical velocity as shown in Fig. 1(b). A simple analytic expression was developed for v_T, the critical velocity for formation of a completely disordered alloy, in both cases. For the second-order case, characteristic of the B2 → bcc transition, the critical velocity is given by a very simple expression for a stoichiometric Bragg-Williams binary alloy:

$$v_T = v_D[(T_C/T_M)-1],\qquad\qquad(1)$$

where v_D is the ratio of the interface diffusivity to the jump distance, T_C is the critical temperature for the order-disorder transition and T_M is the solidus temperature of the compound. Only phases that are ordered up to their solidus point are considered; i.e., T_C/T_M >1. Large values of T_C, which relate directly to the strength of ordering, require large velocities to trap complete disorder.

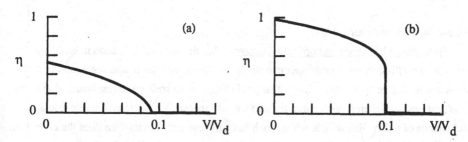

Figure 1. Predicted long range order parameter of the growing solid at the interface, η, as a function of dimensionless growth velocity, v/v_D, for a congruently melting stoichiometric AB compound. Predictions are made for cases where order-disorder transition is thermodynamically (a) second order; (b) first order.

(2) Deviations from stoichiometry can be trapped in by rapid solidification. For solidification from an off-stoichiometric melt, the solid's deviations from stoichiometry increase with increasing solidification velocity. The details depend on the free energy of the solid as a function of composition and order parameter, which is not yet known for most compounds but could, in many cases, be calculated by the cluster variation method [5].

(3) The solid/liquid interface is highly mobile during the solidification of completely disordered or completely ordered compounds, but is much more sluggish at intermediate velocities where partial disorder trapping or partial solute trapping is occurring. This implies the possibility of a competitive growth situation in which a metastable disordered phase nucleates and out-races a growing, partially-ordered phase.

EXPERIMENTAL TECHNIQUE

Sample preparation

<u>Thin film samples</u>. Thin films on insulators are used to facilitate the transient electrical resistance measurement of the interface velocity while avoiding electrical short-circuiting from the substrate. In order to ensure plane-front melting and subsequent solidification (necessary for a quantitative measurement of the interface velocity) it is essential that we have starting materials with surfaces that are smooth to a few hundred Å, providing uniform absorption of

laser irradiation and uniform heat flow into the substrate. Two substrates that we have used successfully are sapphire and thermally oxidized Si wafers. Thin films are prepared by vacuum evaporation or by ion beam sputtering, using a multiple-target sputtering system or a dual-hearth e-beam evaporator. Thin film samples are photolithographically patterned into a geometry for optimizing signal-to-noise during ultrafast electrical measurements.

Bulk Specimens. Initial experiments on Ni_2TiAl were done on bulk specimens. They were prepared by arc melting, diamond sawing and spark cutting.

Time-resolved measurements

Transient conductance and optical reflectance. Melting and solidification induced by pulsed laser melting of semiconductors can easily be monitored with a time resolution of a few nanoseconds. Large optical reflectivity changes indicate when melting or resolidification occur in the first few nanometers adjacent to the surface. By measuring the transverse electrical conductance of a thin film sample whose melt has a different resistivity than does the crystal, as shown in Fig. 2, the melt depth and interface velocity can be measured [6] over depths greater than half a micron and with time resolution of down to one nanosecond, depending upon the bandwidth of one's oscilloscope. The changes in electrical conductivity and optical reflectance upon melting of many elemental metals and alloys are sufficiently large to allow the same transient electrical conductance measurement of melt depths and velocities as has been used

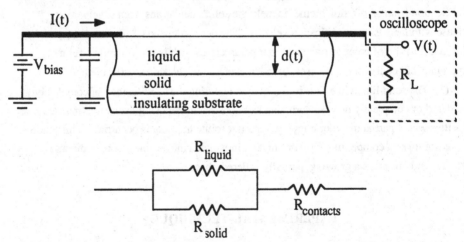

Figure 2. Transient conductance measurement and equivalent DC circuit.

successfully on semiconductors. We have made these measurements on several elemental metallic thin films [7,8], as well as on intermetallic compounds [3,4]. The accuracy is not as great as for Si, but any measurement at all of the interface velocity is a vast improvement over the alternative, which is a mere estimate of the growth rate from heat flow calculations.

Pulsed Laser Melting. Laser irradiation is performed with an ultraviolet (308 nm) pulsed excimer laser. The laser pulse is rendered spatially uniform by passing it through an external beam homogenizer. The pulse is roughly 30 nanoseconds in duration, routinely yielding solidification rates of a few m/s in metal films on SiO_2 substrates.

Post-irradiation analysis

We have used plan-view transmission electron microscopy (TEM) to examine the structure and crystallinity of laser irradiated material, on a Philips 420 TEM-STEM. TEM is usually sufficient to detect superlattice reflections and characteristics of the microstructure such as grain size, anti-phase boundary configuration, and precipitation. Rutherford Backscattering Spectrometry is used to obtain film thicknesses and compositions. Quantitative measurements of the LRO parameter are made with X-ray diffraction, performed on a GE $\theta-2\theta$ diffractometer using Cr K_α radiation. In the $\theta-2\theta$ geometry, only crystals having atomic planes parallel to the film surface contribute to diffraction. Monitoring a superlattice and a fundamental reflection pair with a scattering direction in common (e.g., 110 and 220) guarantees that the same crystals are responsible for diffraction into both peaks, eliminating any problem arising from possible texturing during any of the processing steps.

RESULTS

Disorder trapping in Ni₃Al

Ni_3Al is ordered to its melting point, above which lies a metastable order-disorder transition from $L1_2$ to fcc that is thermodynamically first-order. Its disorder-trapping behavior is predicted to behave as in Fig. 1(b). It is found to form a completely disordered fcc phase upon rapid solidification at velocities of 2 meters per second and above, and to re-order into the $L1_2$ phase during subsequent furnace annealing. The LRO was monitored by an X-ray diffraction measurement of the ratio of the intensity of the 110 superlattice peak to that of the 220 fundamental peak in $\theta - 2\theta$ geometry. The 110 superlattice peak completely disappeared following rapid solidification, and returned after furnace annealing. This result confirms the prediction of the model that complete disorder trapping is possible by rapid solidification.

Disorder trapping in Ni₂TiAl

The Heusler alloy Ni_2TiAl, with the $L2_1$ (DO_3) structure, can have two levels of disorder, characterized by two order parameters. If the order parameter for the Ti-Al sublattice goes to zero, then Ti and Al randomly populate this sublattice and the structure becomes B2; if the other order parameter subsequently vanishes it becomes bcc. Each of these transitions has a unique T_C, so the application of equation (1) indicates that as the solidification velocity is increased, first $L2_1$, then B2, then bcc will form from the melt.

We have laser-melted bulk specimens of Ni_2TiAl. The melt duration was measured and used to calibrate heat-flow calculations, which estimated the velocity during subsequent solidification to be approximately 8-20 m/s. TEM analysis of the solidified material revealed the $L2_1$ phase with a fine dispersion of anti-phase boundaries. The boundaries all had <100> displacements, indicating that they result from a B2 $\rightarrow$ $L2_1$ transition. Since no boundaries were found with <111> displacements, which would be characteristic of the bcc $\rightarrow$ B2 transition, we inferred that the B2 phase was formed directly from the melt and that it ordered to the equilibrium $L2_1$ structure during subsequent cooling of the solid from the melting point to room temperature.

Rapid solidification of thin films of Ni_2TiAl on SiO_2 at velocities measured to be ~4 m/s yielded altogether different results. Only bcc fundamental reflections are always visible in TEM diffraction patterns, B2 superlattice reflections are missing from some grains, and $L2_1$ superlattice reflections are always absent. Furnace annealing recovers the B2 and $L2_1$ reflections in all cases. This result indicates that completely disordered bcc Ni_2TiAl can be solidified directly from the melt and retained during cooling to room temperature.

DISCUSSION

These experiments reviewed here demonstrate that intermetallic compounds can be completely disordered by rapid solidificaton. Yavari and Bochu had formed partially disordered Ni_3Al by melt spinning [9], but were only able to form completely disordered compounds by adding enough Fe to depress the critical temperature below the solidus. In the latter case, rather than forming a metastable material directly from the melt, they were forming an equilibrium disordered solid solution from the melt and suppressing solid-state ordering during rapid solid-phase cooling. Quantitative estimates of the critical velocity for complete disorder trapping depend in detail on the thermodynamics of Ni_3Al and on the magnitude of v_D, the "diffusive speed". For a range of thermodynamic parameters tinkered with while exploring the model, the critical velocity seems to range from near v_D to velocities well below v_D, the latter being the case when the driving force for ordering is very slight. Preliminary experiments on solute trapping in disordered metallic alloys [8] yield values of v_D on the order of 1 m/s, which lies in the realm of pulsed laser melting. As solidification and cooling rates from melt spinning are several orders of magnitude less than those due to pulsed laser melting, complete disorder trapping by rapid solidification following pulsed laser melting and partial disorder trapping by melt spinning seem to be near extreme cases for the range of possibilities. A more complete and quantitative test of the model, by measuring the LRO parameter over a range of velocities, has yet to be performed.

The formation of *unstable* disordered Ni_2TiAl by suppressing second order transitions during rapid solidification may come as a surprise to some. To form such a material, one must score a kinetic "bull's eye"; i.e., forming a phase with a configuration close to the target state is insufficient, because relaxation will occur in a direction away from the target state. Apparently, the kinetics of rapid solidification lead directly to that bull's eye. The solidification and cooling rate must be so high that virtually no time is available for the barrierless relaxation to an ordered structure. The necessary solidification velocity in this second-order case is estimated using eq. (1). Burton *et al.* [10] have estimated the critical temperatures for the $L2_1 \rightarrow B2$ and the $B2 \rightarrow$ bcc transitions as 1900 K and 3300 K, respectively. With a melting point of about 1800 K for ordered Ni_2TiAl and a v_D of 1 m/s, eq. (1) predicts B2 formation above 0.6 m/s and bcc formation above 9 m/s. This is roughly consistent with the experimental velocities. We had expected that solidification of thin films on SiO_2 would occur more slowly than on bulk samples, resulting in a more highly ordered solid. However, it is possible that the thin film samples undercooled significantly prior to solidification in the absence of a crystal seed, and then grew very rapidly once nucleation occurred. There might have been some amount of lateral growth of crystalline nuclei, this would prevent us from interpreting the transient conductance measurement in terms of an interface velocity. Extreme undercoolings from pulsed laser melting are not uncommon; we have hypercooled liquid Ni ($\Delta T \approx 512$ K) by this method [7].

CONCLUSIONS

Rapid solidification forms metastable compositionally-disordered fcc Ni_3Al and unstable compositionally-disordered B2 and bcc Ni_2TiAl, as predicted by the model for disorder trapping. The observed velocities for solidification of disordered alloys are roughly consistent with theoretically estimated critical solidification velocities for complete disorder trapping.

ACKNOWLEDGMENTS

Some of the work reported here is the result of a collaboration with W.J. Boettinger, L.A. Bendersky, and J. Cline at NIST. We are grateful to J.T. Manos and P.M. Smith at Harvard for technical assistance. Work at Harvard has been supported by N00014-88-K-0548.

REFERENCES

[1] W.J. Boettinger and M.J. Aziz, Acta Metall. **37**, 3379 (1989).

[2] W.J. Boettinger, L.A. Bendersky, J.A. West, M.J. Aziz and J. Cline, Mat. Sci. Eng. **A133**, 592 (1991).

[3] J.A. West, J.T. Manos, M.J. Aziz, Mat. Res. Soc. Symp. Proc. **213**, (in press, 1991).

[4] J.A. West, Ph.D. thesis, Harvard University, in preparation (1991).

[5] B.P. Burton, J.E. Osburn, and A. Pasturel, Mat. Res. Soc. Symp. Proc. **213**, (in press, 1991).

[6] M.O. Thompson, J.W. Mayer, A.G. Cullis, H.C. Webber, N.G. Chew, J.M. Poate and D.C. Jacobson, Phys. Rev. Lett. **50**, 896 (1983).

[7] H.A. Atwater, J.A. West, P.M. Smith, M.J. Aziz, J.Y. Tsao, P.S. Peercy, and M.O. Thompson, Mat. Res. Soc. Symp. Proc. **157**, 369 (1990).

[8] P.M. Smith, J.A. West, and M.J. Aziz, Mat. Res. Soc. Symp. Proc. **205**, in press (1991).

[9] A.R. Yavari and B. Bochu, Phil. Mag. A **59**, 697 (1989).

[10] B.P. Burton, A. Pasturel, J.E. Osborn, and W.C. Carter, these proceedings.

KINETIC STATES OF ORDER IN HIGHLY NONEQUILIBRIUM MATERIALS

Brent Fultz
California Institute of Technology, 138-78
Pasadena, Calif. 91125, USA

Abstract

Two concepts from the kinetic theory of disorder→order transformations are presented. Kinetics provides control over the path taken *en route* to equilibrium, permitting different types of microstructures to be synthesized. Variations in microstructural evolution can be parameterized by two or more order parameters. "Kinetic paths" through two or more order parameters are convenient for experimental study because they are independent of the vacancy concentration in the material. Saddle points in the free energy surface have little consequence to the thermodynamic state, but if an alloy approaches a saddle point its kinetic evolution tends to stall. This stalled state is an example of a "pseudostable" state.

Introduction

The thermodynamic state of a system is reached in sufficient time after the system passes through a chain of kinetic states. In this sense the thermodynamic state is a subset of the kinetic states, so perhaps it is not surprising that the science of kinetics remains less developed than thermodynamics. Many fundamental phenomena of the thermodynamics of materials are well-established, such as the forms of free energy surfaces and phase diagrams. These phenomena are also relevant to the kinetics of phase transformations since they must affect the kinetic chain of states, at least near the end of this chain. Our concern here, however, is with kinetic phenomena that are not necessarily related to thermodynamics. Such kinetic phenomena are expected to be important in materials far from thermodynamic equilibrium.

Since time is a kinetic variable but not a thermodynamic variable, understanding the time scale of a phase transformation has been the main goal of most kinetic theories. Especially for the synthesis of nonequilibrium materials, however, kinetics provides a strong control over the internal structures of a material. It is well-known that thermodynamics predicts only general features of a microstructure such as crystal structures and their compositions, and kinetics is often responsible for details such as grain size and morphology. What is less-known, however, is that in materials far from thermodynamic equilibrium the kinetic states of a material can be quite unlike the thermodynamic state.

Since combinatoric enumerations of states can be performed directly for disorder→order transformations, ordering in highly nonequilibrium materials is well-suited for study by "statistical kinetics", which is the analog of statistical mechanics for equilibrium states [1-6]. When two or more variables are available to describe the state of the material, it may be possible to obtain many different combinations of state variables during ordering. "Kinetic paths" are trajectories taken by a material through a space of state variables. The kinetic paths at different temperatures may be quite different [7-15]. The state variables parameterizing the kinetic paths specify microstructural features of a material, so here kinetics is used to predict different microstructures rather than different time dependences of phase transformations. It is interesting to ask if through kinetics we can predict and control a useful range of states in nonequilibrium materials.

With kinetic theory it has been observed that some special long-lived states appear in the chain of states along kinetic paths. Some such states occur when the materials reached a saddle point in its free energy surface. We propose to call these long-lived, but transient states "pseudostable states" [16], and prospects for their influence on the kinetics of real materials are discussed.

B2 Ordering in the Pair Approximation

Activated state rate theory forms the basis for our statistical kinetic theory. In activated state rate theory it is assumed that at a fixed temperature, equilibrium exists between an initial state of the system and activated states of the system. We seek the rates at which the various final states are achieved after activation. Gibbs or Boltzmann factors are employed to obtain the relative probabilities of the initial and activated states*. The energy of the final state may not affect the activation probability; it does not for a vacancy mechanism for atom movements, for example. In all cases the transition probabilities between initial and final states require precise consideration of the number of ways by which the initial and final states are linked. This connection between states is a feature of kinetics that is not part of thermodynamics, and causes the kinetic rate equations to depend on the kinetic mechanism, *i.e.* the mechanism of atom movement. Here we assume that movement occurs by atom motion into an adjacent vacancy. A detailed accounting of the various atom movements during B2 ordering in the pair approximation was developed recently [12,17] so only general concepts are presented here.

In the pair approximation, the state of the alloy is specified by a set of "pair variables". The pair variables for a binary A-B alloy with two interpenetrating sublattices (the A-rich α-sublattice, and the B-rich β-sublattice) are: $\{p_{AA}^{\alpha\beta}, p_{AB}^{\alpha\beta}, p_{BA}^{\alpha\beta}, p_{BB}^{\alpha\beta}\}$. The pair variable $p_{AB}^{\alpha\beta}$, for example, is the probability that a nearest-neighbor pair chosen from the α-sublattice to the β-sublattice comprises an A on α and a B on β. The conservation of atoms and the conservation

---

* From these Boltzmann factors a partition function can be constructed for the activated state, and from the partition function we can learn all thermodynamic information about the activations of the system. The problem is that this thermodynamic information is not interesting; the heat capacity associated with the activations is rarely of interest, for example.

of bonds reduces to two the number of independent pair variables in this problem. To study the kinetics of ordering, however, it is also necessary to parameterize the pair variables involving vacancies. The pair variables $\{p_{VA}^{\alpha\beta}, p_{AV}^{\alpha\beta}, p_{VB}^{\alpha\beta}, p_{BV}^{\alpha\beta}, p_{VV}^{\alpha\beta}\}$ are similarly defined. Because the number of vacancies is small, we do not expect the pair variables involving vacancies to affect thermodynamic aspects of the problem such as the critical temperature for ordering.

Nevertheless, an atom-atom pair variable cannot change unless a vacancy is located next to at least one of the atoms. The rates of change of the atom-atom pair variables are always directly proportional to the number of vacancies in the alloy. For this reason it is meaningful to recast the predictions of kinetic theory from the temporal dependence of the state variables, such as predictions of $\{p_{AB}^{\alpha\beta}(t), p_{BA}^{\alpha\beta}(t)\}$, to predictions of the interdependence of the state variables: $p_{AB}^{\alpha\beta}(p_{BA}^{\alpha\beta})$. A graph of $p_{AB}^{\alpha\beta}$ versus $p_{BA}^{\alpha\beta}$, here termed a "kinetic path", will be conveniently independent of the vacancy concentration. This independence of the kinetic path on vacancy concentration does not depend on the order of the approximation (such as point, pair, tetrahedron...), but only on the assumption that the vacancy concentration is small so that vacancy-vacancy interactions are not important. Vacancy concentrations in highly non-equilibrium materials are often poorly controlled, and during low temperature annealings we expect annihilations of many quenched-in vacancies. Fortunately none of these phenomena affect the shape of the kinetic path (although the time required to traverse the path will be affected). Interpretations of experimental kinetic paths of ordering [8,10,13,14,18] are therefore much more straightforward than interpretations of the temporal dependences of the order parameters. Kinetic path experiments require that at least two independent parameters are measured simultaneously, a single Bragg-Williams long-range order (LRO) parameter, L, is inadequate, for example. (Here $L \equiv 2(R - W)/N$, where in an alloy with N sites R is the number of A-atoms on the α-sublattice and W is the number of B-atoms on the α-sublattice.)

Fig. 1. Kinetic paths through the SRO pair variable, $p_{AA}^{\alpha\beta}$, and the conventional LRO parameter, L, for a binary equiatomic bcc alloy at a temperature of 0.7 of the critical temperature for B2 order. Path from disorder to order is indicated with the arrows on the pair approximation result. Monte Carlo result includes data points (crosses) with interpolated curves.

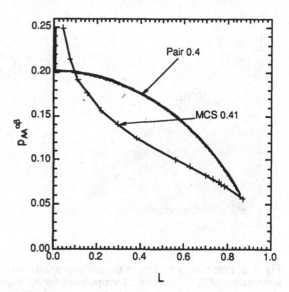

After quenching to below the critical temperature for long-range ordering, B2 ordering of a binary alloy illustrates some important features of a kinetic path. At a temperature of about 0.7 of the critical temperature, the pair approximation predicts the kinetic path of a disorder$\rightarrow$

order transformation shown in Fig. 1 [18]. The initial state of disorder is at the upper left corner of the figure. The initial state of disorder includes a wide variety of local atomic environments, and atoms in the more unfavorable environments tend to quickly exchange sites with a vacancy. The short-range order (SRO) relaxes partly towards equilibrium in this way, but soon there is insufficient variation in the local atomic environments to overcome the return jumps. Once LRO develops, however, there is a growing energy difference for the species occupancies on the two sublattices, and the high energy environments become increasingly less favorable. In this second stage of ordering the SRO and LRO evolve together. A distinct break in the kinetic path occurs between these two stages of ordering. (For the pair approximation results in Fig. 1 the break is at $p_{AA}^{\alpha\beta} = 0.2$, $L = 0$.) This break is a distinct prediction of kinetic theory that can be tested experimentally.

In previous work we developed Monte Carlo algorithms to simulate diffusion and ordering with activated state rate theory for vacancy jumps [19-21]. The assumptions about how the local chemical environment affects the probability of a vacancy jump are the same as in the analytical theories [3-7,10,12,17], but the Monte Carlo simulations are exact for a periodic crystal of finite size. Presented in Fig. 1 is a kinetic path obtained with a Monte Carlo simulation of ordering with vacancies. The simulation was performed at a temperature of 0.77 of the critical temperature. Monte Carlo simulations permit a more physical transition from SRO to LRO because they include correlations over intermediate distances. (Finite size effects modify determinations of the initial state of disorder and the later part of the kinetic path, but the middle part of the path should be reliable.) The sharp break in the kinetic path predicted with the pair approximation is mollified in the Monte Carlo simulations, but increased curvature is still found for the kinetic path after an early relaxation of SRO.

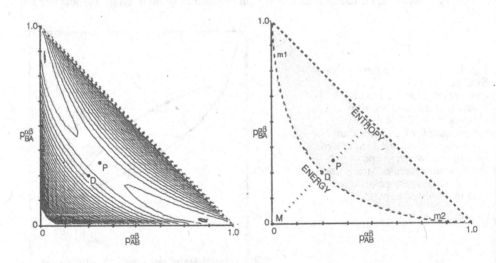

Fig. 2. a) Free energy surface in the pair approximation for a binary equiatomic bcc alloy at a temperature of 0.7 of the critical temperature for B2 order. Conservation of bonds confines the alloy to states within a triangle. b) Features of the free energy surface. The curved, dashed line bounds the states with SRO and LRO. The disordered state is labeled "D", the absolute minima by "m1" and "m2", the maximum by "M", and the saddle point by "P".

The free energy surface in the pair approximation, $F(p_{AB}^{\alpha\beta}, p_{BA}^{\alpha\beta})$, corresponding to the alloy of Fig. 1 is presented in Fig. 2. Note that the sharp break in the kinetic path occurs at a saddle point in the free energy surface. It is reasonable that a saddle point singularity in F will affect the kinetics. A saddle point in $F(\{\xi_i\})$, where ξ_i are independent variables, is characterized by:

$$\frac{\partial F}{\partial \xi_i} = 0 \quad \text{for all } \xi_i \text{, and} \qquad\qquad 1)$$

the eigenvalues of $M_{ij} = \frac{\partial^2 F}{\partial \xi_i \partial \xi_j}$ have mixed signs. $\qquad\qquad$ 2)

Equation 1, which is also true at absolute and relative minima of $F(\{\xi_i\})$, shows that at the saddle point there exists no thermodynamic driving force for ordering. If the rate of change of a state variable depends on the sensitivity of F to the state variables,

$$\frac{d\xi_i}{dt} = \sum_j \gamma_{ij} \frac{\partial F}{\partial \xi_j} , \qquad\qquad 3)$$

where γ_{ij} is a matrix of kinetic coefficients, then ordering will be slow when the state variables of the alloy are near those of the saddle point. It may be possible, however, to devise kinetic mechanisms to circumvent this thermodynamic argument. Nevertheless, since:

$$\frac{d}{dt} F(\{\xi_i\}) = \sum_i \frac{\partial F}{\partial \xi_i} \frac{d\xi_i}{dt} , \qquad\qquad 4)$$

at the saddle point the free energy must be static:

$$\frac{d}{dt} F(\{\xi_i\}) = 0 . \qquad\qquad 5)$$

The free energy function of the pair approximation must have a saddle point, as illustrated with Fig. 2b. Along any path connecting the two minima, m1 and m2, there is a relative maximum whose height depends on its position with respect to the two dashed lines in Fig. 2b. If this path passes close to the straight dashed line, the free energy will be large because of entropy. At the straight dashed line there are zero A-A and B-B pairs, and the entropy rises with a logarithmic singularity. The maximum along the path can also be reduced by avoiding close contact with the curved dashed line. The states along the curved dashed line have LRO but no SRO, and the internal energy does not favor these states. If the maximum along the path can be reduced by avoiding the two dashed boundaries of the ordered region, between these boundaries there will be (at least) one path connecting m1 and m2 that has the smallest relative maximum. This smallest relative maximum is a saddle point. By symmetry, we expect the saddle point to be located along the diagonal line in Fig. 2b. Similar arguments can be applied to free energy functions of more than two pair variables. Because of the equivalence of the different sublattices, the ordering problem always has more than one absolute minimum of its free energy surface, so the existence of saddle points on the free energy surface should be quite common.

It is perhaps surprising that the kinetic path of an alloy should approach a saddle point. Since the curvature of the free energy surface near the saddle point must be negative for at least one of the state variables, the path into the saddle point is unstable against infinitesimal perturbations of at least one state variable. It would seem that the importance of the saddle point to kinetic paths would be highly sensitive to the choice of initial conditions. Nevertheless in our work with the pair approximation, we found states of SRO with unchanged LRO persisted for modest times even when the initial value of L was as high as 0.2. The reason that the kinetic effects of the saddle point are so durable is that the time scale for SRO is so much shorter than the time scale for LRO. Even though the duration of the state of SRO with unchanged LRO is affected by the initial conditions, owing to the quick SRO relaxation this state appears, and remains unchanged, for a significant time. Because such a state falsely appears to be stable, we have called it "pseudostable" [16].

Although we may locate saddle points in F in approximations beyond the pair, and even in the hypothetical "exact" expression for F, these saddle points may not cause such a distinct impact on the kinetic path as in the pair approximation. This is a consequence of the variety of time scales in the more sophisticated problems. Intermediate-range atomic correlations are expected to relax at intermediate rates between the pair variables and L. Consequently small perturbations in the initial conditions will cause the intermediate-range state variables (those for which F is negatively curved) to quickly divert the kinetic path away from the path into the saddle point. The Monte Carlo results in Fig. 1 suggest that the clarity of the pseudostable state becomes blurred, but pseudostability remains a qualitatively useful concept for describing the ordering process. An experimental test of pseudostability in FeCo is presented below.

Experiments

Fe_3Al

In thermodynamic equilibrium, alloys with the Fe_3Al stoichiometry have B2 LRO below a critical temperature of about 800 °C. Below 550 °C, Fe_3Al undergoes a secondary ordering transformation where DO_3 LRO appears within the B2 ordered structure. Here we are not concerned with the equilibrium state of order in Fe_3Al, but rather with the nonequilibrium states of order that evolve as a rapidly quenched alloy is annealed at low temperatures. In an x-ray diffractometry study of Fe_3Al [10], we found that the rate of growth of B2 LRO was more temperature-sensitive than that of DO_3 LRO; B2 order evolved relatively more rapidly than DO_3 order at low temperatures. Equilibrium thermodynamics predicts only a weak variation in the "kinetic path" of the alloy through these two order parameters, and predicts that the rate of growth of DO_3 order is more temperature-sensitive than the rate of growth of B2 order (because the critical temperature for DO_3 ordering is lower than that for B2 ordering). The observed behavior must have a kinetic origin since equilibrium thermodynamics predicts the opposite trend.

Very recently we used Mössbauer spectrometry to study the kinetic paths of short range ordering in Fe_3Al [11,14]. Further work on this problem is presented here. Alloys of Fe-25 at% Al and Fe-27 at% Al were prepared from materials of 99.99 % purity by arc-melting under argon atmospheres, and the ingots were inverted and remelted several times to insure homogeneity. The chemical compositions were checked by atomic absorption spectrometry. Disordered specimens were prepared by piston-anvil quenching with an Edmund Bühler ultra-rapid quenching apparatus, which has a characteristic cooling rate of $10^5 - 10^6$ °C/sec. To induce ordering, samples were annealed in evacuated quartz ampoules for various times at 185 °C, 250 °C, and 300 °C. X-ray diffractometry was performed with several instruments, including an Inel CPS-120 diffractometer with Co Kα radiation and a curved position-sensitive detector.

Mössbauer spectra were obtained in transmission geometry at 295 K with a radiation source of 10 mCi ^{57}Co in a Rh matrix. The distribution of ^{57}Fe hyperfine magnetic fields (HMF) in Fe-Al alloys can be interpreted with a simple model, where the HMF at a particular ^{57}Fe atom is reduced in proportion to the number of Al atoms in its first nearest neighbor shell [22,23]. These systematics are modified somewhat by the presence of 2nn Al atoms and other effects, but in the HMF distributions of Fe-Al alloys we expect a series of distict peaks. These peaks are arranged with decreasing HMF, corresponding to an increasing number of Al atoms in the 1nn shell of the ^{57}Fe atom. These peaks in the HMF distribution are seen when the experimental spectra are processed by the method of Le Caër and Dubois [24] (see Fig. 3). In Fig. 3, the peaks in the HMF distributions are labeled by the numbers of Al first nearest neighbors about the ^{57}Fe atom. The intensities of these peaks correspond approximately to the probability of each 1nn environment. Note the clarity with which the different 1nn environments are resolved in Fig. 3. To quantify these intensities, the peaks in the HMF distribution were fit to a set of seven Gaussian functions.

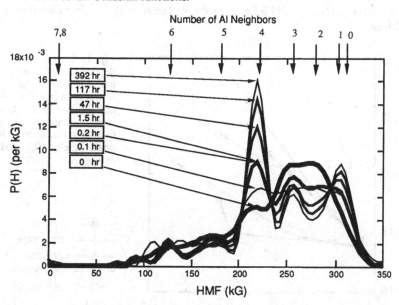

Fig. 3. Hyperfine magnetic field distributions for Fe-25at.%Al alloy as piston-anvil quenched, and after annealing at 300 °C for various times. Numbers at the top of the figure identify the resonance peaks from ^{57}Fe atoms with different numbers of Al neighbors. Note the growth of the peak (4) and (0,1), which are the two Fe environments in the D0$_3$ structure.

Five kinetic paths of short-range ordering in Fe$_3$Al are presented in Fig. 4. The "SRO parameters" in Fig. 4 are the two Fe environments of the D0$_3$ structure; one with 4 Al neighbors, and one with 0 Al neighbors that cannot be resolved from the environment with 1 Al neighbor (denoted (4) and (0+1), respectively). The state of disorder is towards the lower left of the graph, and perfect D0$_3$ order corresponds to Intensity(0+1) = 0.33 with Intensity(4) = 0.66. The change in slope of the kinetic paths between 200 °C and 300 °C is about a factor of four, so significantly different kinetic paths are followed by Fe₃Al at different temperatures. Temperature is an effective means of kinetic control over the state of order in Fe₃Al .

At lower temperatures we see (Fig. 4) that the (0,1) 1nn Al peak grows more rapidly relative to the (4) 1nn Al peak. This temperature dependence agrees with that of the LRO parameters, which showed a more rapid relative growth of B2 order *versus* DO₃ order at low temperatures [10]. To show this agreement, we make the following argument based on the four-sublattice model of the bcc structure [10]. The formation of B2 order requires a segregation of Al atoms off the α and β sublattices. The Fe atoms on the γ and δ sublattices will then have only Fe atoms as nearest neighbors. The degree of B2 order is therefore related to the intensity of the (0) 1nn Al peak, which is from ^{57}Fe atoms that are on the γ and δ sublattices. The differentiation between B2 and DO₃ order requires examination of the neighborhoods of Fe atoms on the α and β sublattices. If there is significant DO₃ order, the Fe atoms on the α and β sublattices will have mostly (4) 1nn Al atoms, but B2 order allows many (3) or (5) 1nn Al atom neighborhoods for the α and β Fe. A strong (4) 1nn Al peak therefore corresponds to strong DO₃ order.

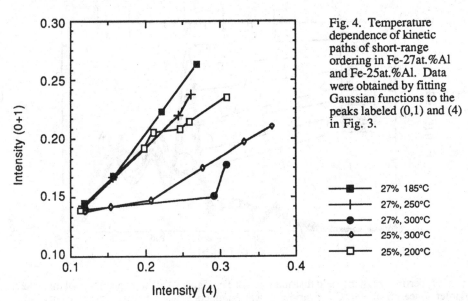

Fig. 4. Temperature dependence of kinetic paths of short-range ordering in Fe-27at.%Al and Fe-25at.%Al. Data were obtained by fitting Gaussian functions to the peaks labeled (0,1) and (4) in Fig. 3.

FeCo

I have recently completed an investigation of the disorder→B2 order transformation in rapidly quenched FeCo [18]. Samples of FeCo were prepared from starting materials of 99.99 % purity, much in the same manner as were the samples of Fe₃Al. X-ray diffractometry was used to measure L in the conventional way by comparing the intensities of B2 superlattice diffractions to the intensities of the fundamental bcc diffractions. Problems with the small difference in x-ray scattering factors of Fe and Co atoms were overcome by using Co Kα radiation to suppress the scattering factor of the Fe atoms, and by using the Inel CPS-120 large angle position sensitive detector. Even with these experimental capabilities, it was not practical to measure the SRO by x-ray diffuse scattering. Mössbauer spectrometry was used to measure the evolution of SRO. Unlike x-ray diffractometry, which makes direct measurements of correlations between atom positions, Mössbauer spectrometry is sensitive to local magnetic disturbances in the material through the ^{57}Fe HMF. Mössbauer spectrometry is a less direct measure of SRO in Fe alloys with high Co concentrations. It is nevertheless understood how variations in local atomic configurations affect the HMF in Fe-Co alloys [25-28], so models of the SRO can be tested against observed HMF distributions.

The Monte Carlo simulations were particularly useful because x-ray diffractometry and Mössbauer spectrometry "experiments" can be performed on the simulated alloys. The x-ray diffraction pattern is easily simulated within kinematical theory by taking the diffracted wave, $\psi(\Delta k)$, as the three dimensional Fourier transform of the alloy, and then calculating $\psi^*\psi$:

$$\psi^*(\Delta k)\,\psi(\Delta k) = \left| \sum_{\text{all sites}} f(\mathbf{r})\,\exp(-i\Delta \mathbf{k}\cdot\mathbf{r}) \right|^2 \qquad 6)$$

The intensities of the diffraction peaks were obtained by integrating $\psi^*\psi$ in a cubical volume around the center of the diffraction peak. The integration volume had a width in Δk that corresponded approximately to our experimental capability. A width in Δk of $0.10\ 2\pi\ a^{-1}$ was used, where a is the edge length of the cubic bcc unit cell. Similar results were obtained when the width in Δk was as large as $0.30\ 2\pi\ a^{-1}$, but when the width in Δk became too large the integration included the diffuse scattering, and there was little structure to the kinetic path of SRO *versus* LRO.

Simulated Mössbauer spectra from FeCo alloys were obtained from the alloys of the Monte Carlo simulations. The average HMF, <H>, was obtained from these spectra. The responses of the ^{57}Fe HMF to the configurations of local magnetic moments [26-28] are too complicated to discuss here. Suffice to say that the 4s conduction electron polarization seen at a ^{57}Fe nucleus owing to the surrounding atoms is included, as is the core s-electron polarization due to the effects on the ^{57}Fe magnetic moment owing to the surrounding Co neighbors. The present work employed the same mechanisms used previously for the ^{57}Fe HMF response to the Co and Fe magnetic moments, and the parameters for these responses and the values of the magnetic moments were the same as used previously [9,25,28]. Again it was necessary to scale linearly the magnitude of the changes in HMF during ordering to correspond to the actual experimental results.

Figure 5 presents the results of simulated experimental kinetic paths of <H> *versus* L, which correspond to SRO *versus* LRO. One set of alloys was obtained by a Monte Carlo simulation of ordering with a vacancy mechanism, and from each state of the alloy the Mössbauer and x-ray diffractometry data were simulated. The second path in Fig. 5 was calculated by assuming a heterogeneous mode of ordering where a region of equilibrium order grows into the region of disorder. The value of <H> and the intensities of the superlattice diffractions were merely the volume-weighted average of the HMF and superlattice intensities of the disordered and ordered states. Figure 6 presents our experimental kinetic paths of <H> *versus* L. The excellent agreement between the kinetic paths at 350 °C and 400 °C in Fig. 6 and the heterogeneous curve in Fig. 5 indicates that B2 ordering in FeCo evolves heterogeneously in a first-order phase transformation at low temperatures. This change in mode from a

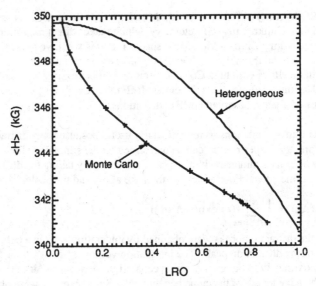

Fig. 5. Simulated and calculated kinetic paths of ^{57}Fe HMF *versus* L for disorder→B2 order transformation in FeCo. The two data sets are:
i. Monte Carlo simulations of ordering with vacancies (same simulations used for the data of Fig. 1). Line is drawn to guide the eye.
ii. Heterogeneous ordering.

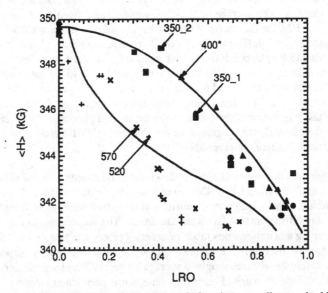

Fig. 6. Experimental kinetic paths of <H> *versus* L for piston-anvil quenched FeCo annealed at 350 °C, 400 °C, 520 °C, and 570 °C for various times. Data set 350_1 was obtained in the present work, data set 350_2 was obtained previously [25]. Data set 400* was obtained by scaling previous data at 77 K [25] to fit an 8 kG range in <H> (this should underestimate slightly the concave downward curvature). Lines are those of Fig. 5.

homogeneous second-order transformation at higher temperatures had previously been reported by Buckley [29]. Evidently this change in mode of ordering has a kinetic origin because at lower temperatures the disordered alloy becomes more unstable against ordering, so thermodynamics does not suggest a first order transformation at low temperature.

More significant from the point of view of statistical kinetic theory is the agreement between the experimental kinetic paths at 520 °C and 570 °C in Fig. 6 and the simulated kinetic path for homogeneous ordering in Fig. 5. As in Fig. 1 the kinetic path begins in the upper left of the figure, and the first four data points (crosses) in Fig. 1 are those of the SRO relaxation. (The same four states of the same simulated alloy were used to obtain the first four data points in Fig. 5.) In the early, independent relaxation of SRO, $<H>$ decreases relatively rapidly with respect to the L, as expected since the HMF is sensitive to atomic arrangements within a few angstroms around the ^{57}Fe atom [25]. The strong difference in the initial slopes of the two kinetic paths in Fig. 6 reflects the early, independent relaxation of SRO, and the good agreement between the kinetic paths of Fig. 5 and Fig. 6 confirms approximately the magnitude of the SRO relaxation predicted by the Monte Carlo simulations. Since the pair approximation and the Monte Carlo simulations showed a qualitatively similar early, independent relaxations of the SRO (Fig. 1), there is qualitative agreement between experiment and statistical kinetic theory in the pair approximation.

The sharp kink in the kinetic path in the pair approximation, associated with a distinguishable pseudostable state of SRO without LRO, is not found in the experiment with FeCo. Nature takes a more gradual transition into long range ordering, much as observed in the Monte Carlo simulations. There is a distinctly large curvature of the kinetic path near the pseudostable state, however, and a change in the time scale of ordering. It seems that the concept of a "pseudostable state" has qualitative usefulness for B2 ordering in FeCo.

Summary and Conclusions

New kinetic phenomena, unrelated to predictions of equilibrium thermodynamics, occur in disorder→order transformations in materials far from equilibrium. As an alternative to predicting the time dependence of ordering, kinetic theory can be used to predict microstructural variations during disorder→order transformations. Formal maps of microstructural evolution are provided by "kinetic paths" through two or more order parameters. Both theory and experiments with Fe_3Al have shown that a material can attain a wide range of different kinetic states *en route* to equilibrium. In Fe_3Al, the rate of B2-like order tends to predominate over DO_3-like order as ordering occurs at lower temperatures.

A saddle point in the free energy surface is of little concern to the thermodynamic state. It can, nevertheless, affect the kinetic path in interesting ways provided that for the variables ξ_i for which $\dfrac{\partial^2 F}{\partial \xi_i{}^2} < 0$ there is a long time scale for relaxation. The state variables associated with the saddle point can then be persistent, and we call this a "pseudostable state" of the material. The effect of the saddle point in the ordering of real materials is mollified by the instabilities against intermediate-length correlations having intermediate relaxation times.

Acknowledgements

I am delighted to acknowledge stimulating conversations with R. Kikuchi, T. Mohri, J. Sanchez, D. de Fontaine, P. Cenedese, L. Anthony, and K. Wada. With such colleagues, science cannot be dull. Theoretical work on pseudostability was supported by a Grant for International Research from NEDO, Japan. This work was supported by the U. S. Department of Energy under contract DE-FG03-86ER45270.

References

1 Kikuchi, R., *Prog. Theor. Phys. Suppl.*, 1966, **35**, 1.

2 Kikuchi, R. and Sato, H., *J. Chem. Phys.* **51** (1969) 161. *ibid* **57** (1972) 4962.

3 Bakker, H., *Phil. Mag.* A, 1979, **40**, 525.

4 Sato, H. and Kikuchi, R., *Acta Metall.*, 1976, **24**, 797.

5 Sato, H., Gschwend, K., and Kikuchi, R., *J. de Physique*, 1977, **C7**, 357.

6 K. Gschwend, H. Sato, and R. Kikuchi, *J. Chem. Phys.*, 1978, **69**, 5006.

7 Fultz, B., *Acta Metall.*, 1989, **37**, 823-829.

8 Fultz, B., Hamdeh, H.H., and Pearson, D.H., *Acta Metall.*, 1989, **37**, 2841-2847.

9 Fultz, B. and Hamdeh, H.H., *Hyperfine Interac.*, 1989, 45, 55-72.

10 Anthony, L. and Fultz, B., *J. Mater. Res.*, 1989, **4**, 1132-1139. *ibid,* 1140-1142.

11 Fultz, B., Gao, Z-Q., and Hamdeh, H.H., *Hyperfine Interac.*, 1990, **54**, 521-526.

12 Fultz, B., *J. Mater. Res.*, 1990, **5**,1419-1430.

13 Fultz, B., *J. Less-Common Metals*, 1991, **168**, 145-157.

14 Fultz, B., Gao, Z-Q., and Anthony, L. *Proc. Mater. Res. Soc.*, 1990, **186**, "Alloy Phase Stability and Design", edited by Stocks, G.M., Pope, D.P., and Giamei, A.F.

15 Anthony, L. and Fultz, B., *Proc. Mater. Res. Soc.*, 1990, **186**, "Alloy Phase Stability and Design", edited by Stocks, G.M., Pope, D.P., and Giamei, A.F.

16 Kikuchi, R., Mohri, T., and Fultz, B., "The Pseudostable Phase", *Proc. Mater. Res. Soc.* Boston, MA, 1990, in press.

17 Fultz, B., "Kinetics of short- and long-range B2 ordering in ternary alloys", to be published.

18 Fultz, B., "Kinetics of short-range and long-range B2 ordering in FeCo", to be published.

19 Fultz, B., *J. Chem. Phys.*, 1987, **87**, 1604.

20 Fultz, B., *J. Chem. Phys.*, 1988, **88**, 3227.

21 Fultz, B. and Anthony, L., *Phil. Mag. Lett.*, 1989, **59**, 237.

22 Wertheim, G.K., Jaccarino, V., Wernick, J.H., and Buchanan, D.N.E., *Phys. Rev. Lett.*, 1964, **12**, 24.

23 Stearns, M.B., *Phys. Rev. B*, 1972, **6**, 3326.

24 Le Caër, G. and Dubois, J.M., *J. Phys. E*, 1979, **12**, 1083.

25 Fultz, B. and Hamdeh, H.H., Phil. Mag. B, 1989, **60**, 601.

26 Fultz, B. and Morris, Jr., J.W., *Phys. Rev. B*, 1986, **34**, 4480.

27 Fultz, B. Hamdeh, H.H., and Okamoto, J., *Hyperfine Interac.*, 1990, **54**, 799.

28 Hamdeh, H.H., Okamoto, J., and Fultz, B., *Phys. Rev. B*, 1990, **42** 6694.

29 Buckley, R.A., *Metal Sci.*, 1975, **9**, 243.

RESISTIVITY VARIATIONS DURING ORDERING AND DISORDERING

WOLFGANG PFEILER
Institut für Festkörperphysik, Universität Wien
Strudlhofgasse 4, A-1090 Vienna, Austria

ABSTRACT

The dependence of electrical resistivity on short-range order (SRO) and long-range order (LRO) is shown in a condensed and simplified way. Experimental observations of recent years on ordering and disordering are reviewed briefly, whereby a special emphasis is laid on the thermal treatment applied. It is shown that a lot of information on the ordering process can be gained concerning ordering relaxation, equilibrium of order and order parameter. For a more complicated behaviour (critical behaviour at T_c or parallel development of SRO and LRO in addition to resistivity) information by other methods would be advantageous.

INTRODUCTION

Atomic ordering in metallic alloys means a certain tendency for the different kinds of atom to correspond to specific lattice sites. This tendency can be just a local deviation from the random value of the occupation probability for the lattice sites by the different atoms (short-range order (SRO)); or certain lattice sites are correlated to one kind of atom over great atomic distances leading to a superlattice (long-range order (LRO)). All these changes in the atomic arrangement are brought about by diffusion processes and therefore are linked to atomic mobility.

Whereas microstructural changes of short-range character need just a few atomic jumps for the new atomic equilibrium distribution to be established, ordering of atoms over large atomic distances obviously is linked to long-distance diffusion resulting in comparably sluggish process kinetics. In addition, the metallurgical state of the sample will be of essential importance in this case.

Atomic ordering in metallic alloys will drastically influence electrical, mechanical and magnetical properties by changing the microstructural state of the sample. In addition, the possibility of technical application of ordered alloys as materials with high mechanical stability at elevated temperatures is the reason for the increased interest in investigations on formation and destruction of order.

It is the aim of the present paper to show that measuring the electrical resistivity in connection with an appropriate temperature treatment gives a lot of interesting information on both the state of order as well as on the ordering kinetics.

RELATION BETWEEN HOMOGENEOUS SRO AND ELECTRICAL RESISTIVITY

In general, the electrical resistivity is a rather complicated quantity, which normally consists of several contributions. For example, scattering of conduction electrons may occur by phonons, lattice defects, internal stresses, dislocations, grain boundaries etc. However, if all these contributions but the one of interest, which in the present case is that resulting from differences in the atomic arrangement, are kept constant, a rather simple relation between SRO and the corresponding resistivity can be derived by usual pseudopotential formalism [1,2]:

$$\rho_{SRO}/\rho_s = \sum_i z_i \alpha_i Y_i \tag{1}$$

where the Warren-Cowley parameter

$$\alpha_i = 1 - (p_{AB}^i/c_B) \tag{2}$$

describes the deviation of local atomic arrangement from the random distribution. ρ_s means the resistivity of a sample without any atomic correlation effects.

The Y_i in eqn.(1) oscillate between positive and negative values depending on Fermi wave number k_f, the radius of the coordination shell R_i and the special shape of the atom potential used. Given a certain atomic correlation of the sample, e.g. a negative and dominating α_1, the sign and magnitude of ρ_{SRO} in addition to α_1 is determined by the coefficient Y_1.

EXPERIMENTAL OBSERVATIONS ON SRO

SRO atomic microstructure depending on temperature via the configurational part of the Gibb's free energy ($G_{conf} = H_{conf} - TS_{conf}$) its investigation requires a certain thermal treatment of the sample.

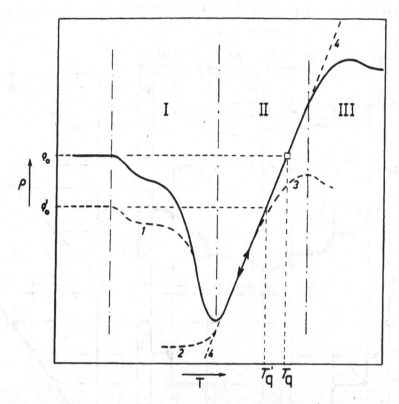

Figure 1. Schematic plot of electrical resistivity as a function of temperature for the case of a decrease of resistivity with increasing degree of order. Part I: increase of SRO due to a growing atomic mobility. Part II: decrease of SRO along an equilibrium line (curve 4). Part III: deviation of measured resistivity from the equilibrium line because of ordering during the quench. Curve 1: influence of thermal and mechanical pre-treatment. Curve 2: decreasing atomic mobility for decreasing temperature. Curve 3: influence of quenching rate during measurement.

Figure 1 gives a schematic plot of electrical resistivity versus annealing temperature as usually observed for isochronal annealing treatment. To exclude the changing phonon contribution to resistivity with changing temperature, which may further couple with changes in the atomic arrangement, the measuring temperature is assumed to be constant. Such an investigation gives an overview over the alloy system with respect to

     (i) the sign and magnitude (step height) of SRO-induced resistivity;

(ii) the onset of atomic mobility, that means the effect of surplus and thermal vacancies;

(iii) the quality of the quenching procedure.

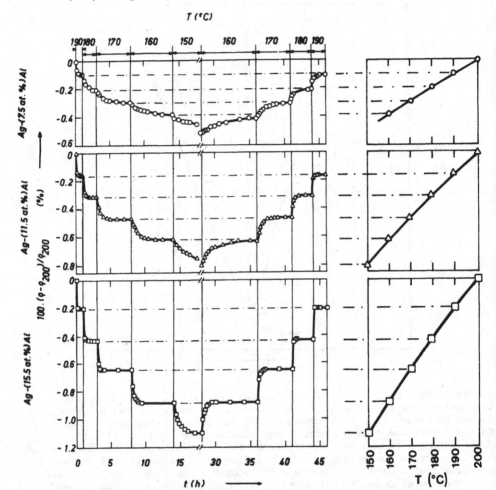

Figure 2. A series of isothermal anneals at falling and rising temperatures (small-step annealing) for the AgAl-system. (o) Ag-7.5at%Al; (Δ) Ag-11.5at% Al; (□) Ag-15.5at%Al. Right side: equilibrium line.

Figure 2 shows a sequence of isothermal anneals (small-step annealing): after a small and sudden change of temperature the system adjusts to the respective new equilibrium values of SRO. From this experimental procedure the following information on SRO can be gained:

(i) Thermodynamic stability: Certain resistivity values belong to specific annealing temperatures as equilibrium values, which are completely independent of the preceding temperature treatment.

(ii) The temperature dependence of SRO-induced resistivity (step height) can be studied in a very accurate way.

(iii) Investigating the step height for a concentration series of an alloy system the SRO-induced resistivity as a function of alloying concentration results.

(iv) The various isothermal annealing curves can easily be analysed with respect to their relaxation behaviour. This yields a SRO-relaxation time.

(v) An Arrhenius-plot of the SRO-relaxation times gives the SRO-activation energy coresponding to a sum of formation and migration enthalpy of vacancies.

For more details of measuring procedure and experimental results see [3].

Results on the temperature dependence of SRO-induced electrical resistivity as well as its concentration dependence can be used to try an evaluation of SRO-parameters by a comparison with model calculations. Good correspondence with experimental results from scattering experiments have been obtained this way [3]. It appears that in spite of the uncertainty with respect to sign and magnitude of the SRO-induced resistivity arising from eqn.(1) several alloy systems show a similar concentration behaviour, if a dominance of the contribution of the first coordination shell to SRO is assumed. Furthermore, a comparison of changes of resistivity as measured in CuNi for appropriate temperature treatment with changes of resistivity as calculated from measured changes in the SRO-parameters by means of diffuse neutron scattering [4] yielded an astonishingly good correspondence already for considering just the first coordination sphere.

Besides the motivation of getting information on SRO-kinetics as well as on SRO-equilibrium values corresponding to rather small changes in the atomic arrangement there arises the possibility to get detailed information about the point defect properties. The diffusion controlled change of atomic correlations usually causes a much larger change of electrical resistivity than the change in defect concentration itself. In this way an investigation of point defect properties can be managed like looking through a magnifying-glass. A special application of this method is the study of point defect kinetics after or during high energy particle irradiation, where diffusion is enhanced due to irradiation produced interstitials and vacancies [5].

RELATION BETWEEN LONG-RANGE ORDER AND ELECTRICAL RESISTIVITY

Whereas SRO will affect the electrical resistivity just by (diffuse) scattering of conduction electrons it is known that LRO in addition will contribute by superlattice Bragg scattering, which changes the effective number of conduction electrons n_{eff} (and maybe their effective mass m^*) by forming new energy gaps in the Fermi surface at the Brillouin zone boundaries of the superlattice [1].

If the LRO structure is described by the Bragg-Williams model it can be shown by pseudopotential formalism [6] that the dependence of conduction electron scattering on the order parameter S for a stoichiometric alloy within the relaxation time approximation can be written as

$$1/\tau(S) = (1/\tau(0))(1-S^2).\tag{3}$$

From a comparison with magnetic systems a plausible assumption can be made for the effective density of conduction electrons as a function of order parameter [7]

$$n_{eff} = n_o(1-AS^2),\tag{4}$$

where n_o is the conduction electron density for the completely disordered state. In A enters the position of Fermi surface relative to the superlattice Brillouin zone.

Both effects can be combined by $\rho=m^*/(n_{eff}e^2\tau)$ to give

$$\rho(S) = \rho(0)(1-S^2)/(1-AS^2).\tag{5}$$

$\rho(0)$ means the resistivity of the disordered state. No contribution of phonon scattering is included in this formula which therefore only is applicable to "quenched results" that means that the actual state of order has to be frozen-in by quenching to a low measuring temperature.

If the measurement is done at the temperature where ordering takes place (at-temperature measurement), an additional term due to the temperature coefficient has to be considered. When the measuring temperature is above the Debye temperature, eqn.(5) can be modified [7]:

$$\rho(S,T) = \rho(0)\frac{1 - S^2}{1 - AS^2} + \frac{TB}{n_o(1 - AS^2)}\tag{6}$$

Here B is the proportionality constant between the temperature coefficient of resistivity and the reciprocal value of the effective density of conduction electrons.

EXPERIMENTAL OBSERVATIONS ON LRO

Experimental investigation of LRO similar to SRO is a question of adequate thermal treatment of the samples. Again an isochronal (or linear) heating and cooling procedure gives an overview over the system with respect to the order-disorder temperature, the atomic mobility and the quality of the quenching procedure. A following series of isothermal annealing experiments at various temperatures yields information on ordering kinetics and if (quasi-) stable states of order are achieved. However, one has to bear in mind that an analysis of ordering kinetics may be complicated by nucleation of ordered domains (homogeneous or inhomogeneous), changes in the domain structure, influence of antiphase boundaries and magnetic scattering, if a magnetic component is present in the alloy.

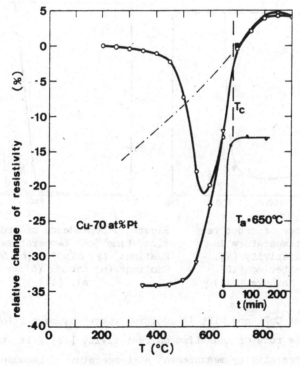

Figure 3. Resistivity change versus isochronal annealing temperature for $Cu_{30}Pt_{70}$ [8]. $\Delta T=50°C$, $\Delta t=20min$. Inset: isothermal annealing at 650°C. Dash-dotted line: hypothetical SRO equilibrium curve. (o) rising, (∇) falling temperatures; ($\blacksquare$) initial value after pre-annealing and quench. (After Banhart et al. [8].)

50

Anyway, such problems do not arise in all cases as is shown in fig. 3 for $Cu_{30}Pt_{70}$. Here the dependence of resistivity measured at 77K is shown as a function of temperature [8]. A picture very similar to figure 1 results for this off-stoichiometric alloy: Below the transformation temperature stable states of LRO seem to be formed, whereas above (and possibly below in parallel) SRO is observed.

It is known that theoretical treatments of order-disorder systems predict a critical slowing down of the relaxation rate of LRO near to the critical temperature. As electrical resistivity is a very accurate tool in determining ordering rates in this way a study of the critical behaviour near T_c can be tried. Figure 4 gives an example for such an investigation on the Ni_3Mn system [9] in comparison with results from neutron scattering [10]. However, a careful investigation of the critical region in $CoPt_3$ (figure 5) shows that the relaxation time of order-disorder is very sensitive to magnitude and direction of the temperature step used [11].

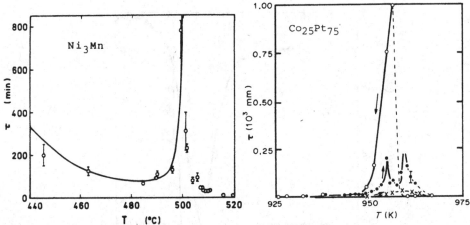

Figure 4. Dependence of order relaxation time on temperature in Ni_3Mn. Points: resistivity [9]; curve: neutron scattering [10]. (After Hatta and Shibuya [9].)

Figure 5. Dependence of order relaxation time on temperature in Co_3Pt. Heating: (x) $\Delta T \simeq 5°$; (o) $\Delta T \simeq 1°$. Cooling: (o) $\Delta T \simeq 5°$. (After Dahmani et al. [11].)

On the systems CoPt and NiPt in addition there has been a lot of work with respect to the role of magnetism on resistivity [12]. It turned out that in CoPt for resistivity measurement at-temperature, because of spin disorder scattering contribution a decrease in resistivity at the order-disorder transition is observed.

In figure 6 it is shown that for certain systems a comparison of the

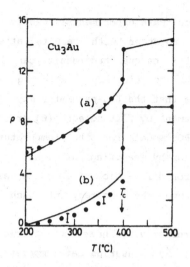

Figure 6. Variation of electrical resistivity with temperature for Cu₃Au.
Points: experiment;curves:calculated.
(a)at-temperature results;(b)quenched
results. (After Rossiter [7].)

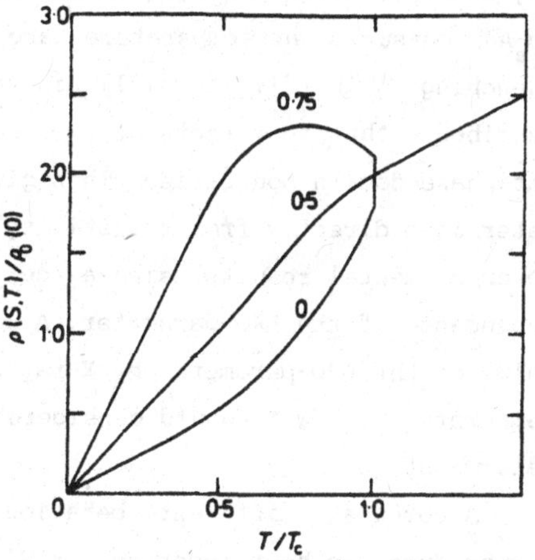

Figure 7b. Normalized resistivity
versus reduced temperature as calculated using eqn.(6) for at-temperature results. Parameter: various values of constant A in eqn.(6).
(After Rossiter [7].)

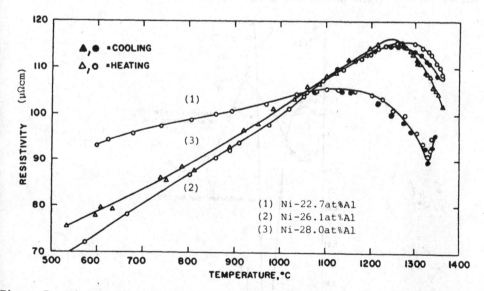

Figure 7a. Electrical resistivity of NiAl-alloys as a function of temperature. (After Corey and Lisowsky [13] and Cahn et al. [14].)

resistivity variations during ordering and disordering as measured and as
calculated from the temperature dependence of the order parameter by eqns.

(5) and (6) yields a very good correspondence. Here experimental results on Cu_3Au measured at-temperature are shown together with results after quenching to 4K [7]. The small discrepancy for the quenched results can be ascribed either to effects of SRO near T_c or to electron scattering at antiphase domain boundaries. This gives hope that the LRO-parameter may be determined directly from resistivity measurement by fitting eqn. (6) or (7) to experimental results using a certain LRO model for the temperature dependence of the LRO-parameter. A calibration by measuring at least one value of the LRO-parameter by X-ray or neutron diffraction for a certain temperature below T_c would considerably improve the accuracy of such a determination.

A somewhat different behaviour of resistivity near the critical temperature has been observed for Ni_3Al [13] by measuring at-temperature (figure 7a). Here after reaching a maximum during the disordering process with increasing temperature the resisitivity decreases up to the transition temperature, where a sharp change to increasing resistivity is observed. This behaviour can be fully understood from equ.(6) if a certain range of values for n_{eff} (constant A) are taken (figure 7b).

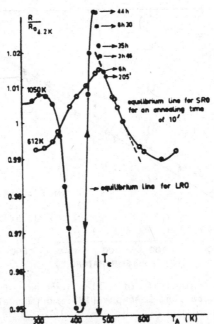

Figure 8. Change of electrical resistivity as a function of isochronal annealing temperature (Δt=10min. and as indicated). Quench temperatures as indicated in the figure. Measuring temperature 4.2K. (After Renaud et al. [15].)

In some systems beside the LRO process SRO seems to proceed rather independently in the disordered matrix and it is then a question of atomic mobility if changes in SRO or LRO are observed. This is shown in figure 8 for Au_9Cu [15]. After quenching from rather low temperatures (612K in fig. 8) due to the correspondingly low atomic mobility only an increase of SRO is observed until an SRO-equilibrium line is reached. The behaviour is similar to that of fig.1 but with an increasing resistivity with increasing degree of SRO. For higher quenching temperatures (1050K in fig.8) the higher atomic mobility enables a drastic decrease of resistivity due to the establishment of LRO between 350 and 400K. The subsequent increase of resistivity is observed to be reversible during anneal cycles at successively increasing and decreasing temperatures. This is interpreted as variation of equilibrium degree of LRO with temperature. Near to the transition temperature at 473K pronounced deviations from the SRO-equilibrium line are observed together with a strong increase of characteristic ordering times. This effect is attributed to the critical slowing down linked to an increasing spatial extension of SRO.

Similar results of a competition between SRO and LRO have been found in off-stoichiometric $Cu_{65}Pt_{95}$ [16], but in this case SRO seems to be not a precursor of LRO, but develops completely independently of the corresponding transition temperature.

CONCLUSIONS AND FUTURE ASPECTS

By measuring electrical resistivity during ordering and disordering it is possible to investigate SRO and LRO ordering processes *time-resolved* as well as *temperature-resolved* with a very high accuracy in comparison to other scattering methods. In this way in several cases relaxation times of the ordering process can be determined, giving insight into diffusion processes and thermally induced changes in materials characteristics with respect to technical application. If stable states of order are observed this may be used for a comparison with model calculations leading to a determination of the order parameter, especially when the resistivity method can be calibrated with other scattering methods.

Further information on the relation between electrical resistivity and atomic ordering is expected from a more rigorous theoretical approach to resistivity. Within the modern theory of alloys several physical quantities (spin-lattice relaxation time, hyperfine field, Knight shift and photo-emission spectra) have recently been calculated *ab initio* by means of the relativistic KKR-CPA and the embedded cluster method [17]. It is hoped that these calculations can be extended to the electrical resistivity in the Kubo formalism in the next future.

REFERENCES

1. P.L.Rossiter, The Electrical Resistivity of Metals and Alloys, Cambridge University Press, Cambridge, 1987, pp. 137-257.
2. P.L. Rossiter and P. Wells, The Dependence of the Electrical Resistivity on Short-Range Order. J. Phys. C: Solid St. Phys., 1971, 4, 354-63.
3. W. Pfeiler, Investigation of Short-Range Order by Electrical Resistivity Measurement. Acta Metall., 1988, 36, 2417-34.
4. W. Wagner, R. Poerschke and H. Wollenberger, Dependence of Electrical Resistivity on the Degree of Short Range Order in a Nickel-Copper Alloy. Phil. Mag., 1981, 43, 345-55.
5. A. Bartels, Study of Point Defect Properties in Concentrated Alloys with Short Range Ordering Methods. Mater. Sci. Forum, 1987, 15-18, 1183-1202.
6. P.L. Rossiter, Long-Range Order and the Electrical Resistivity. J. Phys. F: Metal. Phys., 1979, 9, 891-901.
7. P.L. Rossiter, Long-Range Order and the Electrical Resistivity. J. Phys. F: Metal. Phys., 1980, 10, 1459-65.
8. J. Banhart, W. Pfeiler and J. Voitländer, Detection of Short- and Long-Range Order in Cu-Pt Alloys. Phys. Rev., 1988, B37, 6027-9.
9. I. Hatta and M. Shibuya, Experimental Study on the Critical Dynamics of the Order-Disorder Transition in Binary Alloys. J. Phys. Soc. Japan, 1978, 45, 487-94.
10. M.R. Collins and H.C. Teh, Neutron-Scattering Observations of Critical Slowing Down of an Ising System. Phys. Rev. Lett., 1973, 30, 781-4.
11. C.E. Dahmani, M.C. Cadeville and V. Pierron-Bohnes, Temperature Dependences of Atomic Order Relaxations in Ni-Pt and Co-Pt Alloys. Acta Metall., 1985, 33, 369-77.
12. C. Leroux, M.C. Cadeville, V. Pierron-Bohnes, G. Inden and F. Hinz, Comparative Investigation of Structural and Transport Properties of $L1_0$ NiPt and CoPt Phases; Role of Magnetism. J. Phys. F: Met. Phys., 1988, 18, 2033-51.
13. C.L. Corey and B. Lisowski, Phase Decomposition in Near Ni_3Al Alloys. Trans. Met. Soc. AIME, 1967, 239, 239-43.
14. R.W. Cahn, P.A. Siemers, J.E. Geiger and P. Bardhan, The Order-Disorder Transformation in Ni_3Al and Ni_3Al-Fe Alloys. I. Determination of the Transition Temperatures and their Relation to Ductility. Acta Metall., 1987, 35, 2737-51.
15. G. Renaud, M. Belakhovsky, J. Hillairet, M. Wuttig, G. Bessenay and S. Lefebvre, J. de Phys., 1987, C8, 519-24.
16. B. Urban-Erbil and W. Pfeiler, Ordering and Disordering in CuPt-Alloys. This conference.
17. P. Weinberger, Electron Scattering Theory for Ordered and Disordered Matter, Clarendon Press, Oxford, 1990, pp. 176-238.

Ackowledgement

The financial support by the Austrian "Fonds zur Förderung der wissenschaftlichen Forschung" grant Nr. 7615 und the "Österreichische Forschungsgemeinschaft" is gratefully acknowledged. The author is indebted to Dr. W. Püschl for a critical reading of the manuscript.

CALORIMETRIC STUDY OF REORDERING OF DISORDERED Ll2, Ni3Al BASED ALLOYS

M.D. BARO, J. MALAGELADA, S. SURINACH, N. CLAVAGUERA* and M.T. CLAVAGUERA-MORA

Física de Materials, Departament de Física,
Universitat Autònoma de Barcelona, 08193 Bellaterra, Spain
* Departament d'Estructura i Constituents de la Matèria, Facultat de Física,
Universitat de Barcelona, Diagonal 647, 08028 Barcelona, Spain

ABSTRACT

Technologically interesting Ni3Al based alloys can be effectively disordered by ball milling. The compound disordered in this way shows, when continuously heated, a broad exothermic transformation after which the ordered Ll2 phase is again obtained. The total enthalpy released in the transformation grows with the milling time.

It is observed that other kinetic processes, apart from the ordering one, are involved in the broad exothermic transformation. This study is devoted to the separation and deeper comprehension of these processes and for that purpose, isothermal measurements coupled with continuous heating measurements were performed. Isothermal measurements give a continuous decay of the calorimetric signal. At intermediate temperatures this decay leads to a state where the sample is still able to transform, as it is evidenced on further continuous heating experiments. Three different processes have been identified, the more energetic corresponding to the reordering process. The kinetic parameters of the reordering process have been obtained and are discussed.

INTRODUCTION

The greatly expanded research on mechanical properties of deformed intermetallic phases shows that the subject has always been of interest in Materials Science. In the last years, a considerable amount of work has been done to investigate the mechanical behaviour of ordered alloys. In particular, it has especially centered upon alloys with very high order-disorder transition temperatures because of their high temperature strength [1-3] and so, with immediate impact on industrial application.

The discovery that permanently ordered phases can be completely disordered not only by intense plastic deformation [4-5], but also by electron irradiation [6,7] and, very recently, by ultra-rapid solidification of sputter-deposited films [8] and by melt-spinning [9], offers the possibility for a better understanding of how the physical and mechanical properties of such artificially disordered alloys change as order is gradually reintroduced.

In the present work we investigate some aspects of the disordering and further reordering of a ball-milled Ni_3Al based alloy. Our study is focused on the way in which the disorder is achieved in this permanently, strongly ordered compound, and on the reordering and grain growth kinetics determined from differential scanning calorimetry (DSC).

EXPERIMENTAL METHODS

The intermetallic atomized powder used for the present work was kindly supplied by J.R. Knibloe and R.N. Wright (INEL-Idaho). An extensive characterization of this powder is reported in [10]. The powder is single-phased, ordered γ', as confirmed by XRD analysis. The ball-milling experiments and the analysis of the structural changes introduced during the milling process and after annealings at properly selected temperatures were performed in the Department of Materials Science and Metallurgy of the University of Cambridge by R.W. Cahn and S. Gialanella, which are kindly acknowledged. Experimental details are reported in [11-12].

The thermal stability of the disordered phases formed during ball-milling was tested by studying their behaviour on heating in a computerized differential scanning calorimeter, type Perkin-Elmer DSC 7, under pure argon atmosphere. The continuous heating experiments were performed at scan rates in the range 5-160 K/min. Isothermal experiments were carried out by heating up to the annealing temperature at a rate of 40 K/min, and holding at that temperature until the signal was no longer measurable. After cooling to room temperature, the sample was run again under identical conditions to establish the baseline, which was always properly optimized and matched the zero of the signal. For simplicity it was assumed that the rate of heat generated during the exothermic process was proportional to the transformation rate, dx/dt. Therefore, the fraction transformed, x, at a given time, t, was determined as the ratio between the area subtended by the signal at that time and the total area of the exothermic signal. Similarly, the transformation rate at time t or/and temperature T was determined as the ratio between the height of the DSC curve at that time or/and temperature and the complete area. The calibration procedure is described elsewhere [13]. The estimated errors were 2% for the enthalpy and less than 0.05% for the temperature. The apparent activation energies of the exothermic processes observed in continuous heating experiences were calculated by Kissinger's shift peak method [14].

RESULTS AND DISCUSSION

Differential scanning calorimetry of the whole transformation

The DSC curve reported in figure 1 shows the result of a continuous heating experiment (40 K/min) on a disordered sample milled for 4 hours. It reveals the existence of a broad exothermic process where two peaks are evident. A shoulder on the high-temperature end of the process which may reflect a third contribution can be seen and experimental results suggest that this shoulder is insensitive to the milling time. All these transformations are irreversible, as proved by repeating a thermal cycle with the same specimen. Similar DSC curves were obtained by Harris et al.[15] for disordered Ni_3Al films synthesized by vacuum evaporation.

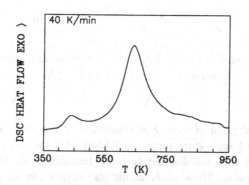

Figure 1. DSC curve from a sample milled for 4 h.

To get more insight into the nature of the ordering process a deconvolution into three exothermic processes has been attempted. The sample chosen was the one ball milled for 20 h (completely disordered). The procedure followed was a combination of isothermal treatments and further continuous heating experiments. The isothermal treatments were performed at temperatures equal to the peak temperatures (measured during continuous heating at 40 K/min). It was proved that these treatments were the best ones to eliminate the contributions of the first and second processes, respectively. Continuous heating experiences (at the heating rate 40 K/min) after every isothermal treatment provided the necessary information to get the results presented in figure 2.

The values of the peak temperatures, as well as the enthalpy released in the individual exothermic processes were measured. Within the experimental error, the peak temperature of the first exothermic event remains unchanged with increasing milling time; however, for the second (large) exothermic process, the peak temperature decreases with increasing milling time down to a well-defined saturation value. The enthalpy values obtained from the integrated area, as a function of the milling time, are illustrated in figure 3 for the first and the second exothermic peaks. The enthalpy release in both peaks increases and for milling times up to 20 h no saturation value is reached, even though XRD measurements show that all long-range order has disappeared after 8 h of milling [12].

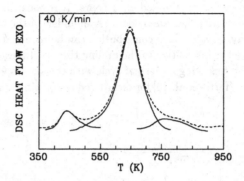

Figure 2. *Possible deconvolution of the exothermic processes carried on the sample milled for 20h.*

From a structural point of view, and since most of the long range order evolves over a relatively narrow temperature range, coincident with the large exothermic peak, this could be associated with the disorder to order transformation and growth of new grains of the ordered phase. These observations also suggest, in an indirect way, that the first exothermic step could be linked with annealing out of point defects [16]. That reordering is accompanied by growth of the nanocrystalline grains is a result that agrees with that of Harris et al. [15].

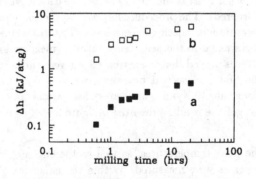

Figure 3. a) *Enthalpy released in the first exothermic process from Ni_3Al milled for different times.* b) *Enthalpy released in the remaining exothermic process from Ni_3Al milled for different times.*

Calorimetric analysis of the reordering kinetics

The calorimetric measurements performed for the study of the reordering kinetics combined both isothermal and continuous heating regimes. This procedure increases, in general, the precision of the determination of the kinetic parameters from DSC results [14,18]. Isothermal measurements as will be presented later, give a continuous decay of the calorimetric signal. As this behaviour is unusual for the classical Johnson - Mehl - Avrami - Erofe'ev (JMAE) analysis, another model predicting it was also assumed. This was the analysis based on a grain growth process, extensely described by Chen and Spaepen in [17,18]. It can be deduced that the rate of transformation is given by an equation of the form:

$$\frac{dx}{dt} = \frac{r_0}{n} \frac{k(T)}{r^{n+1}}$$ (1)

where x is the transformed fraction at time t and temperature T, r_0 is the initial grain radius, n is an exponent, empirically between 1.5 and 4 and $k(T)$ is the reaction rate given by

$$k(T) = \frac{k_0}{T} exp(-E/RT)$$ (2)

E is the apparent activation energy of a singly activated process and k_0 is related to a frequency factor that can be obtained from the standard transition state rate theory [19]. Besides, r is the grain size at a given temperature or time. In an isothermal experiment this is given by

$$r^n(t) = r_0^n + k(T) t$$ (3)

and in a continuous heating experiment at a constant rate, β, it follows the equation

$$r^n(t) = r_0 + \frac{1}{\beta} \int_0^T k(T) \, dT$$ (4)

In the classical nucleation-and-growth process, as depicted by the JMAE analysis [20], the rate of transformation is given by

$$\frac{dx}{dt} = K(T) \, f(x)$$ (5)

Here, the reaction rate, $K(T)$, is generally given by the Arrhenius expression

$$K(T) = K_0 \, exp(-E/RT)$$ (6)

where E is also the apparent activation energy and K_0 the pre-exponential factor. The function $f(x)$ is given by the expression

$$f(x) = p\,(1-x)[-ln(1-x)]^{(p-1)/p} \tag{7}$$

where p is the kinetic exponent that, depending on the nucleation mechanism and growth morphology, varies between 0.5 and 4 [19]. Under continuous heating conditions, equation (5) can be rewritten as

$$\frac{dx}{dT} = \frac{1}{\beta}\,K(T)\,f(x) \tag{8}$$

There are many different types of processes whose thermal behaviour can be described by the general kinetic equations (5) to (7) in a certain range of temperature. On the other hand, equations (1) to (4) were developed to provide a criterion to distinguish between microcrystalline and amorphous materials and have been especially applied to the study of grain growth processes.

The study of the reordering kinetics was carried on a sample ball-milled for 20 h and the values of the kinetic parameters E, k_0 and n (or K_0 and p) were obtained. The peak method [14] was used to obtain the apparent activation energy, E, for both analyses. As it is proved in [18], the very similar form of equations (2) and (6) is the main reason why the Kissinger analysis of nucleation-and-growth processes applies to grain growth processes as well. The values of E obtained in this way for the reordering process, remain constant irrespectively of the milling time, within the experimental errors [11]. The other parameters have been determined by fitting the simulated calorimetric signals that can be deduced from both models to the experimental DSC signal of the reordering process of the sample studied. The results are quoted in table I.

The value of r_0 was taken directly as the mean particle size given by XRD results. As it is shown in Table I, the value of k_0 for isothermal measurements which better reproduces the experimental results is different to that for continuous heating experiments. The fact that some parameters are not the same for iso- and non-isothermal simulations was already indicated in [18] and it was suggested that it could be due to the temperature dependence of some of the parameters assumed to be constant in the grain growth theory. Making a comparison between the pre-exponential factors k_0 and K_0 is possible by considering for instance, $n = 2$ [21] and then replacing r in equation (1) by its expression as a function of x, that is,

$$\frac{dx}{dt} = k(T)\,\frac{(1-x)^3}{2r_0} \tag{9}$$

Using equations (2) and (6) the following approximate expression relating K_0 and k_0 is obtained

$$K_0 \simeq k_0/2Tr_0^2 \qquad (10)$$

provided that the values of the function $(1 - x)^3$, for values of x next to the peak value, are similar to the corresponding ones of the function $f(x)$ of equation (7) for $p = 0.6$. The values found for $k_0/2Tr_0^2$ are: isothermal conditions, $2\ 10^{13}s^{-1}$; continuous heating regime, $3\ 10^{10}s^{-1}$. The later value scales as expected from equation (10). However the values for isothermal and continuous heating conditions are very disimilar, and this strongly suggests that the grain growth theory cannot be applied.

TABLE I.

Value of the kinetic parameters used in the simulation

	E (kJ/mol)	r_0 (nm)	n, p	$k_0\ (Km^2s^{-1})$ non-iso	iso	K_0 (s^{-1})
GG	150	20	2	$1.5\ 10^{-3}$	10	
JMAE	150		0.6			$3\ 10^{10}$

From the results obtained, it can be deduced that the JMAE analysis is the one giving a better fit. This can be seen in figure 4, where the DSC curve, at 40 K/min, corresponding to the reordering peak obtained for the sample studied is shown together with the best fits obtained by the grain growth theory and by the JMAE theory.

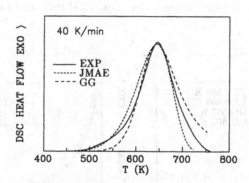

Figure 4. *Enlarged view of the deconvoluted reordering peak from a sample milled for 20 h ($\beta = 40$ K/min). The dashed lines are simulations corresponding to Eqs. (1,4) and Eqs. (7,8).*

Further confirmation of the ability of the considered models to reproduce the experimental behaviour is obtained by comparing continuous heating experiments, performed at different heating rates, to the corresponding simulations; the same must be done for isothermal experiments. An example of this is given in figures 5 and 6. Figure 5 shows the DSC curves taken at two different heating rates. Superimposed to them there are the signals predicted by the grain growth theory (left) and by the JMAE model (right). The shift of the peak temperature is well reproduced in both models because they use a very similar temperature dependence of the rate constant (equations (2) and (6)). In figure 6 the isothermal DSC curves obtained at two different temperatures after heating from room temperature at a scanning rate of 40 K/min are shown. The curves simulated by the grain growth theory and by the JMAE model are also shown in figure left and right graphs, respectively. The rate of decrease of the signal is better reproduced by the JMAE model.

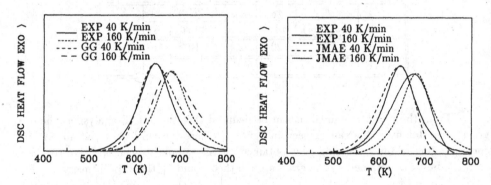

Figure 5. Scanning calorimetry curves, at 40 and 160 K/min, from Ni₃Al milled for 20 h. The dashed lines are simulations using: Eqs. (1,4) (left) and Eqs. (7,8) (right).

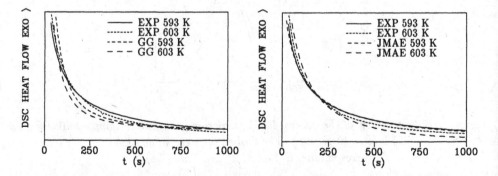

Figure 6. Isothermal DSC curves from Ni₃Al milled for 20 h compared to the calculated ones using Eqs.(1,3) (left) and Eqs.(5,7) (right).

As it is well known, the value of $p = 0.6$ in the JMAE model can correspond to a one- or two- dimensional growth of ordered regions, diffusion-controlled, from pre-existing ordered nuclei, the ordered regions consuming the surrounding disordered material until regions impinge at antiphase domain boundaries. Simultaneously with the growth of ordered domains here is also a coarsening of the particles, as seen by XRD. The enthalpic contribution coming from grain growth, from an initial value of $r_0 = 20$ nm may be obtained by assuming an antiphase boundary energy ranging from 30 to 150 mJ/m^2 and it is of the order of 0.1 to 0.5 kJ/mol. Therefore, almost all of the peak enthalpy (about 7 kJ/g· at) can be associated to the ordering energy.

In order to get indirect information about the mechanisms taking part in the reordering process, the measured value of the apparent activation energy can be compared to those of the literature. The apparent activation energy value is similar to that of the vacancy migration in Ni, which is about 1.5 eV [22] and slightly higher than that measured for Ni vacancy migration in Ni$_3$Al, which is 1.2 eV [23] or the one calculated by the embedded atom method [24] for the migration energy of a Ni vacancy into an adjacent Ni site, which is 1.02 eV. The pre-exponential factor K_0 is two orders of magnitude lower (and the activation energy also lower) than the values assigned in [25] for the resistivity variation under isothermal conditions (862, 880, 925 and 973 K) in Ni$_3$Al rapidly cooled from high temperature, but the reason for these differences may be that other processes are also activated when annealing at higher temperatures.

SUMMARY OF THE RESULTS

- Disordering by ball-milling of Ni$_3$Al based intermetallics involves not only the destruction of chemical order, but even the reduction of crystallite size and an increase of lattice strain.

- It is observed that the reordering process is not unique but there are many processes involved. From the calorimetric measurements three exothermic steps were evidenced. The small low-energy heat release at about 400 K does not correspond to re-ordering but may be linked with annealing out of point defects. Instead during the second exothermic process long range order and grain growth were found to evolve simultaneously and it seems that reordering was always accompanied by growth of the nanocrystalline grains.

- The ordering kinetics has been studied calorimetrically. The experimental results have been fitted assuming two different mechanisms: a grain growth process and the Johnson - Mehl - Avrami - Erofe'ev analysis. The last one gives a unique set of parameters that is able to reproduce both continuous heating and isothermal measurements and, therefore, it seems more suitable to interpret the ordering process. The values obtained for the kinetic exponent, p, and for the activation energy, E, strongly suggest that the ordering process is a diffusion-governed one- or two-dimensional growth.

ACKNOWLEDGMENTS

The research reported in this paper was supported by the EEC (contract No. SCI-0262) under the SCIENCE programme and by the CICYT (project No. MAT90-0454), which are acknowledged. We thank the MEC for a FPI researh studentship to J.M.

REFERENCES

1. Stoloff N.S., Davies R.G., The mechanical properties of ordered alloys. Prog. Mater. Sci., 1966, 13, 1-86.

2. Cahn R.W., Takeyama M, Horton J.A., Liu C.T., Recovery and recrystallization of the deformed, orderable alloy $(Co_{78}Fe_{22})_3V$. J. Mater. Res., 1991, 6, 57-70.

3. Cahn R.W., Recovery, strain-age-hardening in deformed intermetallics. In High Temperature Aluminides and Intermetallics, eds. Whang S.H., Liu C.T., Pope D.P., Stiegler J.O., TMS, Warrendale, PA, 1990, pp. 245-70.

4. Hellstern E., Fecht H.J., Fu Z., Johnson W.L., Structural and thermodynamic properties of heavily mechanically deformed Ru and AlRu. J. Appl. Phys., 1989, 65, 305-10.

5. Jang, J.S.C., Koch, C.C., Amorphization and disordering of the Ni_3Al ordered intermetallic by mechanical milling. J. Mater. Res., 1990, 5, 498-510.

6. Carpenter, G.J.C., Shulson, E.M., Effect of 1 MeV electron irradiation on Ni_3Al. Scripta Met., 1981, 15, 549-54.

7. Liu, H.C., Mitchell, T.E., Irradiation induced order-disorder in Ni_3Al and NiAl. Acta metall., 1983, 31, 863-72.

8. West, J.S., Manos, J.T., Aziz, M.J., Formation of metastable disordered Ni_3Al by pulsed laser-induced rapid solidification. Presented at MRS Symposium on High Temperature Intermetallic Alloys IV, Boston, November 1990; to be published in Mat. Res. Soc. Symp. Proc. Vol.213.

9. Yavari, A.R. and Bochu, B., $L1_2$ ordering in Ni_3Al-Fe disordered by rapid quenching. Phil. Mag. A, 1989, 59, 697-705.

10. Knibloe, J.R., Wright, R.N., The influence of alloying on the microstructure and mechanical properties of P/M Ni_3Al. Acta metall. mater., 1990, 38, 1993-2001.

11. Malagelada, J., Suriñach, S., Baró, M.D. , Gialanella, S. and Cahn, R.W., Kinetics of reordering in Ni_3Al based alloys disordered by ball milling. To be published in

Materials Science Forum, Aedermannsdorf, Switzerland, Trans. Tech. Publications, 1991.

12. Gialanella, S. and Cahn, R.W., Order parameter and lattice parameter in $L1_2$ ordered alloys. Yavari, A.R. ed., Proc. European Workshop on Ordering and Disordering, Grenoble, July 1991, to be published by Elsevier.

13. Suriñach, S., Baró, M.D., Clavaguera-Mora, M.T., Clavaguera, N., Kinetic study of isothermal and continuous heating crystallization in $GeSe_2$-$GeTe$-Sb_2Te_3 alloy glasses. J. Non-Cryst. Solids, 1983, 58, 209-17.

14. Kissinger, H.E., Reaction kinetics in differential thermal analysis. Anal. Chem., 1957, 29, 1702-06.

15. Harris, S.R., Pearson, D.H., Garland, C.M. and Fultz, B., Chemically disordered Ni_3Al synthesized by high vacuum evaporation. J. Mater.Res., 1991, 6, in press.

16. Cahn, R.W., Kinetics of ordering and disordering of alloys. Yavari, A.R. ed., Proc. European Workshop on Ordering and Disordering, Grenoble, July 1991, to be published by Elsevier.

17. Chen, L.C. and Spaepen, F., Calorimetric evidence for the microquasicrystalline structure of 'amorphous' Al/transition metal alloys. Nature, 1988, 336, 366-8.

18. Chen, L.C. and Spaepen, F., Analysis of calorimetric measurements of grain growth. J. Appl. Phys., 1991, 69, 679-88.

19. Christian, J.W., The theory of transformations in metals and alloys. 2nd ed., Pergamon, Oxford, 1975.

20. Avrami, M., Kinetics of phase change I, II and III. J. Chem. Phys., 1939, 7, 1103-12, ibid., 1940, 8, 212-24, ibid., 1941, 9, 177-84.

21. Atkinson, H.V., Theories of normal grain growth in pure single phase systems. Acta metall., 1988, 36. 469-91.

22. Schumacher, D., Schule, W. and Seeger, A., Untersuchung atomarer Fehlstellen in verformtem und abgeschrecktem Nickel. Z. Naturforsch., 1962, 17a, 228-35.

23. Wang, T., Shimotomai, M. and Doyama, M., Study of vacancies in the intermetallic compound Ni_3Al by positron annihilation. J Phys. F, 1984, 3 37-45.

24. Foiles, S.M. and Daw, M.S., Application of the embedded atom method to Ni_3Al. J. Mater. Res., 1987, 2, 5-15.

25. Kozubski, R. and Cadeville, M.C., In situ resistometric investigation of ordering kinetics in Ni₃Al. J. Phys. F, 1988, 18, 2569-75.

ORDER PARAMETER AND LATTICE PARAMETER IN L1$_2$ ORDERED ALLOYS, WITH SPECIAL REFERENCE TO Ni$_3$Al-BASED ALLOYS.

STEFANO GIALANELLA*, SIMON B. NEWCOMB, ROBERT W. CAHN
Department of Materials Science and Metallurgy, University of Cambridge, Pembroke Street,
Cambridge, CB2 3QZ, U.K.
*permanent address: Dipartimento di Ingegneria dei Materiali, Università di Trento, Mesiano di
Povo, 38050, Trento, Italy.

ABSTRACT

Ordered alloys which cannot be disordered by heat treatments and quenching can still be effectively disordered by irradiation or by ball-milling. The disordered state is achieved during this extreme form of cold-work by the creation of several kinds of defects; in particular vacancies and dislocations. The increase of anti-phase boundary density and the consequent reduction of ordered domain size are other aspects of the disordering process. The enhancement of the free energy can destabilize the crystalline structure and amorphous regions can appear.
We present some results concerning the characterization of ball-milled Ni$_3$Al based intermetallic powders, in the state which follows the complete destruction of long range order, as verified by x-ray diffraction. We believe that the observed continued lattice expansion is due to the combined effect of lattice strain, increase of defect concentration, especially vacancies and dislocations, and elastic mismatch. The lattice parameter is strongly influenced by the local atomic configuration, so that another contribution can come from the reduction of short-range order.
A comparison with some other L1$_2$ disorderable alloys is also presented.

INTRODUCTION

The degree of long-range order (LRO) in metallic alloys can be quantified by defining a suitable order parameter. Of course it is a macroscopic parameter, and to each value will correspond a series of possible microscopic configurations.Therefore other parameters that are needed for a complete description of the ordered state of an alloy were taken into account, such as lattice parameter and elastic modulus [1]. The lattice parameter is strongly influenced by the local atomic configuration; as such it is particularly sensitive to variations of the degree of order of an alloy, both long- and short-range. Several studies on the lattice parameter of ordered alloys have been carried out to establish a relationship between this quantity and the short-range order (SRO) parameter. In one of the first systematic studies, Betteridge [2] showed that a linear relationship can be assumed between the lattice parameter and SRO parameter in Cu_3Au. It was observed that the progressive thermal disordering of the material, achieved with quenchings from temperatures above and below the critical temperature for the order-disorder transformation, is accompanied by an increase in the lattice parameter. The LRO parameter, S, as defined in the Bragg-Williams theory, is related to the way in which the superlattice sites are occupied over distances large compared with the interatomic spacings. The SRO parameter, σ, is related instead to the local atomic arrangement of a limited number of atoms. A number of different SRO states correspond to each value of S and only in the case of perfect order are the two parameters both equal to 1. A more complete analysis of the interrelationships between S and σ can be found, for example, in [3]. Above the critical temperature the ordered alloy still retains a certain degree of SRO. In the cited study a simple model was proposed, taking into account the effects of the reduction in that SRO. The lattice expansion observed upon increasing the quenching temperature above T_C was ascribed to the variation of the average bond length. This was expressed as a linear combination of the bond lengths in the pure elements, weighted with factors proportional to the SRO parameter. Figure 1 is the graph a_0- vs -annealing temperature for Cu_3Au, taken from [2]. The full line is the variation of the lattice parameter predicted by the Peierls theory [4]. The circles are experimental determinations of a_0 at different temperatures Some values of σ, theoretically evaluated from the Peierls's model, are also reported. Theory and experiment show a sharp rise in the lattice parameter across the order-disorder transition temperature, which corresponds to the sharp fall in the LRO parameter. The SRO parameter is not zero above the critical temperature: this explains the further increase of a_0.

A more accurate model was proposed by Dienes [5]. A relation between the nearest-neighbour distance, the composition and the SRO parameter was established. He assumed, as starting point for his model, that the bond energies between two atoms in a binary alloy, characterized by a certain value of the SRO parameter, can be expressed in terms of the square of the difference between the nearest-neighbour distance of the atoms in the alloy and in the elemental materials. The model, applicable to both ordered and random alloys, gives results in good agreement with the experimental data for random alloys and satisfactory qualitative predictions about the variations upon disordering of the nearest-neighbour distance in ordered alloys. In [1] an approximately linear dependence of the lattice parameter on the LRO parameter is reported, again for thermally disordered Cu_3Au but only for a limited range of S, from 1 to 0.8 approximately. This is the range of S accessible by changing the temperature

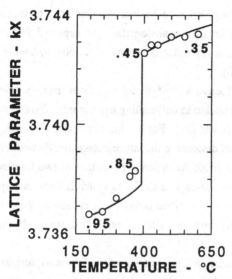

Figure 1. Lattice parameter of Cu_3Au- vs -annealing temperature (from [2]). <u>Full line</u> calculated from Peierls's theory; <u>circles</u>: experimental values. Some values of the short-range order parameter are reported.

below the transition point. Other interesting researches [6,7], again on thermally disordered Cu_3Au, show an increase of a_0 as the degree of order decreases. There is a good agreement in the values reported for the lattice expansion : 0.13 - 0.16%.

A few more data on lattice expansion upon disordering are available for other ordered compounds which have been thermally disordered: for Ni_3Fe, +0.08% [8]; for Nb_3Sn, 1.25% [9]; for Mg_3In, 0.13% [10]; and 0.09% was reported for Cu_3Pt [11].

The samples disordered by quenching contain a certain concentration of non-equilibrium defects which, vacancies in particular, will affect the lattice parameter. This aspect has been considered in irradiation experiments of intermetallics with high energy electrons and ions. Lattice swelling is one of the main effects of irradiation, during which both disordering and ordering have been reported to occur. For comprehensive reviews of irradiation-induced phenomena in intermetallics the reader is referred to [12,13]. The lattice expansion caused by irradiation can be mainly ascribed to the increasing concentration of point defects and to the creation of anti-site defects. The observed extent of the expansion depends on the material and on the irradiation conditions, bombarding particles included. For electron-irradiated Zr_3Al $\Delta a/a_0$ is about +0.55%; the change is $+0.7\pm0.4$ when the alloy was irradiated with 0.5 MeV Ar^+ [14], which does not disorder it completely, and +0.775% when irradiated, and completely disordered, with 1 MeV Kr^+ ions [15]. In this material the lattice expansion leads to an elastic instability, similar to the one observed in the pre-melting stage of a crystalline solid [16], supporting the proposed model about the similarity of melting and solid state amorphization [17,18]. No lattice expansion data are reported for electron-irradiation experiments with another strongly ordered intermetallic, Ni_3Al, having the same ordered structure $L1_2$ as Zr_3Al [19,20].

An alternative and effective way of disordering even strongly ordered compounds is with heavy mechanical deformation in ball-milling experiments. Several materials have been successfully disordered in this way [21] . For this kind of disordering, too, some studies have been done to relate the state of disorder to the lattice expansion. Particularly interesting is the work recently published [22] about the opposite behaviour of two intermetallics, Nb_3Sn, in which the lattice parameter increases, and CoGa, in which it decreases upon disordering. In this work it is pointed out that the observed behaviour is caused by the particular structure of the point defects (vacancies and anti-site defects) introduced during the mechanical disordering.

It is clear that an exhaustive interpretation of the lattice expansion upon disordering, whatever the disordering technique used, must take into account several different factors. In the present work we present some results concerning the disordering induced by ball-milling in a Ni_3Al-based alloy, and the associated variations of its lattice parameter. This is the first study of the variation of lattice parameter of Ni_3Al with the state of order.

EXPERIMENTAL DETAILS

The atomized intermetallic powder used for the present work was kindly supplied by J.R.Knibloe and R.N.Wright (INEL-Idaho). An extensive characterization of this powder is reported in [23]. The powder is single phase, ordered, γ', as confirmed also by our x-ray diffraction (XRD) analysis, and has the following composition (in at.%): Ni-76.6, Al-23.3, B-0.097. The ball-milling experiments were carried out in a planetary ball mill using tempered stainless chrome steel balls and vials. They were loaded and sealed under argon atmosphere, after degassing the powder for several hours at a pressure of about 10^{-3} torr in a desiccator. XRD analysis of the specimens were performed with a wide angle diffractometer, using monochromated $CuK\alpha$ radiation. The conditions were selected to achieve a high sensitivity in detecting the superlattice peaks, even at intermediate and low levels of order. The variations of the relative degree of long-range order were measured from the integrated intensity ratios of the $(110)_S$ and $(220)_F$ peaks, calibrated with the same ratio obtained from an unmilled powder, having the maximum degree of order compatible with the sample composition. The integral breadths of the diffraction peaks, corrected for the instrumental broadening, were used for evaluating the average size of the coherent diffracting domains, D, and the lattice rms strain,

$\sqrt{\langle \varepsilon^2 \rangle}$ following the procedure described in [24]. The diffractometer alignment was checked with a polycrystalline Si sample. The lattice parameter was evaluated by the standard Riley-Sinclair extrapolation method using the five fundamental peaks in the 2θ range $40°$-$100°$.

For TEM observations the powders were embedded in a Ni electroplated matrix. The sheet obtained was mechanically thinned and spark-eroded 3 mm disks were electropolished to electron transparency.

RESULTS

Ball-milling of Ni_3Al based alloy, as already reported [25,26], leads to a progressive reduction of the coherent scattering domain size; the rms strain, after a sharp increase, achieves a steady value of about 0.5%. Another aspect of the disordering process is a consistent lattice expansion. The increase of the lattice parameter with milling time is illustrated in figure 2. The lattice parameter increases to a maximum value of 0.79%, which is very close to 0.775%, reported for the lattice expansion of Zr_3Al upon irradiation with 1 MeV Kr^+ ions [15].

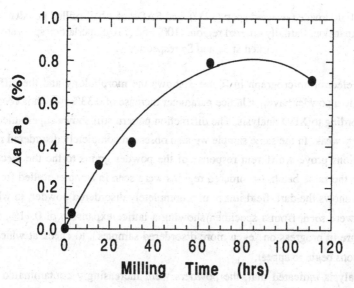

Figure 2. Lattice expansion- vs -milling time for Ni_3Al-based alloy.

The substantial difference is that in our ball-milled powder the disappearance of the superlattice peaks, measured by XRD (in the cited study on Zr_3Al the disordering process was followed by quantitative electron diffraction), occurs at earlier stages. From experiments carried out with the specific aim of pinpointing the critical expansion, we found a value of about 0.25% corresponding to S=0. These results were checked with TEM observations.

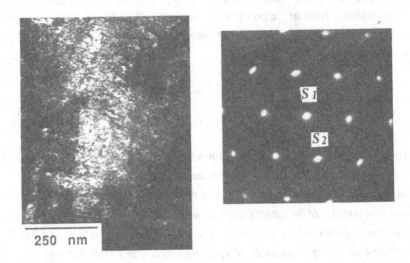

Figure 3. Dark-field image (selected spot: (111)), and SADP of a ball-milled powder showing 0.33% lattice expansion. Partially ordered region. (100) and (110)Superlattice spots are marked at S_1 and S_2 respectively.

The dark field electron micrograph in figure 3 shows the morphology and the diffraction pattern relating to a powder having a lattice parameter increase of 0.33%, which is completely disordered according to XRD analysis. The diffraction pattern still shows superlattice spots, even though very weak. In the same sample we also observed completely disordered regions. These observations prove a different response of the powder grains to the the mechanical attrition, within the same batch. No ordered regions were seen in samples milled for longer times. Figure 4 shows the dark-field image of a completely disordered powder, in which no ordered zones were seen, from a specimen showing a lattice expansion of 0.41%. Further investigations are in progress on "even more disordered samples", to check at which stage amorphous regions begin to appear.

EDS analysis indicated that, the powders were increasingly contaminated by the grinding media. For longer milling times further evidence came from XRD results, with the appearance of extra peaks, due to the Cr-steel contamination, and from microscopical observations. Figure 5 is the micrograph of an iron-based contamination particle, found in the sample of figure 4. Fe and Cr were both found to be present here and since both are soluble in Ni_3Al, and it was then important to check the possible effect of this contamination on the observed lattice expansion. The data available in literature on the lattice expansion observed in

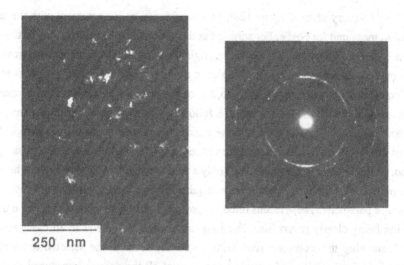

Figure 4. Dark-field image (selected rings (111) and (200)) and SADP of a ball-milled powder showing 0.41% lattice expansion. Completely disordered region.

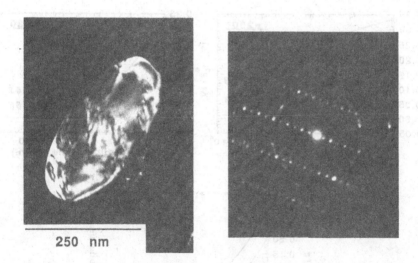

Figure 5. Iron-based contamination particle found in the sample of fig.4. Dark-field image and SADP.

Ni₃Al when ternary elements, Fe and Cr, are added is quite controversial. A theoretical treatment, based on the cluster variation method [27], predicts a lattice expansion of Ni₃Al both for Fe and Cr alloyings. A lattice contraction is reported instead in experimental work on the Ni-Fe-Al ternary phase diagram [28]. This would also be in agreement with the smaller value of a_0 measured for Ni_3Fe. For sorting out this problem we annealed some of the milled samples with the idea of reestablishing the original condition. The isothermal anneals were carried out in a furnace, keeping the samples in an Ar flux. The reordering process of these milled powders, as already reported [25,26,29], occurs over a wide range of temperature and includes several processes. We selected the following three temperatures for the treatments: 250°, 300° and 350° C. After each step the lattice parameter was measured. In figure 6 the series of graphs relative to the three different temperatures and to two different samples is reported. At the lowest temperature (250°) only a slight recovery of a_0 took place, whereas the anneals performed at higher temperatures were quite effective in reestablishing the initial values of the lattice parameter. These results ruled out any significant contamination effect, the lattice expansion being clearly reversible . The long-range order parameter in all cases remained below 1, showing that complete reordering could not occur at the selected temperatures, because of the impossibility of complete recovery of all the defects introduced during the disordering treatment.   The complete restoration of full LRO was obtained after annealing heavily

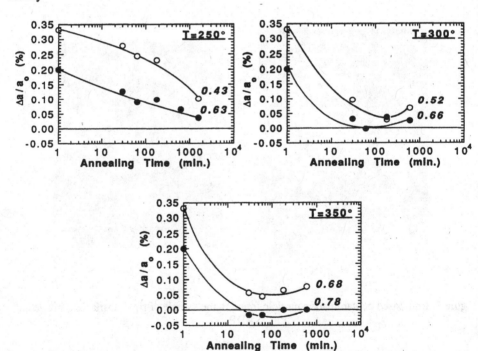

Figure 6. Isothermal curves showing the recovery of the lattice parameter in samples having two different initial values of the order parameter: open circles $S_0=0$; full circles $S_0=0.18$. The final values of the relative LRO parameter are reported.

disorderd powders for 2hrs at 500 °C. This treatment also resulted in the complete diffusion of the contaminating elements, coming from the grinding media, into the powder: XRD analysis revealed the disappearance of the extra peaks. This could also explain why such reordering annealing did not lead to any reduction of lattice parameter, which, on the contrary, increased slightly. We suspect that this same mechanism can explain the increase of a_0, after the initial decrease, in the course of the annealings at 300° and 350° C (figure 6). In the attempt to eliminate, or at least reduce, the possible contamination effects, we decided to perform some experiments with different grinding media. Some of these are already under way, but no results can be reported yet.

DISCUSSION

No experimental data are available on the lattice parameter behaviour of the fcc disordered phase of the permanently ordered Ni_3Al intermetallic. Experiments with Cu_3Au, in agreement with the theoretical prediction of Dienes model [5], show that a lattice expansion accompanies the reduction of LRO and SRO.

In a recent publication [30], based on the application of the Finnis-Sinclair potential to the description of bonds in several fcc binary alloys, the calculated trend of lattice parameter-vs-concentration for the disordered Ni-rich primary phase of the Ni-Al system shows linear behaviour, in accord with Vegard's law. The value of a_0 given by Vegard's law is larger than that experimantally measured for our unmilled powder (3.5645 Å) so that, on this basis, we can expect an expansion to occur also upon disordering Ni_3Al-based alloys. We calculated the lattice parameter for a completely disordered alloy, of the same composition as ours, just applying the Betteridge formula [2] for a face-centered cubic alloy A_3B with degree of SRO σ, as defined by Bethe [31]:

$$d_\sigma=1/8(3+\sigma)d_{AB}+1/16[(9-\sigma)d_{AA}+(1-\sigma)d_{BB}]$$

in which d_{AA}, d_{BB}, d_{AB} are the lengths of the bonds between the three possible kinds of pair of atoms present in a binary alloy. For our purpose σ was set at zero. This calculation was done in order to get a rough a priori extimate of $\Delta a/a_0$. In our calculation we assumed for the bond lengths the values for the pure elements for bonds between the same species, and their average, weighted with the concentrations of each element (Vegard's law), in case of bonds between different ones. We found a value of the lattice parameter corresponding to an expansion of about 0.90% with respect to the lattice parameter of the unmilled, ordered powder. This value is not so far from the observed value of 0.79% In principle this is an expansion just due to the chemical disorder, i.e., to the destruction of the LRO and of the SRO. At the moment we do not have any experimental evidence that after the superlattice spots disappear (S=0), some degree of SRO still remains. Also there are certainly other defects, vacancies and interstitials, introduced during the milling process, which may be responsible for part of the observed supplementary dilatation.

In the light of the reported results, some generalizations can be advanced concerning the disordering and amorphization conditions of strongly ordered intermetallics. The lattice expansion, analogous to that observed in irradiation experiments, seems to be a general condition in the disordering and amorphization of these materials. On the other hand we believe that the microscopic defect configuration is determined by the specific disordering techniques: irradiation, mechanical attrition, rapid quenching, etc. In ball-milling experiments, the increase of plastic lattice strain and the reduction of domain size are aspects of the disordering process. The rms strain achieves a steady value during disordering; in fact an increasing stress, which is the larger the smaller is the domain size, is required for deforming the material. In this respect the results of an investigation of the amorphization conditions of Si sputter-deposited films are of some interest [32]. They show in fact that the Si crystalline structure is destabilized by the combined effect of lattice expansion and reduction of crystallite size.

Acknowledgements. This work was supported by the European Community, in the framework of the SCI-0262-C project "Ordering Kinetics in Alloys".

REFERENCES

1. Feder, R., Mooney, M. and Nowick, A.S., Ordering kinetics in long-range ordered Cu3Au, Acta Met., 1958, 6, 266-77.

2. Betteridge, W., Relation between the degree of order and the lattice parameter of Cu3Au, J. Inst. Met., , 1949, 75, 559-70.

3. Gordon, P., Principles of Phase Diagrams in Materials Systems, McGraw-Hill, New York, 1968, pp.113-15.

4. Peierls, R., Statistical theory of superlattices with unequal concentrations of the components, Proc. Roy. Soc., 1936, A154, 207-22.

5. Dienes, G.J., Lattice parameter and short-range order, Acta Met., 1958, 8, 278-82.

6. Jones, F.W. and Sykes, C., Atomic rearrangement process in copper-gold alloy Cu3Au. II, Proc. Roy. Soc., 1938, 166A, 376-90.

7.	Poquette, G.E. and Mikkola, D.E., Antiphase Domain Growth in Cu3Au, Trans. AIME, 1969, **245**, 743-51.

8.	Leech, P. and Sykes, C., The evidence for a superlattice in the nickel-iron alloy Ni3Fe, 1939, **27**, 742-53

9.	Skvortsov, A.I., Shemeljov, Y.V., Klepatski, V.E. and Levitski, B.M., An x-ray study of fission fragment induced structural damage in Nb3Sn, L. Nucl. Mat., 1978, **72**, 198-202.

10.	Konishi, H. and Noda, Y., The remarkable incubation time on the order-disorder phase transition of Mg3In, in Dynamics of Ordering Processes in Condensed Matter, Komura, S. and Furukawa, H. eds., Plenum Press, New York, 1989, pp. 309-14.

11.	Mitsui, K., Mishima, Y. and Suzuki, T., Hetereogeneous ordering and antiphase domain morphology in Cu3Pt (19 at.% Pt), Phil. Mag., 1986, **A53**, 357-76.

12.	Schulson, E.M., The ordering and disordering of solid solutions under irradiation, L. Nucl. Mat., 1979, **83**, 239-64.

13.	Russell, K.C., Phase stability under irradiation, Progr. Mat. Sci., 1985, **28**, 229-434.

14.	Schulson, E.M., Carpenter, G.J.C. and Howe, L.M., Irradiation swelling of Zr3Al, L. Nucl. Mat., 1979, **82**, 140-47.

15.	Okamoto, P.R., Rehn, L.E., Pearson, J., Bhadra, R. and Grimsditch, M., Brillouin scattering and transmission electron microscopy studies of radiation-induced elastic softening, disordering and amorphization of intermetallic compounds , J. Less-Comm. Met, 1988, **140**, 231-44.

16.	Rehn, L.E., Okamoto, P.R., Pearson, J., Badra, L. and Grimsditch, M., Solid-state amorphization of Zr3Al: evidence of an elastic instability and first-order phase transformation, Phys. Rev. Lett., 1987, **59**, 2987-90.

17.	Johnson, W.L., Thermodynamic and kinetic aspects of the crystal to glass transformation in metallic materials, Progr. Mat. Sci., 1986, **30**, 81-134.

18.	Wolf, D., Okamoto, P.R., Yip, S., Lutsko, J.F. and Kluge, M., Thermodynamic parallels between solid-state amorphization and melting, J. Mat. Res., 1990, **5**, 286-301.

19.	Liu, H.C. and Mitchell, T.E., Irradiation induced order-disorder in Ni3Al and NiAl, Acta Met., 1983, **31**, 863-72.

20.	Carpenter, G.J.C. and Schulson, E.M., Effect of 1 MeV electron irradiation on Ni3Al, Scripta Met., 1981, **15**, 549-54 .

21.	Koch, C.C., Mechanical milling and alloying, Processing of Metals and Alloys, ed. R.W. Cahn, VCH, Weinheim, 1991, pp.193-287.

22.	Di, L.M., Loeff, P.I. and Bakker, H., Atomic disorder in intermetallic compounds by mechanical attrition, Mat. Sci. Engin., 1991, **A134**, 1323-25.

23.	Wright, R.N. and Knibloe, J.R., The influence of alloying on the microstructure and mechanical properties of P/M Ni3Al, Acta Met. Mat., 1990, **38**, 1993-2001.

24. Klug, H.P. and Alexander L.E., X-Ray Diffraction Procedures, 2nd ed., New York: Wiley, 1974.

25. Malagelada, J., Suriñach, S., Barò, M.D., Gialanella, S. and Cahn, R.W., Kinetics of ordering in Ni_3Al based alloys disordered by ball-milling, International Conference on Mechanical Alloying, Kyoto, Japan, May 1991. Proceedings to be published in Materials Science Forum, Aedermannsdorf, Switzerland: Trans. Tech. Publications.

26. Jang, J.S.C. and Koch, C.C., Amorphization and disordering of the Ni_3Al ordered intermetallic by mechanical milling, J. Mater. Res., 1990, 5, 498-510.

27. Enamoto, M. and Harada, H., Analysis of γ / γ' equilibrium in Ni-Al-X alloys by the cluster variation method with the Lennard-Jones potential, Met. Trans. A, 1989, 20A, 649-64.

28. Masahashi, N., Kawazoe, H., Tagasugi, T. and Izumi, O., Phase relations in the section Ni_3Al-Ni_3Fe of the Al-Fe-Ni system, Z.Metallkde, 1987, 78, 88-794.

29. Yavari, A.R., Barò, M.D., Fillion, G., Surinach, S., Gialanella, S., Clavaguera-Mora, M.T., Desrè, P. and Cahn, R.W., $L1_2$ ordering in disordered Ni-Al-Fe alloys, presented at the MRS Symposium on High Temperature Ordered Intermetallic Alloys IV, Boston, November 1990, Mat. Res. Soc. Proc. Vol. 213.

30. Rafii-Tabar, H. and Sutton, A.P., Long-range Finnis-Sinclair potentials for f.c.c metallic alloys, Phil. Mag. Lett., 1991, 63, 217-24.

31. Bethe, H.A., Statistical theory of superlattices, Proc.Roy.Soc, 1935, A150, 552-75.

32. Veprek, S., Iqbal, Z. and Sarott, F.A., A thermodynamic criterion of the crystalline-to-amorphous transition in silicon, Phil. Mag., 1982, 45, 137-45.

ORDERING KINETICS IN SUPERALLOY TYPE INTERMETALLIC COMPOUNDS IN RELATION WITH DIFFUSION MECHANISMS

M.C. CADEVILLE, V. PIERRON-BOHNES
Groupe d'Etude des Matériaux Métalliques, IPCMS, Université Louis Pasteur,
4 rue Blaise Pascal, 67070 Strasbourg, France

R. KOZUBSKI
Institute of Physics, Jagellonian University,
Reymonta 4, 30-059 Krakow, Poland

ABSTRACT

The investigation of ordering-ordering kinetics (LRO or SRO) in inter-metallic compounds yields informations on the activation energy of the ordering process that can be compared to the energy of the interdiffusion. A systematic study of aluminide phases built either on a fcc ($L1_2$) or bcc (DO_3) lattice shows that quite different behaviours are observed in the two types of structure. In $L1_2$ phases, the role played by the ordering energy has been tested in the pseudo-binary $Ni_3Al_{1-x}M_x$ (M = Fe, Mn, Cr) system which displays a strong decrease of the $L1_2$ phase stability when the transition metal M is increasingly substituted to aluminium. The activation energy (E_A) of the ordering process is found to increase with the phase stability, in agreement with theoretical predictions. In binary bcc $Fe_{1-x}Al_x$ phases, the result is different since the activation energy strongly decreases correlatively to the formation of long range ordered DO_3 structure. Such an unexpected behaviour can be attributed to a large decrease of the migration energy through a lattice dynamics effect, as recently proposed in models which explain the enhanced diffusion observed in some bcc transition metals and alloys. Such a comparative experimental study shows that, depending on basic lattices, either thermodynamical or vibrational properties are determinant factors of the activation energy for the ordering kinetics.

INTRODUCTION

The understanding of ordering mechanism in metallic systems is a still largely opened question. Many phenomenological and statistical models as well as computer simulations have been and are still being developed. Some of them deal with far from equilibrium problems, starting from a quenched disordered state. Other approaches consider the near equilibrium states, i.e. the ordering-ordering kinetics either in a solid solution phase (SRO kinetics) or in a long range ordered structure (LRO kinetics) whereas some models such as the Path Probability Method (PPM), initially developed by Kikuchi et al. [1] consider all the situations in bcc alloys.

Most of these models recognize the important role played by the vacancies, either explicitly, or undirectly via the diffusion coefficient (D), in determining the time scale of the kinetics and its temperature dependence.

As noticed by de Fontaine [2], most of the kinetic models can be handled either by a cluster approach, or by a wave approach. The models lying on a cluster approach [1] emphasize the role played by the local binding energies (ε_{ij}) which represent an additional activation energy needed to break bonds when an atom jumps. It is the amplitude of these pair interaction energies which determines, in a pair approximation, the order-disorder transition temperature (T_{oD}). It has been shown recently by Fultz et al. [3] that, in a bcc lattice, these pair interaction energies and their relative strengths between first and second nearest neighbours are determinant factors of the kinetic paths encountered by the alloy en route to the thermal equilibrium.

In the models lying on a wave approach as for example the model of continuous ordering [4] the temperature dependence of the kinetics is mainly driven by the diffusion coefficient and is not directly correlated to the order state, or to the ordering energy.

The intricated problem of the diffusion in concentrated alloys and in ordered compounds has been reviewed by Bakker [5] from both theoretical and experimental points of view. As long as only thermal vacancies are present in the compounds, a lot of experimental results indicate that, being given an alloy system and a structure (fcc or bcc) the activation energy of the diffusion is enhanced in compounds which display the highest phase stability. Other theoretical models such as those developed by Schoijet and Girifalco [6] show that, besides the role played by the type of the LRO structure and by each atomic species, the LRO parameter (η) and the ordering energy (i.e. the pair interaction potentials) are determinant factors of both the activation energy of the self-diffusion of each atomic species, and of the preexponential factor.

More recently, a comparison between diffusivities and ordering kinetics in some $L1_2$ alloys has been discussed by Cahn [7]. He showed that there is a correlation between the relaxation time of the ordering kinetics, the order-disorder transition temperature, and the self-diffusivity of the slowest diffusing species.

In order to pursue in that direction we undertook a systematic experimental investigation of the atomic mobility in transition-metal aluminides. The pseudo-binary $Ni_3Al_{1-x}M_x$ system with a $L1_2$ structure (γ' phase) is a good candidate for such a study because it presents a continuous series of $L1_2$ ordering solid solutions for a large variety of transition metals and a broad range of x values. Changing M and x greatly affects the γ' phase stability. The data were obtained up to now for three transition metals (M = Fe, Mn, Cr) and several values of x. Partial results with M = Fe have been already published [8]. A detailed investigation of the ternary $Ni_{75}Al_{25-x}Fe_x$ ($2 \leq x \leq 15$) is presented in this conference [9].

Previously another systematic study of ordering kinetics has been achieved in binary $FeAl_x$ [10], which present DO_3 and B_2 structures ordered on the bcc lattice, the B_2 phase stability being maximum at the equiatomic composition.

This paper aims in presenting a general overview and a comparative discussion of all the results obtained up to now in these aluminides.

LRO and SRO kinetics were measured by means of resistometry. It is clear that such a technique which is very easy to handle and very sensitive to small changes of order parameters does not give a detailed information on the ordering mechanism. It only yields the average thermodynamic quantities such as the activation energy and the preexponential factor that drive the atomic mobility.

In the next section, we recall briefly the relations between resistivity, order parameters and kinetics. Also the main features of the experimental procedure are given, and a comparison is made between the results of kinetics deduced from resistometry and those using diffraction techniques. It is shown that as long as the relaxation times of the ordering process and their T-dependence are the desired quantities, resistometry yields a reliable information.

The last section is devoted to the results obtained in the ternary $Ni_3Al_{1-x}M_x$ and in the binary $Fe_{1-x}Al_x$ and to their discussion. It is shown that, if in the $L1_2$ phases one observes the expected correlation between the activation energy of the ordering process and the γ' phase stability, in the binary Fe-Al a quite different behaviour is observed which cannot be explained by any model of ordering kinetics developed up to now.

ORDERING KINETICS, ORDER PARAMETERS AND RESISTIVITY

It is well known that resistometry is a sensitive tool to investigate SRO or LRO kinetics in alloys or compounds, as recently reviewed by Pfeiler for the SRO case [11].

Various phenomenological models have been proposed to account for the experimental kinetic behaviour as measured by resistivity. According to the authors, ordering-ordering kinetics can be described either by a single exponential, a sum of 2 or 3 weighted exponentials, an infinite sum of exponentials with a gaussian distribution of the relaxation times (τ), or by a power law. Considering the theoretical models that have been developed to describe the ordering kinetics, two of us have shown in [10] that the approach using a sum of exponentials with a lognormal distribution of τ seemed the most suitable way to account for the microscopic models such as the PPM [1] and the most of the encountered experimental situations.

In this approach, the relaxation of order after a small departure from its equilibrium value, as furnished for example by a small temperature step, and measured by the sample resistivity variation, is represented by:

$$\frac{\Delta\rho(t)}{\Delta\rho(t = \infty)} = \int_0^\infty P(\ell n\tau) \left[1 - \exp\left(- t/\tau\right) \right] d\tau \tag{1}$$

and

$$P(\ell n\tau) = \frac{1}{\beta\sqrt{\pi}} \exp\left[- \left(\frac{\ell n\tau/\bar\tau}{\beta} \right)^2 \right] \tag{2}$$

where $\Delta\rho(t) = \rho(t) - \rho(t = 0)$.

One determines at a given temperature T an average relaxation time $\bar{\tau}$ which is a predominant one, and the half-width β of the τ distribution.

Over the relatively small temperature range where the measurements of τ are possible (about 100°), the temperature dependence of $\bar{\tau}$ can generally be approximated by an Arrhenius law:

$$\bar{\tau} = \tau_0 \exp E_A/kT \tag{3}$$

Consequently, the physical quantities which are deduced from these resistivity experiments, are first E_A, the average activation energy of the ordering process, which can be compared to an effective diffusion energy, and $\bar{\tau}_0$ the preexponential factor which determines the time scales of the kinetics and is inversely proportional to the atomic jump frequency at infinite temperature. τ_0 contains entropic terms.

Similarly to the diffusion process, the vacancies are considered as the main vectors of the atomic mobility which allows the ordering process. Thus the ordering rate (ν) can be written in terms of vacancy jump frequency (ν_v) and vacancy concentration (C_v), giving:

$$\nu = \tau^{-1} = \zeta\, C_v\, \nu_v = \zeta\, C_0\, \nu_0 \exp(-E_F/kT) \exp(-E_M/kT) \tag{4}$$

ζ being an efficiency factor which is generally of the order of the unity.

E_F and E_M, which are respectively the formation and migration energies of vacancies, contain, in the frame of the PPM, terms related to interatomic interaction energies that determine the order-disorder transition temperature (T_{OD}), as well as the kinetic paths [3]. In the frame of the diffusion models in ordered compounds as developed by [6], in addition to the interaction energy, the SRO and LRO parameters modify the activation energy for the different kinds of atomic jumps, this effect being different for minority or majority atoms in an AB_3 structure for example.

In order to test the ability of the resistometry in giving reliable values of $\tau(T)$, it is interesting to compare with data deduced from other techniques such as diffraction techniques for example. This is possible for two phases. The first one is the DO_3 Fe-Al around 30 Al% (figure 1). The activation energy deduced by Chen and Cohen [12] from the measurement of the [1/2, 1/2, 1/2] superlattice Bragg peak intensity is 1.44 ± 0.03 eV. This value is very close to that of 1.41 ± 0.11 eV deduced from resistivity in $Fe_{73}Al_{27}$ [10]. The relaxation times are a little bit higher as measured by resistometry than by diffraction. This is partially due to the fact that (i) resistivity measurements are carried out over very long time ranges, and (ii) our analysis method integrates over the longest relaxation times.

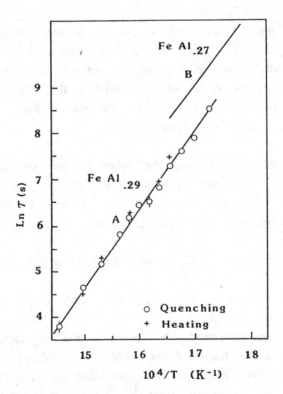

Figure 1. Comparison of Arrhenius plots in $FeAl_{.29}$ (curve A) deduced from X-ray diffraction [12] and in $FeAl_{.27}$ (curve B) from resistometry [10].

The second phase is $CoPt_3$ that has been investigated by two of us some years ago [13] in its $L1_2$ structure, and by Chen and Cohen in its disordered state [12] using X-ray diffuse scattering. The activation energies and the relaxation times are very close from each other, both quantities being a little bit higher in the LRO structure than in the disordered phase as expected from the PPM [1] (figure 2).

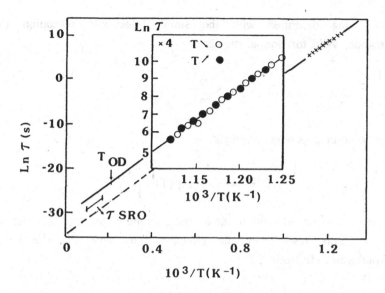

Figure 2. Comparison of Arrhenius plots determined in CoPt₃ by resistometry (●, ○, ×) in the LRO state [13] and by diffuse scattering in the SRO state [12] as indicated by τ SRO (small full line).

It thus appears from these two examples that, considering the orders of magnitude of the relaxation times and their temperature dependence, the resistivity measurements and their analysis using a lognormal distribution of τ yield reliable values of $\bar{\tau}$ and E_A.

The relations between resistivity and order parameters have been established about ten years ago by Rossiter [14]. In non-magnetic alloys, for the case of a stoichiometric composition, and above the Debye temperature θ_D, neglecting the contribution of short range correlations, the Rossiter's approach yields:

$$\rho(\eta,T) = \rho_0^D \left[1-\eta^2(T)\right]/\left[1-A\eta^2(T)\right] + BT/n_0\left[1-A\eta^2(T)\right] \qquad (5)$$

where T is the measuring temperature, ρ_0^D, n_0 and B/n_0 are respectively the residual resistivity, the density of conduction electrons, and the coefficient temperature of the phonon contribution in the disordered state.

The relation (5) was obtained by considering the effect of long range atomic ordering on the scattering of conduction electrons in an appropriate pseudo-potential model treated in the Bragg-Williams approximation, the resistivity being described with the simple electronic relaxation time (τ^e) approximation, valid for normal metals:

$$\rho = \frac{m^*}{n_{eff}} \frac{1}{e^2 \tau^e}$$

where τ^e is order dependent through:

$$\tau^e = \tau_0^e \left[1 - \eta^2(T) \right]^{-1}$$

τ_0^e is the electronic relaxation time corresponding to the disordered state, m^* and n_{eff} are respectively the effective mass and the effective density of the conduction electrons.

As the long range atomic ordering may introduce new gaps into the Fermi surface at the superlattice Brillouin zone boundaries, Rossiter has shown that the value of n_{eff} and/or m^* will be order dependent according to the following expression:

$$\frac{n_{eff}}{m^*} = \frac{n_0}{m_0^*} \left[1 - A\eta^2(T) \right]$$

The coefficient A depends upon the relative positions of the Fermi surface and the superlattice Brillouin zone boundaries. Its sign is determined by the evolution of the electronic band structure near the Fermi level with the formation of the LRO structure. This electronic structure effect is very important in determining both the variation of the resistivity with $\eta(T)$ and the amplitude and sign of the resistivity jump at the order-disorder transition, as discussed in [15] in the case of Fe-Al phases.

Whereas an increase of the resistivity when η decreases is often observed, in some systems such as in Fe-Al DO_3 [15,16] and in CoPt $L1_0$ [17] ρ decreases at T_{OD}. The relation (5) has as consequence that, at a given temperature, the sign of $\partial\rho/\partial\eta$ can be positive, negative or close to zero and moreover it can change with T as observed for example in Ni_3Al [18].

In consequence, when measuring through in situ resistivity the time dependence of the order restauration after a departure from its equilibrium value, it is necessary in some cases to perform rather large temperature steps in order to get measurable resistivity changes. As a consequence of such a procedure, the presence of non-equilibrium vacancies can modify the beginning of the kinetics, introducing a small difference between the relaxation times determined at increasing or decreasing temperatures as observed for example in some DO_3 Fe-Al [10] or Ni_3Al-Fe series [9].

RESULTS AND DISCUSSION

Details on experimental procedure have been given in previous papers, mainly in [10] and [13]. Details on Fe-Al preparation and phase diagram will be found in [10], those on $Ni_3Al_{1-x}M_x$ for M = Fe in [8,9], whereas those concerning M = Cr and Mn will be given in a forthcoming paper.

$Ni_{75\pm\varepsilon}(Al-M)_{25\mp\varepsilon}$ Results

As an example, results obtained in $Ni_{74}Al_{21}Cr_5$ are shown in figure 3 [19]. Figure 3a displays some isothermal curves set up at decreasing T ($\Delta T = -100°$) and their simulation with relations (1) and (2) calculated for $\beta = 1$. Figure 3b collects all the data of the isotherms on an Arrhenius plot that shows, in that compound, the relaxation times measured at decreasing or increasing T determine a unique Arrhenius straight line whose slope $E_A = 3.11 \pm 0.07$ eV is well defined.

The overall results obtained up to now in paramagnetic $Ni_{75}(Al-M)_{25}$ for M = Fe, Mn and Cr are collected in figure 4 which represents the dependence of the activation energy E_A with the temperature Tc of the γ' phase stability limit, as read on the vertical cross-sections of the pseudo-binary lines Ni_3Al-Ni_3M of the ternary phase diagrams [20,21,22]. In addition to our own determinations of E_A, we have plotted the value for Ni_3Mn we have deduced from a neutron diffraction study by Wakabayashi [23].

The error bars of E_A values are very different for each sample. This is firstly due to the various sensitivities of the total resistivity variation to a small change of η. In some phases, even in the paramagnetic state, magnetic effects can still reduce $\partial\rho/\partial\eta$. Secondly the error bar of E_A

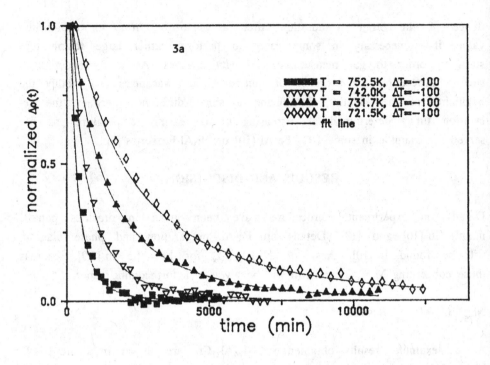

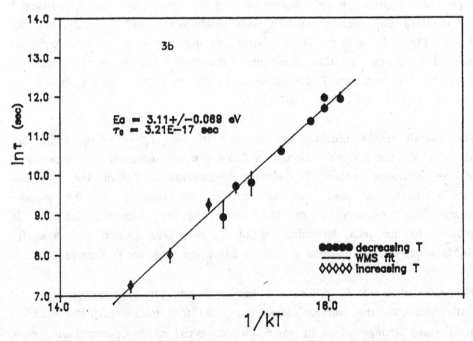

Figure 3. LRO kinetics in $Ni_{74}Al_{21}Cr_5$ as measured by resistivity from [19].

depends also on the number of isotherms that it is possible to set up over the small temperature range where the measurements of τ is accessible, on account of the phase diagram shape. Thus in the case of Ni_3Al [18], only an inferior limit of E_A could be deduced from resistivity measurements. For $Ni_{75}Al_{15}Fe_{10}$ two values of E_A have been obtained using two different resistometric methods. In situ resistometry yields a low value of E_A with a large error bar (2.4 ± 0.4 eV), but the measurement range is close to the Curie temperature, whereas 4K resistometry yields a higher value with a better accuracy (2.9 ± 0.1 eV).

In spite of some unaccuracies in the determination of E_A, figure 4 displays the expected trend, i.e. an increase of the activation energy of the ordering process with Tc, i.e. with the strength of the pair interaction energies. As discussed in [9] in the frame of the Schoijet and Girifalco's model, the order parameters can also play a role in the observed dependence of E_A with Tc and explain the dispersion of the experimental values.

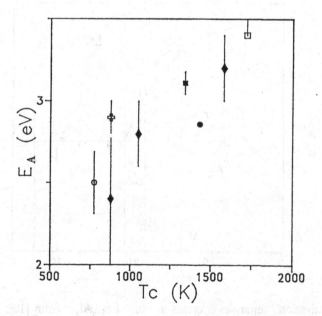

Figure 4. Activation energies versus Tc in Ni_3Al (□), Ni_3Mn (○), $Ni_{75}Al_{25-x}Fe_x$ (♦ and +), $Ni_{74}Al_{21}Cr_5$ (■) and $Ni_{75}Al_{20}Mn_5$ (●).

Ordering kinetics are also being investigated in ternary Ni_3Al-M (M = Ti, Ta) by Dimitrov et al. [24] using resistometry. Preliminary results indicate the same trend.

Fe-Al Results

SRO kinetics have been investigated in two $FeAl_x$ alloys (x = .10 and .17) and LRO kinetics in five alloys (x = .19, .23, .27, .30 and .33) using either 4K or in situ resistometry. The last four compositions have the DO_3 structure over the T-range investigated, whereas the first one is located at the limit between the solid solution phase A2 and the two-phase region A2 + DO_3. All the measurements were carried out in the ferromagnetic state, except for $FeAl_{.30}$. For sensitivity reasons, we do not succeed in getting results beyond 33 Al % in the B_2 phase, the temperature dependence of η being too flat in the interesting τ measurement range.

The concentration dependence of E_A is shown in figure 5 and compared with other data from the literature obtained using quite different techniques.

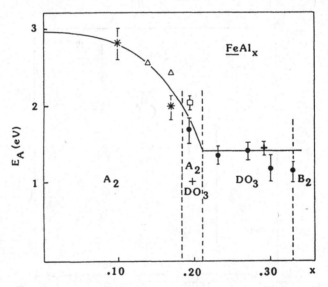

Figure 5. Activation energies versus x in $Fe_{1-x}Al_x$ from [10]: 4K resistometry (*), in situ resistometry (•), Zener relaxation (△), surface segregation (□) and x-ray diffraction (+).

The agreement between the different determinations of E_A is very good. So, the strong decrease of E_A with the Al content increase and with the formation of the DO_3 structure can be considered as a significative result.

Such a behaviour is quite different from that we expect from the above mentioned theories [1,6]. A recent investigation of positron-lifetime measurements in $Fe_{.763}Al_{.237}$ [25] yielding an effective vacancy formation energy of 1.18 ± 0.04 eV, we therefore conclude that a very small value of the migration enthalpy (0.2 ± 0.1 eV) would originate the low values of the total activation energy for $.20 \lesssim x_{Al} \lesssim .33$. A simple vacancy mechanism cannot explain so low values of E_M (or E_A). Considering the models recently developed by Herzig and Köhler [26] or Petry et al. [27] to explain the enhanced diffusion in some bcc group IV metals or alloys, we believe that a similar mechanism could explain the present data. In this model, two peculiar phonon modes, a $2/3 <111>$ longitudinal phonon and a $1/2 <110>$ transverse phonon are thought to play an important role in lowering the atomic migration potential barrier for the diffusing atom in the nearest neighbour direction. There is some evidence in the literature [28,29] for such low energy phonons in $Fe_{75}Al_{25}$, especially the transverse $<110>$ mode which is the most sensitive to the order state according to Robertson [29].

Such a mechanism would be valid for all bcc metals as indicated by a strong correlation of the bcc phonon anomalies and the self-diffusion properties, the higher the $1/2 <110>$ frequency the higher the activation energy. The representative point of Fe_3Al works well in such a scheme as previously shown [30].

Thus in Fe-Al, the expected increase of the activation energy of the ordering process with the formation of LRO structures is not observed, this effect being probably hidden by the role played by the vibrational properties in enhancing the atomic mobility.

CONCLUSION

An overview of the investigation of the ordering kinetics in two series of aluminides, $Ni_3(Al-M)$ with $L1_2$ structure, and $Fe_{1-x}Al_x$ with bcc or DO_3 structures, has been presented. We have shown that resistometry is a good method for investigating ordering kinetics in alloys and intermetallic compounds, and that it yields reliable information on the average activation energy of the ordering process, in agreement with those deduced from more sophisticated experiments.

A correlation between the activation energy E_A and the temperature of the γ' phase stability limit is well established in some pseudobinary $Ni_3Al_{1-x}M_x$ for M = Fe, Mn and Cr, in agreement with theoretical predictions.

In the DO_3 phase of Fe-Al alloys, low values of E_A were measured, that would be mainly due to very low values of the migration energy, induced by some "damped" phonon modes. This effect seems to be specific of bcc structures and of phases ordered on the bcc lattice.

REFERENCES

1. Sato, H. and Kikuchi, R., Kinetics of order-disorder transformations in alloys. Acta Metall., 1976, 24, 797-809.

   Gschwend, K., Sato, H. and Kikuchi, R., Kinetics of order-disorder transformations in alloys. II. J. Chem. Phys., 1978, 69 (11), 5006-5019 and J. Phys. (Paris), 1977, 38, C7, 357-362.

2. De Fontaine, D., Introductory talk at the International Conference on Order and Disorder in Solids. J. Phys. (Paris), 1977, 38, C7, 1-4.

3. Fultz, B., Kinetic paths in two order parameters: Theory. Acta metall., 1989, 37, 823-829, Kinetics of short- and long-range B2 ordering in the pair approximation. J. Mat. Res., 1990, 5, 1419-1430, and Anthony, L. and Fultz, B., Kinetic paths of B2 and DO_3 order parameters: Theory, J. Mater. Res., 1989, 4, 1132-1139.

4. Cook, H.E., Continuous transformations. Mater. Sci. Eng., 1976, 25, 127-134.

5. Bakker, H., Tracer diffusion in concentrated alloys. In Diffusion in Crystalline Solids, ed. G.E. Murch and A.S. Novick, Academic Press, Orlando, 1984, pp. 185-256.

6. Schoijet, M. and Girifalco, L.A., Diffusion in order-disorder alloys. The face centered cubic AB_3 alloys. J. Phys. Chem. Solids, 1968, 29, 911-922.

7. Cahn, R.W., Ordering kinetics and diffusion in some $L1_2$ alloys. Mat. Res. Symp. Proc., 1987, 57, 385-404.

8. Kozubski, R., Soltys, J. and Cadeville, M.C., Long-range order kinetics in $Ni_3Al_{0.4}Fe_{0.7}$. J. Phys. Condens. Matter, 1990, 2, 3451-3458.

9. Kozubski, R., Soltys, J., Cadeville, M.C., Pierron-Bohnes, V., Kim, T.H., Schwander, P., Hahn, J.P. and Kostorz, G., Long-range ordering kinetics in Ni_3Al-based alloys with various ordering energies. This conference.

10. Vennégues, P., Cadeville, M.C., Pierron-Bohnes, V. and Afyouni, M.,

Strong decrease of the activation energy as a function of Al content in $FeAl_x$ alloys ($x \leq 30$ at%) deduced from kinetic measurements of ordering. Acta metall. mater., 1990, **38**, 2199-2213.

11. Pfeiler, W., Investigation of short-range order by electrical resistivity measurements. Acta metall., 1988, **36**, 2417-24 .

12. Chen, H. and Cohen, J.B., A comparison of experiment and the theory of continuous ordering. J. Phys. (Paris), 1977, **38**, C7, 314-327.

13. Dahmani, C.E., Cadeville, M.C. and Pierron-Bohnes V., Temperature dependences of atomic order relaxations in Ni-Pt and Co-Pt alloys. Acta metall., 1985, **33**, 369-377.

14. Rossiter, P.L., The Electrical Resistivity of Metals and Alloys, Cambridge University Press, Cambridge, 1987, pp. 137-169.

15. Cadeville, M.C., Sanchez, J.M., Pierron-Bohnes, V. and Moran-Lopez, J.P., to be published in Proceedings of the Adriatico Conference on Structural and Phase Stability of Alloys, ICTP, Trieste, May 1991.

16. Rossiter, P.L., Long range order and the electrical resistivity. J. Phys. F: Metal Phys., 1980, **10**, 1459-1465.

17. Leroux, C., Cadeville, M.C., Pierron-Bohnes, V., Inden, G. and Hinz, H., Comparative investigation of structural and transport properties of $L1_0$ NiPt and CoPt phases; the role of magnetism. J. Phys. F: Met. Phys., 1988, **18**, 2033-2051.

18. Kozubski, R. and Cadeville, M.C., In situ resistometric investigation of ordering kinetics in Ni_3Al. J. Phys. F: Met. Phys., 1988, **18**, 2569-2575.

19. Kim, T.H., Private communication, Strasbourg, 1991.

20. Masahashi, N., Kawazoe, H., Takasugi, T. and Izumi, O., Phase relations in the section Ni_3Al-Ni_3Fe of the Al-Fe-Ni system. Z. Metallkd., 1987, **78**, 788-794.

21. Masahashi, N., Takasugi, T., Izumi, O. and Kawazoe, H., The phase diagram of the pseudobinary Ni_3Al-Ni_3Mn system. Z. Metallkd., 1986, **77**, 212-217.

22. Merchant, S.H. and Notis, M.R., A Review, Constitution of the Al-Cr-Ni Systems. Mater. Sci. Eng., 1984, **66**, 47-60.

23. Wakabayashi, N., Neutron-diffraction study on the kinetics of the atomic order in Ni_3Mn. Phys. Rev. B, 1986, **33**, 6441-6445.

24. Dimitrov, C. and Tarfa, T., Equilibrium and kinetics of thermal ordering or disordering in Ni_3Al. This conference.

25. Schaefer, H.E., Würschum, R., Sob, M., Zak, T, Yu, W.Z., Eckert, W. and Banhart, F., Thermal vacancies and positron-lifetime measurements in $Fe_{76.3}Al_{23.7}$, Phys. Rev. B, 1990, **41**, 11869-11874.

26. Herzig, C. and Köhler, U., Anomalous self-diffusion in BCC IVB metals

and alloys, <u>Mater. Sci. Forum Volumes</u>, 1987, **15-18**, 301-322.

27. Petry, W., Heiming, A., Trampenau, J. and Vogl, G., On the diffusion mechanism in the bcc phase of the group 4 metals. <u>Dimeta 88, Proc. Int. Conf. on Diffusion in Metals and Alloys</u>, Balatonfured, Hungary. Trans Tech Publ., Switzerland, 1989, **66-69**, p. 157.

28. Van Dijk, C., Lattice dynamics of Fe_3Al. <u>Phys. Lett.</u>, 1970, **324**, 255-256.

29. Robertson, I.M., Phonon dispersion curves in ordered and disordered Fe_3Al. <u>Solid State Commun.</u>, 1985, **53**, 901-904.

30. Cadeville, M.C., Pierron-Bohnes, V., Vennégues, P., Afyouni, M. and Kim, T.H., Kinetics of SRO and LRO in FeAl alloys with A_2 and DO_3 structures. To be published in <u>MRS Symposium Proceedings: Kinetics of Phase Transformations</u>, 1991.

Phase Transformation and Order–Disorder in Al–Li observed by High Resolution Microscopy

G. SCHMITZ and P. HAASEN

Institut für Metallphysik, Univ. Göttingen
and SFB 345
Hospitalstr. 3/5
D-3400 Göttingen

ABSTRACT

The decomposition of the supersaturated Al-7at%-Li α phase was studied by high resolution microscopy. Investigating the spatial arrangement of the metastable δ' precipitates a further hint of congruent ordering prior to decomposition is obtained. During examination of δ' precipitates in an 120 kV TEM a strain contrast appeared which increased with irradiation time (typically 30 min). In 400 kV electron microscope these effects came up faster and could be better resolved. As the $L1_2$ particle disappeared Moiré fringes became visible and also new lattice fringes. From these and the diffraction pattern the new phase could be identified as the stable δ-Al-Li phase (B32 with a lattice parameter of $a_0 = 0.641$ nm)

INTRODUCTION

Because of technologically important features, such as high stiffness and low density, the Al–Li–system has been the object of intensive research in the last ten years. A look at the phase diagram (see fig. (1)) shows that besides the technical applications a lot of interesting questions concerning phase stability and phase decomposition arise from the various ordered intermetallics which can be found in this system. The stable phase diagram is mainly determined by the ordered δ phase with about 50at% Li. The phase has a complicated structure (B32) with 16 atoms in the cubic unit cell. Because of a large lattice misfit ($\approx 11\%$) to the α matrix the formation of the stable δ phase is kinetically suppressed. Due to difficulties in preparation of specimens suitable for electron microscopy, the details of the precipitation of this phase are not known.

At lower aging temperatures ($T < 513$ K), instead of the stable δ phase first a metastable phase (δ') with a $L1_2$ structure forms by homogenous nucleation within the α matrix. The Li concentration of this metastable phase is nearly the stoichiometric concentration of 25 at%. The δ'particles are important precipitates for age hardening of commercial Al–Li alloys. Therefore

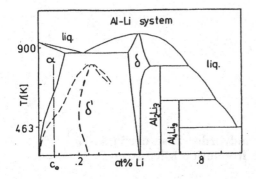

Figure (1):
phase diagram of Al–Li system. The Li concentration c_0 of the specimen used is marked as a vertical line.The aging treatment was done at 463 K. Dashed lines: metastable solubility and phase field of δ' phase.

the solubility of Li in the α matrix and the extent of the δ' phase field have been measured often and are known quite well. However the precipitation mechanism of δ' from the supersaturated α matrix is still controversial. Here, besides the Li concentration, one has to consider the order parameter as a second variable describing the δ' phase. The precipitation path in this two–dimensional parameter space is still an open question. Recently spinodal ordering before decomposition was proposed [3] and since then some researchers found experimental hints of $L1_2$ ordered particles with a Li concentration far below the stoichiometric 25 at% [6] whereas others reject the existence of such particles [5].

The aim of our paper is twofold. In a first part we want to present experimental results about the phase stability and the order in the metastable δ' phase. Some further hints for congruent ordering prior to decomposition are obtained. In a second part we describe the formation of the stable δ phase under electron irradiation. High resolution TEM proves to be a powerful tool as small $L1_2$ ordered particles ($r \approx 0.4$ nm) can be identified. At the same time the microscope supplies a suitable source of electron radiation.

EXPERIMENTAL

The ingots of a binary Al–7at%Li alloy, supplied by PECHINEY, were cold rolled to sheets of 0.23 mm in thickness. From these sheets, discs of 3 mm diameter were punched. These specimens were solution treated at 773 K for 45 min in an argon atmosphere and finally quenched into iced brine. The following aging treatment was done in an oil–bath at 463 K ± 1.5 K for various times. We choose aging conditions to allow for the homogenous formation of metastable δ' particles but to suppress the stable δ phase. Standard jet–polishing with HNO_3/methanol (1:4) at 243 K yielded suitable HREM foils, which could be examined at ≈ 23 nm foil thickness. To avoid large effects of room–temperature decomposition, the specimens were examined within a period of 24 h after the solution treatment. High resolution images of the specimen's microstructure were taken by a multiple beam technique in axial illumination. For this purpose two different microscopes were used: a PHILIPS EM 420 ST (120 kV, point–to–point resolution 0.33 nm) and a JEOL 4000 EX (400 kV, point–to–point resolution 0.17 nm). With $L1_2$ ordered domains, the {100} and the {110} superstructure reflections are excited. Choosing the central beam and the four {100} beams for image formation, one can easily distinguish the $L1_2$ ordered areas with orthogonal superlattice fringes, spaced ~ 0.4 nm, from the disordered α matrix (fig.(2)). The resolving power of the EM 420 ST is sufficient to use this contrast technique. With the 4000 EX, also the lattice fringes of the α matrix, spaced ~ 0.2 nm, can be resolved. The irradiation experiments were done with 120 kV electrons as well as with 400 kV electrons during the microscopic investgation. The typical irradiation time was about 30 min.

DECOMPOSITION OF THE SUPERSATURATED α-PHASE

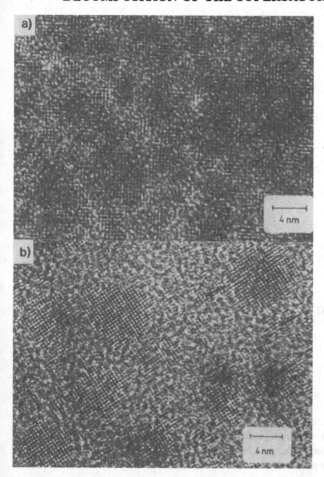

Figure (2):
high resolution images of Al-7at%-Li taken with an 120 kV electron microscope. $L1_2$ order is clearly identified by the superstructure fringes spaced ≈ 0.4 nm. a) microstructure of the alloy after quenching. Ordered domains with a complicated morphology and diffuse interface are observed. b) microstructure after 5 min aging at 463K. Large, nearly spherical δ' particles are nucleated. Between these particles small ordered domains are found (some of them marked by an arrow).

Using the order contrast technique described above the decomposition of the supersaturated α matrix was studied. Two examples of high resolution images obtained are given in fig.(2). Already in the as-quenched specimen a lot of small $L1_2$ ordered domains with a somewhat diffuse interface are observed (fig.(2)a). During the first minutes of aging this diffuse domain structure is preserved. After about 4 minutes suddenly a lot of larger nearly spherical δ' particles with a clear interface are nucleated (fig.(2)b). At further aging these spherical precipitates grow whereas the small ordered domains, formed during quenching, disappear. Using a digitizer, high resolution images like those shown in fig.(2) are evaluated. The size distribution of the ordered particles and their spatial arrangement within the specimen area were measured. Three typical size distributions are shown in fig.(3)a)...c). Here, the decomposition behaviour described above results in a bimodal size distribution (fig.(3)b)defined by one peak at r $\approx$ 1 nm belonging to the small domains and a second peak at larger radii belonging to the clearly interfaced spherical δ' particles which are nucleated during aging. Up to this day the nature of the small ordered domains and the mechanism of their formation is not clear. A congruent ordering mechanism which should take place during quenching was proposed by Khachaturyan in 1986 [3] (see also [2]).

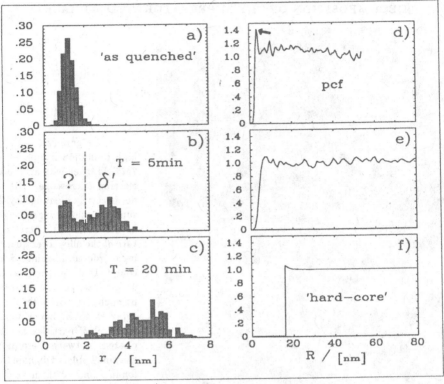

Figure (3):
a)...c) size distributions of $L1_2$ ordered particles: a) as–quenched state, b) 5 min aging: bi-modal size distribution , c) 20 min aging. d) pair-correlation function (pcf) calculated from the arrangement of the ordered particles in the as–quenched state. The large first peak marks a pre-ferred distance of 3.4 nm between the ordered domains. e) pcf calculated for the microstructure formed during 5 min aging. f) theoretical pcf of a 'hard–core' process. A similar pcf is expected for a microstructure formed by a nucleation mechanism.

In principle these congruently ordered domains should have a Li concentration similar to the initial one (7at%). However, due to decomposition following directly such congruent ordering, every concentration between the initial and that of metastable equilibrium seems possible[1]. Consequently for deciding between congruent ordering and a mechanism based on nucleation and growth, it is not sufficient to measure the Li concentration of the small ordered domains. In the hope of learning something about the decomposition mechanism we investigated the spatial arrangement of the ordered particles. For that purpose the pair–correlation function $f(R)$ of the centers of the particles was calculated, which is defined by the probability of finding in two volume elements dv_1 and dv_2 spaced by a distance R at least one particle respectively:

$$p \cdot dv_1 dv_2 = N_v^2 \cdot f(R) \cdot dv_1 dv_2$$

(with N_v : volume density of the particles [1]). For a decomposition mechanism based on nucleation and growth the pair–correlation function of a so–called 'hard–core' process is expected

---

[1]In fact, atom probe measurements recently done in Göttingen find small particles with about 20 at% Li already in the as–quenched state (T. Al-Kassab, private communication).

(fig.(3)f). The particles are randomly distributed but due to the competition of solute atoms the probability for distances smaller than a certain 'hard–core' distance goes to zero. In fact, such a correlation function is found for the later aging states, when the small ordered domains disappeared (fig.(3) e). On the other hand an example of the correlation function typically found for the as–quenched specimens is given in fig.(3)d). Here, the function is quite different from that of the later aging states. A certain distance of ≈ 3.4 nm between the ordered domains is clearly prefered. Considering nearly strain–free precipitates, such a regular arrangement of the precipitates is incompatible with a nucleation and growth mechanism!

Therefore there are good reasons to assume, that the decomposition of the α phase (under the chosen experimental conditions!) takes place as a two stage process. First during quenching small (≈ 3 nm in diameter) ordered domains form by congruent ordering. Maybe, already during quenching the decomposition starts following a spinodal mechanism. The partly unmixed specimen is then heated to the aging temperature and the further decomposition takes place by a nucleation and growth mechanism.

IRRADIATION INDUCED DECOMPOSITION — RESULTS

During the examination of Al–Li specimens with 120 kV electrons, irradiation effects were observed. Fig.(4) shows an image series of a specimen with large δ' particles formed during aging at 463K for 24 h . (The Li concentration within the particles is assumed to be that of the metastable equilibrium, $c_{Li} = 20.5 at\%$). The images were taken using the order contrast technique, i.e. compared with usual high resolution work, a relative small objective aperture was used. The essential effects are:

- Within the δ' particles, areas of clear distortion contrast are recognized, which grow with irradiation time.

- Often this contrast looks like a "coffee bean", a contrast which is known for isotropic distortion.

- The line–of–no–contrast usually lies in the (110) direction of the fcc matrix.

- Between the δ' particles, small domains with $L1_2$ order about 2–3 nm in diameter grow, sporadically also distorted areas are found.

After an irradiation time of 30 min, no further changes are observed. Similar irradiation effects are observed using 80 kV electrons — however the formation rate of the defects is much smaller. Taking into account the axial illumination used, an anisotropic distortion contrast is only expected if the defects are anisotropic too. Looking at the specimens from another direction confirms this assumption. Fig.(5) shows an image of similar δ' particles after 10 min irradiation time, using a (011) beam direction. Under these illumination conditions, the irradiation defects grow very fast; it was hardly impossible to take an image of δ' particles without some distortion contrast. We noticed:

- After an irradiation time of 10 min under the already described distortions, line contrasts appear.

- These line contrasts are found in two different orientations (see inlay). Selected area diffraction (obtained with 5 min illumination time) shows clearly streaks in (111) directions. These streaks, together with the trace analysis, point towards plate–like defects with {111} habitus .

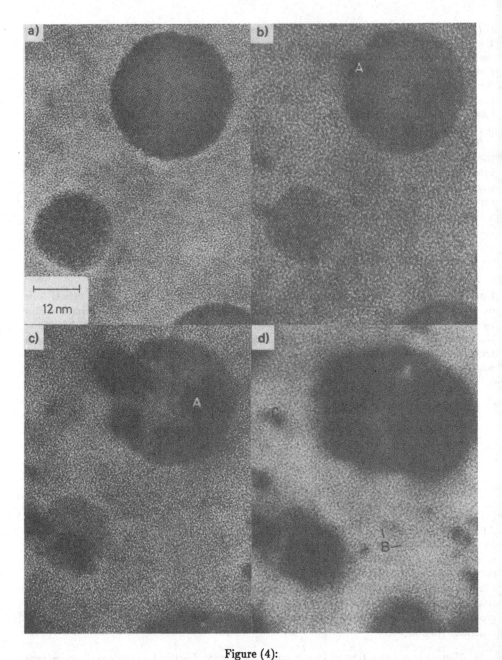

Figure (4):
Specimen with large δ' particles (24h aging) during irradiation with 120 kV electrons. Irradiation times: a) $\approx$ 0 min, b) 3 min, c) 10 min, d) 30 min. Growing areas with distortion contrast within the δ'particles (A), small ordering domains (B) and small distortions (C) in the matrix are observed.

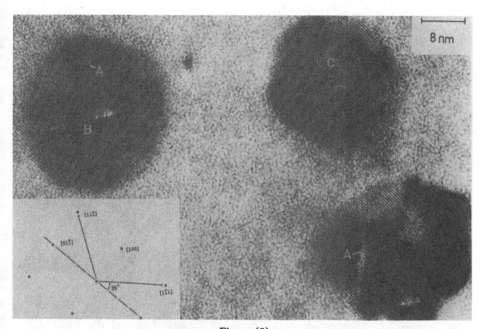

Figure (5):
same specimen as in fig.(4) but irradiated with 120 kV electrons for ten minutes along (011) direction. Only the direct beam and two {001} beams are used for imaging. Therefore super-structure fringes are only observed in one direction. Under the distortions line–contrasts are formed (A), sometimes with new fringes within these lines (B). Looking at the details distorted fringes are found (C). The trace analysis points toward two–dimensional defects with {111} habitus.

The irradiation effects were tested also with 400 kV electrons. Two typical images are shown in fig.(6). They were taken from the same specimen area after $\approx$ 0 min (a) and 30 min (b) irradiation time. Due to the higher resolution of the instrument used, the lattice fringes of the fcc matrix are also resolved. On the other hand, because of the larger objective aperture, a much lower contrast results from similar lattice distortions than in the 120 kV case. With customary laser diffractometry, patterns from the δ' area were taken. These patterns survey the space frequencies which are present in the negative. The main results are:

- The ordered δ' particles and the corresponding superstructure reflections in the diffraction pattern disappear nearly completely during the irradiation.

- Instead of the superstructure fringes of the former δ' particles, Moiré fringes appear. At some positions even new lattice fringes and the corresponding reflections in the diffraction pattern can be observed.

DISCUSSION

Considering these experimental results, we address the following questions: Does a phase trans-formation take place during irradiation? Can the new phase be identified from the reflections? What are the mechanisms of such a transformation?

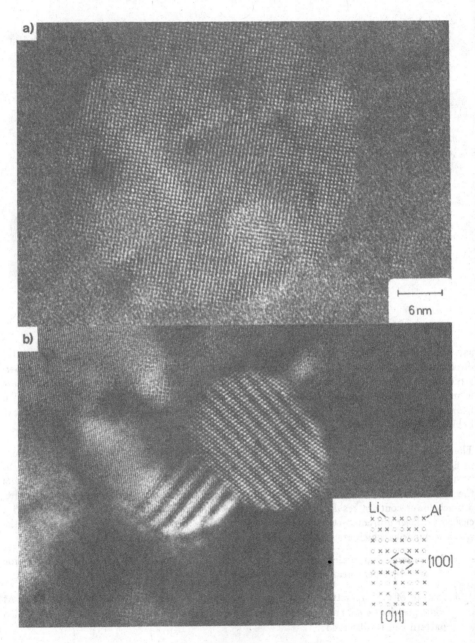

Figure (6):

Irradiation of a δ' particle with 400 kV electrons. irradiation times: a) $\approx$ 1 min, b) 30 min. Both images show the same specimen area. Due to the higher resolution power of the microscope the matrix lattice is imaged also. During irradiation the δ' particle disappear completely. Instead Moiré fringes and a new lattice structure (A) appear. The small inlay shows the positions of the atoms for a B32 structure projected in (011) direction. Obviously the lattice structure at (A) is closely related to this projection.

The results of the 120 kV irradiation give no direct evidence for the existence and structure of the new phase. Neither are Moiré fringes found, nor give any new reflections hints of a new structure. Streaks in <111> direction are also expected for prismatic dislocation loops on {111} planes. Surely there are some fringes within the plate–like defects shown in fig.(5). But those fringes change their orientation on changing the defocus. Therefore they cannot be interpreted without simulation, which we have not yet done. On the other hand the results of the 400 kV irradiation prove the formation of a new structure within the δ' particles. Considering the strong distortions and the {111} habitus, we assume that the new structure is the stable δ phase. Some data on the stable phase [9] are given in table (1). A detailed analysis of the diffraction pattern fig.(7)a), comparing the new reflections with those expected from the B32 structure in a (011) direction fig.(7)b), prove, that the new lattice structure and the Moiré fringes can be explained by the B32 structure, superposed to the fcc lattice of the α phase. Using the lattice parameter of the α phase for calibration, one gets the lattice parameter of the δ phase as:

$$a_0 = (0.641 \pm 0.005)nm$$

This value agrees very well with that known in the literature [9]. Moreover the new structure (see area A in fig.(6)b), which forms during irradiation, is closely related to the columns of equal atoms which are be seen looking at the B32 structure along a (011) direction. The projected positions of the atoms are shown in fig.(6) (small inlay). The projection consist of two by two identical columns with a distance of 0.16 nm, each, which cannot be resolved. Only one point, somewhat smeared, is imaged, which marks either the position of an Al column or of a Li column depending on defocus. Finally, the orientation relation between the α phase and the δ phase, which can be calculated from the diffraction pattern, is compared with that known in literature. Good agreement is found. Considering all the facts stated above, the following conclusion is drawn: There is an irradiation induced transformation of the metastable δ' phase to the stable δ phase during irradiation with 400 keV electrons!

How can we explain the quick formation of the stable δ phase which would form during aging only at increased temperatures ($\approx$ 573 K) in 20 h by heterogeneous nucleation? A heating of the specimen during the microscope investigation can be estimated to be smaller than 65 K (for details see [7]). Consequently, ordinary thermal decomposition can be excluded. In the literature, typical irradiation effects which form in a high voltage electron microscope (U > 600 kV) are reported [10]. Strong effects, like a phase transformation, were only observed if Frenkel defects could be generated by irradiation.Caused by the increased concentration of vacancies and by the diffusion of interstitials, the diffusion coefficient can be increased by about 5 orders of magnitude. To generate a stable Frenkel defect, the so–called Wigner energy has to be transferred from the electron to the target atom during the collision. Because of the large mass difference between electron and target atom, for such events typically an electron energy of >300 keV is required. However, the following estimate shows that in the case of Al–Li a surprisingly small electron energy is sufficient to generate stable Frenkel defects. In a first approximation for the Wigner energy E_d, the value of pure Al is used. Including relativistic effects, the maximum energy, which can be transfered by collision is given by [4]

$$E_{max} = \frac{2E(E + 2mc^2)}{Mc^2}$$

Assuming $E_{max} = E_d = 14eV$ (T=300 K) [10], $M_{Li}c^2 = 6.46 \cdot 10^9 eV$, one gets a lower limit for the acceleration voltage $U_{min} = 43kV$. All the experiments reported above were therefore done with electron energies sufficient for Frenkel defect generation!

The results of a more detailed estimate using a Kinchin–Pease model (after [4]) are given in fig.(8). There, the total cross section $\sigma_d^{(tot)}$ of atom displacements (defined by $N_d = In_v\sigma_d^{(tot)}$; N_d: displacements generated in unit volume and in unit time; I: electron current; n_v:volume density of target atoms) is shown for a lattice occupied with Li atoms or with Al atoms. In the

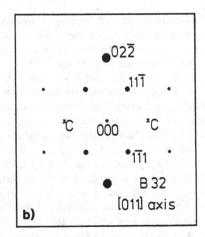

Figure (7):
a) diffraction pattern of the negative in Fig.(6) b) obtained with ordinary laser diffractometry. b) diffraction pattern expected for the B32 structure of the stable δ phase. Interpreting the diffraction pattern one must take into account the particular physical features of the laser technique. Not all of the reflections are due to a certain periodicity in the specimen. Some of the outer reflections may result from interference of two inner ones. For example, every strong reflection in a) is surrounded by some Moiré satellites (B). The 'forbidden' {200} reflections (C) can be explained by addition of two {111} reflections. All of the remaining 'significant' reflections can be explained by the diffraction pattern of the fcc α matrix superposed with two diffraction patterns expected for the B32 structure. The reflections marked (A) are related to the lattice structure (A) in Fig.(6)b).

- plate–like precipitates, {111} habit plane
- B32 structure, (NaTl) type
- lattice parameter $a_0 = 0.637nm$
- orientation relation $(010)_\delta \parallel (011)_\alpha$; $(101)_\delta \parallel (1\bar{1}1)_\alpha$; $(10\bar{1})_\delta \parallel (21\bar{1})_\alpha$
- lattice misfit between α and δ phase: $\frac{\Delta a_0}{a_0} = 11\%$

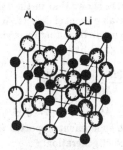

B32 structure

Table (1): properties of stable δ phase

$U_A/[kV]$	electron current/$[m^{-2} \cdot s^{-1}]$	dpa in α matrix	δ' particle
120	4.2 $\cdot 10^{22}$	0.01	0.03
400	36 $\cdot 10^{22}$	2.8	2.6

Table (2): displacements per atom after 30 min irradiation time

case of 120 keV electrons, only the Li target atoms contribute to the irradiation effects. Nearly all interstitials are generated during the primary event. Consequently, only Li interstitials can be expected. Since the Li concentration within the metastable δ' phase is considerably higher than the Li concentration of the matrix, the major irradiation effects are expected within the δ' precipitates. Using 400 keV electrons, the main contribution to the irradiation effects is due to Al atoms. (Keeping in mind that even in the Li rich δ' particles most of the lattice positions are occupied by Al atoms.) Therefore many Al interstitials will be found! The absolute numbers of Frenkel defects generated by 30 min irradiation are given in table (2).   There, the different

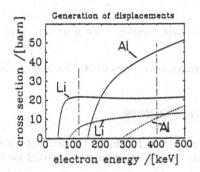

Figure (8):
cross sections for generation of displacements calculated using a Kinchin–Pease model. Al target atoms and Li target atoms are given separately. Solid line: total cross section, dashed line: part of the cross section due to collision cascades. The acceleration voltage used are marked by vertical lines.

electron currents of the microscopes and the different Al and Li target densities were taken into account. In the 400 kV experiments reported above, the rate of displacement was about two orders of magnitude higher than in the 120 keV experiments. The densities of defects in the α matrix and in the δ' precipitates were of the same magnitude. The rates of defect generation and recombination and the elimination rate of defects at sinks like free surfaces and dislocation loops, can be described by rate equations. Generally, after some short irradiation time ($\approx$ 30 s), a steady state with constant defect densities is reached. This steady state depends sensitively on the electron current, beam direction, temperature, dislocation density, and similar factors. For the experimental situation reported here, the rate equations were not formulated yet. Nevertheless, according to [8], we propose the following mechanism to explain the phase transformation observed:

• Due to the electron irradiation, the density of vacancies and interstitials is increased. Consequently, the diffusion coefficient is increased by several orders of magnitude compared with thermal equilibrium. Prismatic dislocation loops form on {111} habit planes.

• During the course of decomposition, the stable δ phase nucleates semi–coherently at these dislocation loops, saving distortion energy.

• Because of the accelerated diffusion, the nucleated δ precipitates grow fast. The required Li atoms are drawn from the metastable δ' phase. Consequently, the δ' particles dissolve.

There are still a lot of open questions. At first sight, the effects observed during 120 kV irradiation are quite similar to those observed at the higher electron energy. Looking at the details, there are some remarkable differences: At the lower electron energy, a phase transformation or a new phase could not be proven yet. The irradiation seems to stabilize the $L1_2$ order (there are ordering domains in the α matrix), whereas during 400kV irradiation, the $L1_2$ order is destroyed. It is a challenging question, whether these differences are only due to the lower generation rate of defects in the case of 120 keV electrons, or whether they are caused by the different interstitial atoms which are generated.

CONCLUSIONS

- During quenching of a supersaturated Al-Li alloy a congruent ordering takes place.

- For the congruently ordered phase a regular arrangement with a preferred mean distance between domain centers was found.

- During electron microscopic investigation irradiation effects were found at low electron energies.

- Under irradiation with 400 kV electrons a phase transformation from the metastable δ' phase to stable δ phase was detected.

- For the transformation a nucleation mechanism starting from dislocation loops is proposed.

ACKNOWLEDGEMENTS

The authors thank Dr. D. Gerthsen, Forschungszentrum Jülich, for the help with the JEOL 4000 EX and Dr. J. Ohser, Bergakademie Freiberg, for providing the mathematical background to calculate the pair–correlation functions.

REFERENCES

[1] T. Ficksel *statistics*, **19** (1988), pp. 67-75

[2] D. de Fontaine, *acta met.*, **23** (1975), pp.553-571

[3] A.G. Khachaturyan, T.F. Lindsey und J.W. Morris jr., *met. trans*, **19A** (1988) pp. 249-258

[4] G. Leibfried,*Bestrahlungseffekte in Festkörpern*, Stuttgart: B.G. Teubner 1965, p.90 ff. and p. 153 ff.

[5] J. Lendvai, H-J. Gudladt, W. Wunderlich und V. Gerold, *Zeitschrift für Metallkunde*, **80** (1989) pp. 310-317

[6] V. Radmilovic, A.G. Fox and G. Thomas *acta met.* **37** (1989), pp. 2385

[7] G. Schmitz,*dipl.thesis*, Univ. Göttingen (1990)

[8] P.S. Sklad und T.E. Mitchell, *scripta metall.*, **8** (1974) p. 1113

[9] K. Welpmann, M. Peters und T.H. Sanders jr., *Aluminium*, **60** (1984) pp. 735-739 / 846-848

[10] K. Urban, *phys. stat. sol.*, **56a** (1979) pp. 157-168

STRUCTURAL INSTABILITIES ASSOCIATED WITH PHASE TRANSITIONS
an electron microscopy study.

G. VAN TENDELOO, D. SCHRYVERS, L.E. TANNER *

University of Antwerp (RUCA) Groenenborgerlaan, 171, B2020 Antwerp, Belgium
* Lawrence Livermore National Laboratory, Livermore, CA 94550, USA

ABSTRACT
Electron microscopy combined with electron diffraction are highly qualified to provide microstructural data on phenomena occurring in the vicinity of phase transitions. Such pretransition effects occur for diffusion controled as well as for diffusionless transformations.
a) for diffusion controled transformations short range order (SRO) is formed when approaching the ordering temperature Tc from above while upon approaching Tc from below, interfaces associated with the ordering show wetting phenomena .
b) for diffusionless transformations the pre-transition modulations associated with the displacive austenite-martensite transformation are considered.

INTRODUCTION

Phase transformations can be subdivided into displacive and non-displacive, or into diffusionless and diffusion controled transformations. In the present contribution we shall concentrate on microstructural changes occurring in the vicinity of these phase transformations. Different aspects can be observed depending on wheather the transformation is approached from higher temperatures or from lower temperatures.

a) for the non-displacive, diffusion controled transformations; we shall investigate the short range order (i.e. approaching Tc from above) in D1a type alloys and the behaviour of ordering defects, such as antiphase boundaries in Co-Pt or Cu-Pd, when approaching Tc from below.

b) for the displacive, diffusionless transformations; we shall treat as an example the austenite-martensite transformation in a nickel rich Ni-Al alloy. Special attention will be focused on the "tweed" microstructure of the premartensitic state.This modulated structure is known to be associated with unusual variations of physical properties of the parent phase.

Microstructural changes associated with these pre-transition effects are recorded using different electron microscopy techniques.

SRO IN FCC BASED ALLOYS

It is common knowledge that above the ordering temperature in diffusion controled transformations short range ordering takes place. In some alloys this ordering state can be retained by quenching the alloy to room temperature. The short range ordered state (or the initial stages of ordering) can then be studied on an atomic scale by high resolution electron microscopy. In Ni_4Mo, which is thouroughly studied in the literature, the SRO of order is hard to retain, but in the isostructural Au_4Cr ordering is slow and the initial stages of ordering can be quenched in. Under the high resolution electron micoscope, viewed along one of the <100> directions, an image of (at first glance) disordered dark and bright dots is observed (see fig.1a). However the optical fourier transform of this area reveals relatively intense short range order reflections (fig.1c) indicating that the SRO state has definitely been imaged. Detailed inspection, with the help of image processing techniques, reveals the presence of square (D1a based) and triangular (DO_{22} based) arrangements in about equal amounts.

The high resolution techniques therefore allow to describe the short range order state or initial ordering stages in <1 1/2 0> type alloys as a mixture of clusters of two different types: D1a type clusters with an ideal composition of 4:1 and DO_{22} type clusters with a composition 3:1 [1,2].

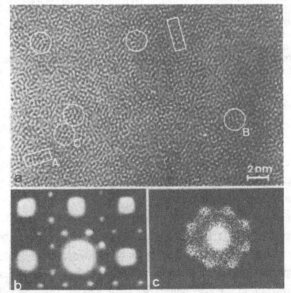

fig.1: a) High resolution [001]-image of short range ordering in Au_4Cr. Different areas showing pronounced square or triangular configuration of bright (or dark) dots are indicated.
b) Corresponding electron diffraction pattern
c) Optical diffraction pattern corresponding with the area shown in a)

INTERFACE WETTING IN ORDERED ALLOYS

Below 755°C the FCC based alloy $Pt_{70}Co_{30}$ becomes ordered with the $L1_2$ structure. Details on the phase diagram are to be found in [3]. Within the ordered phase, antiphase boundaries with a displacement vector of the type $R = 1/2<110>$ appear as a consequence of the decrease in translation symmetry. When the antiphase boundary is imaged under high resolution conditions as in fig.2a this displacement vector can immediately be identified. Moreover it becomes clear that the structure is perfect up to the boundary plane and that the width of the interface is restricted to a single atomic plane. Note that in this image the bright dots correspond to the minority atom configuration, i.e. they figure the projected Co-atom configuration when projected along one of the cube axes [4,5]. When annealing the material at temperatures between Tc - 17° and Tc, the width of the interface gradually increases (see fig.2b). The antiphase boundary is now 2 to 4 nm wide and a slab of disordered material has formed between the two ordered regions. This disordered region is imaged in fig.2b as a square lattice of 0.19 x 0.19 nm.

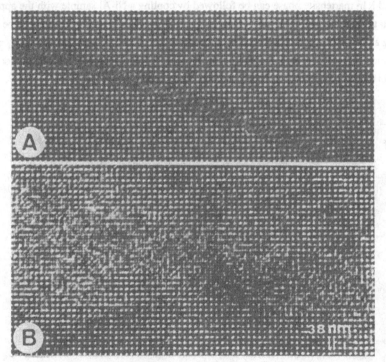

fig.2: High resolution image along [001] of antiphase boundaries in the $Pt_{70}Cu_{30}$ alloy.
    a) when annealed well below Tc
    b) when annealed just below Tc

In the Cu-17% Pd, the low order-disorder transition temperature has allowed us to perform highly controled heat treatments in the range between Tc-10° and Tc with temperature steps of 0.2° [6]. This experimental study has been undertaken in order to characterize the nature of the divergence of the width of the domain walls. In order to do so a number of precautions have to be taken but some problems persist: (a) the width of the interface is always measured at room temperature after quenching, (b) high resolution images invariably show a sharper interface than conventional dark field images. The preliminary results show a good agreement with statistical mechanics calculations [7] which predict that the width of the interface logaritmically diverges as log(Tc-T).

PRE-TRANSITION EFFECTS IN DIFFUSIONLESS TRANSFORMATIONS

We will focus on the austenite - martensite transition in Ni-rich Ni-Al because strong diffuse scattering is associated with this transition. The transition from the bcc based B2 austenite phase to the fct based L1$_0$ martensite phase can be followed by cooling a Ni-Al sample with the appropriate composition through its thermoelastic M$_s$ temperature. However a similar transition can be induced by applying a proper stress configuration [8]. In practice, when a Ni-Al sample with an M$_s$ close but still below room temperature is thinned for observation in a HREM instrument, strong but local stress fields around micro-cracks in the thin foil can induce the transformation.

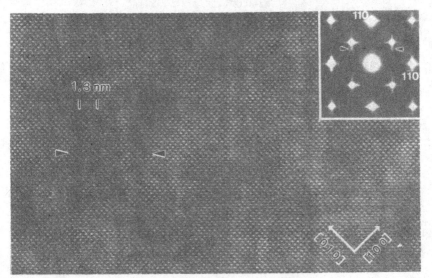

fig. 3: [001] HREM image of the precursor β' austenite phase.

In the latter case the different transition stages can be observed under HREM conditions. In fig.3 a typical [001] HREM image of the precursor austenite phase, away from the crack and labelled β', is presented: a patchwork of contiguous domains of approximately 3-5nm diameter can be observed, the contrast inside each domain being modulated parallel with (110) or (1$\bar{1}$0) and with a wavelength slightly incommensurate with the bcc lattice and corresponding with the satellites observed in the diffuse streaks. It was concluded that the distortion of the perfect bcc lattice can best be described by superimposing a transverse (110)[1$\bar{1}$0] sine modulation onto a uniform shift of the same type [9].

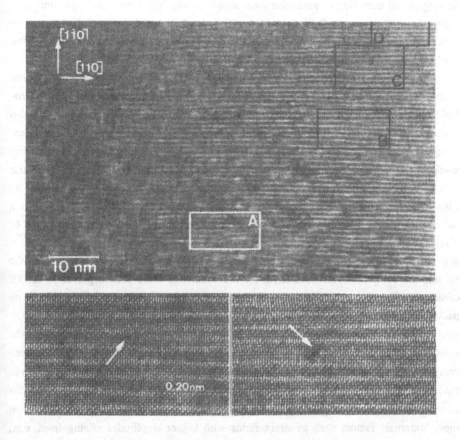

fig. 4: [001] HREM image of the long period transition structure with indications of some specific defects. Top (a), bottom left (b), right (c)

The present multiple-beam phase contrast modulation is called "micro-tweed" [9] and should not be confused with the original two-beam tweed contrast. The origin of this patchwork of small distorted domains can most probably be related to the existence of a large number of point-like defects inducing a local strain field with tetragonal symmetry to which the surrounding matrix responds following its soft modes [10].

Closer to the crack where the amount of stress is larger, out of all variants of the micro-modulation existing in the precursor phase, one is chosen which gradually increases its correlation both along and perpendicular to the corresponding wave-vector when moving closer to the micro-crack (fig.4a)

From these experimental results it can be concluded that when moving from the premartensitic region into the long period transition structure the local atomic arrangements must undergo some well defined alterations. The stress configuration near the micro-crack apparently dedicts which, out of the six possible transverse {110}<1$\bar{1}$0> distortions will be preferred for the formation of the long period modulated structure with the increased correlation lengths.

It can thus be expected that the domains belonging to the (110)[$\bar{1}$10] variant will grow at the expense of domains distorted in one of the other variants. Since the uniform distortions inside the micro-modulated domains are small, those domains belonging to the (110)[1$\bar{1}$0] variant will probably also grow. Since there is no correlation between domains of the same variant in the precursor phase some conflicting situations will arise when domains with out-of-phase micro-modulations connect during this growth: an example hereof is seen at the marker A in fig.4. The periodicity of the modulation also decreases when moving into the long period modulated structure. Two different mechanisms by which the structure achieves this decrease in wavelength are observed in fig.4. The first is a bifurcation at B while the second is a dislocation type arrangement at C, both enlarged in 4b and 4c respectively.

Only at the tip of the crack the amount of stress is sufficient to fully transform a small portion of the material into the 7R phase. The wavelength (or the number of close packed planes in the 7R monoclinic unit cell) has to increase from 5.5 to 7 planes and the modulation wave has to lock in on close packed planes so as to form the stable (5$\bar{2}$) stacking.

In conclusion we can say that from the present HREM investigation new detailed information on the atomic structures of precursor and transition states of martensitic transition in Ni-Al has been collected. It can be concluded that the precursor phase consists of contiguous domains (3-5nm in diameter) with each domain being modulated by an incommensurate transverse {110}<1$\bar{1}$0> displacement wave superimposed onto a uniform shift of the same type. As the transformation continues, "external" factors such as strain fields with higher amplitudes arising from, e.g., dislocation tangles, grain boundaries etc., determine which of the six variants will grow and develop the micro-twinned martensite structure. Before this phase is formed, an intermediate long period modulated phase with a shorter wavelength appears. Its modulation will then lock in on the fcc-like close packed planes to construct the 7R stacking or multiply twinned L1$_0$.

ACKNOWLEDGEMENTS

The authors (L.T, N.S and G.V.T.) like to thank the National Centre for Electron Microscopy at the Lawrence Berkeley Laboratory, Berkeley, (U.S.A.) for the use of the facilities. Part of this work was supported by the U.S. Department of Energy contract no. W-7405-ENG-48.

REFERENCES

[1] Van Tendeloo,G., Amelinckx,S., de Fontaine D.(1985) Acta Cryst B41,281
[2] Banerjee,S., Urban, K., Wilkens,M (1984) Acta Metall.32, 299
[3] Leroux,C., Cadeville, M.C., Kozubski, R. (1989) J.Physics: Condensed Matter,1,6403.
[4] Broddin,D., Leroux,C., Van Tendeloo G.(1991) Proceedings of MRS Spring Meeting, San Fransisco 1990, in the press.
[5] Broddin,D., Van Tendeloo,G., Van Landuyt,J., Amelinckx,S., Portier,R., Guymont,M, Loiseau,A. (1988) Phil. Mag. A54,395.
[6] Ricolleau ,C., Loiseau, A., Ducastelle,F. (1991) Phase Transitions, in the press
[7] Finel,A., Mazauric,V., Ducastelle,F. (1990) Phys.Rev.Lett. 65, 1016
[8] Martynov,V.V., Enami,K., Khandros, L.G., Nenno,S., Tkachenko,A.V.(1983) Phys. Met. Metall. 55, 136
[9] Schryvers, D., Tanner,L.E., (1990) Ultramicroscopy 32, 241
[10] Tanner,L.E., Schryvers,D., Shapiro,S.M., (1990) Mat. Sci. and Eng. A127, 205
[11] Shapiro,S.M., Yang,B.X., Shirane,G., Noda,Y., Tanner,L.E. (1989), Phys. Rev. Lett., 62, 1298

ELECTRON MICROSCOPY INVESTIGATION OF WETTING PHENOMENA IN L1$_2$ ALLOYS

Ch. RICOLLEAU[1,2], A. LOISEAU[1,2] and F. DUCASTELLE[1]
[1]ONERA, Laboratoire de Physique des Solides
[2]CNRS/ONERA, Laboratoire d'Etude des Microstructures
29 Avenue de la Division Leclerc B.P. 72 92322 Châtillon Cedex, France

ABSTRACT

We have investigated the behaviour of antiphase boundaries as a function of temperature in A$_3$B binary ordered alloys using transmission electron microscopy. The study shows a typical wetting phenomenon at the order-disorder transition. At low temperature, the domain walls are perfectly thin and micro-faceted. Complete interfacial wetting by the disordered phase, characterized by a logarithmical divergence of the width of the walls, is shown to occur when approaching the order-disorder transition from below.

INTRODUCTION

Numerous A$_3$B binary compounds are ordered at low temperature on the FCC lattice according to the simple cubic L1$_2$ structure (Cu$_3$Au-type) and recover at high temperature the full FCC symmetry (A1) through an order-disorder transition. Dividing as usual the FCC lattice into four interpenetrating simple cubic sublattices, the perfect L1$_2$ structure is obtained by placing B atoms on one sublattice and the A atoms on the three others (this structure is shown in figure 1a).

In agreement with Landau-Lifshitz arguments, the L1$_2 \rightarrow$A1 order-disorder transition is generally found to be of first order with (i) a discontinuity of the order parameter at the transition and (ii) the presence of two-phase regions (in figure 1b, T_{c1} and T_{c2} are respectively the lower and the upper limits of the two-phase region) except at the congruent point (labelled T_c in figure 1b).

The loss of translational symmetry characterizing this transition implies the formation, during the ordering process, of translational variants (or domains), related one to the other through a/2<110> type translations where a is the lattice parameter, separated by domain walls or so-called antiphase boundaries (APB).

At low temperature, the domain walls are perfectly thin. When approaching the order-disorder transition, they are shown to be progressively wetted by the disordered phase with premonitory effects starting a few 10$^{-2}T_c$ below T_c. This phenomenon is characterized by a divergence of the width of the APB. Qualitatively, the boundary extends over a few atomic planes which become partially disordered : the width of this disordered film increases when increasing the temperature but remains microscopic compared to the domain size. At T_c a macroscopic layer of disordered phase takes place giving rise to two new order-disorder interfaces. Evidence for this phenomenon has been qualitatively shown in $CoPt_3$ [1] and in Cu_3Au [2] [3].

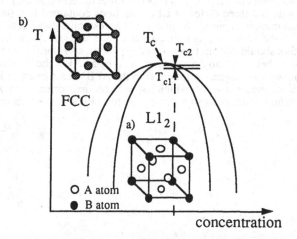

Figure 1. a) $L1_2$ structure. b) Typical phase diagram of an A_3B alloy which undergoes a $L1_2 \rightarrow A1$ phase transition.

We present here a quantitative characterization of the domain walls as a function of temperature using conventional and high resolution electron microscopy. These techniques provide a convenient tool for studying the nature and behaviour of these defects.

From a theoretical point of view, mean field approaches [4] predict a logarithmical divergence of the APB width. The theoretical models consider the case of a single APB between two variants respectively at $+\infty$ and $-\infty$. Experimentally this constraint imposes to study samples with large domains, bounded by boundaries of low curvature and to analyse the APB profile perpendicularly to their plane.

In order to characterize quantitatively the nature of the divergence, we have chosen to investigate the Cu-17%Pd alloy. A preliminary study [5] has shown premonitory effects of wetting starting about 10°C below T_{c2}, the order-disorder transition temperature. The low value of T_{c2} (T_{c2} = 506.6°C) has allowed us to perform highly controlled heat treatments in the range between T_c-10°C and T_c with temperature steps of 0.2°C.

EXPERIMENTAL PROCEDURES

The Cu-17%Pd samples were prepared according to the experimental procedure given in [5]. In order to obtain time stability and high accuracy of temperature, the salt bath used for the thermal treatments was isolated from the environment fluctuations and highly controlled by a P.I.D. type regulation. We were then able to perform heat treatments with a precision of ± 0.05°C during 20 hours.

All the samples were first ordered as perfectly as possible as follows : (i) annealing treatment a few degrees below the transition temperature during one week in order to obtain large ordered domains and (ii) cooling treatment down to 450°C for 24 hours to ensure a high long range order parameter within the ordered domains. This ordering treatment is our initial state. Samples were then aged 16 hours at different temperatures in a small temperature range just below the order-disorder transition : 506.55, 506.3, 506.1, 505.8, 505.6, 505.4, 505.2, 505, 504.7, 504.3, 503.85, 503.3, 502.6, 501.4, 500 and 498.6°C.

Figure 2 shows the initial microstructure of the Cu-17%Pd alloy i.e after the ordering process described previously. The APB contrast clearly obeys the extinction rules expected for perfect APB in two beam condition with a superstructure reflection [1,5,6]. The dark field (DF) images realized with the three different $L1_2$ reflections contained in a (001) zone axis thus reveal different but complementary informations. Two essential peculiarities can be noticed in the micrograph : (i) the APB usually terminate at a node necessarily formed of three different APB whose vectors should add up to zero modulo a lattice vector. In some case, in order to ensure the continuity between two domains separated by an APB, the interface is bounded by a partial dislocation with a Burgers vector equal to the APB vector (mark A in fig. 2a). (ii) The APB appear with an apparent width which depends on their orientation with respect to the electron beam. The minimum contrast width occurs for APB seen edge on (mark B in fig. 2a).

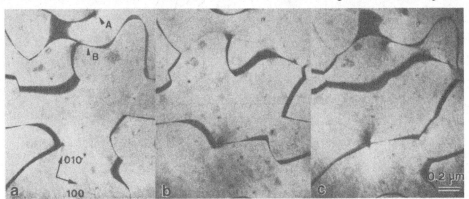

Figure 2. Dark field images of Cu-17%Pd aged at 450°C. a) 110 DF - b) 120 DF - c) 210 DF.

On the other hand, high resolution imaging allows to study the domain walls on an atomic scale. APB were imaged under the high resolution imaging conditions detailed in [5,6]. In these conditions, the $L1_2$ structure appears as a square pattern of bright dots figuring the Pd configuration projected along a [001] direction. An APB is directly visualized through the shift of certain rows of bright dots from one side of the boundary to the other. For perfect APB, this shift is restricted to one single atomic plane (see figures 7,9 and 10).

EVIDENCE OF DOMAIN WALL WETTING

Boundaries have been studied in the different aged samples and observed in dark field mode according to the method previously described. As a general feature, the APB contrast does not obey any more the extinction rules working for perfect APB when the temperature is increased. The contrast modifications are illustrated in figure 3 for three temperatures : 498.6°C (Fig. 3a), 503.3°C (Fig. 3b) and 505.2°C (Fig. 3c). At a given APB node, the APB which was previously out of contrast in a given DF is now seen with a residual contrast which becomes more pronounced as the temperature increases.

Up to 502.6°C, the residual contrast remains very weak but still permits an unambiguous characterization of the different APB. In the vicinity of T_{c1} (T=505.4°C) the different APB appear with almost the same dark contrast in the three DF images which indicates that the initial APB is now transformed into a thin film of disordered phase.

To characterize the divergence of the disordered layer width, width measurements have been performed, for each temperature, on numerous APB seen edge on along a [001] projection and viewed in the same dark field mode in order to make a meaningful statistical analysis and to eliminate aberrant values [7].

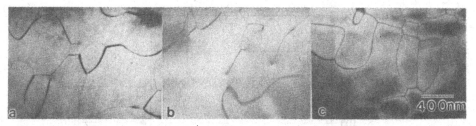

Figure 3. Dark field images showing the evolution of wetting phenomenon with temperature.

Figures 4a and 4b show the evolution of the domain walls width with temperature. We can notice a clear-cut difference between two regimes : APB wetting region and two-phase regime. Up to 505.4°C, the APB width increases but remains below 7 nm and at 505.6°C it abruptly jumps to 18 nm : this corresponds to a significant scale change between a microscopic regime and a macroscopic one which allows us to differentiate a two-phase situation from a wetting phenomenon occurring below the coexistence temperature : T_{c1} = 505.6°C. Typical two-phase microstructures are shown in figure 5.

This scale change is also characterized by important fluctuations of APB width. In fact, in the vicinity of the lower limit of the two-phase region, the system easily switches from one regime into the other because of concentration fluctuations ($\approx$ 0.3 to 0.5% Pd). The T_{c1} temperature being thus well defined, we were able to plot the APB width evolution as a function of - Log (T_{c1} - T). Although experimental evidence of logarithmical divergence was very difficult because of the weak singularity of this function, the results presented in figure 4c are remarkably consistent with such a logarithmical law.

The results presented here do not include the lowest temperatures. Their analysis is more delicate, since the disorder at the APB being weaker, the APB contrast is no more that of a disordered layer but is a coupling of the APB fringe contrast and of the disorder contrast. The latter one is the only observed in the extinction condition of the APB contrast but is very faint and fluctuating. High resolution imaging is necessary to achieve precise measurements, which is currently in progress. We now present results concerning the atomic structure of perfect boundaries observed in high resolution imaging.

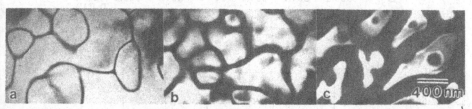

Figure 5. Two-phase microstructure in (110) dark field : a) 505.6°C - b) 506.1°C - c) 506.55°C

ATOMIC STRUCTURES OF PERFECT APB IN Cu₃Pd AND IN CoPt₃

Morphology on a macroscopic scale

We have performed a comparative study of the perfect APB morphology in the Co-Pt and in the Cu-Pd systems. From a macroscopic point of view, whereas in CoPt₃ the perfect APB are isotropically oriented [6], in Cu-17%Pd they are preferentially oriented in (210) type planes at low temperature. Figure 6a shows such APB type configuration (arrows indicate normals to the APB planes). When the temperature increases (fig. 6b,c), the APB are then more frequently contained in (100) and (110) planes (mark 1 and 2 respectively in figures 6b and 6c).

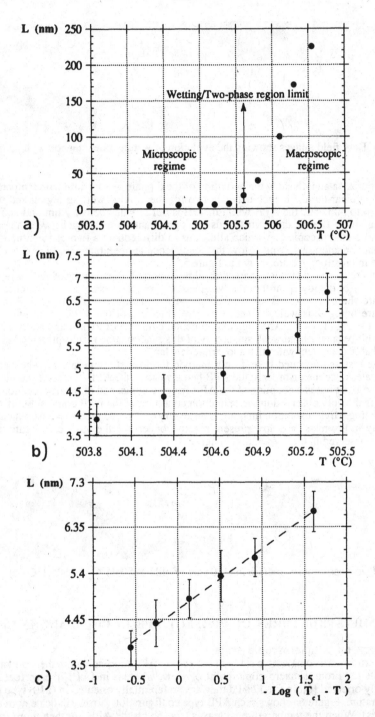

Figure 4. a) Evolution of APB width as a function of temperature. b) Enlargement in the microscopic regime. c) Evolution of the APB width as a function of $- \text{Log} (T_{c1} - T)$.

Figure 6. Three different geometrical configurations in 110 DF imaging at 498.6°C (a) and at 505.2°C (b and c).

Atomic scale morphology

Let us now study the microscopic structure of perfect APB in both systems. As a general feature, boundaries in the Cu_3Pd alloy are micro-faceted in (100) and (110) type planes whatever their orientation, whereas in $CoPt_3$ they are more or less rough depending on their local orientation with respect to a simple cristallographic plane (see figures 7b, 9b and 10b). In the following, typical examples of APB in Cu-17%Pd are analyzed and compared to APB in $CoPt_3$ having the same average orientation.

In all cases, APB seen edge on in a [001] projection consist of a succession of terraces in (100) and (110) planes, of length n and m respectively, n and m being integers. When the APB is strictly contained in a (h,k,0) plane, the micro-faceting is almost periodic of period p, and we can write :

$$p (h,k,0) = n (1,0,0) + m (1,1,0) \qquad (1)$$

with :

$$\begin{cases} m = p.k \\ n = p.(h - k) = p.h - m \end{cases} \qquad (2)$$

According to the previous relations the steps length in (100) and in (110) planes are respectively equal to $p.(h - k).a$ and $p.k.\sqrt{2}.a$.

Typical encountered APB configurations observed at 450°C are described in figures 7 to 10. Figure 7b shows the example of an APB lying in (210) type plane : it is constituted by (100) and (110) facets whose length varies between 5a and 7a ($5\sqrt{2}a$ and $7\sqrt{2}a$ respectively). Therefore that case can be described by the relation (1) with m = n = p where p = 6 ± 1. In figure 8 we have modelled the (210) APB for p=6. The solid line is the APB real position and the dashed line is the perfect (210) plane in absence of micro-faceting, which corresponds to the real position observed in $CoPt_3$ (figure 7a).

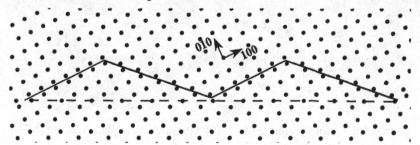

Figure 8. Schematic representation of a micro-faceted (210) APB. The APB displacement vector is equal to 1/2[011].

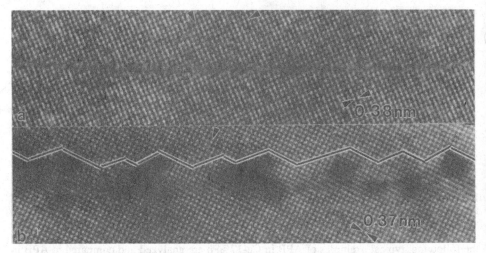

Figure 7. High resolution image of a (210) APB in CoPt$_3$ (a) and in Cu$_3$Pd (b).

The two simplest cases occur when APB are nearly lying in the (100) and (110) planes. In both these extreme situations, the interfaces are constituted of very large terraces (several atomic distances) in the main plane connected one to the other by small ledges on a single atomic plane rotated by 45°. Figure 9 is an example of the (110) orientation. In Cu$_3$Pd (figure 9b), the deviations from the cristallographic plane are locally accommodated by small (100) ledges, indicated by the arrows. On the other hand, the APB in CoPt$_3$ (figure 9a) is not constrainted to lie in simple planes and it freely fluctuates around the (110) plane.

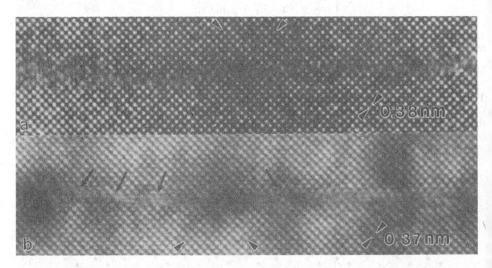

Figure 9. High resolution image of a (110) APB in CoPt$_3$ (a) and in Cu$_3$Pd (b).

121

Finally, when the APB are not contained in a well defined plane, the (100) and (110) facets are not regularly arranged but their relative lengths depend on the local curvature of the boundary. As a general feature, the terrace lengths become smaller in the areas of curvature changes. This last point is well shown in figure 10b. The APB orientation evolves from the (310) plane (left side of the picture) to the (100) plane (right side). In these two extremities, the local orientation can be analyzed as previously. In between, the terrace length arrangement is completely random and determined by the local curvature. For comparison, figure 10a shows that in $CoPt_3$ the boundary position continuously fits the local curvature without terraces or ledges.

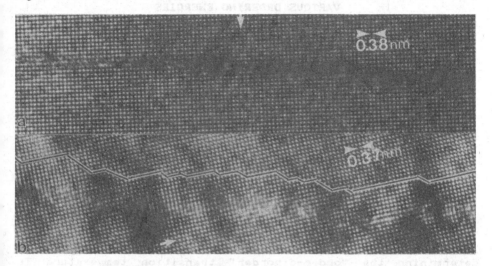

Figure 10. APB with curvature change in $CoPt_3$ (a) and in Cu_3Pd (b).

REFERENCES

1. Leroux, C., Loiseau, A., Cadeville, M.C. and Ducastelle, F., Wetting of antiphase boundaries by the disordered phase in $CoPt_3$. Europhys. Lett., 1990, 12, 155-160.

2. Morris, D.G., Disordering study of copper-gold alloys with compositions close to 25% Au. Phys. Stat. Sol., 1975, 32, 145-156.

3. Tichelaar, F.D., Schapink, F.W. and Xiaofeng, L., In situ TEM observations of the order-disorder transition at interfaces in Cu_3Au. Phil. Mag. A, 1991, in press.

4. Binder, K., Critical behaviour at surfaces. In Phase Transitions and Critical Phenomena, ed. C. Domb and J.L. Lebowitz, Academic Press, New York, 1983, 8, 2-144.

5. Ricolleau, C., Loiseau, A. and Ducastelle, F., Electron microscopy study of domain wall wetting in ordered copper-palladium. Phase Transitions, 1991, 30, 243-254.

6. Leroux, C., Loiseau, A., Cadeville, M.C., Broddin, D. and Van Tendeloo, G., Order-disorder transformation in CoPt alloy : evidence of wetting from the antiphase boundaries. J. Phys. : Condens. Matter., 1990, 2, 3479-3495.

7. Ricolleau, C., Loiseau, A. and Ducastelle, F., to be published.

LONG-RANGE ORDERING KINETICS IN Ni₃Al-BASED ALLOYS WITH VARIOUS ORDERING ENERGIES

R. KOZUBSKI, J. SOŁTYS
Institute of Physics, Jagellonian University,
Reymonta 4, 30-059 Kraków, Poland
M. C. CADEVILLE, V. PIERRON-BOHNES, T. H. KIM
Groupe d'Étude des Matériaux Métalliques, I.P.C.M.S.,
4, rue Blaise Pascal, 67070 Strasbourg, France
P. SCHWANDER, J. P. HAHN, G. KOSTORZ
Institut für Angewandte Physik, ETH-Hönggerberg,
8093 Zürich, Switzerland

ABSTRACT

According to theoretical models, the value of ordering energy determining the "order-disorder" transition temperature T_t should affect the kinetics of long-range ordering (LRO). The kinetics of $L1_2$ ordering was studied by means of in situ resistometry within the γ'-phase of a quasi-binary system $Ni_{75}Al_{25-x}Fe_x$, where T_t drops drammatically with x.

Experimental evidence of an increase of the LRO activation energy with increasing T_t has been achieved. The result was discussed in terms of a classical model of diffusion in $L1_2$ superstructure proposed by Schoijet et al.

Low temperature resistometry and high resolution electron microscopy have been used to clarify some irreversibilities remarked in disordering kinetics in alloys with low Fe content.

INTRODUCTION

Goal of the study
Atomic mobility in ordered crystalline structures has for a long time been a subject of numerous studies. Experimentally, it has been shown that establishing of chemical long-range order generally slows diffusion and diffusion-controlled processes e.g. LRO kinetics. In particular, it has been found out that the kinetics generally slows down while the ordering advances (see e.g. [1]).

The experimentally observed effects have been reproduced by computer simulations, the results of which were summarised by Bakker several years ago [2].

Model approaches to atomic mobility in LRO structures and, specifically, to LRO kinetics (see e.g. the reviews of Yamauchi [3] and Bakker [1]) can be developed much more easily when applied to B2-type superstructures than in the case of the fcc-based L1o/L12 ones. This is mostly due to a lack of full equivalency of sublattices in fcc-based superstructures, which considerably complicates their configurational thermodynamics. Consequently, while the B2 long-range ordering is now quite well described within the Path Probability Method (PPM) as a collective process composed of elementary jumps of atoms to neighbouring vacancies [4], no such approach to L1o/L12 ordering has up to now been fully elaborated.

In the latter case, the still quoted classical model of Schoijet et al.[5] is focused on the evaluation of activation energies of particular, structurally non-equivalent jumps of atoms to nearest neighbour (nn) vacancies, the concentration of which is determined within equilibrium thermodynamics.

When considering LRO kinetics, it is useful to concentrate on the activation energy of the process, which, provided a relaxation between two well determined equilibrium states is in question, is a sum of the activation energies of vacancy formation and migration. While the model of Schoijet et al. directly yields this quantity, it also appears within the PPM, provided the initial departure of the system from equilibrium is small.

The LRO activation energy E_A results from the models generally as a function of atomic pair interaction energies V_{ik} and of the degree of atomic order. It is, however, shown that a term proportional to the ordering energy v ($v=2V_{ik}-V_{ii}-V_{kk}$) may always be separated. Consequently, if local-order effect is neglected, as v is proportional to the "order-disorder" transition temperature T_t, E_A should increase both with T_t and the LRO degree η. While the influence of η on E_A has been quite widely evidenced, no systematic experimental study of the effect of T_t has up to now been performed. The present work reports the first results of such a study.

Required experimental conditions and specimens

As all the above models describe homogeneous LRO structures, the experiments should be carried out using homogeneous, monophase samples preferably free from structural defects such as grain and antiphase boundaries, precipitates etc.

Because of a general complexity of the V_{ik} dependence of E_A (terms other than the v-dependent one are also present), it is not reliable to compare E_A evaluated for completely different systems with various T_t. Another variant of examining LRO solid solutions A-B differing in component concentrations is also inadequate because of the expected strong effect of the maximum attainable LRO degree dependent of the sample composition.

Appropriate experimental conditions were found in the case of quasi binary systems $Ni_{75}Al_{25-x}M_x$ ($0 < x < 25$) where M = Fe, Mn, Cr. The systems form continuous series of L12 ordered solid solutions — the γ' phase, whose stability limit T_t drammatically decreases with x. The phase diagram of $Ni_{75}Al_{25-x}Fe_x$, a system up to now investigated the most extensively , was established by Masahashi et al.[6].

As the Fe atoms occupy preferentially (though not exclusively) the minority Al sublattice (the cube corners) [7], the alloys $Ni_{75}Al_{25-x}Fe_x$ may be treated as $Ni_{75}X_{25}$ $L1_2$-ordered ones, X denoting an "average" atom whose properties — mostly interaction energies with nn atoms, depend on the Fe content x. Thus, the considered system assures both a wide range of T_t variation (Table 1) and a continuity of energetics in the crystalline lattice.

MATERIALS AND METHODS

Five alloys representing the compositional range from 0 to 15 at.%Fe were produced from 5N Al, 4N Ni and 4N Fe. The alloys were cast in a high-frequency furnace under Ar and rapidly solidified in quartz capillary tubes which resulted in needle-like samples 25 mm long and 1 mm in diameter. In order to attain a γ' homogeneity as high as possible and large antiphase ordered domains, all the samples were annealed for 1 week at temperatures T_A below the corresponding γ' phase boundary (Table 1). The alloys $Ni_{75}Al_{10}Fe_{15}$ and $Ni_{75}Al_{15}Fe_{10}$ were, however, first homogenised at 1223 K, as the corresponding temperatures T_A were supposed to be too low to eliminate the casting defects.

The homogeneity, effective compositions and microstructures of the specimens were tested by X-ray diffraction, SEM and TEM using JEOL JSM 840 and JEM 100 C devices. The TEM specimens, spark-eroded discs ca. 0.1 mm thick, were first annealed at T_A and then jet-polished at 230 K in a methanol solution of HNO_3.

DTA tests were carried out by heating the specimens water quenched from 1300 K. The samples were sealed under Ar in quartz ampoules together with Al_2O_3 powder added in order to improve the thermal contact. The heating rates were 2, 5, 10 and 20 K/min.

Resistometry was performed by a four-probe method using a computer controlled device. The first tests consisted of measuring the temperature dependence R(T) of the electrical resistance of the samples quenched from T_A in order to detect any possible transformations below T_t. Isothermal relaxations of LRO towards the equilibrium state were monitored by measuring time evolutions R(t) of the resistance at temperatures stabilised better than 0.1K. After a constant electrical resistance of the specimen had been reached at a given temperature (always below T_t), the subsequent measurement was carried out after the temperature had been rapidly decreased by a certain decrement ΔT. The ΔT value was kept constant all over the examination of each alloy.

RESULTS

Preliminary tests of samples annealed at T_A

X-ray diffraction examinations showed the presence of $L1_2$-type LRO in all the alloys and a decrease of a lattice parameter with increasing nominal Fe content (Table 1), in agreement with literature data [6].

SEM observations of the specimens indicated a lack of precipitates and an average grain size of 100 μm.

TEM dark field images of the $Ni_{75}Al_{20}Fe_5$ and $Ni_{75}Al_{23}Fe_2$ taken using the (100) and (110) superstructure reflections showed a homogeneous $L1_2$ superstructure with no antiphase boundaries (Fig.1a). This was, however, not the case of $Ni_{75}Al_{10}Fe_{15}$ and $Ni_{75}Al_{15}Fe_{10}$, where $L1_2$-ordered domains of a size 0.1 μm ($Ni_{75}Al_{10}Fe_{15}$) and 0.5 μm ($Ni_{75}Al_{15}Fe_{10}$) were separated by regions ca. 10 nm thick containing fine $L1_2$ domains dispersed in a disordered matrix (Fig.1b).

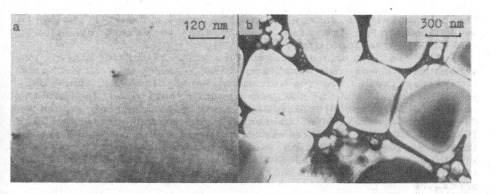

Figure 1. Microstructures of $Ni_{75}Al_{23}Fe_2$ (a) and $Ni_{75}Al_{15}Fe_{10}$ (b) after annealing at T_A. TEM dark field images at ⟨001⟩ zone axis using {100} superlattice reflection.

The $R(T)$ curves showed anomalies between 700 and 800 K. This effect, generated by an activation of vacancy mobility and the resulting recovery of LRO, indicated a temperature range suitable for isothermal experiments.

DTA tests of $Ni_{75}Al_{10}Fe_{15}$ and $Ni_{75}Al_{15}Fe_{10}$ resulted in exothermal peaks appearing systematically between 750 and 850 K and associated with the activation of vacancy mobility reflected by $R(T)$. Kissinger analysis of the heating-rate dependence of peak positions yielded an activation energy of about 1 eV for both examined alloys.

In situ LRO kinetics studies

Isothermal relaxations of the sample electrical resistance were recorded at temperatures between 700 and 830 K, provided a suitable step ΔT was applied. While $\Delta T=-10K$ was sufficient for $Ni_{75}Al_{10}Fe_{15}$, the other alloys required $\Delta T=-20K$. An exception was pure Ni75Al25, where, due to extremely fine effect of the LRO relaxation, ΔT was taken equal to -200K [8].

The relaxation curves were fitted with superpositions of single exponentials with log-normal distributions of relaxation times:

$$R(t,T) = R_{eq}(T) + [R(t=0)-R_{eq}(T)] \times \int_o^\infty d\tau P(\tau) \exp(-t/\tau) \qquad (1)$$

where:
$$P(\tau) = const. \times \exp[-(\ln\tau - \ln\tau_{av}(T))^2/\beta^2] \qquad (2)$$

Each one of the fits was χ^2 tested. In several cases also a single exponential was employed. However, the χ^2 was always ca. 10% higher than when using the formula (1) with $\beta \neq 0$. In no case did the half-width β exceed unity.

Assumming that τ_{av} obeys the Arrhenius law:

$$\tau_{av} = \tau_o \times \exp(E_A/kT) \qquad (3)$$

the LRO activation energy E_A and the pre-exponential factor τ_o were (except for the case of Ni75Al25) evaluated by fitting lines to $\ln(\tau_{av})$ against T^{-1}.

The results are displayed in Table 1 and in Fig.2. The errors of E_A and τ_o, resulting from χ^2 analysis, were 1% and 10% respectively. Two special cases were: Ni75Al25, where only an inferior limit of E_A was roughly estimated by putting $\tau_o = 10^{-14}$ s [8], and Ni75Al10Fe15, where two different values of E_A were obtained within the ferro- and paramagnetic phase [9]. The E_A value for Ni75Al10Fe15 displayed in Table 1 and related to the paramagnetic region was evaluated from the disordering kinetics as, being the Curie point in this case close to T_t, the region was not accessible by means of the step-cooling procedure.

TABLE 1

Nominal compositions, lattice parameters, "order-disorder" transition temperatures [6], annealing temperatures and LRO kinetics parameters of the examined alloys.

Composition (at.%)			a	T_t	T_A	E_A	τ_o
Ni	Al	Fe	(nm)	(K)	(K)	(eV)	(s)
75	25	0	0.35675	1720	1200	>3.4	$\leq 10^{-14}$
75	23	2	0.35660	1580	1200	3.2	4×10^{-17}
75	20	5	0.35640	1050	1000	2.8	7×10^{-15}
75	15	10	0.35608	877	770	2.3	2×10^{-11}
75	10	15	0.35594	825	770	2^P	
						3.1^f	1×10^{-16}

^{f} ferromagnet, ^{P} paramagnet

Disordering kinetics and irreversibility problems
Disordering kinetics were investigated by applying positive ΔT

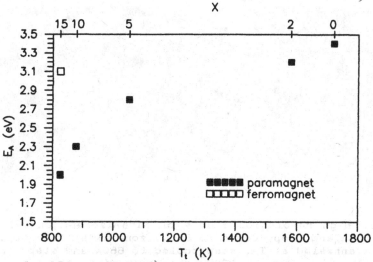

Figure 2. LRO activation energies in Ni₇₅Al₂₅₋ₓFeₓ alloys against the "order-disorder" transition temperature.

after having finished each step-cooling series. A satisfactory reversibility of the electrical resistance relaxations was obtained in the case of Ni₇₅Al₁₀Fe₁₅ [9] and Ni₇₅Al₁₅Fe₁₀. Because of the non-perfect LRO homogeneity shown by TEM, the latter alloy was isochronally annealed at temperatures T (T<Tt), step-decreased and then increased; each step was followed by quenching the sample in liquid He for measuring its residual resistivity, ρ_{res}, at 4K. Gradual increase of ρ_{res} with increasing T confirmed that the in situ measured R(t) relaxations were, at least predominantly, due to LRO evolution. The ρ_{res}(T) curves yielded a LRO activation energy of 2.8 eV. A slight difference between ρ_{res} measured before and after the experiment revealed the occurrence of some irreversible processes.

No reversibility of R(t) was found in the case of Ni₇₅Al₂₀Fe₅ and Ni₇₅Al₂₃Fe₂. In the former alloy regular R(t) relaxations, with ΔT>0, took place below 800 K; the resulting activation energy of disordering being, however, more than two times lower than that corresponding to ordering. At higher temperatures completely irregular isotherms were registered. Similar effects, inverted in sequence, occurred in the second alloy.

TEM samples of both these two alloys were step-cooled, subsequently step-re-heated, in analogy to the in situ resistometry cycles, and then examined by HREM using a Philips CM30 microscope. Two kinds of structural features appeared at the thermal-treatment stages corresponding to particular R(t) anomalies: a homogeneous distribution of regions showing Moiré fringes (Fig.3a) coincided with irregular R(t) isotherms, while the presence of localised particle-like domais, also showing Moiré fringes (Fig.3b) was observed during the regular R(t) relaxations with low activation energies.

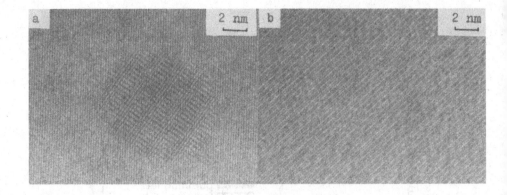

Figure 3. High resolution images of Ni75Al20Fe5: annealed at
TA and step-cooled to 660K, zone axis <110> (a),
annealed at TA, step-cooled to 660K and step-heated
to 820K, zone axis <100> (b).

DISCUSSION

The extensive tests of the γ'-phase homogeneity in the
examined alloys Ni75Al25-xFex, performed by means of a number
of complementary techniques, indicated that, after the
annealing at TA, the samples contained either entirely or
predominantly large γ' domains. It is thus possible to discuss
the observed Fe-content dependence of EA for ordering in terms
of the models reviewed in the first paragraph. The model of
Schoijet et al.shows that, in the case of an L12 ordered AB3
alloy, both short- and long-range order affect specifically
the total activation energy for atoms jumping between various
sublattices, when exchanging lattice sites with nn vacancies.
In particular, LRO causes an increase of the activation energy
of the jumps of A (minoritary) atoms while the corresponding
energy of the B (majoritary) atoms is decreased. After having
extensively discussed the model in the light of the
experimental data for Cu3Au, the authors concluded that the
rate of experimentally observed LRO relaxation is determined
by the A-atom migration, the activation energy of which
increases both with v (hence with Tt) and η.
        If all of the measurements had been performed in a fully
ordered γ'-phase (η=1), a linear rise of EA with Tt might have
been expected. Different behavior of the experimental data
(Fig.2), i.e. the higher slope of EA(Tt) in the range of low
Tt,is explained by the temperature range selected for the
resistometric tests. This was in fact, for the alloys
containing 10 and 15at.%Fe, close to Tt (Table 1), where η was
apparently lower than 1. Fig.2 thus illustrates both the Tt
and η effect on the activation energy of LRO kinetics to be in
good qualitative agreement with the model of Schoijet et al.

The difference between the values of E_A resulting from the in situ and 4K resistometry of $Ni_{75}Al_{15}Fe_{10}$ can be ascribed to the presence of excess vacancies quenched-in during the step-cooling of the samples. In such a case all the activation energies evaluated by the in situ experiments would be underestimated. A similar effect was due to the inhomogeneities revealed by HREM. Thus, as both the encountered perturbations of ideal experimental conditions acted in the same direction - i.e. towards lowering the effective LRO activation energy, they were not supposed to affect the qualitative behavior of E_A against T_t.

Currently, the investigations of the remaining quasi-binary systems $Ni_{75}Al_{25-x}Mn_x$ and $Ni_{75}Al_{25-x}Cr_x$ are being carried out.

REFERENCES

1. Benci, S. and Gasparrini, G., The activation energy for self diffusion in the Cu_3Au alloy. J.Phys.Chem.Solids, 1966, 27, 1035-39.

2. Bakker, H., Tracer diffusion in concentrated alloys. In Diffusion in Crystalline Solids, ed. G.E.Murch and A.S.Novick, Orlando, Academic Press, New York, 1984, p.185.

3. Yamauchi, H. and de Fontaine, D., Kinetics of order-disorder. In Order-Disorder Transformations in Alloys, ed. H.Warlimont, Springer Verlag, Heidelberg, New York, 1974, p.148-78.

4. Sato, H. and Kikuchi, R., Kinetics of order-disorder transformations in alloys. Acta Metall., 1976, 24, 797-809.

5. Schoijet, M. and Girifalco, L.A., Diffusion in order-disorder alloys. The face centered cubic AB_3 alloys. J.Phys.Chem.Solids, 1968, 29, 911-22.

6. Masahashi, N., Kawazoe, H., Takasugi, T and Izumi, O. Phase relations in the section Ni_3Al-Ni_3Fe of the Al-Fe-Ni system. Z.Metallkde., 1987, 78, 788-94.

7. Shindo, D., Kikuchi, M., Hirabayashi, M., Hanada, S. and Izumi, O., Site determination of Fe, Co and Cr atoms added in Ni_3Al by Electron Channelling Enhanced Microanalysis. Trans.JIM, 1988, 29, 956-61.

8. Kozubski, R. and Cadeville, M.C., In situ resistometric investigation of ordering kinetics in Ni_3Al. J.Phys.F:Met.Phys., 1988, 18, 2569-75.

9. Kozubski, R., Sołtys, J. and Cadeville, M.C., Long-range order kinetics in $Ni_3Al_{0.4}Fe_{0.6}$. J.Phys.:Condens.Matter, 1990, 2, 3451-58.

EQUILIBRIUM AND KINETICS OF THERMAL
ORDERING OR DISORDERING IN Ni₃Al

Colette DIMITROV, Tahar TARFA and Omourtague DIMITROV
CNRS, C.E.C.M.,
15 rue G. Urbain, 94407 Vitry-sur Seine Cedex, France

ABSTRACT

The high sensitivity of residual electrical resistivity to structural modifications in metallic materials was used for studying the composition and temperature dependence of the equilibrium and kinetics of atomic order, in binary Ni₃Al intermetallic compounds with 24.6 to 26.5 at% Al and in ternary materials with 4 at% Ti or Ta. In the binary alloys, the resistivity characterizing the equilibrium long-range order exhibits a positive temperature dependence, which is quantitatively consistent with changes of the LRO parameter calculated by the Cluster Variation Method. The analysis of the isothermal ordering (or disordering) kinetics shows that two first order processes are involved. The order change-rates are dependent on composition ; they are minimum at stoichiometry and increase with off-stoichiometry. The presence of a third element (Ti or Ta) results in a slower diffusion.

INTRODUCTION

The high-temperature properties of intermetallic structural materials are directly controlled by the characteristics of atomic order and by the atomic mobility resulting from diffusion. In the L1₂ Ni₃Al alloys the order-disorder transition temperature is near to, or above, the melting point. Nevertheless, changes of the equilibrium long-range order of these materials at lower temperatures have been detected by resistivity variations determined either in situ [1,2] or at 4.2 K [3]. In the present paper, the high sensitivity of residual electrical resistivity was used for studying, in Ni₃Al-based intermetallics, the temperature and composition dependence of the equilibrium atomic order and of its evolution kinetics. Some preliminary results have been reported in [4].

MATERIALS AND METHODS

The materials investigated were binary Ni₃Al compounds with 24.6, 25.0 and 26.5 at% Al, and ternary intermetallics Ni-21at%Al-4at%Ti and Ni-21at%Al-4at%Ta. They were prepared by melting together wheighed quantities of high-purity Ni, Al, Ni₃Ti or Ni₃Ta in an inductive plasma furnace, then annealed for 24 hours at 1273 K in a vacuum of 10^{-6} Pa. Rods ($1 \times 1 \times 10$ mm³) were cut out from the ingots with a diamond saw. Thermal treatments were performed either under a reduced pressure of pure helium gas in a stainless steel tube immersed in silicon oil or molten salt baths, or in a silica tube under a vacuum of 10^{-4} Pa. Electrical resistivities were measured at 4.2 K.

EQUILIBRIUM RESIDUAL RESISTIVITIES

Isochronal anneals were performed, at increasing or decreasing temperatures in the range 360-1260 K, after a preliminary treatment of 10 hours at 1173 K followed by slow furnace cooling. Results are shown in figure 1 for the binary Ni₃Al, and in figure 2 for the ternary materials. The evolution may be divided into three stages : i) no important change is detected up to about 800K ; ii) a large reversible resistivity variation was found in the ~ 800-1000 K range. Its amplitude is composition dependent : it is maximum in the stoichiometric binary alloy and decreases progressively when the composition deviates from stoichiometry ; it remains important when Al is partially substituted by Ta or Ti ; iii) above ~1000 K, a weaker variation (positive or negative) was observed.

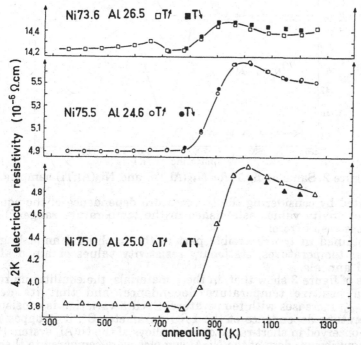

Figure 1. Residual resistivity changes during isochronal anneals (Δt = 40 min), in Ni₃Al.

The origin of the reversible resistivity increase has been previously discussed in the case of the binary Ni₃Al compounds [3]. It was shown that it could not be assigned to the elimination of antiphase boundaries or to the resistivity of equilibrium vacancies. It was concluded that the variation had an amplitude consistent with a change of the equilibrium degree of long-range order with temperature. In the present paper, this assignement is further

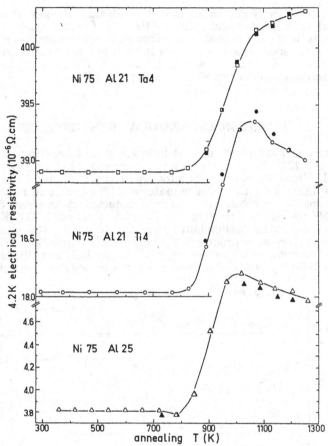

Figure 2. Same as fig.1, for $Ni_3(Al,Ta)$ and $Ni_3(Al,Ti)$ samples.

substantiated by considering the temperature dependence of the equilibrium residual resistivity values established in the temperature range 710-910 K. These values result from:
i) data obtained in the reversible part of the isochronal annealing curves,
ii) at lower temperatures, stationary resistivity values obtained after long isothermal anneals.
The curves of figure 3 show that, in these materials, the equilibrium resistivity exhibits a positive temperature dependence, and that its derivative $d\rho_{eq}/dT_{anneal}$ increases with temperature. Such a variation is consistent with the temperature dependence of the LRO parameter S ; it is opposite to the behaviour observed in short-range ordered alloys of the Ni(Al) system [5].

The equilibrium data of the Ni_3Al samples were compared to theoretically calculated variations by writing

$$\rho_{eq} = \rho_{ord} + \rho(S)$$

where ρ_{ord} is the resistivity of the fully ordered material, including any contribution of impurities or slight heterogeneities, and $\rho(S)$ is the LRO-dependent resistivity. According to Rossiter [6], in a stoichiometric long-range ordered alloy, the order-dependent resistivity is

$$\rho(S) = \rho_{dis}(1-S^2) / (1-AS^2) \qquad (1)$$

where S is the LRO parameter, ρ_{dis} is the resistivity of the disordered alloy and A is a parameter depending on the electronic structure of the material. In the case of off-stoichiometric materials, the maximum value of S, S_{max}, is smaller than 1 and $\rho(S)$ in equation (1) will not vanish in a fully ordered alloy. Also, the maximum change in $\rho(S)$ will then be (ρ_{dis} - ρ_{ord}). Therefore, equation (1) was used in a generalized form as :

$$\rho(S) = (\rho_{dis} - \rho_{ord})[1 - (S^2/S^2_{max})] / [1 - A(S^2/S^2_{max})] \qquad (2)$$

The temperature dependence of S was considered according to two theoretical calculations : i) the Bragg-Williams approximation, ii) the Cluster Variation Method (CVM).

i) In the Bragg-Williams theory [7], the temperature dependence of S is scaled on the critical order-disorder temperature T_c ; therefore, the experimental data were fitted with relationship (2) by adjusting the three parameters ρ_{ord}, A and T_c. A good fit is obtained in the stoichiometrie Ni₃Al compound and the values of the parameters : ρ_{ord} = 2.91 ± 0.01 µΩ.cm, A = 0.70 ± 0.04, T_c = 2210 ± 50 K, are of a reasonnable order of magnitude. However the transition temperature is substantially higher than the value (1723 ± 10 K) estimated by Cahn et al [8] on the basis of dilatometric studies. In the case of the off-stoichiometric alloys, the Bragg-Williams theory is known to yield a wrong composition dependence of the critical ordering temperature.

ii) The temperature dependence of the long range order parameter S was calculated by Cenedese [9] by the cluster variation method in the Tetrahedron-Octahedron approximation, with the pair interaction potentials given in [10]. These calculations yield T_c = 1744 K, which is in good agreement with the above cited value of 1723 K. The two parameters adjusted in relationship (2), ρ_{ord} and A, are given in table 1. The experimental data were satisfactorily fitted both for the stoichiometric and the off-stoichiometric alloys (figure 3).

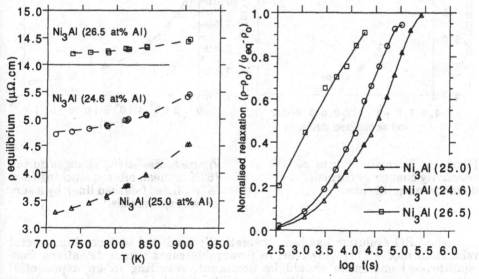

Figure 3. Equilibrium residual resistivities in Ni₃Al. Dashed lines : theoretical variations (CVM method)

Figure 4. Normalised variation of residual resistivity during anneals at 737 K (ρ_0 is the initial resistivity).

TABLE 1
Parameters ρ_{ord} and A obtained by fitting relation (2), with S calculated by the CVM method, to the equilibrium resistivity data of figure 3

Alloy	$Ni_{73.5}Al_{26.5}$	$Ni_{75.0}Al_{25.0}$	$Ni_{75.4}Al_{24.6}$
ρ_{ord} ($\mu\Omega$.cm)	14.20 ± 0.01	2.96 ± 0.01	4.72 ± 0.01
A	0.66 ± 0.02	0.628 ± 0.004	0.68 ± 0.01

ISOTHERMAL KINETICS OF ORDER RELAXATION IN Ni₃Al COMPOUNDS

The isothermal kinetics of ordering or disordering were studied in the temperature range 713-818 K, after temperature changes $\Delta T = T_a - T_q = \pm 25$ K, where T_a is the annealing temperature and T_q the temperature at which the sample was previously equilibrated. Figure 4 shows an example of the resistivity changes (normalised to the total variation amplitude) as a function of annealing time at 737 K, for the three compositions of the Ni₃Al compound. The variation rate, resulting from self-diffusion, is minimum for the 25.0 at% alloy, and increases with increasing deviations

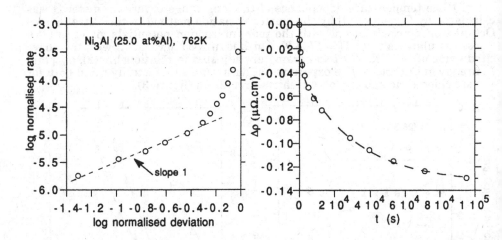

Figure 5. Normalised rate of resistivity change vs normalised distance to equilibrium.

Figure 6. Resistivity changes during 762K anneal, after quenching from 787K , fitted (dashed line) by a sum of two exponentials.

The LRO evolution has been theoretically predicted to depend on several relaxation modes [11]. However, in favourable cases (small deviations from equilibrium) one mode should be dominant, resulting in an exponential relaxation. In order to check the applicability of such kinetics, a log-log plot of the normalised relaxation rate $-d\rho/dt \,/\, (\rho_0 - \rho_{eq})$, as a function of the normalised deviation from equilibrium $(\rho - \rho_{eq}) \,/\, (\rho_0 - \rho_{eq})$, is given in figure 5.

A straight line of slope 1 is expected for an exponential relaxation, i.e. for first-order kinetics. It can be seen that such kinetics is actually followed for the last half of the evolution. The first part of the relaxation takes place with a much larger rate. This cannot be due to the presence of excess of vacancies at the beginning of the anneal, since such a behaviour is observed both for down-quenching ($\Delta T < 0$) and for up-quenching ($\Delta T > 0$) experiments. It suggests that the resistivity changes result from at least two processes, the second of which follows first-order kinetics. Therefore, the isothermal resistivity variations were analysed as a sum of two first order processes of amplitudes $\Delta \rho_1$ and $\Delta \rho_2$, with time constants τ_1 and τ_2. An example of the resulting fit is shown in figure 6.

Relaxation time-constants were obtained by this analysis in the binary Ni_3Al compounds, for up- and for down-quenching experiments. The corresponding relaxation rates (τ^{-1}) are shown in the Arrhenius plot of fig.7. An effect of aluminium content is clearly present. The ordering is generally faster (higher relaxation-rates) in the off-stoichiometric alloys than in the stoichiometric one ; however the fast process (τ_1^{-1}) seems to be comparable in the low-Al and in the stoichiometric material. The temperature dependence of the relaxation rates appears complex (fig.7) : the Arrhenius plot is curved, so that it is difficult to derive a meaningful activation enthalpy. A linear fit in the restricted range of the data yields a rather low value (~ 2.3eV).

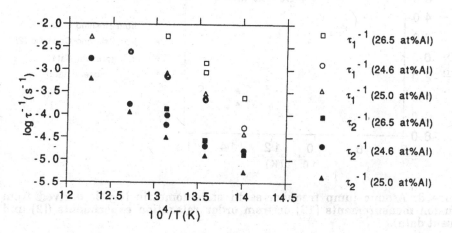

Figure 7. Arrhenius plot of the relaxation rates obtained by a 2-exponential analysis.

The present relaxation rates can be compared, in stoichiometric Ni_3Al, to the results obtained by Kozubski and Cadeville [2] by in-situ resistivity measurements. A large discrepancy exists between the two sets of data (figure 8). Atomic jump frequencies derived from the ^{63}Ni self-diffusion measurements of Hoshino et al.[12] by the relation $\nu_{at} = 12D/a^2$ (where D is the diffusion coefficient and a the lattice parameter) are also shown. It can be seen that these results are consistent with the present data, taking into account the different temperature range investigated. They also agree with the composition dependence observed in our experiments since, below 1000°C, diffusion was found to be slower at stoichiometry.

A possible interpretation of the presence of two relaxation processes was previously proposed [4] on the basis of the high ordering energy of Ni_3Al and of the specific geometry of the $L1_2$ structure. It consists in assigning the two processes respectively to the recovery of Ni-atom antisites (for the faster relaxation rate) and of Al-atom antisites (for the slower one). Both processes depend on the same fundamental mechanism (the diffusion of Ni-vacancies) but involve very different numbers of jumps. Quantitative calculations of the resulting kinetics should be performed for substantiating such an interpretation.

In the case of compounds in which a part of aluminium has been substituted by titanium or tantalum, no detailed isothermal kinetic studies are presently available. However, some indication on the atomic mobility may be derived from the isochronal curves of fig.2. In these curves, the intercept between the initial plateau and the rising part of the curves corresponds to the onset of sufficient atomic mobility for order changes to take place. This temperature increases when going from binary Ni3Al (795 K) to the Ti-substituted material (875 K) and to the Ta-substituted one (960 K). This suggests that the presence of Ti and particularly Ta leads to a slower diffusion, and thus should have a favourable effect on the high-temperature properties of Ni3Al-based materials.

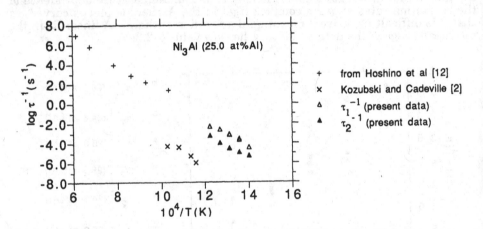

Figure 8. Atomic jump frequencies in stoichiometric Ni3Al, derived from diffusion measurements [12] or from order relaxation experiments ([2] and present data).

CONCLUSIONS

The structural evolution in Ni3Al intermetallics, during thermal treatments in the range 710K-910K, has been investigated by means of residual electrical resistivity measurements. The temperature-dependence of the equilibrium resistivity values is quantitatively consistent with changes of the equilibrium degree of long-range order. The analysis of the isothermal order relaxation shows that two kinetic processes are involved. In the binary Ni3Al compounds, the corresponding diffusion-rates were smaller at the stoichiometric composition. The presence of titanium or of tantalum atoms results in a slower diffusion.

REFERENCES

1. Corey, C.L. and Lisowsky, B., Trans. Metall. Soc. AIME, 1967, **239**, 239-43.
2. Kozubski, R. and Cadeville, M. C., J. Phys. F : Metal Phys., 1988, **18** , 2569-75.
3. Sitaud, B., Zhang, X., Dimitrov, C. and Dimitrov, O., Advanced Materials and Processes, eds. H. E. Exner, V. Schumacher, DGM, Oberursel, FRG, 1990 pp. 389-94.
4. Dimitrov, C., Zhang, X., Tarfa, T. and Dimitrov, O., Proceedings of International Symposium on Intermetallic Compounds, JIMIS-6 ; ed. O. Izumi, The Japan Institute of Metals, Sendai, 1991, pp. 75-79.
5. Sitaud, B. and Dimitrov, O., J.Phys.: Condens. Matter., 1990, **2**, 7061-75.
6. Rossiter, P.L., J. Phys. F : Metal Phys.,1980, **10**, 1459-65.
7. Bragg, W.L. and Willams, E.J., Proc. Roy. Soc. A, 1935, **151** 540-66.
8. Cahn, R.W., Siemers, P.A., Geiger, J.E., Bardhan, P., Acta Metall., 1987, **35** , 2737-51.
9. Cenedese, P., 1991 (private communication)
10. Cenedese, P., Marty, A. and Calvayrac, Y., J. Phys. France, 1989, **50**, 2193-2200
11. Gschwend, K., Sato, H. and Kikuchi, R., J. Chem. Phys., 1978, **69**, 5006-19.
12. Hoshino, K., Rothman, S.J. and Averback, R.S., Acta Metall., 1988, **36** 1271-9.

Acknowledgements
The authors are grateful to Dr P. Cenedese for the CVM and to Dr P. Bellon for the Bragg-Williams calculations of the temperature dependence of the long range order parameter in Ni3Al.

INVESTIGATION OF DEFORMATION INDUCED DISORDER AND RE-ORDERING BY SUBSEQUENT ANNEALING IN Ni_3Al+B

J. BALL, R. MITTEAU, and G. GOTTSTEIN,
Institut für Metallkunde und Metallphysik, RWTH Aachen,
Kopernikusstr. 14, W-5100 Aachen, FRG

ABSTRACT

The change of the degree of order with increasing rolling reduction was determined by X-ray diffraction. During a subsequent isochronal annealing treatment between 273 K and 1473 K of 40% rolled samples, hardness, yield-stress and the intensities of the superlattice reflections were measured and the microstructural evolution was investigated by TEM. The correlation between the measured property changes are discussed.

INTRODUCTION

The current study addresses the effect of plastic deformation and annealing on the long range order (LRO) parameter S of Ni_3Al and its correlation to hardness, flow stress and microstructure.

The incentive for this investigation is the frequently reported change in the degree of LRO and microhardness of cold worked intermetallic compounds, in particular Ni_3Al in the temperature range between 673 K and 923 K [1,2]. For more information on the dislocation structure and the state of recrystallization also yield-stress measurements as well as TEM investigations were conducted.

EXPERIMENTAL PROCEDURE

The nickel aluminide under investigation had the composition $Ni_{76}Al_{24}$ + 0.24at% B. The initial grain size of the fully ordered alloy, attained after a sequence of thermomechanical treatments, was approximately $30\mu m$. Mechanical properties and order parameter S were investigated as a function of the rolling degree and for 40% rolled specimens during isochronal annealings.

The LRO was determined by means of X-ray diffraction. To account for the influence of texture development on diffracted intensities during rolling, the specimen was tilted to obtain the best signal to noise ratio for most precise measurements. For the same reason the specimen was laterally oscillated to maximize the illuminated specimen area.

The LRO parameter S was determined from the ratio of the intensities of {100} superlattice reflection and {200} fundamental reflection as obtained from a θ-2θ scan in an X-ray diffractometer. After correcting for the background the peak intensities were integrated to take eigen-stresses into account. For comparison of measurements under different tilt angles the integrated intensities were corrected for absorption by means of a standard sample. Finally the LRO parameter S was calculated according to

$$S^2 = \left[\frac{I_S}{I_F}\right]_{\alpha=const} \cdot \left[\frac{I_F^{St}}{I_S^{St}}\right] \qquad (1)$$

where I_S, I_F correspond to the integrated intensities of {100} and {200} reflections, respectively. The supercript St indicates the measurements on a standard sample, α denotes the tilt angle. The reproducibility of the measured intensity ratio I_S/I_F was determined by conducting several measurements on the same deformation sample. A very small standard deviation of $\sigma=0.01$ was obtained. The standard deviation from the results on three different specimens was determined to be $\sigma=0.02$.

To investigate the influence of annealing on the degree of LRO, microhardness and yield stress the recrystallized bulk material was cold rolled to a reduction of 40%. From these rolled sheets flat tensile specimens were machined with a gauge size of $0.5\times5\times22mm^3$. The samples were annealed for 30 min. at various temperatures between 273 K and 1473 K under Argon gas atmosphere. Tensile tests were conducted at room temperature in an Instron mechanical testing machine.

To properly account for absorption and the influence of texture all measurements on the annealed specimen were performed at $\alpha=30°$. The tilt angle $\alpha=30°$ was selected because the rolling texture contains a strong brass component and thus would yield maximum intensity [2]. For the recrystallized state of the specimen the choice of α is uncritical since Ni_3Al develops a random texture upon recrystallization [2].

Microstructural investigations were performed in a TEM (JEOL 2000FX) operating at 200 kV. TEM studies were carried out on specimens annealed in the temperature interval between 673 K and 873 K where primary recrystallization would take place.

EXPERIMENTAL RESULTS

On specimens subjected to rolling reductions between 10% to
95%, X-ray scans were performed and the LRO parameter S was
calculated according to equation (1) (Fig.1). A continuous
degradation of LRO is observed from S=1 prior to deformation,
to S≈0.2 after a rolling reduction of 95%.

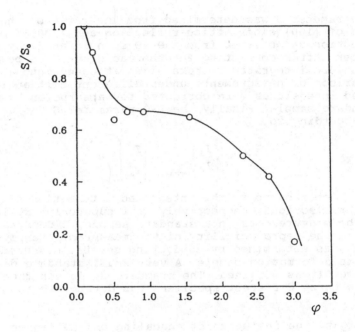

Figure 1. Change of the LRO as a function of rolling reduction
of initially fully ordered boron-doped Ni_3Al

A comparison of the development of LRO, microhardness and
yield stress during annealing is illustrated in Fig.2. Below
673 K there is neither a change in microhardness nor in the
degree of LRO. Above 673 K the LRO parameter S increases
steadily while the hardness peaks at 773 K and thereafter decreases until it reaches the same value as prior to deformation.

Also the yield stress $\sigma_{0,2}$ does not show any significant change
after annealing below 673 K (Fig.2). It is noted that all tensile specimens fractured after some plastic deformation, although the plastic stress was small at low annealing temperatures (Fig.3). Annealing above 673 K results in a smooth decrease of $\sigma_{0,2}$ with a noticeable increase in ductility.

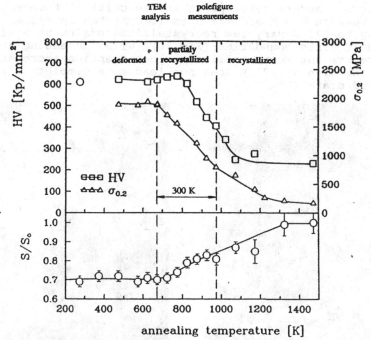

Figure 2. Influence of the annealing temperature on LRO, microhardness and yield-stress on 40% cold-rolled boron-doped Ni₃Al

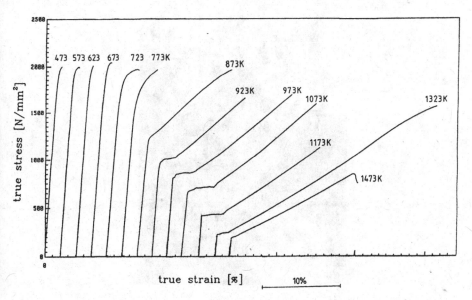

Figure 3. The stress strain curves of 40% cold-rolled boron-doped Ni₃Al annealed at various temperatures between RT and 1473K

The microstructure development in the critical temperature range between 673 K and 973 K is illustrated in Fig.4. After heating to 673 K very few recrystallized grains have already formed at grain boundaries (Fig.4a). With increasing annealing temperature the recrystallized volume fraction increases but annealing even at 873 K does not lead to a completely recrystallized state (Fig.4b).

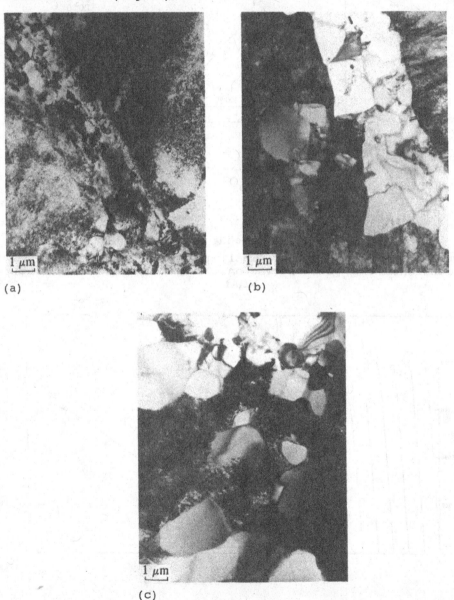

(a)

(b)

(c)

Figure 4. Microstructure development of 40% cold-rolled boron-doped Ni_3Al after annealing at (a) 673 K, (b) 873 K and (c) 973 K

Even after annealing at 973 K the apparently fully recrystallized samples exhibit high densities of stacking faults and microtwins. Locally even high dislocation densities are evident (Fig.4c). No antiphase domain configurations could be observed.

DISCUSSION

The degradation of order during deformation has been reported before by other authors [3,4]. The most conspicuous result is the drastic reduction of S to almost complete loss of order after 95% rolling reduction. This can not be understood in simple terms. Under the assumption that the disordered volume in a deformed material is confined to the atomic layer between the partials of an extended superdislocation, that the dislocation density is as high as $10^{16}m^{-2}$, usually an upper limit for heavily deformed metals, and that the antiphase boundary energy amounts to $\gamma_{APB}=110$ mJ/m^2, a LRO parameter S=0.93 [Eq. (1)] is obtained. This is obviously very different from the observed value. Apparently the disordered volume is substantially underrated in this considerations and we have to assume that more than a single atomic layer will contribute to the weakening of the superlattice reflection if disordering is due to dislocation core structures. Moreover, other imperfections like stacking faults may further degrade the intensity of the superlattice reflection.

The increase of S with progressing recrystallization is understandable in terms of restored order during the removal of dislocations in the course of nucleation and growth of recrystallized grains. The same holds for the recovery of the yield stress. The behavior of hardness, however, is contrary to expectations. The increase of hardness to attain a maximum during incipient stages of recrystallization can only be understood as a hardening of the deformed structure during annealing. With progressing recrystallization softening will finally overcompensate the hardening so that the measured hardness will go through a maximum.

The hardness increase can be rationalized on the basis of a relaxation process of the dislocation cores during annealing. We propose that during annealing the superdislocations relax by cross slip from {111} to {100} planes, and thus forms "Kear-Wilsdorf locks" in analogy to models discussed for the anomalous yield behavior of Ni$_3$Al. This process will neither be observed in the degree of order, nor in the dislocation arrangement. Such a hardening, however, ought to affect the yield stress as much as it affects hardness, contrary to experimental results. We presume that this discrepancy is due to morphological reasons of recrystallized grains at grain boundaries and microbands or shear bands. The microbands / shear bands are oriented under some angle ±α to the rolling direction. This morphology of intersecting systems of microbands and shear bands offers an explanation for the observed discrepancy between the behavior of hardness and yield stress. With recrystallization being confined initially to prior grain boundaries and deformation inhomogeneities very thin and progressingly contiguous layers of recrystallized material develops through the sheet cross section and correspondingly through the cross section of the tensile specimen machined from this sheet. These thin layers represent a very small volume fraction of the recrystallized material and will not substantially contribute to the hardness value which is averaged over a plastically deformed area much larger than the layer thickness. A continuous layer through the cross section of a tensile sample will have a drastic effect on the yield stress, however, because yielding will primarily be confined to this cross section followed by rapid hardening until the flow stress of the unrecrystallized volume is attained. Any progress of recrystallization towards a more contiguous network will therefore lower the yield stress despite the additional hardening of the bulk material. This concept is further supported by the observed ductility increase during recrystallization up to 10% while the hardness still increases. In summary the property correlations are far from being unique. While the hardness yields information on the mechanical properties of the bulk, it does not correlate to the degree of order or recrystallized volume fraction during recrystallization. The yield stress, however, is a very sensitive measure of the recrystallized volume and therefore indirectly of the degree of order but the correlation is only provided through the particular morphology of the recrystallized microstructure, which avoids the interference of hardness effects in the bulk.

CONCLUSIONS

1. The long range order parameter exhibits a strong decrease during rolling, much larger than expected from disordering by dislocation core structures.

2. Yield stress decrease, gain in ductility and increase of LRO change monotonically during progressing recrystallization.

3. The hardness exhibits a maximum during incipient stages of recrystallization. This is attributed to Kear-Wilsdorf lock formation during annealing.

4. The different behavior of yield stress and hardness is attributed to the morphology of the recrystallized microstructure.

REFERENCES

1. Cahn, R.W., in <u>High temperature Aluminides and Inter-
    metallics</u>, ed. S.H. Whang, C.T. Liu, D.P. Pope and J.O.
    Stiegler, The Minerals, Metals & Materials Society, 1990.

2. Ball, J., Zeumer, B., Gottstein, G., Proceedings
    ICOTOM 9, Avignon, 1990, in press

3. Jang, J.S.C. and Koch, C.C., <u>J. Mater. Res.</u>, 1990, **5**,
    498.

4. Clark, J.P. and Mohanty, G.P., <u>Scripta metall.</u> 1974, **8**,
    959

DISORDERING BY PLASTIC DEFORMATION OF SHORT-RANGE AND LONG-RANGE ORDERED Ni-BASED ALLOYS

M. VILTANGE and O. DIMITROV
Centre d'Etudes de Chimie Métallurgique du C.N.R.S,
15 rue Georges Urbain, F-94407 Vitry-sur-Seine Cedex, France

ABSTRACT

The change of atomic order and of defect concentration during plastic deformation was investigated in short-range ordered austenitic Fe-Cr-Ni alloys and in long-range ordered substoichiometric boron-doped Ni3Al alloys, by measuring electrical resistivity changes. Analytical relationships were used for representing the respective contributions of these two processes to the total resistivity variation. Disordering efficiency was found to be very different in short-range and long-range ordered materials.

INTRODUCTION

Plastic deformation affects the microstructure of alloys by :
- producing lattice defects (dislocations, point defects ...)
- changing atomic order : the neighbourhood of a given atom is modified during dislocation glide.
    Electrical resistivity, which is widely used to characterize the deviations from an ideal periodic lattice, is sensitive to both these processes. Lattice-defect production contributes positively to resistivity. This is similar to the case of pure metals, for instance nickel (1) ; however, very little information is available on the magnitude of this contribution in alloys. Disordering increases the resistivity of long-range ordered alloys. In short-range ordered alloys, it may increase or decrease resistivity, according to the relative positions of Fermi surface and Brillouin zone boundaries (2,3).
    We have investigated the influence of plastic deformation on the atomic order of two families of concentrated crystalline nickel-based alloys :
    1) short-range ordered austenitic Fe-Cr-Ni alloys with ~17 at.% chromium and 19 to 73 at.% nickel
    2) long-range ordered substoichiometric Ni3Al doped with 0.1 weight % boron.

The alloys were prepared from high-purity metals by plasma-furnace melting.After cold-rolling and specimen fabrication, standard anneals produced reproducible atomic structures with well-defined degrees of order ; these were characterized by resistivity measurements.The preparation technique and the compositions have been described elsewhere (4,5). After deformation, resistivity was again determined in order to obtain the modifications introduced by cold-work.

DEFORMATION OF SHORT-RANGE ORDERED FeCrNi

Four alloys were studied with nickel contents of 19, 24, 43 and 73 at.% Ni. They are face-centered cubic and short-range ordered (6,7). It has been found that their resistivity increases with the degree of short-range order, produced by thermal or irradiation treatments (8-10).

Tensile Deformations (11)

Two sets of resistivity measurements were performed :
- at 4,2K before and after deformation
- at room temperature, during the tensile tests : a direct current was fed through the electrically insulated grips, and the potential drop was measured between two nickel wires on the specimen.
1) Measurements at 4.2 K before and after deformation : Four series of resistance measurements were performed.

α) after a standard anneal at 1273 K and slow cooling to reach a reproducible degree of order (normalized state), of known electrical resistivity, ρ_N (8,9) :
$$R^4_N = \rho_N . F_0$$
where F_0 is the shape-factor of the undeformed sample,

β) after a second anneal and adequate thermal treatment : fast cooling (Tr) or thermal ageing (Vth) at successively decreasing temperatures between 800 and 700 K, to get either a lower degree of order or a higher degree of order (4) :
$$R^4_0 = \rho_0. F_0$$

γ) after a tensile deformation ε :
$$R^4_\varepsilon = (\rho_0 + \Delta\rho_{pl}).F = (\rho_0 + \Delta\rho_{pl}) . F_0 . (1 + \varepsilon)^2$$

δ) after a new standard anneal to return to the same structural state as in (α) :
$$R'^4_N = \rho_N . F_0 . (1 + \varepsilon)^2$$
These four measurements give access to the resistivity ρ_0 , the effective strain ε and the resistivity change due to plastic deformation $\Delta\rho_{pl}$.
2) Room-temperature measurements during extension : They were used to continuously measure the change in room-temperature resistivity caused by the elastoplastic strain $\Delta\rho_{el + pl}$; after stopping the test and unloading, the resistivity change due to plastic deformation $\Delta\rho_{pl}$ is determined.

Results obtained by both methods are shown in Fig. 1 for fast-cooled and aged $Fe_9Cr_{18}Ni_{73}$.Weak positive changes of resistivity occur during the first few percents of deformation ; afterwards, a general decrease of resistivity is observed, greater for thermally-aged specimens than for fast-cooled ones.

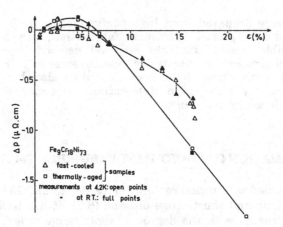

Figure 1. Electrical resistivity changes produced by tensile deformation of the
Fe₉Cr₁₈Ni₇₃ alloy.

The effect of composition appears in Fig. 2. The values of the resistivity
changes, found for the four fast-cooled or thermally-aged materials are
compared for the same strain value of 0.15. Changes of resistivity are negative
and their magnitude increases :
- with nickel content,
- with the thermal-ageing treatment (the cross-over around 25 % Ni might be
due to the the uncertainty on small resistivity changes).

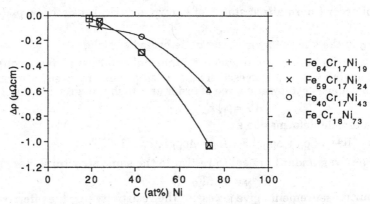

Figure 2. Resistivity changes of FeCrNi alloys, for an extension ε = 0.15. Plain
symbols : fast-cooled samples ; squared symbols : thermally-aged samples.

The relation between Δρ (at ε = 0.15) and the order resistivity, i.e. the
difference between initial resistivity (ρ_0(Tr) or ρ_0(Vth)) and the resistivity of the
entirely disordered state (ρ_{dis}) is shown in Fig. 3. Values of ρ_{dis} have been
determined by extrapolating, to infinite temperature, the temperature

dependence of the equilibrium resistivity of the alloys $\rho_{eq} = a.(1 + b / T)$ (12). It can be seen that $\Delta\rho$ varies approximately linearly with ($\rho_0 - \rho_{dis}$).

The whole set of results is consistent with deformation-induced decreases of the short-range order of these alloys.

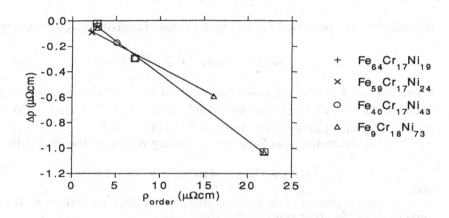

Figure 3. Resistivity changes of FeCrNi alloys at $\varepsilon = 0.15$, as a function of the initial order resistivity. Plain symbols : fast-cooled samples ; squared symbols : thermally-aged samples.

Deformation by Cold-rolling (13)

Samples of an $Fe_{40}Cr_{17}Ni_{43}$ alloy, in the normalized state, were cold-rolled at room temperature to reductions in thickness (eo - e) / eo up to 90 %, or D = Ln(eo / e) up to ~2,3. Fig.4 shows that resistivity decreases with increasing deformation.

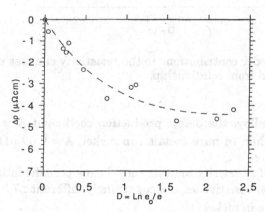

Figure 4. Values of resistivity changes of cold-rolled normalized $Fe_{40}Cr_{17}Ni_{43}$

150

In the relationship :

$$\Delta\rho = \Delta\rho_{def.} + \Delta\rho_{order}$$

- the defect contribution $\Delta\rho_{def.}$ was assumed to be proportional to deformation, $\Delta\rho_{def.} = A \cdot D$. In nickel, linearity has been observed up to D values of 2 (1).
- the increment of order resistivity $\delta\rho_{order}$ was assumed to be proportional to

the deformation increment δD and to the order resistivity ρ_{order}, resulting in :

$$\Delta\rho_{order} = \rho^0_{order} \cdot (\exp(- C \cdot D) - 1) \qquad [1]$$

when ρ^0_{order} is the initial order contribution to resistivity, i.e. the difference between the resistivity of the normalized state ($\rho_N = 95.6\ \mu\Omega cm$) (9) and the resistivity of the disordered state ($\rho_{dis} = 89.43\ \mu\Omega cm$) (12).
    This relationship was found to adequately represent the data, with :

$$A = 0{,}50 \pm 0.04\ \mu\Omega cm$$
and $$C = 1.18 \pm 0.11$$

In the deformation range investigated, the defect contribution is always smaller than the order contribution (figure 5).

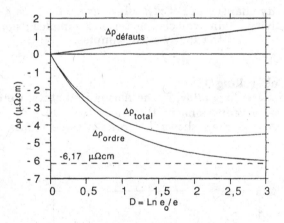

Figure 5. Defect and order contributions to the resistivity changes of cold-rolled $Fe_{40}Cr_{17}Ni_{43}$, deduced from relationship.

In the present alloy, the defect production coefficient A is found to be significantly greater than in pure metals : in nickel, $A = 0.09\ \mu\Omega cm$ (1). This difference is not due to :
- a higher sensitivity of resistivity to deviations from a periodic lattice, since the room-temperature ideal resistivities are not strongly different : 7.5 $\mu\Omega cm$ in the alloy (9) and 6.85 $\mu\Omega cm$ in nickel (1).

- a higher specific defect resistivity : the Frenkel-pair resistivity is 8.6 $\mu\Omega$cm/% in the alloy and 6.7 $\mu\Omega$cm /% in nickel (15).
- the nature of created defects, since in both cases deformation occurs by slip and vacancies are immobile at room temperature.

The higher value of A may be assigned to a larger dislocation resistivity, due to a more important dislocation dissociation in the alloy : the stacking fault energies are estimated to be approximately 50 mJ m^{-2} in the alloy (16), as compared to 250 mJ m^{-2} in pure nickel (17).

The value of the disordering parameter C (C = 1,18) shows that local order is destroyed by a deformation somewhat larger than unity.

ROLLING DEFORMATION OF Ni$_{76}$Al$_{24}$(+B)

Deformation studies were performed on two single-phase substoichiometric Ni$_3$Al alloys (doped with 0.1 weight % boron to prevent intergranular brittleness), prepared from two different grades of nickel :
Ni$_{75.8}$Al$_{24.2}$ (+ B)   (from 99.994 % Ni)
Ni$_{76}$Al$_{24}$ (+ B)   (from 99.9 % Ni)

The two materials, in the normalized state, were deformed by cold-rolling. 4,2 K resistivity changes were determined, using either ρ_N values measured on each individual sample, or the average ρ_N of all samples.

Fig. 6 shows the experimental increase $\Delta\rho$ in the 24.2 at. % Al compound. In this material, the resistivity increase is a little slower than in the impure 24.0 at. % Al material (the $\Delta\rho$ values were 32.6 and 40 $\mu\Omega$cm for D = 2.16, while the initial resistivity values were 21.9 and 24.38 $\mu\Omega$cm). Nevertheless, the relative values $\Delta\rho / \rho_N$ are very close (Fig. 7).

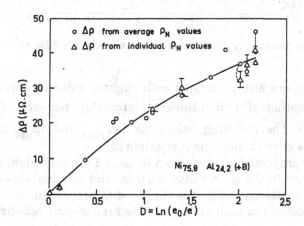

Figure 6. Resistivity changes values of cold-rolled normalized Ni$_{75.8}$Al$_{24.2}$(+B)

In long- range ordered materials, a decrease of the degree of order induces an increase of resistivity (5).A mathematical representation taking into account the two contributions to the resistivity change was established :

$$\Delta\rho_{total} = A . D + \rho_{dis} . [1 - \exp(-2\,C.D\,)] / [1 - a . \exp(-\,2\,C.D)] \qquad [2]$$

The change in order resistivity (second term in right-hand side of equation [2]) has been deduced from the relationship between order resistivity ρ_{order} and

long-range order coefficient S (18) :

$$\rho_{order} = \rho_{dis} . (1 - S^2) / (1 - a.S^2)$$

where a is a coefficient depending on the electronic structure of the material. We have taken a value of a = 0.75, which is consistent (19) with the temperature dependance of the electrical resistivity of Ni_3Al. The value of ρ_{dis} was deduced from (20). The resulting values of A and C are given in table 1.

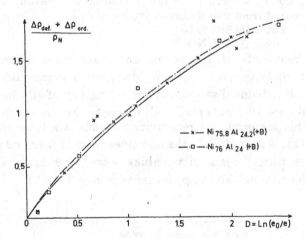

Figure 7. Relative resistivity changes of the two cold-rolled Ni aluminides

Other fits were also performed with slightly different values of a, and with the assumption of a non-linear relationship between $\Delta\rho_{def}$ and D ($\Delta\rho_{def} = A' . D^n$). The resulting values for $\Delta\rho_{order}$ and $\Delta\rho_{def}$ are not very different from the ones obtained from equation [2].

The defect production coefficients A in table 1 are very high, as compared to the one obtained in the short-range ordered solid solution. However, such a difference is consistent with the large differences in the resistivity contributions of defects in such materials. The Frenkel-pair resistivity in Ni_3Al was estimated from irradiations studies to be of the order of 100 $\mu\Omega cm$ / % (5), much higher than the value of 8,6 $\mu\Omega cm/\%$ for the $Fe_{40}Cr_{17}Ni_{43}$ alloy.

TABLE 1
Defect production and disordering coefficients, resulting from the
relationships between electrical resistivity and deformation (see text)

	Defect production coefficient A ($\mu\Omega$.cm)	Disordering parameter C
Nickel	0.09	
Short-range ordered alloy $Fe_{40}Cr_{17}Ni_{43}$	0.50 ± 0.04	1.18 ± 0.11
Long-range ordered alloys $Ni_{75.8}Al_{24.2}(+B)$ $Ni_{76}Al_{24}(+B)$	9.1 ± 4.0 7.8 ± 1.2	0.07 ± 0.05 0.10 ± 0.02

The low value found for the disordering coefficient C (0,07 to 0,1) , as
compared to C = 1,2 in the short-range ordered alloy points to a low efficiency of
disordering by dislocation glide. This might be due to the fact that, at least at
small deformations, the disorder created by the glide of a dislocation in the
long-range ordered alloy will be restored by the passage of a second dislocation
in the same glide plane. Therefore the total disordering of Ni_3Al will require a
much larger deformation than for a short-range ordered alloy.

CONCLUSION

In these two examples, we have established quantitative relationships between
deformation and the resistivity changes related to the decrease of order during
cold-rolling, both in the short-range and the long-range ordered materials In
the case of short-range ordered alloys, high tensile deformations yield results
in qualitative agreement with cold-rolling. However, at small strains, a
different process appears to take place. In the long-range ordered alloys,
the disordering efficiency of cold-rolling was found to be much smaller than in
the short-range ordered one.

REFERENCES

1. Merklen, P., Thesis, Doctorat d'Etat ès Sciences-Physiques, Paris 1968.
2. Gibson, J.B., J. Phys. Chem. Solids,1956, 1 , 2.
3. Rossiter, P.L., Wells, P., J. Phys. C : Solid St. Phys. , 1971 ,4,354.
4. Aïdi, B., Viltange, M., Dimitrov, O., J. Nuclear Mater., 1990, 175, 96.

5. Zhang, X., Thesis, Doctorat, Spéc. : Science des Matériaux, Caen, 1989.

6. Bley, F. Cénédèse, P., Lefebvre, S., A.I.P. Conf. Proc. n⁰ 89, Neutron Scattering, New-York, 1981, p. 276.

7. Cénédèse, P., Bley, F. Lefebvre, S., Acta Cryst., 1984, A 40, 228.

8. Dimitrov, C., Tenti, M., Dimitrov, O., J. Phys. F.: Metal Phys., 1981, 11, 753.

9. Dimitrov, C.,Dimitrov, O. J. Phys. F.: Metal Phys. , 1984, 14, 793.

10. Dolique, G., Report C.E.C.M.n⁰ 82A3, 1982.

11. Aïdi, B., Thesis, Doctorat Univ. Paris VI, Spéc. : Métallurgie, 1990.

12. Dimitrov, O., Dimitrov, C. J. Phys. F : Metal Phys., 1986, 16, 969.

13. Viltange, M., Dimitrov, O., J. Nuclear Mater., 1991 (in press).

14. Dimitrov, C., Huguenin, D., Moser, P., Dimitrov, O.,, J. Nucl. Mater., 1990, 174, 22.

15. Dimitrov, O., Dimitrov, C. , Rad. Effects, 1985, 84, 117.

16. Rhodes, C.G., Thompson, A.W., Met. Trans., 1977, 8 A, 1901.

17. Gallagher, P.C.J., Met. Trans., 1970, 1, 2429.

18. Rossiter, P.L., J. Phys. F : Metal Phys., 1980, 10, 459.

19. Cahn, R. W., Riemers, P. A., Geiger, J. E., Bardhan, P., Acta Met., 1987, 35, 2737.

20. Sitaud, B., Dimitrov, O., J. Phys.: Condensed Matter, 1990, 2, 7061.

DEFECT RECOMBINATION PHENOMENA IN MELT-SPUN ORDERED ALLOYS OF THE Fe-Al SYSTEM.

F. MARINO°, S. GIALANELLA*, R. DELORENZO+.

°Dipartimento di Ingegneria dei Materiali, Università di Trento, 38050, Mesiano di Povo, Trento, Italy.
*Department of Materials Science and Metallurgy, University of Cambridge, Pembroke Street, Cambridge, CB2 3QZ, U.K., *perm. address* (°)
+Dipartimento di Scienza dei Materiali e Ingegneria Chimica, Politecnico di Torino, Corso D. degli Abbruzzi 24, 10129, Torino, Italy.

ABSTRACT

Aluminides having B2 ordered structure can accomodate in their lattice very high concentrations of vacancies. In the present work we study the effect of the different defect concentrations on the recovery and reordering phenomena of some rapidly solidified iron aluminides ordered alloys.
Several series of melt spun ribbons of binary, FeAl, and ternary, Fe-Al-Cr, compositions were prepared.
The kinetics of the annealing out of the frozen-in defects was investigated by Differential Scanning Calorimetry, and the variations in the microstructural characteristics before and after the heat treatments were analysed by X-ray Diffraction
The observed kinetics show single or multi-step recombination processes, depending on the composition of the samples.

INTRODUCTION

As reported in several studies, B2 phase intermetallic compounds cannot be disordered by thermal quenching. One of the first investigations on β-brass [1] showed, through accurate measurements of superlattice x-ray peak intensities, intrinsically very weak in this material, no sensible variations after quenchings from different temperatures. On the other hand analogous heat treatments, that is anneals followed by rapid quenches, are very effective in freezing in these class of materials high concentrations of lattice defects.

The annealing out of these defects was studied for β-brass [2] and other alloys with the same B2 structure: NiAl [3], AuZn [4], AuCd [5].

FeAl is another system widely studied, initially just for its crystallographic and microstructural properties [6], more recently also for its interesting characteristics as structural material for high temperature applications [7].

In this material also, which can exist at room temperature over a wide composition range (23-55 at.%), high concentrations of non-equilibrium defects, vacancies and anti-site defects, can be hosted without decreasing consistently the degree of long range order. One of the reasons of this behaviour relies in the particularly low energies of vacancy formation, E_f [3]. It is worth noting that, differently from elemental metals and ordinary random alloys, in B2 intermetallics the ratio E_m/E_f, between the energy of migration and of formation of vacancies, is >1 (see table in ref. 3).

A recent investigation [8] proved how different concentrations of vacancies can significantly change the mechanical properties, in particular the hardness, of this ordered compound.

A thorough characterization, using mainly resistometric techniques, of the processes accompanying the annealing out of vacancies in FeAl (Fe-40at.% Al) was carried out on samples quenched from high temperatures [9,10] and also neutron-irradiated at low temperature, 20 K, both in the quenched and untreated state [11].

An increase in resistivity was observed upon increasing the quenching temperature: this effect was explained in terms of an increase of extra-vacancies. The extimated value of energy of formation is E_f=.91 eV, and their migration energy E_m=1.61 eV: this values give a ratio E_m/E_f>1, in agreement with other values reported for other B2 alloys.

In this work we present some results on the disordering and reordering aspects of rapidly solidified samples of the Fe-Al system, in the stability region of the B2 phase, with and without ternary alloying of Cr.

EXPERIMENTAL DETAILS

Binary and ternary ingots, Fe-Al and Fe-Al-Cr, were prepared by arc melting pure element pellets in an Ar atmosphere. Rapidly-solidified ribbons were spun using the apparatus described elsewhere [12]. Two spinning rates were selected for each composition: 42m/s (6000rpm) and 28 m/s (4000rpm).The ribbon compositions were measured with Energy Dispersive X-ray Spectroscopy (EDXS), using virtual standards and conventional correction programs. The values, calculated from five measurements on each sample, are reported in table1.

X-Ray Diffraction (XRD) analysis were carried out with a wide angle diffractometer, operated in step-scan mode and using monochromated Cu-Kα radiation. From the integral breadth

values of the diffraction peaks, corrected for the instrumental broadening, the average values of
the particle size and lattice rms strain were calculated.

TABLE 1			
Chemical compositions of melt-spun ribbons (at.%) by EDXS			
	Fe	Cr	Al
FeAl-4000rpm (FA-4)	52.5	-	47.5
FeAl-6000rpm (FA-6)	51.2	-	48.8
$Fe_{50}Cr_8Al_{42}$-4000rpm (F5CA-4)	49.7	8.6	41.7
$Fe_{50}Cr_8Al_{42}$-6000rpm (F5CA-6)	48.3	8.6	43.1
$Fe_{42}Cr_8Al_{50}$-4000rpm (FCA5-4)	42.5	9.0	48.5
$Fe_{42}Cr_8Al_{50}$-6000rpm (FCA5-6)	43.3	9.2	47.5

The annealing kinetics of the defects trapped in during the melt-spinning process were studied
with a Differential Scanning Calorimeter (DSC), Perkin Elmer DSC7, operated in a continuous
heating mode.

RESULTS

All the samples resulted single phase and very well ordered, irrespective of their compositions
and cooling rates. This confirms the findings of a previous work on rapid solidification of Fe-
Al binary alloys, which showed the inefficiency of this prepartion techniques in freezing in an
appreciable degree of disorder [13]. Figure 1 is the typical XRD pattern of one of the sample,
F5CA-6.

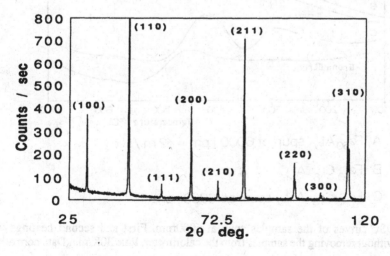

Figure 1. XRD pattern of the sample $Fe_{50}Cr_8Al_{42}$-6000rpm, showing the ordered B2 phase.

The samples were strain free and with very large particle size. The average particle size, as measured from XRD peak broadening, was >2500Å. The same characteristics were found in all the other investigated samples, both the binary and ternary ones.

The calorimetric tests revealed the existence of several irreversible exothermic peaks. In figure 2 are reported the thermograms of consecutive runs relative to the series melt-spun at 6000 rpm. The sample FCA5-6 clearly shows two distinct signals. In the binary FA-6 the lower temperature peak is extremely weak: this did not allow a reliable calculation of its apparent activation energy to be done. Finally the thermograms of the sample F5CA-6 show only the high temperature signal. After the first cycle the ternary sample still shows a weak low temperature exothermic peak, most likely due to the defects trapped in the specimen during the intrinsic cooling down of the calorimeter.The second and the following runs on the same sample coincided exactly.

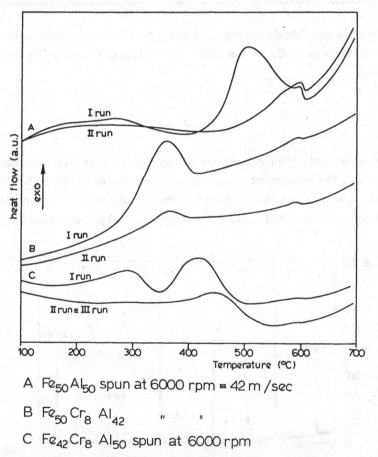

A $Fe_{50}Al_{50}$ spun at 6000 rpm ≡ 42 m/sec

B $Fe_{50}Cr_8Al_{42}$ " "

C $Fe_{42}Cr_8Al_{50}$ spun at 6000 rpm

Figure 2. DSC curves of the samples spun at 6000rpm. First and second heatings were performed without removing the samples from the calorimeter. Rate 30K/min.Data normalized.

The thermograms relative to the samples spun at 4000rpm, not reported, have essentially the same behaviour, though with some variation in the peak intensities and positions.

XRD analysis, performed before and after the DSC runs, do not show any significant change in the microstructural properties.The relative long-range order parameter, S, estimated from the ratio of the integrated intensities of four superlattice and five fundamental peaks, was the same before and after the heat treatments. Slight lattice contractions were observed in the ternary samples : 0.06% in the sample FCA5-6, 0.08% in the sample F5CA-6. An expansion of 0.15% was observed instead in the binary sample FA-6. No data are yet available for the specimens melt-spun at 4000rpm. Different heating rates, Φ (10, 30, 50, 100 K/min) were used in order to determine whether or not the observed process were thermally activated and to calculate their apparent activation energies, E_a. For this purpose the Ozawa method was used [14]. The curves for the FCA5-6 sample, obtained at the four mentioned rates, are shown in figure 3 and in table 2 are reported the apparent activation energies for all the series of samples spun at 6000rpm.

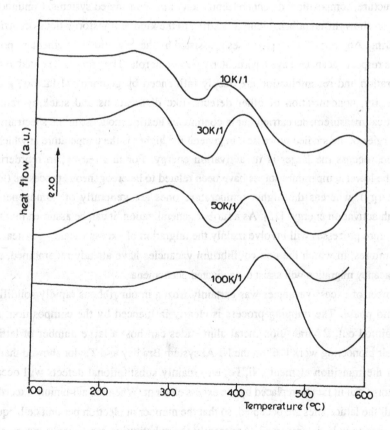

Figure 3. DSC curves for the sample $Fe_{42}Cr_8Al_{50}$-6000rpm, taken at different heating rates: 10, 30, 50 and 100 K/min. All data normalized.

TABLE 2		
Apparent activation energies (eV) from DSC analysis for the samples spun at 6000rpm		
	E_{a1}	E_{a2}
FeAl-6000rpm	(*)	1.30
$Fe_{50}Cr_8Al_{42}$-6000rpm		1.60
$Fe_{42}Cr_8Al_{50}$-6000rpm	0.60	1.20
(*) a reliable analysis of this very weak peak could not be carried out.		

DISCUSSION

As reported [15 and references there cited], there are common characteristics in the reordering processes in alloys, appearing at different extents and temperatures, according to the specific crystalline structure, composition, defect concentration of the considered systems. Continuous heating runs can show different behaviour, according to the kind of transformations occurring during reordering. Any one of these processes is assisted by the migration of elementary point defects. In this respect vacancies have a particularly important role. The processes related to the vacancy migration and recombination are deeply influenced by geometrical factors, grain boundary density, concentration of other defects, like dislocations and stacking faults. Thermoanalytical measurements carried out in continuous heating mode, scan the spectrum of activation energies of the studied processes: in general the higher is the temperature at which a transformation occurs the larger is its activation energy. For this reason, in reordering phenomena, the lower-temperature stages have been related to hetereogeneous processes (low activation energy), whereas the higher-temperature ones are generally of homogeneous character (high activation energy) [15]. As a further generalization, it can be assumed that the lower-temperature processes will involve mainly the migration of excess vacancy ; instead at higher temperatures, in which the non-equilibrium vacancies have already recombined, the equilibrium vacancy migration will assist the activated phenomena.

A certain number of excess vacancies was certainly frozen in our ribbons rapidly solidified from the liquid phase. The trapping process is clearly influenced by the composition. As previously pointed out, B2 transition metal aluminides can host a large number of lattice defects. In their pioneering work [16] on the Ni-Al system Bradley and Taylor showed that in alloys rich in the transition element (Ni, Fe, etc) mainly substitutional defects will occur. Aluminium atoms can in fact be replaced by the excess element. When the aluminium exceedes 50 at.%, not all the lattice sites are occupied, so that the number of electron per unit cell, equal to 3 in these compounds, is preserved. Apparently the substitution rate is only marginally influenced by the effective atomic size: Al atoms have in these ordered alloys an effective size comparable to the size of the transition metal atoms, and so they could fit in their lattice sites quite comfortably [17]. On this basis the samples having a larger concentration of aluminium are expected to have a higher concentration of equilibrium vacancies. This involves also the possibility of trapping a larger number of non-equilibrium vacancies, as experimantally tested for other FeAl-based alloys [18].

The calorimetric curves show two different behaviours: a single peak transformation, and a multi-step one, in which at least two processes are evident. In the first case, reported for the sample F5CA-6 only (see figure 2), the measured apparent activation energy (1.60 eV) is in very good agreement with the vacancy migration energy reported in [9] for the ordered alloy Fe-Al 40 at.%. Its aluminium content is close to the one of our sample, so that is quite plausible that also our observations are related to a process assisted by vacancy migration. The activation energy of this process is lower in the samples having a larger concentration of aluminium. On the other hand, in agreement with the previous discussion, higher aluminium concentrations, by generating an higher number of empty sites, render more efficient the mechanisms of excess vacancy trapping, occurring during rapid quenches. We believe that the first stage peak, the one observed at lower temperatures, is due to the recombination of these defects. The annihilation of excess vacancies is certainly facilitated by the presence of other defects, such as grain boundaries, acting as sinks. The grain size of the rapidly solidified ribbons ranges from 5 to 20 μm: the grain boundary concentration is therefore quite high (figure 4).

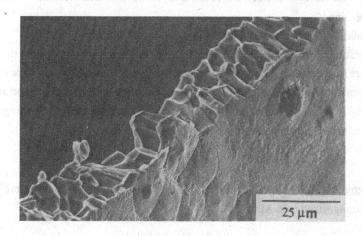

25 μm

Figure 4. SEM micrograph of the fracture surface of the sample FeAl-6000rpm.

Of course also the presence of the ternary element must play an important role, as proved by the fact that in the binary samples the first stage peak is extremely weak. In the light of the present results we are not in the position of giving a complete explanation of this and other aspects, such as the lattice expansion observed in the binary samples after annealing in the DSC. For this reason further investigations are needed. On the other hand we can already put forward some hypotesis for the interpretation of the observed phenomena. The as-spun ribbons were only slightly disordered: no sensible variations of S, the long-range orderd parameter, were measured after annealings. For this reason we can tentatively associate the higher temperature process to the coarsening of ordered domains. Even though quite large, they could not achieve a complete growth because of the very rapid solidification. Moreover the temperature range of this process would be suitable for this kind of processes: $T \cong T_m/2$, where T_m is the melting temperature of the binary FeAl compound. The increasing value of the activation energy of this process with the decreasing concentration of aluminium can still be ascribed to the lower concentration of equilibrium vacancies, the lower is the aluminium concentration, and to the consequent slowing down of any diffusive process.

CONCLUSIONS

Several series of melt-spun ribbons of binary, FeAl, and ternary, Fe-Al-Cr, compositions were prepared, and characterized via calorimetric and x-ray diffraction measurements. The following results were obtained:

1) the rapid solidification is not capable of inducing chemical disorder;

2) the concentration of frozen-in vacancies depends on the chemical composition; the samples having a higher Al content are able to retain a larger number of non-equilibrium vacancies.

3) the experimental results suggest that excess-vacancy recombination (lower temperature) and the coarsening of the ordered domains (higher temperature), assisted by the migration of equilibrium vacancies, are the principal recovery phenomena .

Acknowledgements. S.G. wishes to thank the EEC for financial support: science project SCI-0262-C, "Ordering Kinetics in Alloy".

REFERENCES

1.   Chipman, D. and Warren, B.E., X-ray measurement of long range order in β-brass, J. Appl.Phys., 1950, 21, 696-97.

2.   Cupschalk, S.G. and Brown, N., Observation of defects in β-brass, Acta Met., 1967, 15, 847-56.

3. Ball, A. and Smallman, R.E., Vacancy defects in the ordered compound NiAl, Acta Met., 1968, **16**, 233-41.

4. Mukherjee, K., Liebermann, D.S. and Read, T.A., Quenched-in defects in beta-phase Au-Zn, J.Appl.Phys., 1965, **5**, 857-58.

5. Wechsler, M.S., The quenching-in of lattice defects in gold-cadmium, Acta Met., 1957, **5**, 150-58.

6. Bradley, A.J. and Jay, A.H., The formation of superlattices of iron and aluminium, Proc.Roy.Soc., 1932, **A136**, 210-32.

7. Baker, I. and Munroe, P.R., Properties of B2 compounds, in High Temperature Aluminides and Intermetallics, ed. by Whang, S.H., Liu C.T., Pope, D.P. and Stiegler, J.O., The Minerals, Metals & Materials Society, 1990, 425-52.

8. Nagpal, P. and Baker, I., Effect of cooling rate on hardness of FeAl and NiAl, Met.Trans., 1990, **21A**, 2281-82.

9. Rivière, J.P. and Grilhè, J., Restauration de defauts de trempe dans un alliage Fe-Al 40 at.% ordonne, Acta Met., 1972, **20**, 1275-80.

10. Rivière, J.P., Zonon, H. and Grilhè, J., Annealing kinetic of quenched-in vacancies in Fe-40 at% Al ordered alloy, Phys.Stat.Sol. (a), 1973, **16**, 545-52.

11. Rivière, J.P., Zonon, H. and Grilhè, Restauration des defaut ponctuels crees par irradiation a 20 K dans un alliage Fe-Al 40 at.% ordonne, Acta Met.,1974, **22**, 929-36.

12. Antonione, C., Battezzati, L. and Marino, F., Formation and stability of quasicristalline and amorphous phases in Al-Mn and Al-Mn-Si alloy, Trans.Jap.Inst. Met., 1988, **S29**, 505-10.

13. Raviprasad, K. and Chattopaday, K., The effect of rapid solidification on the order disorder transformation in iron-based alloys, Mat.Sci.Eng., 1988, **98**, 281-84.

14. Ozawa, T., Kinetic analysis of derivatives curves in thermal analysis. J.Therm.An., 1970, **2**, 301-24.

15. Mitsui, K., Mishima, Y. and Suzuki, T., Heterogenous ordering and antiphase domain morphology in Cu_3Pt (19 at.%), Phil.Mag., 1986, **A53**, 357-76.

16. Bradley, A.J. and Taylor, A., An x-ray analysis of the nickel-aluminium system, Proc.Roy.Soc., 1937, **A159**, 56-72.

17. Ridley, N., Defect structures in binary and ternary alloys based on CoAl, J.Inst.Met., 1966, **94**, 255-58.

18. Paris, D. and Lesbats, P., Vacancies in Fe-Al alloys, J.Nucl.Mat., 1978, **69-70**, 628-32.

ORDERING AND DISORDERING IN CuPt-ALLOYS

BIRGÜL URBAN-ERBIL and WOLFGANG PFEILER
Institut für Festkörperphysik, Universität Wien
Strudlhofgasse 4, A-1090 Vienna, Austria

ABSTRACT

Changes of resistivity during ordering and disordering have been
investigated for $L1_1$-ordered CuPt alloys. Whereas for stoichiometric
$Cu_{50}Pt_{50}$ a drastic influence of LRO only is observed, two contributions
to resistivity of opposite sign were detected for off-stoichiometric
$Cu_{65}Pt_{95}$. These two processes are attributed to LRO and SRO, respectively,
being in competition: Short annealing times support SRO and the formation
of LRO is suppressed. Prolonged annealing times yield a corresponding
increase of LRO. These findings are confirmed by X-ray diffraction at wide
and small angles.

INTRODUCTION

Apart from those intermetallic compounds which show a minor change of order
with temperature right up to the melting point, there are binary alloy
systems where the degree of order depends strongly on temperature, with a
critical disordering temperature quite below the melting point (reversible
ordering [1]).

In the present study, ordering and disordering processes in CuPt are
investigated as a function of annealing time by means of electrical
resistivity measurement. This investigation is accompanied by X-ray
measurements (Bragg scattering and small angle scattering) to support the
interpretation of results.

EXPERIMENTAL

Foils of 0.2 mm thickness were prepared by cold rolling with intermediate
anneals in a purified argon atmosphere. All annealing treatments were done
in a normal furnace with the samples dismounted from the sample holder.

After each annealing interval the foils were quenched in cold water with a cooling rate of about 1000K/s.

The resistivity measurements were performed at liquid nitrogen temperature relative to a dummy specimen by the usual potentiometric method. Before the actual measuring and annealing series, the samples were preannealed for one hour just above the phase boundary.

X-ray diffraction measurements were done using a point-focus camera with Cu-K$_\alpha$ radiation from a 12 kW rotating anode X-ray generator. For the wide-angle diffraction a conventional two-circle diffractometer was used; the data for small-angle measurements were collected using a linear position sensitive detector.

RESULTS OF RESISTIVITY MESUREMENTS

For stoichiometric Cu$_{50}$Pt$_{50}$ a strong influence of LRO is observed. This can be seen from figure 1, where the relative change of resistivity of Cu$_{50}$Pt$_{50}$

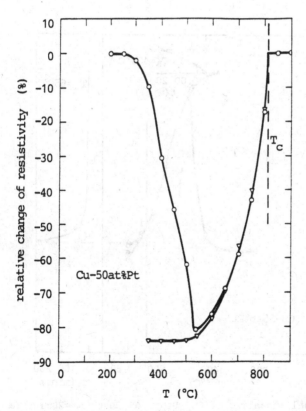

Figure 1. Isochronal annealing of Cu$_{50}$Pt$_{50}$. $\Delta T = 50°C$, $\Delta t = 20$min. (■) initial value after quenching from 850°C.

is plotted against temperature during isochronal annealing (increasing and decreasing temperatures; ΔT=50°C, Δt=20min.) after quenching from 850°C. A drastic decrease of resistivity at about 300°C clearly indicates the formation of LRO. Above the critical temperature at about 820°C the resistivity remains constant and exactly equals the starting value. This gives evidence that there is no falsification of results by the quenching procedure. The good correspondence of resistivity values at rising (o) and falling (∇) temperatures suggests the establishment of equilibrium values of LRO as long as the atomic mobility is high enough (above 650°C).

A significantly different behaviour is observed for the off-stoichiometric alloy $Cu_{65}Pt_{95}$. During the first run (figure 2, dashed line) the resistivity during isochronal annealing first increases up to about 450°C then decreases below the starting value up to about 600°C. Above the phase boundary, the resistivity again increases towards the starting value. During the subsequent annealing at falling temperatures as

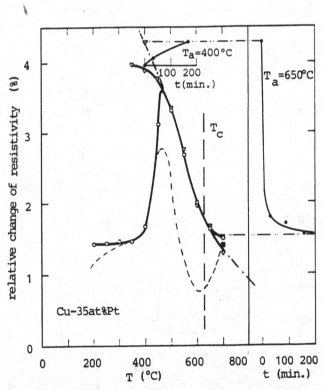

Figure 2. Isochronal annealing of $Cu_{65}Pt_{95}$. ΔT=50°C, Δt=20min. (■) initial value after quenching from 700°C. Insets: isothermal annealing at 400° and 650°C. Dashed line: first run after cold rolling and 0.5h at 720°C.

well as during a second run of isochronal annealing at rising and falling temperatures after a re-homogenization treatment (full line in figure 2) just one process could be detected, leading to a resistivity decrease with increasing annealing temperatures as long as the atomic mobility was high enough to enable complete microstructural rearrangement within the 20 minutes of annealing at each temperature of isochronal treatment. Additional isothermal annealings at 400°C (upper inset) and at 650°C (right inset) confirm these findings and show an "equilibrium line" of this process (dash-dotted line).

A possible interpretation given already earlier [2] is that there are two processes in competition: SRO leading to an increase of resistivity with increasing degree of order and LRO leading to a decrease of resistivity with increasing order. Since for the establishment of SRO only a few atomic jumps are necessary, this process is rather fast, whereas the establishment of LRO needs diffusion over long distances (several atomic diameters). In addition the change from cubic to rhombohedral symmetry gives rise to lattice distortions between matrix and ordered phase. Therefore changes in the degree of LRO are thought to be rather sluggish and essentially dependent on the metallurgical state of the sample.

To check this interpretation, isochronal annealing experiments at increasingly prolonged annealing times were performed. In figure 3 the results of this third run are shown. The previous results during a 20 min. isochronal annealing treatment at rising and falling temperatures are reproduced: just one process leading to higher resistivities at lower temperatures is observed (SRO-process, upper equilibrium line). Prolonged annealing times (1h, 2h, 4h, 8h, 16h) between 450° and 600°C obviously, owing to a strong second process of opposite sign, more and more lead away from the initial upper equilibrium line to lower values. For long annealing times there seems to exist a lower equilibrium resistivity line for this second process, which for lower temperatures, in correspondence with the previous interpretation as LRO formation, leads to a great reduction of resistivity. It may be noted that an asymmetry is observed between rising and falling temperatures (disordering and ordering), which is at maximum for four hours annealing time. Annealing at 600°C for all annealing time intervals leads to the same value which coincides with an annealing time of only 20 minutes.

To check the lower equilibrium line of the second process (LRO) additional isothermal annealing experiments have been done at 450°, 500°

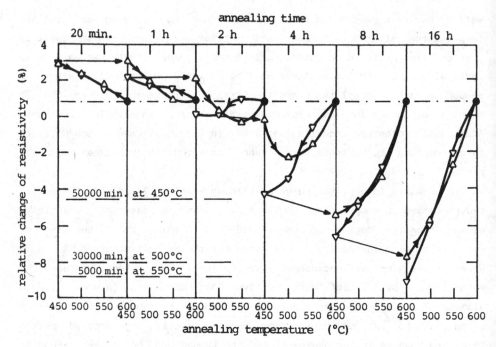

Figure 3. Isochronal annealing of Cu$_{65}$Pt$_{95}$ with increasingly prolonged annealing times. Annealing intervals as indicated in the figure.

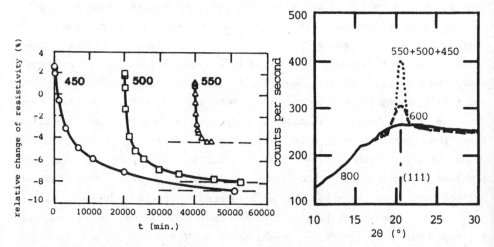

Figure 4. Long-time isothermal annealing of Cu$_{65}$Pt$_{95}$ at 450°,500° and 550°C.

Figure 5. X-ray intensity as a function of scattering angle 2⊖. Annealing of 20min.,respectively, at indicated temperatures (in °C).

and 550°C as shown in figure 4; the final values are given in figure 3 by dashed lines. It seems that stable values are adjusted for 500° and 550°C, whereas for 450°C an annealing time longer than 50000min. would be necessary. Before each of this isothermal measurements a 90min. anneal at 600°C was followed by 20min. anneals at falling temperatures down to the starting temperature of the isotherm. This was done in order to use the corresponding value of the SRO-equilibrium line as a starting value.

RESULTS OF X-RAY MEASUREMENTS

First we tried to look for a superstructure peak near to the (111) ordering line ((1/2 1/2 1/2) disordered position) during 20 min. annealings at falling temperatures. Figure 5 gives the intensity as a function of scattering angle 2θ for a 20 min. annealing at 800°, 600° and 550°+500°+450°C respectively. For 800°C no indication for a superstructure is observed, whereas the start of ordering can be recognized for 600° and more developed for 550°+500°+450°C. This gives evidence that there actually a LRO-process takes place, which may already start at 600°C.

The remaining question is: is the upper resistivity equilibrium line, previously attributed to SRO, in fact rather a consequence of a LR ordering process? To check this point, we tried to destroy LRO completely by an annealing treatment of 15h at 800°C. Subsequently an isothermal annealing treatment was applied immediately at 450°C without any intermediate 20 min. treatment. Figure 6 shows the result of X-ray scattering measurements around the (111) superstructure peak for 100 and 120 min. at 450°C. The intensity after the annealing treatment of 15h at 800°C is given for a comparison. No indication for a peak can be found for either annealing time at 450°C[*]. On the other hand figure 7 gives the result of resistivity measured in parallel for this annealing treatment at 450°C, yielding the expected increase to the upper equilibrium line. An additional annealing treatment of 20 min. at 550°C indicates the beginning ordering process, whereas the resistivity decreases exactly along the upper equilibrium line.

To get information if the early stages of ordering may be connected with homogeneously distributed ordered particles giving rise to the observed resistivity increase of the upper equilibrium line, small-angle X-ray scattering (SAXS) was investigated during an annealing treatment at 450°C after a pre-annealing during 15h at 800°C. Because of the difference

---

[*] A small increase of background scattering with increasing annealing time at 450°C may be caused by SRO diffuse scattering.

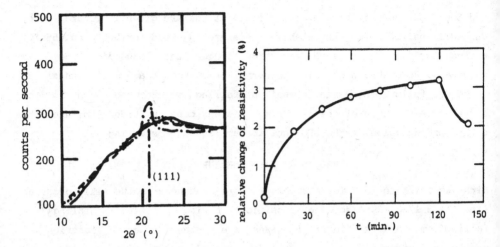

Figure 6. X-ray intensity as a
function of scattering angle 2Θ.
Isothermal annealing of 100 (--)
and 120min. (· · ·) at 450°C after
a treatment of 15h at 800°C (—).
(—·—) additional 20min. at 550°C.

Figure 7. Resistivity change for the
same thermal treatment as in
figure 6.

in concentration between matrix and ordered phase of about 10at% a corresponding small angle signal could be expected. The q-range investigated was
0.02-0.3Å^{-1} corresponding to a mean particle distance of 20-300Å. The SAXS
curves remained completely flat and unchanged during the annealing
treatment, giving no indication for any small particles homogeneously
distributed within the matrix. For this reason the increase in resistivity
cannot arise from early LR ordering.

DISCUSSION AND CONCLUSION

For $Cu_{50}Pt_{50}$ a strong LRO-process is observed without indication to
SRO-processes from resistivity measurement.

The experimental results for $Cu_{65}Pt_{95}$ show that there are two
contributions to electrical resistivity during ordering and disordering
(upper and lower equilibrium line). The slow process decreasing the
resistivity unequivocally has to be attributed to LRO leading for long
enough annealing times to the lower equilibrium line. This is shown by the
existence of superstructure lines after longtime annealing below the phase
boundary. In principle, the second process increasing the resistivity with
decreasing temperature could be due to the nucleation process of the new
LRO phase if the decomposition process would start with small nuclei,

homogeneously distributed within the matrix. This origin of the upper equilibrium line, however, can be excluded because of the negative result of the SAXS experiment. The previous interpretation of the resistivity increase with decreasing temperatures as arising from changes in the degree of SRO therefore seems to hold.

Obviously during the 20 min. anneals from 600° to 450°C, LRO already starts, although the electrical resistivity continues to change reversibly along the upper equilibrium line. Therefore the resistivity change seems not to be influenced by the early stages of LRO.

We come to the following conception of microstructure during ordering and disordering in off-stoichiometric $Cu_{65}Pt_{95}$-alloys:

SRO and LRO develop essentially independent of each other.

SRO seems to be a homogeneous property of the matrix and each change in temperature is followed by a rather rapid change in the degree of SRO towards its equilibrium value.

On the other hand LRO may be predominantly generated at the grain boundaries [3]. During too short annealing times just this nucleation process is enabled and further ordering towards the actual equilibrium degree of order, which may proceed by a growth of ordered domains into the interior of the grains, is suppressed. It therefore is a question of annealing time, annealing temperature and the metallurgical state of the sample if a development of LRO towards a certain final state corresponding to the actual annealing temperature is observed.

A more detailed future investigation of the diffuse scattering intensity near to the (111)-ordering peak should give more information on short-range ordering of (1/2 1/2 1/2)-type within the matrix.

REFERENCES

1. Cahn, R.W., Ordering Kinetics and Diffusion in L1$_2$ Alloys. In <u>Phase Transitions in Condensed Systems - Experiment and Theory</u>, eds. G.S. Cargill, F. Spaepen and K.-N. Tu, Mat. Res. Soc. Symp. Proc., 1987, 57, 385-404.
2. Banhart, J., Pfeiler, W. and Voitländer, J., Detection of short- and long-range order in Cu-Pt alloys. <u>Phys. Rev.</u>, 1988, B37, 6027-29.
3. Irani, R.S. and Cahn, R.W., The mechanism of crystallographic ordering in CuPt. <u>J. Mater.Sci.</u>, 1973, 8, 1453-72.

Acknowledgement

The financial support by the Austrian "Fonds zur Förderung der wissenschaftlichen Forschung" grant Nr. 7615 and the "Österreichische Forschungsgemeinschaft" is gratefully acknowledged. We thank Dr. P. Fratzl and Mag. F. Langmayr for their assistance during the X-ray measurements and Dr. Fratzl in addition for a critical reading of the manuscript.

STOCHASTIC DESCRIPTION OF ORDERING IN DRIVEN COMPOUNDS

P. BELLON, G. MARTIN

CEREM/SRMP, CEN Saclay F91191 Gif-sur-Yvette, France

ABSTRACT

The relative stability of phases in driven systems such as alloys under irradiation cannot be assessed from thermodynamic potentials. Nevertheless, starting from a mesoscopic stochastic description, one can define a potential which governs the probability distribution of macrovariables (for instance the composition and order parameters). As examples, we recall here the results obtained, in the simplest mean-field approximation, for order-disorder transitions on the BCC lattice (B2 ordered phase) and on the FCC lattice ($L1_2$, $L1_0$, $D1_a$, $LiFeO_2$-type ordered phases). Inversion of phase stability, occurrence of tricritical lines, sensitivity of phase boundaries to " cascade size effect " are predicted.

INTRODUCTION

When a compound is maintained in far-from-equilibrium configurations by a dynamical external forcing (e.g. nuclear collisions induced by irradiation or the disordering induced by ball milling), the steady-state properties of the system can no longer be predicted by minimization of the classical thermodynamic potentials (Gibbs or Helmholtz free energy).

This problem is mapped onto the kinetic Ising model with two competing exchange dynamics : thermally activated jumps and forced atomic jumps (the latter result e.g. from nuclear collisions). Despite the fact that such systems lack a partition function for the micro-configurations, a stochastic potential can be derived for the probability of macro-configurations [1,2]. The respective stability of competing crystallographic structures can thus be assessed as a function of the external forcing intensity : dynamical equilibrium phase diagrams can be predicted [3-6].

Here we briefly recall the technique to build such a stochastic approach for alloys exhibiting order-disorder reactions ; we then summarize the results for the different cases already treated : systems undergoing phase transitions with or without composition changes, where the ordered structures are described by one-dimensional or by multidimensional order parameters, and with ballistic jumps occurring individually or by bursts.

MASTER EQUATION FOR ORDER-DISORDER TRANSITIONS UNDER IRRADIATION

Both for taking into account the fluctuations of order and to assess the relative stability of several ordered structures under irradiation, the kinetic equation needed to describe the time evolution of the system has to deal with the probability $P(\underset{\sim}{N},t)$ of finding the system in a state $\underset{\sim}{N}$ at time t. Here the system is described at a mesoscopic level : we consider finite systems and choose to represent the system by the number of clusters of a given type. For instance, in a A-B alloy on a BCC lattice described in the simplest mean-field approximation (Bragg-Williams), $\underset{\sim}{N}(N_1,N_2)$ represents the number of A-atoms on sublattice 1 and 2. If the concentration of the alloy is kept constant, only one independent variable remains. For Markovian processes, the time evolution of $P(\underset{\sim}{N},t)$ is governed by a Master Equation (ME) [7,8] :

$$\partial P(\underset{\sim}{N},t)/\partial t = -\sum_{\underset{\sim}{\epsilon}} W(\underset{\sim}{N},\underset{\sim}{\epsilon})P(\underset{\sim}{N},t) + \sum_{\underset{\sim}{\epsilon}} W(\underset{\sim}{N}-\underset{\sim}{\epsilon},\underset{\sim}{\epsilon})P(\underset{\sim}{N}-\underset{\sim}{\epsilon},t) \qquad (1)$$

where $W(\underset{\sim}{N},\underset{\sim}{\epsilon})$ is the transition rate from a state $\underset{\sim}{N}$ to a state $\underset{\sim}{N}+\underset{\sim}{\epsilon}$; in Eq. (1), $\underset{\sim}{\epsilon}$ runs over all accessible states.

Without irradiation, the thermal transition rates correspond to A-B pair permutations between neighbouring sites in order to model the thermal diffusion in the alloy : the number of such permutations per unit time must obey the detailed balance. This number is given by the product of the number of available configurations times a thermally activated term. The latter has an activation energy chosen to be equal to the increase in internal energy required to displace these atoms from their lattice site to some saddle point configuration [9,10]. Because of the detailed balance condition, Eq. (1) is easily solved analytically under steady-state conditions. We showed previously [3] that with transition rates built in a mean-field approximation one recovers a steady-state solution of the ME in full agreement with the equilibrium thermodynamics, i.e. :

$$P(\underset{\sim}{N}) \propto \exp\{-\Omega \mathscr{F}(\underset{\sim}{n})/kT\} \tag{2}$$

where Ω is the number of lattice sites in the system, $\underset{\sim}{n}$ the intensive variable defined by $\underset{\sim}{n} = \underset{\sim}{N}/\Omega$, and $\mathscr{F}(\underset{\sim}{n})$ the free energy function per site computed in the same mean-field approximation.

Under irradiation, the ballistic jumps induce transitions which are independent of the state of the system (corresponding to an infinite temperature dynamics), the rate of such jumps being proportional to the replacement cross-section times the irradiation flux. Thus in Eq. (1), a transition rate is now the sum of two terms, one is the above thermal transition rate and the second is the ballistic transition rate. While thermal transitions connect only nearest neighbouring states ($|\xi| = 1$), bursts of ballistic jumps as occur in cascades connect many states together ($|\xi| \geq 1$ and $|\xi|_{max} = b$).

In the following, we focus on the steady-state solutions of the ME. Contrarily to equilibrium systems, the detailed balance is, in general, not fulfilled in non-equilibrium systems, such as the alloy with the thermal and forced jumps acting in parallel. Therefore the solution of Eq. (1) can not always be found analytically. Nonetheless, Kubo et al [1] and Suzuki [2] have shown that under some general conditions, in the limit of large system size ($\Omega \longrightarrow \infty$), the probability distribution exhibits an extensive property and is still governed by a potential :

$$P_{irr}(\underset{\sim}{N}) \propto \exp\{\Omega \Phi(\underset{\sim}{n})\} \tag{3}$$

where the subscript specifies that the relation holds for systems under irradiation.

Our main goal is to compute the stochastic potential $\Phi(\underline{n})$ for given irradiation conditions. Indeed the knowledge of this potential allows the construction of dynamical equilibrium phase diagram, i.e. to predict the most stable state of the system as a function of the irradiation parameters.

DYNAMICAL EQUILIBRIUM PHASE DIAGRAMS

We now present several examples of this potential for models with increasing complexity.

Dynamical equilibrium between phases at the same composition

One-dimensional order parameter [3,4,5]: as an example, we consider an A_{50}-B_{50} alloy on a BCC lattice with nearest neighbour interactions exhibiting a A2-B2 order-disorder equilibrium transition. Here, in a Bragg-Williams approximation, the state of the system is given by a scalar, N, the number of A atoms on the sublattice 1.

First, ballistic jumps are assumed to occur individually, which may correspond to the case of an alloy under electron irradiation. In this case the detailed balance for the macrovariable n is fulfilled, and an analytical expression if found for $\Phi(n)$. We define a reduced parameter γ, which scales the intensity of the irradiation, as the ratio of the forced to an average thermal A-B pair permutation frequencies. When the dominant point defect annihilation mechanism is the interstitial-vacancy recombination, it is shown that γ writes :

$$\gamma = \gamma_0 \ \exp(E_v^m/2kT) \text{ and } \gamma_0 \propto \frac{\sigma_r}{f \ \sigma_d} \sqrt{f \ \sigma_d \phi} \qquad (4)$$

where E_v^m is the vacancy migration energy, σ_r and σ_d the replacement and displacement cross-sections respectively, f the fraction of freely migrating defect and ϕ the irradiation flux ; the potential is thus a function of the variable n, the irradiation parameters being the temperature T and the reduced parameter γ_0. The dynamical equilibrium phase diagram (Fig. 1) is displayed in the $T \otimes \gamma_0$ plane (γ_0 is temperature independent). Notice the occurrence of a tricritical point : the order-disorder transition which is second order at equilibrium becomes first order below a tricritical temperature. A similar description [3] allows to rationalize the stabilization of metastable phase and the bistability reported in Ni4Mo under 1 Mev electron irradiation [11].

Under ion or neutron irradiation, when cascades are operating, ballistic jumps occur by burst (of size b) : in this case the detailed balance for the macrovariable is no longer fulfilled and an analytical expression cannot be found. Kubo at al [1] showed how to transform the original ME in the large system size limit : one gets a non-linear partial differential equation (the Kubo equation) for the potential Φ :

$$\partial\Phi(\underset{\sim}{n},t)/\partial t = \sum_{\underset{\sim}{\varepsilon}} w(\underset{\sim}{n},\underset{\sim}{\varepsilon}) \{ \exp[-(\underset{\sim}{\varepsilon}.\underset{\sim}{\nabla}) \Phi] - 1 \} \qquad (5)$$

For driven systems, no analytical solution is known for Eq. (5), thus, in the most general case, the solution has to be found numerically.

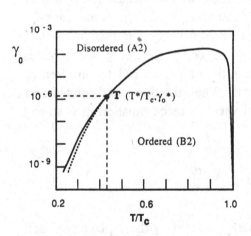

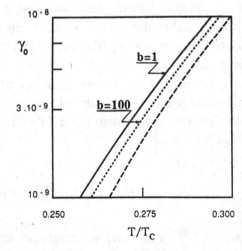

Fig. 1 : Dynamical equilibrium phase diagram for a BCC alloy. The dashed line is for the spinodal ordering line.

Fig. 2 : Cascade size effect on the low temperature part of the dynamical phase diagram of Fig. 1. b measures the cascade size. (Ref. [4]).

The ME can also be approximated by a Fokker-Planck equation (this technique is however only valid for scalar order parameter [6]) : in this case an analytical expression is found for a potential which approximates Φ. We checked the validity of this approximation by comparison with numerical solution of the Kubo equation (5). From the analytical expression of the potential governing the Fokker-Planck equation one notices that at a given temperature, the frequency of replacement per atom determines the possible steady-state value of n, while this frequency *and the size of the cascades* determine their relative stability. As a result, when the transition is 1st order, increasing the cascade size shifts the transition line towards its spinodal (Fig. 2).

Multidimensional order parameter : AB_3 alloy on an FCC lattice was considered [6], and the states of the alloy are described by the degree of A-atomic occupation on the four simple cubic lattices. This alloy presents an $L1_2$ equilibrium ordered phase.

In the case of individual ballistic jumps, the detailed balance is in general not fulfilled, so that no analytical expression is available for Φ. However symmetry properties may guarantee the detailed balance along symmetry axes : for example along the axes corresponding to $L1_0$ and $L1_2$ ordered structures the detailed balance is fulfilled by reasons of symmetry. In the case where all the possible steady-states lie on these axes, one can thus build a dynamical equilibrium phase diagram. It is shown that *metastable ordered structures* can be stabilized by irradiation, depending on the model for the thermal kinetics (Fig. 3).

Multiple simultaneous ballistic jumps have not yet been incorporated in this case, although as shown above, we expect a cascade effect on the 1st order order-disorder transition line.

Dynamical equilibrium between phases at different compositions

Here we have proposed two types of approach :(i) an homogeneous grand canonical description for driven systems, (ii) an heterogeneous canonical description. These descriptions were exemplified on the A_cB_{1-c} BCC alloy already discussed above.

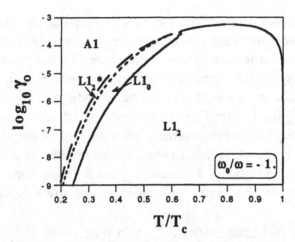

Fig. 3 : Dynamical equilibrium phase diagram for an AB$_3$ FCC alloy. Here the saddle point for the thermal diffusion depends on the neighbours of the jumping atoms (see ref. 6). In addition to the equilibrium L1$_2$ ordered structure, irradiation can stabilize L1$_0$ and A$_3$B L1$_2$ (here referred to as L1$_2$*) metastable ordered phases.

Homogeneous grand canonical description [12] : the composition of the system may evolve because the system is interacting with a particle reservoir. For this grand canonical description to be consistent with a canonical one, the dynamics for chemical exchanges, which will enter the master equation, have been derived by considering two canonical systems in interaction and then letting one system becoming very large with respect to the other -thus acting as a particle reservoir.

Furthermore, by the use of symmetry arguments, we show that the grand canonical potential is related to the canonical one through an analytical formula. As in this simple example the canonical potential is known analytically (see above), the grand canonical potential is also known analytically, and a dynamical equilibrium phase diagram is computed in the $T \otimes \gamma \otimes C$ space. It is shown that, beyond a critical forcing, the transition becomes first order (Fig. 4), and when $C \neq 0.5$ we predict a possible *phase coexistence* between an ordered phase and a depleted disordered solid solution (Fig. 5). When the ballistic jumps will occur by cascade, instead of the uncorrelated jumps considered here, this should affect this coexistence field : this is kept for future work.

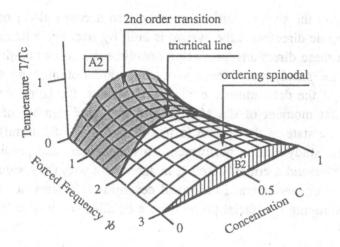

Fig. 4 : Dynamical equilibrium phase diagram for a BCC alloy in T,γ,C
space. For clarity, when the transition is first order, only the
spinodal is displayed (from refs. [12, 14]).

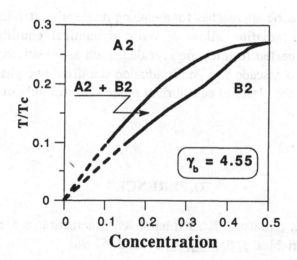

Fig. 5 :The two-phase field computed from the stochastic
potential for an irradiation intensity beyond the tricritical line
(from ref. [12]).

<u>Heterogeneous canonical description</u> [13] : here only the A-B exchange
dynamics introduced in the beginning is kept, but spatial dependence of the A

occupation on the two sublattices is taken into account along one or two crystallographic directions : the system is built by stacking cells of the BCC structure in these directions, producing one-dimensional or two-dimensional systems. The steady-state degrees of order are obtained by numerical integration of the deterministic evolution equations, the latter are obtained from the first moment of the ME. Due to the large number of variables describing the state of the system (a few tens to a few thousand) only this deterministic study can be performed. This technique also predicts *phase coexistence* beyond a critical forcing, in agreement with the results obtained from the homogeneous grand canonical description. It turns out to be very useful for studying interfacial properties : a detailed study is to be found in refs. [13,14].

CONCLUSION

The above stochastic approaches for assessing the relative stability of ordered phases under irradiation allow to build dynamical equilibrium phase diagrams. Unexpected features are revealed, such as sensitivity of 1st order transition lines to cascade size and irradiation stabilized two phase structures in alloys where the classical equilibrium phase diagram only displays single phase field.

REFERENCES

1.   Kubo, R. , Matsuo, K., Kitihara, K., Fluctuation and relaxation of macrovariables. J. Stat. Phys., 1973, **9**, 51-96.

2.   Suzuki, M., Statistical mechanics of non-equilibrium systems. II, Prog. Theor. Phys., 1976, **55**, 383-399.

3.   Bellon, P., Martin, G., Irradiation-induced formation of metastable phases : a master-equation approach, Phy. Rev. B, 1988, **38**, 2570-2582.

4. Bellon, P., Martin, G., Cascade effects in a nonequilibrium phase transition with metallurgical relevance, Phy. Rev. B, 1989, **39**, 2403-2410.

5. Bellon, P., Martin, G., Stochastic description of cascade size effects on phase stability under irradiation, Mat. Res. Soc. Symp., 1989, **138**, 15-28.

6. Haider, F., Bellon, P., Martin, G., Non equilibrium transition in driven AB_3 compounds with FCC lattice : a multivariate master equation approach, Phy. Rev. B, 1990, **42**, 8274-8281.

7. Nicolis, G., Prigogine, I., Self-organization in nonequilibrium systems, Wiley, New-York, 1977.

8. Haken, H., Advanced synergetics, Springer, Berlin, 1983.

9. Fultz, B., Kinetics of short- and long-range B2 ordering in the pair approximation, J. Mater. Res., 1990, **5**, 1419-1430.

10. Martin, G., Atomic mobility in Cahn's diffusion model, Phys. Rev. B, 1990, **41**, 2279-2283.

11. Banerjee, S.,Urban, K., Wilkens, M., Order-disorder transformation in Ni_4Mo under electron irradiation in a high-voltage electron microscope, Acta Metall., 1984, **32**, 299-311.

12. Bellon, P., Grand canonical stochastic potential for ordering and phase coexistence in driven systems, submitted to Phys. Rev. Lett.

13. Soisson, F., Bellon, P., Martin, G. , Interface stability in driven compounds, same volume.

14. Soisson, F., Bellon, P., Martin, G. , to be published.

THERMODYNAMICAL PROPERTIES OF ANTIPHASES IN FCC ORDERED ALLOYS.

A. FINEL

ONERA, BP 72, 92322 Chatillon Cedex (France)

ABSTRACT

We discuss the thermodynamical behaviour of APB's on the FCC lattice. Wetting and layering mechanisms are discussed in relation with pinning effects. We show, in particular, that only high order mean field theories, such as the Cluster Variation Method in the tetrahedron-octahedron approximation, predict the correct thermodynamical properties of surfaces and interfaces on the FCC lattice.

INTRODUCTION

Planar defects in ordered systems play a significant role in many physical properties. As example, when an ordering process starts in different locations of a disordered phase, the thermodynamical properties of the interfaces between the ordered domains and the disordered phase govern the shape of the precipitates and the nucleation rate. Another important situation concerns the deformation properties of ordered alloys. In that case, antiphase free energies determine the dissociation width of superdislocations.

We first propose a general discussion of the quantitative behavior of antiphases in ordered systems. Then, we proceed to a quantitative analysis of a particular situation, using the Cluster Variation Method (CVM) and Monte Carlo simulations.

PHENOMENOLOGICAL DISCUSSION: WETTING EFFECTS.

Due to the symmetry breaking associated to an order-disorder transition, an ordering process leads to the coexistence of different variants (or domains) of the same ordered phase. An antiphase boundary (APB) forms wherever two such domains contact.

Whether the ordering transition is first or second order is a very important factor for the qualitative behaviour of the APB. At the critical temperature of a second order transition, the ordered phase become identical to the disordered one and, consequently, the APB disappears.

On the other hand, for a first order transition, the ordered domains retain always a certain degree of long-range order and the APB's exist up to the transition temperature T_C. In that case, the free energy of the APB (i.e. the excess free energy with respect to a situation with no APB's) does not vanish at T_C. However, it may happen that, precisely at T_C and in order to minimize the excess free energy, the APB takes a particular configuration: a large layer of disordered phase develops in-between the two ordered domains. In other words, the APB splits into two order-disorder interfaces : this is the complete wetting of the APB by the disordered phase.

We present briefly a simple argument which gives the general behaviour of an APB in case of wetting. Consider the situation where the ordered phase has only two variants. Moreover, suppose that these variants are characterized by two opposite values of a *scalar* order parameter (η , -η.) In that case and for a first order transition, the free energy, as a function of η and for T just below T_c, has the qualitative behaviour presented in Fig.1. To first order, the difference δF of free energy between the disordered phase (η=0) and the ordered one ($\eta \neq 0$) (see Fig.1) is proportional to T_c- T .

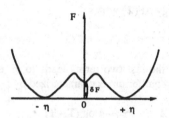

Figure 1. *Free energy as a function of the order parameter η for a symmetric first order transition*

Consider now an inhomogeneous situation with domain 1 (η<0) at z=-∞ and domain 2 (η>0) at z=+∞ (Fig. 2a). Obviously, the order parameter η, as a function of z, has to go through zero. Suppose (this is a qualitative discussion !) that the APB is sufficiently large and, consequently, that η stays in the neighborhood of zero in a finite region. Then the order parameter profile should look as in Fig.2a and the excess free energy as in Fig.2b

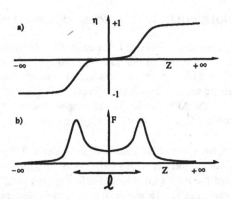

Figure 2. *Example of inhomogeneous system: a) order parameter and b) excess free energy profiles*

The APB free energy ΔF, given by $\Delta F = \int_{-\infty}^{+\infty} F(z)\, dz$, can be approximately analysed as the sum of three terms :

a) a constant term, equal to twice the energy of an order-disorder interface,

b) an attractive term, due to the excess free energy in the central region, and proportional to ℓ $(T_c- T)$, where ℓ is the size of disordered region (see Fig. 2b),

c) and finally a repulsive term due to the overlap of the two interface tails, proportional to $\exp(-a\,\ell)$[1].

In short, the APB free energy $\Delta F(\ell)$ writes :

$$\Delta F(\ell) \;\sim\; 2\,C + b\ell\,(T_c- T) + \exp(-a\,\ell) \tag{1}.$$

The competition between the last two terms leads to an equilibrium width ℓ given by $\partial\Delta F(\ell)/\partial\ell = 0$. Hence, the equilibrium is characterized by :

$$
\begin{aligned}
\ell \;&\sim\; -\mathrm{Log}\,(T_c- T) \\
\Delta F \;&\sim\; -(T_c- T)\,\mathrm{Log}\,(T_c- T) + Ct \\
\Delta S \;&\sim\; -\mathrm{Log}\,(T_c- T) \\
C/k \;&\sim\; (T_c- T)^{-1}
\end{aligned}
$$

where $\Delta S = -\partial\Delta F/\partial T$ is the APB entropy and C/k the excess specific heat.

Therefore, this crude approach predicts that, in case of wetting and with a *scalar* order parameter, the width ℓ and the APB entropy diverge logarithmically as the temperature T increases up to T_c (for a recent review of wetting phenomena, see [1], see also [2]).

---

[1] *It is easy to show, in a continuous Landau theory, that the tails of the excess energy profile of a single order-disorder interface is exponentially decaying.*

QUANTITATIVE EXAMPLE: ANTIPHASES IN $L1_2$

We now present a quantitative study of a particular APB in the FCC lattice. On a microscopic scale, ordering effects on fixed lattices are described by an Ising model. In the present case, we use an Hamiltonian with positive first neighbor interactions :

$$H = J \sum_{(n,m)}' \sigma_n \sigma_m - h \sum_n \sigma_n \qquad (2)$$

where the prime means that the sum is over the first neighbour pairs. The spin-like variables σ_n take values ± 1 depending on the type A or B of the atom at site n. The corresponding phase diagram is now fairly well known [3] and schematically represented in Fig.3 a. We shall focus here on the $L1_2$ ordered phase.

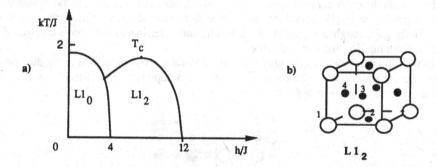

Figure 3. a) *Schematic phase diagram of the FCC lattice with positive first neighbor interactions*; b) $L1_2$ *structure*.

In $L1_2$, one simple cubic sublattice of the FCC structure is predominantely occupied by B atoms and the three other by A atoms (see Fig.3b). More precisely, if 1,2,3 and 4 denote the four simple cubic sublattices and c_i the concentration in minority atoms of sublattice i, a perfectly ordered $L1_2$ structure is defined, say, by $c_1=1$, $c_2=c_3=c_4=0$. More generally, at finite temperature, an homogeneous $L1_2$ may be described by:

$$c_1 = (1+3\eta)/4$$
$$c_2 = (1-\eta)/4$$
$$c_3 = (1-\eta)/4$$
$$c_4 = (1-\eta)/4$$

where η, the order parameter, is equal to one at zero temperature. However, this point of view doesn't reflect the degeneracy of the $L1_2$ phase. Indeed, we have four choices for labelling the sublattice which carries the minority atoms. Hence, $L1_2$ has four possible variants. Therefore, besides the mean concentration $(c_1+c_2+c_3+c_4)/4$, a *three*-dimensional order parameter $\eta = (\eta_1 \eta_2 \eta_3)$ is required to reflect this fourfold degeneracy. This yields to:

$$c_1 = (4 + \eta_1 + \eta_2 + \eta_3) / 4$$
$$c_2 = (4 - \eta_1 - \eta_2 + \eta_3) / 4$$
$$c_3 = (4 - \eta_1 + \eta_2 - \eta_3) / 4$$
$$c_4 = (4 + \eta_1 - \eta_2 - \eta_3) / 4$$

The four $L1_2$ variants are then represented respectively by $\eta = (1\,1\,1)$, $(1\,\bar{1}\,\bar{1})$, $(\bar{1}\,1\,\bar{1})$ and $(\bar{1}\,\bar{1}\,1)$. In other words, they represent the four vertices of a regular tetrahedron in the three-dimensional order parameter phase space.

This point of view is perfectly adapted to the description of inhomogeneous systems. Indeed, consider a situation where two variants are simultaneously present. Then, the ternary axis, which makes equivalent three sublattices in an homogeneous $L1_2$ phase, is lost. The four sublattices are then independant of each other and the long range order is locally represented by the three-dimensional order parameter $\eta = (\eta_1\,\eta_2\,\eta_3)$. More precisely, provided we can define local average values for the order parameter[2], we have to find a path, in the order parameter phase space, which relies the vertices associated with the two variants. Obviously, as we are in a three-dimensional space, this path does not necessarily go through the origine. As a result, and referring to the above discussion, complete wetting is no more unavoidable.

In the following, we consider only the case of a non-conservative APB along direction [100] ; i.e. we assume that minority atoms principally occupy sublattice 1 (or 4) at $z = -\infty$ and 2 (or 3) at $z = +\infty$ (see Fig.4).

Figure 4. *Non-conservative* [100] *APB in the* $L1_2$ *phase. Full (empty) circles represent B (A) atoms. Large (small) circles are the sites on* [n 0 0] *planes with n integer (half integer)*

CVM results in the tetrahedron approximation.

In short, the Cluster Variation Method is a mean field theory in which the correlations inside a given basic cluster are treated correctly, the others being neglected. Hence, we have a level of approximation for each selected basic cluster and the quality of the method increases with the size of that cluster. Of course, the use of the CVM in inhomogeneous systems is much more heavier than in homogeneous ones.

Within the tetrahedron approximation, the first detailed study is due to Kikuchi and Cahn [4]. However, the accuracy of their calculations was not sufficient for studying in details what happens close to the order-disorder tansition.

Recent improved studies [5] have shown that wetting indeed does occur: as the temperature increases up to T_c, the APB splits into two order-disorder interfaces whose

---

[2]*This is possible, through some sort of coarse grained average, if the local concentrations vary smoothly with respect to the lattice spacing.*

separation diverges logarithmically with $(T_c - T)$. But this wetting process is not continuous: there is an infinite series of first order layering transitions associated with entropy jumps. In other words, as the order-disorder interfaces move away, they rub against the lattice. This layering effects is clear in a temperature range ΔT of the order of $\Delta T/T_c = 10^{-1}$ J (Fig.5). We note, for latter use, that the size of the disordered layer in-between the two order-disorder interfaces increases by jump of two (100) disordered planes at each layering transitions.

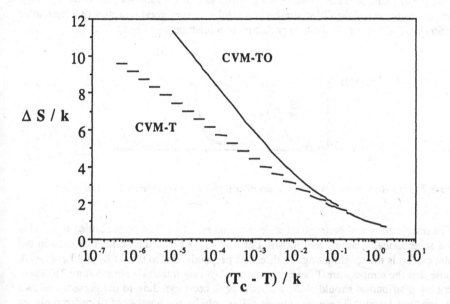

Figure 5. *Excess entropy per site $\Delta S/k$ for a non-conservative (100) APB in $L1_2$: CVM tetrahedron (CVM-T) and tetrahedron-octahedron (CVM-TO) results.*

CVM results in the tetrahedron-octahedron approximation.

In order to test the valadity of the previous CVM results, we have used an approximation of higher order, namely the CVM tetrahedron-octahedron (CVM-TO). The calculations become heavy, but the computation can still be done with a very high accuracy.

We have first investigated the wetting process as T increases up to T_c, and found no layering effects anymore. More precisely, with an accuracy of 10^{-10} J on the excess free energies per site, the wetting is continuous: each order-disorder interfaces moves away with no friction on the lattice. Consequently, the excess entropy is now a continuous function of $T_c - T$, as can be seen in Fig.5. We are then faced with an important contradiction between the two CVM approximations.

Monte Carlo simulations

A most direct way for testing the previous analytical approximations is obviously to use Monte Carlo simulations. Provided the number of sites and the number of Monte Carlo steps are large enough, the calculation is in principle exact and quite straightforward. However, as we are dealing with the thermodynamical properties of a localized defect, the computation demands a very high accuracy, especially as this defect is characterized by a wetting mechanism with a divergent specific heat.

This remark can be made more precise. Suppose that we look for first order layering effects, as suggested by the CVM tetrahedron results. A very good indicator of a first order transition is a double peak in the energy distribution function (see Fig.6).

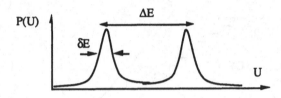

Figure 6. *Energy distribution near a first order transition;* U *represents the total energy.*

This double peak will be resolved under very severe conditions on the lattice sizes. This can be schematically shown as follows. Let $N_v=4 L_z * L^2$ be the number of sites in the simulation (L_z is the lenght along the direction perpendicular to the APB, and $L_x=L_y=L$). Suppose that the temperature T is very close to a layering transition temperature T_0. Then, the energy distribution should show a double peak corresponding to the presence of two states, labelled 1 and 2. These two phases differ only by the number of disordered planes inside the APB, i.e. two (see the CVM tetrahedron investigation). Hence, if Δu is the energy difference, per site , between $L1_2$ and the disordered phase, the energy difference ΔE between the states 1 and 2 is approximately given by:

$$\Delta E \approx 4 L^2 * \Delta u \qquad (3)$$

(there is 2 L^2 sites per (100) planes). On the other hand, the size δE of *each* peak is driven by the specific heat C, which measures the fluctuations of internal energy in each of the two phases:

$$\delta E = kT \sqrt{C/k} \qquad (4)$$

with

$$C = (<U^2> - <U>^2) / kT.$$

In the present case, the specific heat is the sum of two contributions: a volume term, proportional to the total number of sites $N_v = 4L_z * L^2$ and a surface term, due to the APB, proportional to $N_s = 2 L^2$. Thus, we have

$$C = N_v\, C_v + N_s\, C_s$$

where C_v and C_s are the specific heat, *per site*, of the bulk and of the APB, respectively. The double peak shape will be resolved if the separation between the two peaks is much larger than their width (see Fig. 6) :

$$\Delta E \gg \delta E.$$

This gives:

$$L \gg \sqrt{4\, L_z\, C_v/k + 2\, C_s/k}\ \ \frac{kT}{4\Delta u} \qquad (5)$$

Hence, we have a relation between L_z, the linear dimension perpendicular to the APB, and L, its transverse size. We note in anticipation of subsequent results that the width of each order-disorder interfaces, which limits the APB in the wetting regime, is of the order of 10 FCC cubes. Consequently, since the boundary conditions must not interfere too much with the APB, we must have $L_z \gtrsim 40$. Thus, with $\Delta u \approx 0.4$ J and $C_v/k \approx 4$ (as determined by homogeneous Monte Carlo studies) and $kT/J \approx 1.8$ J, i.e. in the wetting regime ($kT_c = 1.815$ J), the first contribution to the right hand side of (5) leads, *alone*, to $L \gg 40$. The situation is even worse if we consider the effect of C_s. Indeed, C_s, the excess specific heat per site of the APB, diverges as ΔT^{-1} (see above).

This shows that a direct observation of layering effects, if any, may need very large simulations.

The simplest way to circumvent this difficulty is simply to look at the origine of the layering mechanism, i.e. the thickness of the order-disorder interfaces with respect to the lattice constant. This is the purpose of the last paragraph.

Pinning effects

The phenomenological approach proposed above is based on a continuous approximation and consequently leads to continuous wetting phenomena. Naturally, the discrete nature of the lattice, which is obviously taken into account in the CVM and Monte Carlo approaches, may change things drastically and may lead to layering transitions. Such pinning effects are encountered in many physical situations (see for example [6] and [7]).

The lattice discreteness plays a significant rôle when the local excess free energy varies significantly in space, that is, in the present case, in each of the two order-disorder interfaces which limits the APB (see Fig.2b). Hence, we just need to estimate the effect of the lattice on the excess energy of an isolated interface.

This interface free energy is no more given by an integral, but by a discrete sum, as:

$$\Delta F^{int} = \sum_n F(n)\,,$$

where $F(n)$ is the local free energy contribution of unit n. Eventhough there are many ways of defining $F(n)$ for the unit cell n, there is no ambiguity concerning what the unit cell must be: as ΔF^{int} must be invariant by a translation of the interface of one FCC cube (because $L1_2$

and the disordered phase are both invariant through this translation), the unit cell is nothing else than the FCC cube. A classic decomposition of a discrete sum into successive terms leads then to:

$$\Delta F^{int} \approx C + A \cos(2\pi x_0) + ...$$

where C correspond to the continuous approximation and x_0 represents the position of the interface on the lattice (in unit of the FCC cube and with respect to some origine). Hence, the first correction to the continuous approach is an oscillatory term, whose period is equal to the FCC cube length, and whose amplitude A is exponentially decaying with the interface width d:

$$A \approx D \, e^{-\alpha d}$$

Concerning the APB free energy and referring to the discussion which gives Eq.1, the cosine term in the interface free energy leads obviously to a similar cosine term in $\Delta F(\ell)$:

$$\Delta F(\ell) \sim 2C + b\ell\,(T_c - T) + \exp(-a\,\ell) + 2A\cos(2\pi\ell)$$

As a result, and in the limit $\Delta T = (T_c - T) \to 0$, the equilibrium APB width ℓ adopts now a staircase behaviour with steps of size one FCC cube. More precisely, these layering transitions develop in a range of temperature ΔT of the order $\Delta T \approx A/b$. The effect of the lattice is then entirely dictated by A, the pinning energy.

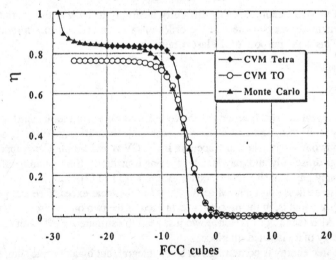

Figure 7. *Interface between* $L1_2$ *and the disordered phase at the transition temperature* T_c, *within two CVM approximations and with Monte Carlo simulations. We plot the order parameter* $\eta_3 = (-c_1 + c_2 - c_3 + c_4)$ *as a function of n, the label of the fcc cubes (see Fig. 4 for definitions).* η_3 *would be equal to one in a perfectly ordered* $L1_2$ *structure, and to zero in an homogeneous disordered phase.*

Now comes the crucial point. The two CVM approximations used above give very different estimates for the interface thickness d, and consequently for the pinning energy A. This can be seen in Fig.7, where we present the order parameter profil of the interface at the transition temperature T_C. We note that the CVM-tetrahedron leads to an interface width (~ 3 fcc cubes) much smaller than the CVM-TO (~ 10 cubes). This discrepancy may be interpreted in terms of bulk properties.

In some way, the interface thickness d is related to the correlation lengths of the *homogeneous* phases which are stabilized on both sides of the interface, say ξ_{L12} and ξ_{dis}. In a mean field scheme, as the CVM approach, these correlation lengths are related to the spinodal concept. More precisely, let T_S^{dis} and T_S^{L12} be the spinodal temperatures of, respectively, the disordered and L1$_2$ phases. These temperatures mark the limits of metastability of these phases: $T_S^{dis} < T_C$ and $T_S^{L12} > T_C$. According to a general property of mean field theories, the correlation lengths are related to the gaps between the spinodals and T_C:

$$\xi_{L12} \sim (T_S^{L12} - T_C)^{-1/2}$$

$$\xi_{dis} \sim (T_C - T_S^{dis})^{-1/2}$$

Table 1

Some characteristics of the order-disorder transition between L1$_2$ and the disordered phase for two different CVM approximations (h/J = 7.5 J, i.e. very close to the congruent point). T_C is the transition temperature, T_S^{L12} and T_S^{dis} are the spinodal temperatures of L1$_2$ and the disordered phase, respectively.

	CVM tetrahedron	CVM -TO
kT_S^{L12}/J	2.009	1.898
kT_C/J	1.943	1.881
kT_S^{dis}/J	1.451	1.809

As shown in Table 1, the two CVM approximations differ notably on the spinodal temperatures: the CVM-TO leads to spinodals which are very close to T_C. We may then argue that the situation may be even different for a higher order CVM. In other words, the question arises of the convergence of the gaps $\Delta T^{L12} = T_S^{L12} - T_C$ and $\Delta T^{dis} = T_C - T_S^{dis}$, with the order of the CVM approximation.

As the order-disorder transition between L1$_2$ and the disordered phases is first order, these gaps must stay finite to *any* order, otherwise the transition would be continuous. Consequently, they must converge to *finite* limit values[3] ΔT_{lim}^{L12} and ΔT_{lim}^{dis}. In order to

---

[3] *As the CVM approximation is expected to converge towards the exact results with the size of the basic cluster [8], this procedure should provide a way of determining "exact" spinodal temperatures. We will not address this general problem here. We just point out that this procedure should be a way of computing the unique analytical continuation of the free energy curve F(T) of a given phase, beyond the singular point T_C.*

decide wether the CVM-TO estimates for ΔT^{L1_2} and ΔT^{dis} are close to these limits or not, we should in principle consider the next order CVM approximation. A more direct way is to compare the CVM-TO results to the Monte Carlo ones. However, this comparison cannot be done readily on spinodal temperatures[4]. We would rather consider the interface profile, especially as this is the point of interest here. As shown in Fig. 7, the profile thicknesses in Monte Carlo and in the CVM-TO are very similar.

We may then conclude that the CVM-TO gives the correct thermodynamical properties of the interface and, consequently, of the APB : the pinning effects on the lattice must be very weak and the layering transitions not detectable, eventhough they will ultimately settle down in a very narrow range of temperature near T_C .

Conclusion

An important conclusion of this study is that, concerning the thermodynamical behaviour of surfaces and interfaces, low order mean field theories may lead to incorrect results. This has been illustrated here in the case of non-conservative (100) APB's in the $L1_2$ phase: the tetrahedron and tetrahedron-octahedron CVM predict very different behaviours as a function of T, eventhough they give similar results for the bulk phase diagram. As explained above, this is due to the fact that the thermodynamical behaviours of defects are governed by the relaxation properties of the *bulk homogeneous* phases, i.e. by the correlation lengths. When the transitions are first order, these correlation lengths play only a minor rôle for the bulk properties. Thus, a mean field theory which underestimates the relaxation effects may lead to an overall correct bulk phase diagram, but will give unreliable results for defects. This is the case for the CVM in the tetrahedron approximation. On the other hand, the CVM-TO predicts the actual relaxation properties of the homogeneous phases. Consequently, as far as we are concerned with surfaces and interfaces on the FCC lattice and near a bulk first order transition, the first reliable mean field theory is the CVM-TO.

Acknowledgements

The author gratefully acknowledge F. Ducastelle and V. Mazauric for very helpful discussions.

---

[4] *From a general point of view, the spinodal concept is related to the hysteresis in the internal energy U(T) as a function of T. Thus, it should be possible, at first sight, to define spinodal temperatures in Monte Carlo simulations and to study their behaviour as a function of lattice sizes. However, due to the very nature of the Monte Carlo method, these spinodals would not behave as in the CVM. More precisely, in the limit of an infinite number of Monte Carlo steps, the hysteresis shrinks to a unique curve U(T) for any finit lattice, and, consequently, there are no spinodals. In fact, as we increase the lattice size and if we want to stabilize an hysteresis, the number of Monte Carlo steps should not increase as fast as exp(NΔE), where ΔE is the internal energy difference, per site, between the two phases in the neighborood of the transition, and N the number of sites in the simulation.*

REFERENCES

1. S. Dietrich, in *Phase Transitions and Critical Phenomena*, eds. C. Domb and J. Lebowitz (Academic Press, London), **12**, 1988, 1

2. R. Lipowsky, Phys. Rev. Let., 1982, **49**, 1575
   R. Lipowsky and W. Speth, Phys. Rev. B, 1983, **28**, 3983

3. A.Finel and F.Ducastelle, Europhys.Letters , 1986, **1**, 135 ; erratum 1986, **1**, 543.
   R.Kikuchi, Progr.Theor.Phys.Supp.,1986, **87**, 69.
   R. Tetot, A. Finel and F. Ducastelle, J. Stat. Phys., 1990, **61**, 121.

4. R. Kikuchi and J.W. Cahn, Acta Metall., 1979, **27**,1337.

5. A.Finel, V. Mazauric and F. Ducastelle, Phys. Rev. Let., 1990, **65**, 1016.

6. Y. Ishibashi and I. Suzuki, J. Phys. Soc. Jap., 1984, **53**, 4250.

7. V. L. Pokrovsky, J. Physique, 198, **42**, 761

8. A. Finel, in *Alloys Phase Stability*, eds. G. M. Stocks and A. Gonis, (Kluwer Academic Publishers) 1989, 269

CVM TREATMENT OF ORDER - DISORDER TRANSFORMATIONS

Cathérine COLINET

LTPCM, URA CNRS no 29, ENSEEG, BP 75, Domaine Universitaire,
F-38402 St. Martin d'Hères, Cédex France

Gerhard INDEN

Max-Planck-Institut für Eisenforschung, Postf. 140444, D-4000 Düsseldorf, Germany

In systems with a body centered cubic crystal structure the Cluster Variation Method (CVM) in the irregular tetrahedron approximation allows to take first and second nearest neighbour pair interactions into account. With these interactions it is possible to obtain the most frequently observed superstructures B2, D0$_3$, B32. For binary systems it could be shown [1] that this approximation yields almost identical phase diagrams to the Monte Carlo simulation technique which is considered to come closest to the real situation.

The presented paper reported on CVM calculations in the irregular tetrahedron approximation of the ternary system Fe-Co-Al. This system is particularly interesting since all three binary subsystems show an ordering tendency which compete in the ternary system. In order to keep with a ternary system we have considered only chemical pair interactions between first and second nearest neighbours. Magnetic effects were thus neglected although we know that magnetic interactions may have important effects on the phase diagram at temperatures close to and below the Curie temperature. A treatment including magnetic effects with spin 1/2 would in fact have implied to treat a six component system. In the Fe-Co-Al system such effects may appear near the binary system Fe-Co where the Curie temperature is high. They are confined to a composition range close to the binary system since the Curie temperature decreases rapidly with the addition of Al to Fe-Co. The numerical values for the interactions were taken from previous studies (Fe-Co[2], Fe-Al [3]-[4], Co-Al [5]-[6]).

Several complete isothermal sections were presented for the temperatures T= 1400, 1000, 800, 700, 600, 300 K. The most striking result of the calculations is the appearence of a ternary

miscibility gap between ordered and disordered phases. This miscibility gap is induced by the competition between the various ordering tendencies of Fe-Co, Fe-Al and Co-Al pair interactions. The shape of the miscibility gap is very asymmetric. This property gives rise to interesting possibilities for materials development as outlined in [7] for analoguous magnetic cases. The miscibility gap closes at multicritical points which are produced at the intersections of second order transition temperature lines with the phase boundaries of the miscibility gap. Good qualitative agreement was obtained with recent experimental data [6].
Several vertical sections were also presented.
The results presented at the meeting will be published in full detail [8].

[1] H. Ackermann, G. Inden and R. Kikuchi, Tetrahedron approximation of the Cluster Variation Method for BCC alloys, Acta Met. 37 (1989) 1-7
[2] G. Inden, Determination of chemical and magnetic interchange energies in bcc alloys, Z. Metallk. 68 (1977) 529-534
[3] G. Inden, The mutual influence of magnetic and chemical ordering, Mater. Res. Soc. Proc. Vol. 19, Elsevier Sci. Publ. Co. 1983, p. 175-188
[4] H. Ackermann, C. Colinet, G. Inden and R. Kikuchi, in preparation
[5] G. Inden and W. Pitsch, Atomic Ordering, in Materials Science and Technology, Cahn R.W., Haasen P. and Kramers E.J. (Eds.), Vol. 5, Phase Transformations, VCH Verlagsgesellschaft, Weinheim 1991, p.497-552
[6] H. Ackermann, Thesis, Technische Hochschule Aachen, 1988
[7] G. Inden, in User Application of Alloy Phase Diagrams, Kaufman L. (Ed.), ASM International 1987, p. 25-30
[8] C. Colinet, G. Inden and R. Kikuchi, in preparation for Acta Met.

A MONTE CARLO STUDY OF THE ANTIPHASE FREE ENERGY IN FCC ORDERING TRANSITIONS

FREDERIC LIVET
Laboratoire de Thermodynamique et
Physico-chimie des materiaux(URA CNRS 29)
ENSEEG-DU-BP75-38402 St Martin D'Heres Cedex

ABSTRACT

The ordering transition of FCC systems is modelized with Ising Hamiltonian (pair interactions).The Monte Carlo method is used to calculate the free energy c the lattice.This permits a discussion of the relative stability of various "long period (LP) structures.This discussion is carried with Hamiltonians close to th degeneracy, where the simplest mean-field theories cannot be used.It appears tha the LP structures are not stable with simple Hamiltonians(antiferromagneti nearest-neighbour interactions).The antiphase surface free energy (σ_a) can then b calculated.A study of the variations of σ_a is carried among variable degenerat Hamiltonians.One shows an increase of σ_a when the Hamiltonian leads to stronger fluctuations close to the ordering temperature.

INTRODUCTION

As in the unmixing phase transitions /1/,the ordering of alloys can b qualitatively described with Ising Hamiltonians.The macroscopic thermodynami properties are thought to be contained in these model Hamiltonians.

In the case of ordering processes,the simplest Hamiltonian has a antiferromagnetic nearest-neighbour interaction.Qualitatively,this Hamiltonian lead to second-oder phase transitions in the case of BCC alloys /2/ and to first-orde transitions in the case of FCC alloys /3/,/4/ .This behaviour is experimentall observed if one compares the Cu-Zn alloy /5/ and the Au-Cu alloy /6/ :the forme alloy has critical exponents which correspond to the results of the Ising spi model , the latter has a first order transition and the fluctuations close to th transition remain of a finite size /7/.

The first order nature of the transition in the FCC antiferromagnetic Isin model has been studied by series expansion /4/,cluster variational methods (CVM /8/ and Monte Carlo calculations /3/ and is well established . It is usually ascribe

, the degeneracy of the ground state of the system . At T=0 , the 50-50 alloy has various structures of the same energy deduced from the L1$_0$ structure (AF1 or ∞>) , by the introduction of periodic antiphases (APB) in the [1 0 0] plane . These structures are here represented by means of the familiar ANNNI /9/,/10/ notations :<i j ...> means that the unit cell has i L1$_0$ cells separated by APB , followed by j cells...

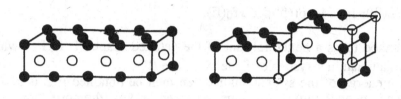

FIG 1 . the < ∞ > (L1$_0$) and the < 2 2 > structures at the 50-50 concentrations

The same occurs in the 25-75 alloy , where the various degenerate structures can be deduced from the L1$_2$ (< ∞ >) one by the introduction of periodic APB(for instance the DO22 structure of "Strukturbericht" is here < 11>) .
The degeneracy of this simple nearest neighbour Hamiltonian can be compared with the various APB strucures observed in the alloys /6/,/11/.

In this paper ,I examine the relative stability of some of the structures already described in the vicinity of degenerate Hamiltonians . This needs precise calculations of the free energy from the MC method.In the first part , the calculations are carried out with the simplest nearest neighbour Hamiltonian . Then these results are extended to the vicinity of that Hamiltonian by an extension of the range of the interactions. Eventually, the degenerate line obtained from a suitable combination of the second and third nearest neighbour pair interactions is explored.

1- The antiferromagnetic nearest-neighbour interaction.

In spin notation,the starting Hamiltonian is:

(1) $H_0 = J \Sigma \sigma_i \sigma_j$, where the sum is over all nearest-neighbour pairs

All MC calculations are carried out on L=8 (2048 sites) and L=16 (16384 sites) CC lattices on canonical ensembles (fixed atomic concentrations).
The basic method used to compute the free energy is the "overlap" method /0/,/12/ :
    (i) At a given temperature ($\beta=1/kT$) and a given Hamiltonian H_0 , the Metropolis algorithm /13/ samples the states of energy E with the probability :

(2) $f_\beta(E)=\Sigma_E g(E)\exp(-\beta E)=\Sigma_{\{c\}}\delta(H_0\{c\}-E)\exp(-\beta H_0\{c\}/Z(\beta)$

where $g(E)$ is the density of states of a given energy and $Z(\beta)$ is the partition function .

(ii) On studying the overlap of the distributions obtained at two temperatures sufficiently close together , one can deduce the ratio of the partition functions :

$$Z(\beta)/Z(\beta')=f_{\beta'}(E)\exp((\beta'-\beta)E)/f_\beta(E)$$

This method takes a better account of the detailed shape of $f_\beta(E)$, which is never trully gaussian in a system of finite size (see /10/).

The free energy of the system studied can then be obtained if Z is known at some reference temperature . Care must be taken here of the corrections arising from calculating in a finite canonical system /10/.

In order to test the results obtained , fig 2 gives the free energies (F ,in J units per atom), obtained in the vicinity of the transition in the 50-50 case. These curves compare , in a system of size L ,the probabilities of the disordered and of the ordered phases.

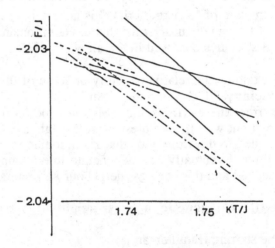

FIG 2. Free energies close to the transition in the L=8 (dash-dots line) and L=16 (dashed line) systems,at the50-50 concentrations , and results from serie expansion (continuous lines) /4/.

The transition temperatures observed here are in close agreement with the results of Polgreen/3/ (kT/J = 1.735) and the small discrepancy with the serie expansion results of Styer /4/ seems mainly due to fluctuations of the disordered state close to the transition.

The same method is used to calculate the free energies of the simple antiphase
tructures in the 25% and in the 50% alloy.The study is here limited to the < 1 1 >,
2 2 >, < 4 4 > in the L=8 case and extended towards < 8 8 > in the L=16 case.
g 3 and fig 4 give $F_{< p\,p >} - F_{< \infty >}$ versus T for the two concentrations studied
the L=16 case .

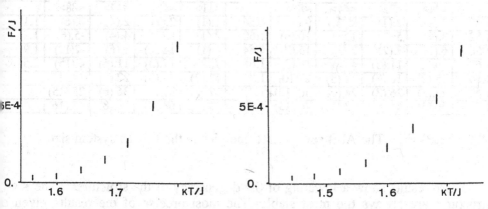

FIG 3. Difference of free energies   FIG 4. Difference of free energies
etween the   < 1 1 > and < ∞ > between the < 1 1 > and < ∞ >
ructures in the 25-75% system (L=16) structures in the 50-50% system (L=16)

From these results , an APB surface energy (σ_a) can be obtained.In this
lculation , a correction has been applied :in the 50-50 % lattice of size L , as the
riodic boundary conditions are used , the < ∞ > phase cannot be distinguished
om the < L L> phase . This means that the free energy of the < ∞ > lattice still
ntains a small amount of antiphase . This free energy has been corrected from the
rface free energy calculated with the < 4 4 > structure.In the case of the L=16
ttice one writes ,F_0 being the non-corrected result:

$F_{< \infty >}=F_{0< \infty >} - (F_{0< 4\,4 >} - F_{0< \infty >}) / 8$ for the 50-50% concentration

d : $F_{< \infty >}=F_{0< \infty >} - 3 (F_{0< 4\,4 >} - F_{0< \infty >}) /16$ for the 25-75% concentration.

Table 1 summarises the results obtained. Each σ_a corresponds to a unit surface
a^2 , the square of the cubic cell , and the energies are given in J units. On Table
are also given the results of the first term of the low temperature expansion
tained by Finel/14/:

$F_{< 1\,1 >} - F_{< \infty >} = (kT / 2J) \exp (-16\beta J)$

50-50					25-75					σ_a /14/
kT/J	σ_a(in J per at)x10^4				kT/J	σ_a(in J per at)x10^4				
	<11>	<22>	<44>	<88>		<11>	<22>	<44>	<88>	
1.48	2.(1)	2.(1)	.5(3.)	9.(5)	1.56	1.(1)	2.(2)	2.(4)	0.(6)	1.1
1.52	3.(1)	5.(2)	1.(3)	10.(6)	1.60	2.(1)	8.(2)	4.(4)	-5.(6)	1.5
1.56	6.(1)	7.(2)	3.(3)	15.(7)	1.64	4.(1)	14.(2)	6.(6)	-2(12)	1.9
1.60	10.(2)	11.(3)	4.(4)	19.(9)	1.68	7.(2)	17.(3)	11.(6)	4.(15)	2.5
1.64	15.(2)	16.(3)	11.(5)	27(10)	1.72	12.(2)	25.(3)	22.(6)	14.(15)	
1.68	23.(2)	26.(3)	24.(6)	50.(12)	1.76	22.(2)	39.(4)	36.(6)	25.(15)	
1.72	36.(3)	44.(4)	42.(7)	76.(15)	1.80	39.(2)	63.(4)	70.(8)	43.(16)	

TABLE 1 . The APB surface free energies in the L = 16 system size.

These results show a lowering of the degeneracy of the structures . The < ∞ structures are always the most stable .The most precise of the results given o Table1 strongly suggest an increase of the antiphase free energy with the distanc between APB.

A few alloys have periodic APB phases appearing or disappearing on varyin the temperature.These alloys can be considered as close to the degeneracy,bu interactions of longer range are certainly involved in the thermodynami properties.For this reason,the results obtained must be extended to othe Hamiltonians.

2.The vicinicity of the antiferromagnetic nearest neighbour Hamiltonian.

a.The method followed.

Higher order pair interactions are added to the Hamiltonian (1):

$$(3)\quad H = H_0 + \quad J_n \; \Sigma \; s_i \, s_j \quad,$$

where J_n refers to a nth nearest neighbour interaction. The partition function ca be written :

$$Z_H(\beta) = Z_{H0}(\beta)\Sigma_{\{c\}}\exp(-\beta(H-H_0)\{c\}) \, \exp(-\beta H_0\{c\})/ \, \Sigma_{\{c\}}\exp(-\beta \, H_0\{c\})$$

This formula shows that $Z_H(\beta) / Z_{H0}(\beta)$ can be calculated by averagir $\exp -(\beta(H - H_0)\{c\})$ with a probability $p\{c\} = \exp(-\beta H_0\{c\}) / Z_{H0}(\beta)$. Th corresponds to the Metropolis sampling with H_0 and β (see formula 2). I ractice, this average can be estimated only if $\beta < (H-H0)\{c\} >$ is not too large. If ne limits to second order :

$$(4) \quad N\Delta F = \Sigma_n <\Delta E_{Jn}> \delta J_n / J_n + \beta \Sigma_{n,m} <\Delta E_{JnJm}{}^2 - <\Delta E_{JnJm}>^2> \delta J_n \delta J_m / (2 J_n J_m)$$

The first order derivatives are deduced from the average energy change bserved on adding an interaction J_n to $H0$, the second order term corresponds to ne fluctuations of the energy change when two interactions J_n and J_m are added, N ; the number of atoms of the sytem.

The first order term is a "potential energy " : it can be written in terms of the air correlation functions (as it can be directly seen from formula 3). The second erm accounts for entropy variations.If the energy distribution were exactly iaussian,formula (4) should be valid in a wide range of Jn (characteristics ieorem).On a finite system,close to the transition,this is not observed /10/,and the second order term is here calculated only to ensure that it can be neglected.

b.The relative stability of the structures studied.

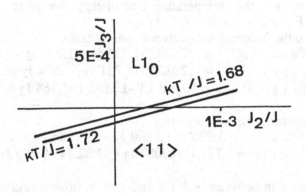

FIG 5. Phase boundaries between the $L1_0$ and the $< 11 >$
structures in the J_2/J - J_3/J plane ,calculated at low temperature.

The free energy variations have been calculated in the 50-50% case (L = 16):
(i) $kT / J = 1.72$:
$F_{<11>} - F_{<\infty>} = -8.7E-4 \ J - 1.681 \ J_2 + 6.835 \ J_3 - 6.800 \ J_4$
$F_{<22>} - F_{<\infty>} = -5.3E-4 \ J - .813 \ J_2 + 3.363 \ J_3 - 3.339 \ J_4$
$F_{<44>} - F_{<\infty>} = -2.5E-4 \ J - .405 \ J_2 + 1.668 \ J_3 - 1.653 \ J_4$

(i i) $kT / J = 1.68$:
$F_{<11>} - F_{<\infty>} = -5.6E-4 \ J - 1.758 \ J_2 + 7.119 \ J_3 - 7.103 \ J_4$
$F_{<22>} - F_{<\infty>} = -3.3E-4 \ J - .879 \ J_2 + 3.537 \ J_3 - 3.524 \ J_4$
$F_{<44>} - F_{<\infty>} = -1.5E-4 \ J - .436 \ J_2 + 1.762 \ J_3 - 1.754 \ J_4$

If the Jn < 2.E-3 , the second order terms are found to be negligible,and main errors come from the free energy calculations of part 1 .

These results provide a simple discussion of the phase diagram around Jn = 0 Fig 5 shows the phase boundaries obtained in the J_2/J , J_3/J plane at the two temperatures .

The results of part 1 strongly suggest an increase of the free energy per APB with the distance between APB. The results hereover given unambiguisly show that the variations of that free energy (per APB) is the faster ,the closest the APB are.In such a case,the phase diagram is simple : the system can be stable only in the < > or in the < 1 1 > structure , and no intermediate longer period structure can be stable , at least among the structures studied.

These results can be compared with the mean-field ones : $F_{< 1 1 >} - F_{< \infty >} =$ $-2J_2 + 8J_3 - 8J_4$ (assuming that the order parameter is 1). In the mean field theory the antiphase energy is constant if : $J_2 - 4 J_3 + 4 J_4 = 0$. Here , at kT/J = 1.72 one obtains : $J_2 - 4.07(1) J_3 + 4.05(1) J_4 = 0$ .The differences are small.

c .Ordering temperatures.

One can add the temperature free energy change to the calculation by the formula : $F / kT = -Ln Z - < E > / kT$

Close to the ordering temperature , one obtains:

$$F_{< \infty >} - F_{dis} = .2753(kT - 1.735 J) + 1.792 J_2 - 3.848 J_3 + 4.667 J_4$$

(5) $$F_{< 1 1 >} - F_{dis} = .2862(kT - 1.731 J) + .024 J_2 + 3.271 J_3 - 2.436 J_4$$

$$F_{< 1 1 >} - F_{< \infty >} = 10.E-3 kT - 17.E-3 J - 1.768 J_2 + 7.119 J_3 - 7.103 J_4$$

and the ordering temperatures are :

$$kT_{< \infty >} = 1.735 J - 6.509 J_2 + 13.98 J_3 - 16.95 J_4$$

(6) $$kT_{< 1 1 >} = 1.731 J - .0087 J_3 - 11.82 J_3 + 8.805 J_4$$

The transition between < 1 1 > and < ∞ > structures occurs if :

$$1.768 J_2 - 7.119 J_3 + 7.103 J_4 = 10.5 E-3 kT - 17.07 E-3 J$$

The results of (6) can be compared with the mean-field ones :

$$kT_{< \infty >} = 4 J - 6 J_2 + 8 J_3 - 12 J_4$$

(7) $$kT_{< 1 1 >} = 4 J - 2 J_2 - 8 J_3 + 4 J_4$$

Fig 6 gives a schematic view of the phase diagram along the J_2/J , $J_3/$ directions.

The comparision between formula (6) and (7) shows that , if the longer range interactions lower the degeneracy , the increase of the transition temperature is faster than the mean-field results. The contrary is observed if the degeneracy remains : if $J_2 = 4 (J_3 - J_4)$, formula 6 gives : $\delta kT_0 = -12.1 J_3 + 9.1 J_4$, and formula (7) gives : $dkT_0 = -16 J_3 + 12 J_4$. In this latter case , the MF result overestimates the variations of the ordering temperatures.

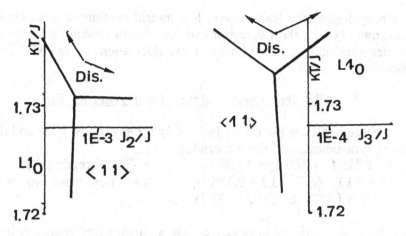

FIG 6 .Phase diagram close to $J_2/J = 0$, $J_3/J = 0$.
(Arrows indicate the MF directions)

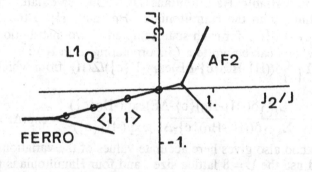

FIG 7 . The T = 0 phase diagram in the J_2/J ,J_3/J plane ,and the line studied

One now focuses interest on the directions of the (J , J_2 , J_3) space where the system remains degenerate between the < 1 1 > and the < ∞ > structure . The direction : J - 4 J_3 , 4 J_3 , J_3 has also a MF constant ordering temperature . From formula (6) , $\delta k T_0 = -5.12 J_3$. The reason of that variation is given on Fig 7: on decreasing J_3 / J along the degenerate line , at T = 0 , from .1 to -.5 , the system starts from the AF2 ordered structure (MnO type of antiferromagnetism) towards the ferromagnetic state . The disorder- AF2 transition is first-order (from group theory) ,and the para-ferromagnetic transition (which corresponds to the unmixing of alloys) is a second-order transition of the ferromagnetic Ising universality class . Along this line , the system passes from a strongly degenerate state ($J_3/J = .1$) to a second order transition with a critical end-point .

The strong degeneracy leads to very low transition temperatures . On the othe side (negative J_3/J) , the fluctuations of the system increase on approaching second-order transition, and the effect of these fluctuations on the APB energy ha to be discussed.

3 . APB free energy along the degenerate line.

The line studied (see Fig 6) is $J_2/J = 4 J_3/J$ .Only the < 1 1 > and the $< \infty$: structures will be compared . From formula (6),

$kT_0 = 1.735 J - 6.605 J_2 + 13.98 J_3$, from formula (5),

$\sigma_a = 42.E-3 kT - 68.3 E-3 J + 0.188 J_3$, and , from these two results :

$\sigma_a = 73.E-3 J T/T_0 - 68.E-3 J - .32 J_3$

This shows that , if J_3/J is negative , the antiphase free energy is increased T/T_0 beeing held constant , close to 1 .This result is nevertheless difficult to exten towards $J_3/J=-.5$: higher order terms of formula(4) become dominant if $J_2/J=2.E-2$

In order to further explore the degenerate line of Fig 7 , the method of part can be extended to variable Hamiltonians /15/ .One calculates by the usua Metropolis algorithm , with the Hamiltonians H_0 and H_1 , the difference o energy : $E = H_1\{c\} - H_0\{c\}$ for each state sampled . Two distributions of ΔE ar obtained , each of which can be written (incorporating β in H) :

$f_H(\Delta E) = \Sigma_{\{c\}}\delta((H_1-H_0)\{c\}-\Delta E)\exp(-H\{c\})/Z(H)$,from which it follows:

$$\frac{Z(H_0)f_0(\Delta E)}{Z(H_1)f_1(\Delta E)} = \frac{\Sigma_{\{c\}}\delta((H_1-H_0)\{c\}-\Delta E)\exp(-H_0\{c\})}{\Sigma_{\{c\}}\delta((H_1-H_0)\{c\}-\Delta E)\exp(-H_1\{c\})} = \exp(\Delta E)$$

and the overlap method also gives here accurate values of the variations of LnZ.Th calculations carried use the L = 8 lattice size , and four Hamiltonians are sufficier to describe the line (circles on Fig 7) . On each Hamiltonian , the temperature i varied .

On Table 2 are given the ordering temperatures : the Hamiltonians used are : ($J (1 + 4x)$, $4xJ$, xJ), in order to have constant MF T_0 . The σ_a of table II ar in the same J units than on Table I , they have been calculated only at some fixe temperatures .

x	kT_0/J	T/T_0	σ_a	T/T_0	σ_a	T/T_0	σ_a
0.	1.735	0.90	6.(2)E-4	0.94	16(2)E-4	0.96	21(2)E-4
-.054	1.95(1)	0.90	16(2)E-4	0.94	28(3)E-4	0.96	37(3)E-4
-.108	2.12(2)	0.88	10(2)E-4			0.94	30(3)E-4
-.162	2.2(?)	0.86	33(4)E-3	0.91	60(4)E-3		

TABLE 2 . The transition temperatures and the APB free energies between the < 1 1 > and the $< \infty >$ structures.

One observe (Table 2) that,at the same T/T_0 , σ_a increases if x decreases.On th same time the fluctuations close to the transition become stronger.Thes luctuations make the T_0 calculated at x=-.162 somewhat unreliable.These results how that ,with negative valuse of x ,much stronger values of J_n are necessary to nake the < 1 1 > structure stable , as compared to Fig 3 and Fig 4.

DISCUSSION

In the case of nearest-neighbour antiferromagnetic interactions, σ_a is very mall.One can estimate the average distance between APB of a finite lattice of size ، ($4L^3$ atomic sites) : d=exp($\sigma_a L^3/kT$) ,in FCC lattice units.If L=50 and $/T_0$=0.9,one obtains d=2.5 with x=0. and no antiphase with x=-.162.

The results given are somewhat decieving :

(i) The σ_a never change sign and the shape of the phase diagram of Fig 3 and 'ig 4 remains the same.

(ii) There are strong evidences that the Hamiltonians studied cannot give table LP structures.The σ_a can be interpreted by a local decrease of the luctuations of the system (entropy effect).If such an interpretation is valid,the ecrease of the fluctuations of a system with two APB is minimum if the two APB re close together (attractive interactions between APB).

(iii)If a long range electronic interaction (as it is suggested in the Au-Cu ystem) stabilizes the LP structure,the only thermodynamic transition should be imilar to the results of Fig 3 or Fig 4 : LP --> $L1_0$ (or $L1_2$) ,on increasing the emperature.This is exactly the contrary which is observed.

The MC method is relatively heavy in calculating free energies.This paper hows that reliable results can nevertheless be obtained.The methods given here can e used in the case of more complex Hamiltonians (tetrahedron interactions ...) and ɔ other structures.

REFERENCES

,Marro,J.,Bortz ,A.B.,Kalos , M.H. and Lebowitz,J.L., Phys.Rev. B12,1975,p2000.

,Zinn Justin,Z., J.Phys.(France) 42,1981,p783.

,Polgreen ,T.L., Phys.Rev.B8,1984,p1468.

,Styer,D.F., Phys.Rev.B32,1985,p392.

,Nix and Shockley,Rev of Mod.Phys.10,1938,p1.

,Hansen,M., Constitution of Binary Alloys,McGraw-Hill, New York,1958.

Keating,D.T. and Warren,B.F., J.Appl.Phys.22,1951 p286.

8.Finel,A. and Ducastelle,F., Europhys. Lett.1-3,1986,p135.

9.Selke,W.,Phys.Rep.170,1988,p213.

10.Livet,F.,J.Phys.(France)50,1989,p2983.

11.Loiseau,A.,van Tendeloo,G.,Portier,R.and Ducastelle,F., J.Phys.(France)46, 1985,p595.

12.Bichara,C.,Gaspard,J.P. and Mathieu,J.C.,Phys.Rev.Lett.A-119,1987,p462.

13.Metropolis,N,Rosenbluth,A.W,Teller,A.H. and Teller,E., J.Chem.Phys.21, 1953,p1087

14.Finel,A.,Thèse Doctorat d'Etat,Université Pierre et 'Marie Curie,Paris(1987).

15.Perini,A.,Jacucci,G. and Martin,G.,Phys.Rev.B29,1984,p2689.

INTERFACE STABILITY IN DRIVEN COMPOUNDS

F.SOISSON, P.BELLON a n d G.MARTIN
Section de Recherches de Métallurgie Physique, CEREM, Centre d'Etudes de Saclay
F-91191 Gif-sur-Yvette Cedex, France

ABSTRACT

To model the behaviour of an ordered alloy under irradiation, we propose a mean-field approximation of the B2-A2 order-disorder transition on a B.C.C. lattice with two atomic exchange mechanisms acting in parallel: thermally activated and forced jumps. By integration of deterministic evolution equations, we show that irradiation can induce the stabilization of a two-phase alloy, while the classical equilibrium phase diagram only displays single phase fields (the B2-A2 transition is second order). The steady-state diagram is computed: in the two-phase region, the lever rule is obeyed and surface tension effects are identified; a new mechanism for precipitate redistribution under irradiation and shearing is discovered.

INTRODUCTION

When a compound is maintained in far for equilibrium configurations by a dynamical external forcing (e.g. the nuclear collisions under irradiation or the disordering process in ball milling), the steady-state properties of the system can no longer be predicted by equilibrium thermodynamics. A stochastic description for such driven systems can nevertheless be developed at a mesoscopic level: stochastic potentials controlling the probability distribution of macrostates can be built, from which the relative stability of competing steady-states can be assessed [1]. However for heterogeneous systems, such potentials are difficult to compute. As an alternative way, we propose here a deterministic kinetic mean-field description which turns out to be very useful for studying interfacial properties in coherent two-phase driven compounds.

DETERMINISTIC KINETIC DESCRIPTION

As an example, we consider a binary $A_c B_{1-c}$ alloy on a rigid B.C.C. lattice and focus on the B2-A2 order-disorder transition. A Bragg-Williams mean-field approximation is then appropriate [2]. For B2 ordering it is convenient to decompose the B.C.C. lattice into two sublattices; then the state of order is described by the A atomic concentrations C_1 and C_2 on both sublattices, or by the concentration C and the degree of order $S=C_1-C_2$. We restrict to nearest neighbour interactions: with V_{ij} the energy of a pair of i-j atoms, we define $V=(V_{aa}+V_{bb}-2V_{ab})/2$ (where V>0 for ordering systems). For the sake of simplicity, we choose $V_{aa}=V_{bb}$. Since we deal with heterogeneous systems, we define a local concentration on each sublattice: if we deal with a one-dimensional problem, the concentration on plane i on the sublattice j, C_1^i is the total number of A atoms on sublattice j in plane i divided by the corresponding number of lattice sites; in a two-dimensional system, the superscript i refers to an atomic row. In the cases reported here we handeled up to 12800 atomic rows.

We model atomic diffusion, by allowing the exchange of nearest-neighbour atom pairs. The exchange of an A atom on the ith site of the sublattice 1 with a B atom on the jth site of the sublattice 2 occurs at the frequency $\Gamma_b + \Gamma_{12}^{ij}$. Indeed two atomic exchange mechanisms operate in parallel: "ballistic" exchanges (Γ_b) induced by nuclear collisions or various shearing mechanisms and thermally actived exchanges (Γ_{12}^{ij}), Γ_{12}^{ij} is proportional to the exponential of the energy necessary to extract an AB pair from its environment and bring it into a saddle point position [3,4].

$$\Gamma_{12}^{ij} = \Gamma^\circ \exp\left(-\frac{E_{12}^{ij}}{k_b T}\right) \tag{1}$$

$$\text{with } E_{12}^{ij} = E_s - \sum_p^{nn(i)} \{C_2^p V_{aa} + (1-C_2^p) V_{ab}\} - \sum_q^{nn(j)} \{C_1^q V_{ab} + (1-C_1^q) V_{bb}\} \tag{2}$$

where the summation in equations (2) are restricted to nearest-neighbour of sites i and j (nn(i) and nn(j) respectively), E_s is the saddle point energy, which for simplicity is chosen here to be independent of the surrounding.

Then the rate of change of concentration on sublattice 1 at the location i is:

$$\frac{dC_1^i}{dt} = \sum_j^{nn(i)} \{-(\Gamma_{12}^{ij}+\Gamma_b)C_1^i(1-C_2^j) + (\Gamma_{21}^{ji}+\Gamma_b)C_2^j(1-C_1^i)\} \tag{3}$$

The similar expression for $\frac{dC_2^i}{dt}$ is as eq.(3) with 1 and 2 interchanged. For thermal systems ($\Gamma_b=0$), the above kinetic equation drives the system to the equilibrium state which is precisely that obtained by minimizing of the mean-field free energy computed in the same Bragg-Williams approximation [4,5].

As can be seen from eq.(1)-(3), the control parameters are: the reduced temperature T/T_c ($T_c= zV/2k_b$), the average composition C and the reduced forced frequency :

$$\gamma_b = \frac{\Gamma_b}{\Gamma° \exp \left\{ -\frac{(E_s-16V)}{k_bT} \right\}} \tag{4}$$

The evolution of the system as well as its steady-states are obtained by numerical integration of eq.(3), using periodic boundary conditions, for different sets of initial conditions $\{C_1^i(t=0), C_2^i(t=0)\}$. The fourth-order Runge-Kutta method with adaptive step-size control was found to be the most efficient [6]. The relative error on the concentration is kept less than 10^{-6} during each integration step, and steady-state is assumed to be reached when all derivatives have absolute values smaller than 10^{-8}.

RESULTS AND DISCUSSION

Steady-states diagrams

In the Bragg-Williams approximation with nearest-neighbour interactions, the equilibrium order-disorder transition is of second kind with a critical temperature $T_c= zV/2k_b$. Under irradiation the transition becomes first order beyond a critical forcing (see fig.1): this result can be obtained either from a homogeneous deterministic description ($C_1^i=C_1, C_2^i=C_2$, for all i) or from a stochastic description [1]. Beyond the tricritical line, various locally stable steady-states may exist for the same values of T_c, C and γ_b: they are identified by varying the initial conditions.

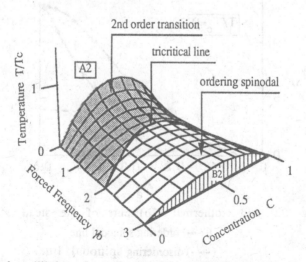

Fig.1: Dynamical equilibrium phase diagram for a BCC alloy in T, γ_b, C space. For clarity, when the transition is first order, only the spinodal is displayed [1].

With homogeneous initial conditions, either ordered or disordered homogeneous solutions (hereafter refereed to as OH and DH) are obtained, defining three domains (fig.2a): on going from high to low temperature, only DH, then both DH and OH (bistability domain) and finally only OH are stable.

With heterogeneous initial conditions, a two-phase system can be stabilized where an ordered phase coexists with a solute-depleted disordered one. We checked that the respective proportion of these two phases fulfils the lever rule. Two examples of steady-state diagrams so obtained are displayed on fig.2a and 2b (one for a given forcing and one for a given temperature).

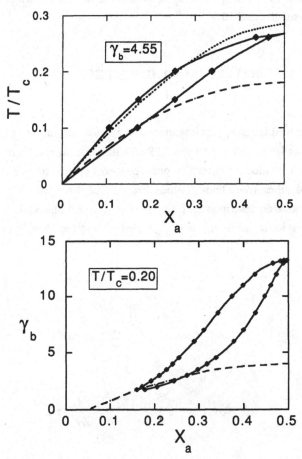

Fig.2. Isoforcing (2a) and isothermal (2b) cuts of the steady-state diagram.
(— —) ordering spinodal line
(- - -) disordering spinodal line

Although the assessment of the relative stability among the various steady-states here obtained requires a stochastic description, the local stability of these steady-states in response to a given "perturbation" can be studied. For instance the introduction of an antiphase boundary in the homogeneous ordered steady-state induces a decomposition when the representative point of the system (T/T_c, C, γ_b) lies in the two-phase field. (Similar effects have been reported for a thermal system by Chen and Khachaturyan [7]) The latter steady-state should therefore be the most stable.

Surface tension effects

In the two-phase region, spherical steady-state precipitates are observed in two dimensional description. A surface tension effect is identified: the A atomic concentration in both phases depends on the precipitate radius and an effective surface tension can be calculated by fitting the composition dependence with a *thermal* regular solution model consistent with the Bragg-Williams approximation (solid and dashed lines on Fig.3 correspond to this fit).

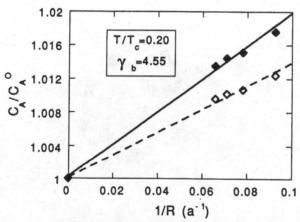

Fig.3. Atomic concentrations in ordered (♦) and disordered (◊) phases vs radius of curvature of the the ordered precipitate.

In the latter, the variation of chemical potential (e.g. in the ordered phase α) is given by:

$$\mu_A^\alpha = \mu_A^{o\alpha} + k_b T \ln C_A^\alpha + \Lambda(1-C_A^\alpha)^2 \tag{4}$$

where C_A^α is the average concentration in α and $\Lambda = T_c/2k_b$. For a two dimensional system, the variation of μ_A^α induced by a variation of pressure δp^α is :

$$\delta\mu_A^\alpha = \frac{\delta p^\alpha}{\Omega_A^\alpha} = \frac{\sigma}{4R\,\Omega_A^\alpha} \tag{5}$$

where $\Omega_A^\alpha = a^2/2C_A^\alpha$ is the A atomic volume in the phase α, σ is the effective surface tension, R is the radius of curvature and a is the lattice parameter.

For $T/T_c=0.20$ and $\gamma_b=4.55$ we obtained $\sigma=0.0337$ RT/a^2. With the typical values $T=1000$ K and $a=3$ Å, this would lead to $\sigma=5.2$ mJ/m^2. The order of magnitude of this value lies in the range of surface tension difference between ordered and disordered phases calculated in a *thermal* Ni_4Mo alloy [8].

Coarsening

Consistently with the surface tension effect, the system described above exhibits coarsening. We build a system containing two ordered precipitates with different curvatur Each of them is surrounded by a disordered phase, the composition of which is the steady-state value obtained with a single precipitate alone.

An interdiffusion flux appears which equalizes the composition in the whole disordered phase and the biggest precipitate is coarsening while the smaller one shrinks and disappears (Fig.4). The system exhibits no tendency for patterning at this scale.

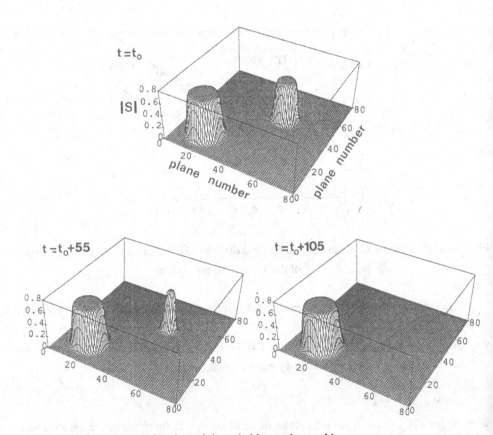

Fig.4. Coarsening of two ordered precipitates(arbitrary time unit)

Precipitate shearing under irradiation

We now study the stability of an ordered precipitate when an antiphase boundary (APB) is introduced, as would be the case after shearing by a dislocation. A disordered region is found to grow at the antiphase and the two antiphased precipitates undergo a diffusive repulsive interaction (fig.5). Finally the smaller shrinks and desappears to the benefit of the larger one.

Under irradiation, dislocation loops formed by point defect agglomeration are growing and thus could shear ordered precipitates: indeed Potter et al. [9] observed such a shearing of γ' precipitates in NiAl (12%at.Ni) under Ni^+ ion irradiation. The shearing was accompanied by a precipitate dissolution which was interpreted in terms of coupling between solute and vacancy fluxes towards the loop[10]. Our model, where point defects are not taken into account, offers an alternative explanation.

It would be of interest to find under which conditions this mechanism is still operating on a two-phase thermal system without external forcing.

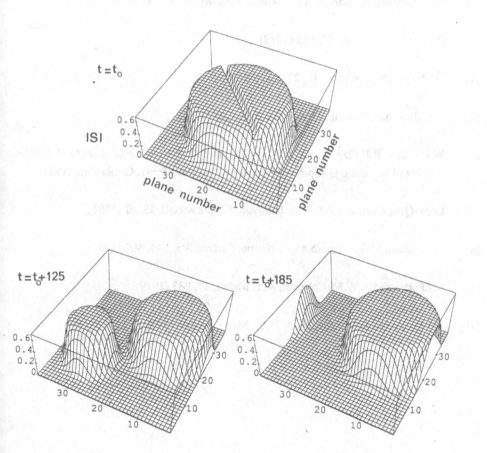

Fig.5. Evolution of an ordered precipitate after APB introduction (arbitrary time unit)

CONCLUSION

The kinetics of an ordered B.C.C. alloy under irradiation has been modeled by numerical integration of deterministic kinetic equations. Several new features have been discovered: irradiation may induce phase separation in a fully ordered compound, coarsening is observed in the two-phase field, an effective surface tension can be measured and it is found that ordered precipitates are unstable with respect to shearing. Further work is required for including vacancies and interstitials in the kinetic theory.

REFERENCES

[1]   P.Bellon and G.Martin, same volume and references therein.

[2]   H.Ackermann, G.Inden and R.Kikuchi, Acta Metall. **37**,1 (1989)

[3]   B.Fultz, Acta Metall. **37**, 823 (1989)

[4]   G.Martin, Phys. Rev. B **41**, 2279 (1990)

[5]   P.Bellon and G.Martin, Phys. Rev. B **39**, 2403(1989)

[6]   W.H.Press, B.P.Flannery, S.A.Teukolsky and W.T.Vetterling, Numerical Recipes: The Art of Scientific Computing (Cambidge University Press, Cambridge, 1986)

[7]   Long-Qing Chen and A.G.Khachaturyan, Scripta Metall. **25**, 61 (1991)

[8]   T.Kingetsu, M.Yamamoto and S.Nenno, Surface Sci. **144**, 402 (1984)

[9]   D.I.Potter and A.W.McCormick, Acta Metall. **27**, 933 (1979)

[10]   D.I.Potter and H.Wiedersich, J. Nucl. Mater.**83**, 208 (1979)

MONTE CARLO STUDY OF THE PHASE STABILITY
OF ORDERED FCC PHASES UNDER IRRADIATION

F. HAIDER

Institut für Metallphysik der Universität Göttingen
Hospitalstr. 3-5, D-3400 Göttingen, Germany
and SFB 345, Göttingen

ABSTRACT

The order-disorder transition in a binary fcc Ising-crystal with competing thermal and ballistic jumps, as is found e.g. under irradiation with high energy particles is studied by a Monte-Carlo technique. The state of order is characterized by a four-dimensional order parameter $\underline{X}$, the relative A-occupation of the four simple cubic sublattices (each consisting of Ω atoms) into which the fcc lattice may be decomposed. In a mean-field approximation, the transition rates $w(\underline{X} \to \underline{X}')$ between neighbouring states of order can be found. By integrating a trajectory in the order parameter space *stochastically*, i.e. with a suitable Monte-Carlo-method, the respective probability $P(\underline{X})$ of each state of order $\underline{X}$ can be computed. For a large system this yields a stochastic potential $\varphi(\underline{X}) = \frac{1}{\Omega} \log P(\underline{X})$. The simulation algorithm is based on a method proposed some years ago by Gillespie [8], where one computes the probability for the system to leave its current state through a "reaction channel" μ, in our case a transition between two special sublattices, after a delay time τ. Integration yields the total time in a certain state $\underline{X}$, i.e. its probability.

The results can be compared to recent analytical results, and it is found, that the cases where an analytical solution is found, are in good agreement with the above simulations. Further, the simulation technique allows for a treatment of more complicated cases, where analytical solutions up to now could not be found. Especially the case of compositions deviating from AB_3 can be treated, and a phase diagram in the (T, γ_0)-plane (T temperature, γ_0 temperature rescaled relative ballistic jump rate) for different compositions can be constructed.

I. INTRODUCTION

In binary alloys with fcc structure, one finds a series of ordered superstructures, the most simple ones being the $L1_2$- and the $L1_0$-phases. Under thermal (i.e. equilibrium) conditions, the relative stability of the respective phases is given by the free energy difference between these phases. Under nonequlibrium conditions, i.e. under an athermal perturbation, the thermal states can loose stability at the cost of other competing phases, namely, but not always the disordered one. Such external perturbations may be irradiation with energetic particles [1] or strong deformation [2]. The stability is no longer governed by the thermodynamic potentials like free energy, but has to be calculated by other methods.

In previous work, Martin and Bellon [3,4] developed a method to construct analytically a stochastic potential, the extrema of which represent the stable structures. They started from a mesoscopic kinetic description of the time evolution of the order parameter, and derived a master equation for the probability of each state of order. In [5] this method was generalized to cases, where one must describe the order by a multidimensional parameter, as is the case for a unique description of different ordered phases on the fcc lattice. The master equation was approximated by a nonlinear partial differential equation, which originally was derived by Kubo et al [6]. For a special case, i.e. an AB_3-stoichiometry, analytic solutions could be found, and it was possible to compute the stability of the $L1_2$-phase for different values of the irradiation parameter and different temperatures. Especially it was found, that under certain conditions explained below, other ordered phases like the $L1_0$- and a nonstoichiometric $L1_2$-phase could be stabilized in a small range of the temperature-irradiation-field.

The main shortcoming of that method was, that only under very special symmetry conditions a solution of Kubo's equation could be constructed and, if those conditions are not fulfilled, a numerical solution of this equation seems to be rather tedious if not impossible. Therefore in the following we present a numerical method which circumvents all these problems. The method is based on a Monte Carlo technique originally proposed by Gillespie [7-9]. The main idea is to calculate the random walk trajectory of a finite system in the order parameter space and to measure the residence time in each state of order to obtain the respecitve probability of the competing phases.

The article proceeds as follows: first we introduce the thermodynamic model, then we construct the kinetic equivalent. In part IV, the simulation method is explained, and some results are presented in part V.

II. THE MODEL

The simplest ordered structures of a binary alloy on the fcc-lattice of composition AB_3, i.e. $L1_2$ and $L1_0$, can be described by the average A-occupation of the simple cubic sublattices, into which the fcc lattice can be decomposed. Using the vector $\underline{X}$ of concentrations x_i of A-atoms on sublattice i:

$$x_i = \frac{A_i}{\Omega}, \qquad i = 1,...,4 \tag{1}$$

(Ω: number of sites per sublattice), as four dimensional order parameter, the free energy in a Bragg-Williams approximation reads [10]:

$$F(\underline{X}) = F_0 - \Omega kT \sum_{i=1}^{4} f(x_i),$$
$$f(x) = 2\frac{\omega}{kT}x^2 - (x\log x + (1-x)\log(1-x)), \tag{2}$$
$$\omega = v_{AA} + v_{BB} - 2v_{AB},$$

where v_{ij} is the contribution of an ij-pair to the ordering energy. The stable or metastable states, the minima of F under the constraint

$$\sum_{i=1}^{4} x_i = 4c \tag{3}$$

are given as solution of

$$4\frac{\omega}{kT}(x_i - \frac{1}{4}) - \log\frac{x_i}{1-x_i} = \frac{1}{4}\sum_{i=1}^{4}\log\frac{x_j}{1-x_j}, \qquad i=1,...,4 \tag{4}$$

III. KINETIC DESCRIPTION

As described in more detail in [5] the thermodynamically stable states can as well be found as the infinite time solutions of kinetic equations. Such a kinetic description allows easily to include nonthermal processes as e.g. forced atomic exchanges under irradiation. To construct a kinetic model we suppose as elementary step an exchange of one A-atom on sublattice i with a B-atom on sublattice k,

$$x_i \rightarrow x_i - \frac{1}{\Omega}, \qquad x_k \rightarrow x_k + \frac{1}{\Omega},$$

in vector notation

$$\underline{X} \rightarrow \underline{X} + \underline{\epsilon_{ik}}, \qquad (\underline{\epsilon_{ik}})_j = \frac{1}{\Omega}(\delta_{kj} - \delta_{ij}).$$

Then the kinetic equations can be written as

$$\frac{\partial}{\partial t} x_i = \sum_{i=1}^{4} w(\underline{X} \to \underline{X} + \underline{\epsilon}_{ki}) - \sum_{i=1}^{4} w(\underline{X} \to \underline{X} + \underline{\epsilon}_{ik}). \tag{5}$$

Starting from a microscopic model for the transition process, and introducing the same mean field approximations as for the thermodynamic model, one can derive the following expression for the transition rates:

$$w(\underline{X} \to \underline{X} + \underline{\epsilon}_{ik}) = \frac{Z}{3} x_i (1 - x_k) \Gamma_{ik}(\underline{X}), \tag{6}$$

where $Z = 12$ is the coordination number of the fcc lattice and $\Gamma_{ik}(\underline{X})$ the atomic jump frequency between lattice sites of sublattice i and sublattice k. If only thermal jumps are allowed, Γ_{ik}^{th} is proportional to the exponential of the energy E_a necessary to lift an A- and a B-atom to the saddle point position, divided by kT:

$$\Gamma_{ik}^{th}(\underline{X}) = \Gamma_0 \exp -\frac{E_a(\underline{X})}{kT}. \tag{7}$$

In [5], an expression for E_a was derived

$$E_a(\underline{X}) = \frac{Z}{6} \omega (x_i - x_k) + 2\omega_0 (1 - x_i - x_k) + 2E_0 \tag{8}$$

with

$$\omega_0 = (v_{AA}^* - v_{AA}) - (v_{BB}^* - v_{BB}),$$

v_{ij}^* denoting interaction energies in the saddle point position. The first part is the energy necessary to extract an $A-$ and a $B-$atom from the lattice, the second term represents the energy of these atoms in the saddle point position. Allowing now for additional forced exchanges, the atomic jump frequency becomes

$$\Gamma_{ik}(\underline{X}) = \Gamma_{ik}^{th}(\underline{X}) + \Gamma^{ball}, \tag{9}$$

the latter being independent of the configuration.

This deterministic description of the evolution of the degree of order leads to unique infinite time solutions only, if no metastable states exist, that is, for a system with first order phase transformations beyond the respective spinodal lines. As soon as one finds more than one stable or metastable state, one has to determine the relative stability of the corresponding phases. While for thermal systems this is easily done by comparing the values of the free energy, for nonequilibrium systems a priori no such function is known. But computing the relative probabilities $P(\underline{X})$ to find the system in state $\underline{X}$,it can be shown [6], that for system size $\Omega \to \infty$, a nonequilibrium potential φ can be found as

$$\varphi(\underline{X}) = -\frac{1}{\Omega} \log P(\underline{X}). \tag{10}$$

$P(\underline{X})$ can be found either by solving directly a master equation

$$\frac{\partial}{\partial t}P(\underline{X},t) = \Omega \sum_{\underline{X}'} P(\underline{X}',t)w(\underline{X}\rightarrow\underline{X}') - P(\underline{X},t)w(\underline{X}'\rightarrow\underline{X}), \tag{11}$$

or approximations [3-5], or it can be computed by a simple numerical procedure to be described in the next chapter.

IV. STOCHASTIC INTEGRATION

Following Gillespie [7-9] we regard the time evolution of the order parameter as a random walk in the order parameter space, i.e. in the space of sublattice occupation. The transition rates per time interval are given by eq. (6), but now one has (a) to choose one transition ϵ_{ik} according to its relative probability, (b) to calculate the residence time in the current state of sublattice until a transition occurs. This means calculating the probability $\Pi_{ik}(\underline{X},\Delta t)$ that, provided the system is in state $\underline{X}$ it leaves this state after a time Δt via a transition $\underline{X}\rightarrow\underline{X}+\epsilon_{ik}$. As shown in [7], Π_{ik} reads

$$\Pi_{ik}(\underline{X},\Delta t) = \Pi_0(\Delta t)\cdot w(\underline{X}\rightarrow\underline{X}+\epsilon_{ik}),$$
$$\Pi_0(\Delta t) = \exp -\sum_{i,k} w(\underline{X}\rightarrow\underline{X}+\epsilon_{ik}), \tag{12}$$

for large enough systems. This method, which is similar to one proposed earlier by Bortz, Lebowitz and Kalos [11], can be implemented easily: choosing an arbitrary initial state $\underline{X}_0$ at time t_0, one finds the time increment as

$$\Delta t = -\frac{1}{\sum_{i,k} w_{i,k}} \log r_1 \tag{13}$$

and the transition channel (i,k) through which the system leaves its current state

$$\sum_{\nu=1}^{n-1} w_{(ik)_\nu} \leq r_2 \leq \sum_{\nu=1}^{n} w_{(ik)_\nu}, \tag{14}$$

r_1, r_2 being two uniformly distributed random numbers from the unit interval. The probability of a certain state $\underline{X}$ is given by the total residence time $\tau(\underline{X})$ in this state, divided by the total integration time:

$$P(\underline{X}) = \frac{\tau(\underline{X})}{\sum_{\underline{X}} \tau(\underline{X})}, \tag{15}$$

provided, the system is ergodic.

V. Results and Discussion

The above described method is a powerful tool to compute nonequilibrium properties of open systems like relaxation behaviour, fluctuation strength, transition times etc. Nevertheless we restrict ourselves to the calculation of steady state probabilities, i.e. phase diagrams in the (T, γ_0)-plane.

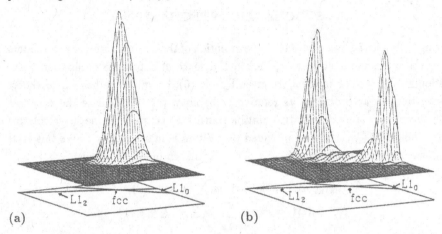

(a)   (b)

Figure (1): Probability surface $P(\underline{X})$ in a twodimensional cut of the order parameter space. $c_A = 30\%, \gamma_0 = 10^{-5}$, (a) $T/T_c = 0.41$ (b) $T/T_c = 0.44$

The simulations were performed for systems with 200 A-atoms for a set of different concentrations. Usually, the typical integration time was between 1 and 10 million steps. Figure (1) shows an example of the obtained probability functions, a cut through the fourdimensional order parameter space at $(x_1 = x_2; x_3; x_4 = 4c_A - x_1 - x_2 - x_3)$ for two temperatures. Below the lower critical temperature only the disordered state is stable, while above that temperature the $L1_2$- and the $L1_0$-phases are stable. Figure (2) shows a set of phase diagrams for A-concentrations $c_A = 15\%, c_A = 25\%, c_A = 35\%$ and $c_A = 50\%$ for two values of the saddle point parameter ω_0.

For $\omega_0 = 0$ at $c_A = 15\%$ and at $c_A = 25\%$ the only ordered structure found is the $L1_2$-phase ($x_1 = x_2 = x_3 < x_4$). For $c_A = 35\%$ additionally an ordered phase with two sublattices equally and the other differently occupied ($x_1 = x_2 < x_3 < x_4$) (here denoted X-phase) appears, and at $c_A = 50\%$ only the $L1_0$-phase ($x_1 = x_2 < x_3 = x_4$) is stable. For $\omega_0 = 0.5$ the behaviour changes drastically: except at $c_A = 15\%$ always new phases appear at low temperature. In all three cases first a phase called $L1_2^*$ forms, a nonstoichiometric $L1_2$-phase with sublattice occupation ($x_1 = x_2 = x_3 > x_4$). Then, depending on the concentration, other phases can be found. This behaviour was already observed in [5], but in the analytical treatment there, it was not possible to consider the X-phase, which arises as a metastable phase above $c_A = 30\%$.

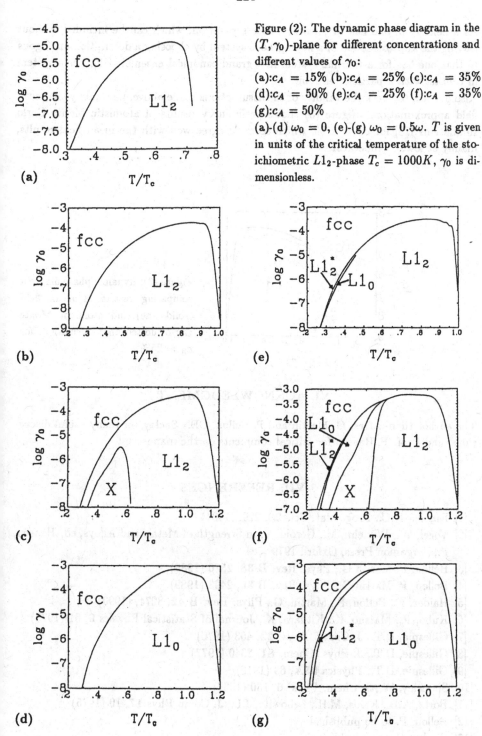

Figure (2): The dynamic phase diagram in the (T, γ_0)-plane for different concentrations and different values of γ_0:

(a):$c_A = 15\%$ (b):$c_A = 25\%$ (c):$c_A = 35\%$
(d):$c_A = 50\%$ (e):$c_A = 25\%$ (f):$c_A = 35\%$
(g):$c_A = 50\%$
(a)-(d) $\omega_0 = 0$, (e)-(g) $\omega_0 = 0.5\omega$. T is given in units of the critical temperature of the stoichiometric $L1_2$-phase $T_c = 1000K$, γ_0 is dimensionless.

As a next step, one has to consider phase separation, which can be included without leaving the approximation of a homogenous system by choosing a description analogous to that one has for a thermal system in a grand canonical ensemble [12], i.e. a system in contact with a particle reservoir.

Also one has to check the validity of the assumptions made above, i.e. mainly the mean field approximation. Figure (3) shows preliminary results of atomistic Monte Carlo simulations [13] for $c_A = 25\%$, which seem to agree well with the mean field results, mainly at lower temperatures.

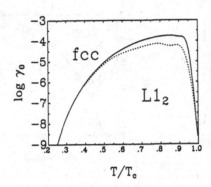

Figure (3): dynamic phase diagram comparing results of mean field (solid line) and atomistic Monte Carlo (dotted line) calculations for $c_A = 25\%$.

VI. ACKNOWLEDGMENT

The author thanks Drs. G. Martin and P. Bellon, CEN Saclay, for many useful discussions, and Prof. P. Haasen for critical comments on the manuscript.

VII. REFERENCES

[1] Russell, K.C., Prog. Mat. Sci. **28**, 229, (1984)

[2] Vogel, W., Wilhelm, M., Gerold, V. in Strength of Metals and Alloys, ed. Haasen, P., Pergamon Press, Oxford 1979

[3] Bellon, P. Martin G., Phys. Rev. B **38**, 2570 (1988)

[4] Bellon, P. Martin G., Phys. Rev. B **39**, 2403 (1989)

[5] Haider, F., Bellon, P., Martin, G., Phys. Rev. B **42**, 8274, (1990)

[6] Kubo, R., Matsuo, K., Kithara, K., Journal of Statistical Physics **9**, 51 (1973)

[7] Gillespie, D.T., J. Comp. Phys. **22**, 403 (1976)

[8] Gillespie, D.T., J. Phys. Chem. **81**, 2340 (1977)

[9] Gillespie, D.T., Physica **95A**, 69 (1978)

[10] Shockley, W., J. Chem. Phys. **6**, 130 (1037)

[11] Bortz, A.B., Kalos, M.H., Lebowitz, J.L., J. Comp Phys **17**, 10 (1975)

[12] Bellon, P. to be published

[13] Haider, F. to be published

LMTO/CVM Calculations of Partial BCC Based Phase Relations in the System Ni-Al-Ti

BENJAMIN P. BURTON
and
W. C. CARTER
Materials Science and Engineering Laboratory NIST, Gaithersburg, MD 20899, USA

ALAIN PASTUREL
Laboratoire de Thermodynamique et Physico-Chimie Metallurgiques E.N.S.E.E.G.
B. P. 75- 38402 St-Martin-D'Heres-Cedex France

ABSTRACT

Linear Muffin Tin Orbital calculations of equations of state were performed for observed and hypothetical BCC based ordered structures in the Ni-Al-Ti system. Total energies were parameterized in the Connolly-Williams and ϵ-G approximations. The resulting effective cluster interactions (ECI) were used to calculate: 1) BCC based ground-states; 2) The A2→B2 and B2→L2$_1$ order-disorder transition temperatures; 3) Quasibinary phase diagrams for the NiAl-NiTi join. The BCC octahedron approximation of the cluster variation method (CVM) was used to calculate $T_c(L2_1 \to B2)$ and $T_c(B2 \to A2)$. The simple cubic "cube approximation" of the CVM was used to calculate quasibinary NiAl-NiTi phase diagrams.

INTRODUCTION

The pseudobinary[1] NiAl-NiTi (1-7) system is interesting from both practical and theoretical points of view. Its high-temperature creep strength (4) is comparable to that of superalloys which makes it potentially attractive for high-temperature applications, and its metastable solid-state phase relations appear to be of the conditional spinodal/tricritical point type (2). The pseudobinary diagram (Fig. 1a) has a central field for the Heusler alloy based on Ni$_2$AlTi (L2$_1$), and broad B2 + L2$_1$ two-phase fields (NiAl + Ni$_2$AlTi, Ni$_2$AlTi

---

[1] The real system is "pseudobinary" which means that the atomic fraction of Ni may depart from $\frac{1}{2}$. The calculated diagram is "quasibinary" because the atomic fraction of Ni is fixed at $\frac{1}{2}$.

+ NiTi) that extend up to the solidus. Experimentally determined tielines (5-7) confirm the presence of broad B2 + L2$_1$ fields, and "QDTA (6)" experiments have located the liquidus and solidus curves in the range 0<X$_{NiTi}$<0.8. Based on rapid solidification experiments, Boettinger et al. (2) concluded that there must be metastable tricritical points above the liquidus (P & Q Fig. 1a), and our calculations support their conclusion.

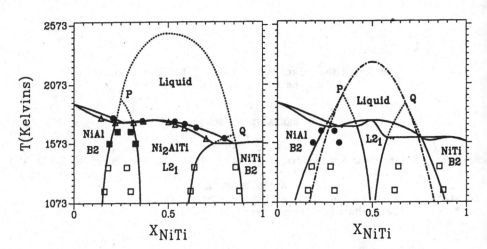

Figure 1: a) A drawing of the NiAl-NiTi pseudobinary phase diagram: Open (5) and filled (7) squares indicate tielines; Liquidus (●) and solidus (△) data are from reference (6); b) Calculated (CVM SC-CA) quasibinary phase diagrams. Dotted and dot-dashed lines indicate metastable phase boundaries.

A previous paper (8) presented quasibinary calculations of the NiAl-NiTi pseudobinary phase diagram that were based on the simple cubic (SC) "cube approximation" (CA) of the cluster variation method (CVM; Fig. 1b). The SC-CA is equivalent to assuming that negligible disordering of Ni occurs up to the L2$_1$→B2 transition in Ni$_2$AlTi. That is, if the Ni atoms are "frozen" on one SC substructure of the B2 superstructure, then the problem reduces to quasibinary ordering of Al and Ti on the remaining SC substructure. This assumption was justified on the basis of the large difference between the calculated formation energy for the L2$_1$ Heusler phase:

$$\Delta E_0(Ni_2AlTi) = \Delta E(NiAl_{BCC}+NiTi_{BCC}\rightarrow Ni_2AlTi_{L2-1}) \approx -13kJ/gram\text{-}atom$$

and the formation energies of the B2 phases from BCC elements:

$$\Delta E_0(NiAl) = \Delta E(Ni_{BCC}+Al_{BCC}\rightarrow NiAl_{B2}) \approx -85kJ/gram\text{-}atom,$$

$$\Delta E_0(NiTi) = \Delta E(Ni_{BCC}+Ti_{BCC}\rightarrow NiAl_{B2}) \approx -53kJ/gram\text{-}atom.$$

$$\Delta E_0(AlTi) = \Delta E(Al_{BCC}+Ti_{BCC}\rightarrow AlTi_{B2}) \approx -21kJ/gram\text{-}atom$$

The critical temperature for the L2$_1$→B2 transition, T$_c$(L2$_1$→B2), is approximately proportional to ΔE_0(Ni$_2$AlTi), and T$_c$(B2→A2) is approximately proportional to [ΔE_0(NiAl)+ΔE_0(NiTi)-2ΔE_0(AlTi)]/4. Therefore, one expects T$_c$(L2$_1$→B2) ≈ 0.54T$_c$(L2→A2) and the long-range order (LRO) parameter that is associated with the disordering between the two SC substructures $[\eta(B2\rightarrow A2)]$ should be almost saturated at the $L2_1\rightarrow B2$ transition.

In this paper we present the results of fully ternary BCC based calculations that include: 1) Formation energies for 11 ternary ordered phases (10 hypothetical + one observed); 2) A BCC based ground-state analysis based on a Connolly and Williams (9) parameterization of the 11 ternary plus 12 binary formation energies; 3) A full ternary BCC based calculation of the order-disorder transitions in Ni_2AlTi.

LMTO ASA CALCULATIONS

Linear Muffin Tin Orbital (LMTO, 10, 11) atomic sphere approximation (ASA) calculations of cohesive energies were done using the code of Van Schilfgaarde et al. (12) with Von Barth-Hedin (13) exchange-correlation energy density as given by Moruzzi et al. (14). We used a uniform mesh of sampling points with at least 12 divisions on each primitive vector. Equal Wigner-Seitz radii were used for all elements, and spherical harmonics up to $1=2$ (d orbitals) were used to construct the basis functions. Equations of State (EOS) were calculated for a total of 26 BCC derived ordered structures: the three BCC elements; DO3, B2, and B32 structures on each binary join; 11 ternary structures. Calculated minimum energies of formation (ΔE_0), and volumes per atom (V_0) are tabulated in Table 1; these were obtained by fitting quadratic polynomials to calculated EOS.

Table 1: Observed and hypothetical BCC based $Ni_{8-m-n}Al_mTi_n$ ordered structures

Formula	Structure type	ΔE_0 kJ/gram--atom	V_0 au³/atom
Ni	A2	0	72.35
Al	A2	0	105.82
Ti	A2	0	110.94
NiAl	B2	-84.80	78.39
NiTi	B2	-53.05	89.10
AlTi	B2	-20.88	105.96
NiAl	B32	-46.64	79.93
NiTi	B32	-42.64	83.78
AlTi	B32	-15.74	104.89
Ni_3Al	DO3	-50.28	73.24
$NiAl_3$	DO3	-21.09	92.02
Ni_3Ti	DO3	-39.07	77.44
$NiTi_3$	DO3	-16.56	96.44
Ti_3Al	DO3	-24.27	107.93
$TiAl_3$	DO3	- 7.02	104.97
Ni_4Al_3Ti	cI16*	-76.75	81.15
Ni_4AlTi_3	cI16	-61.79	85.54
Ni_3Al_4Ti	cI16	-52.57	84.92
$NiAl_4Ti_3$	cI16	-31.27	98.97
Ni_3AlTi_4	cI16	-40.56	92.12
$NiAl_3Ti_4$	cI16	-24.67	101.85
Ni_8Al_7Ti	cP16**	-81.84	79.89
Ni_8AlTi_7	cP16	-58.10	87.38
Ni_2AlTi	$L2_1$	-81.46	83.25
$NiAl_2Ti$	$L2_1$	-40.44	90.69
$NiAlTi_2$	$L2_1$	-30.02	97.49

In Pm3m: cI16* has 8Ni at $(\frac{1}{4},\frac{1}{4},\frac{1}{4})_{8g}$; 3Al at $(0,\frac{1}{2},\frac{1}{2})_{3c}$ + 3Al at $(\frac{1}{2},0,0)_{3d}$; Ti at $(0,0,0)_a$ + $(\frac{1}{2},\frac{1}{2},\frac{1}{2})_b$. In cP16** occupancies are the same as for cI16 above except that the Ti at $(\frac{1}{2},\frac{1}{2},\frac{1}{2})$ is replaced by Al (8).

TERNARY ECI AND THE BCC CONVEX HULL

The formation energies listed in Table 1 were used to calculate ternary effective cluster interactions (ECI; Table 2) for the first three near neighbor (nn) pairs plus the triangle that includes two first-nn pairs plus one second-nn pair. The ν-matrix coefficients were calculated according to the procedure outlined in Kikuchi et al. (15). That is, one writes elements of the ν-matrix as products of the operators B_i and C_i where: $i = -1, 0, 1$ for components Al, Ni, Ti respectively; $B_i = i/2$; $C_i = (3/2)i^2 -1$. For example, a nn pair in configuration i--j has ν-matrix elements of the form B_iB_j, $(B_iC_j+C_iB_j)$, and C_iC_j. The specific set of parameters listed in Table 2 was obtained by a Connolly and Williams type parameterization (9) which implies that volume, and concentration dependence of the ECI are ignored.

From the parameters listed in Table 2 a BCC-based convex hull analysis was performed. [If only BCC derived phases were stable this would be equivalent to a ground-state (GS) analysis]. A total of 84 candidate phases at 63 distinct composition points in the Ni-Al-Ti space were considered. The results are plotted in Figures 2a and 2b which are projections on a plane perpendicular to the formation energy axis. Figure 2a includes only those structures for which LMTO cohesive energies were calculated. Figure 2b includes all structures that were considered in the GS analysis. Large circles that are connected by solid lines indicate vertices of the convex hull. Smaller circles indicate points (or superimposed stacks of points) that lie above the convex hull. Note that the L2$_1$ phase is the only *ternary* BCC derived ordered phase that lies on the convex hull.

ORDER-DISORDER TRANSITIONS IN Ni$_2$AlTi

Figure 3 was calculated by using the ECI in Table 2 as input for an octahedron approximation (OA) CVM calculation. Squares (■) indicate calculated values of the long-range order parameter [η(L2$_1 \rightarrow$B2)] for the L2$_1 \rightarrow$B2 order→disorder transition, and circles (●) indicate calculated values of the LRO parameter [η(B2→A2)] for the B2→A2 transition. The experimentally determined melting temperature is also indicated (T$_m$). Very little disordering of Ni is predicted up to T$_c$(L2$_1 \rightarrow$B2)$\approx$0.57T$_c$(B2→A2), where η(B2→A2)$\approx$0.96 which suggests that the SC quasibinary approximation should be reasonably good. Note however, that the value of T$_c$(L2$_1 \rightarrow$B2)$_{BCC-OA}$$\approx$1900K predicted by the ternary BCC-OA is significantly lower than that predicted by the quasibinary SC-CA, T$_c$(L2$_1 \rightarrow$B2)$_{SC-CA}$$\approx$2430K. Apparently, the SC-CA introduces a significant quantitative error even though η(B2→A2) is nearly saturated at T$_c$(L2$_1 \rightarrow$B2).

DISCUSSION AND CONCLUSIONS

In the SC-CA (8) the quasibinary ECI that are associated with second and third nn Al-Ti pair interactions (corresponding to Al——Ti and Al------Ti respectively in Table 2) actually have signs that are opposite to those in Table 2. This apparent contradiction emphasizes the *effective* character of the ECI, and the importance of volume dependence in them. In (8) it was argued that both elastic contributions (volume dependence) and repulsive third nn Al------Ti pair interactions promote immiscibility between B2 and L2$_1$ phases. The results presented here, however, suggest that this conclusion should be revised to: a combination of elastic (volume dependent) and many-body interactions are responsible for the observed immiscibility between B2 and L2$_1$ phases.

Table 2: Ternary ECI for the BCC based Ni-Al-Ti system.

i	J(i)	Cluster
1	14.61	Ni—Al
2	6.02	Al—Ti
3	10.53	Ni—Ti
4	7.00	Ni——Al
5	-1.08	Al——Ti
6	-1.65	Ni——Ti
7	-0.43	Ni------Al
8	1.39	Al-----Ti
9	-3.01	Ni------Ti
10	-5.75	(B_i)——(B_i) \ (B_i)
11	-4.93	(B_i)——(B_i) \ (C_i)
12	1.44	(C_i)——(Bi) \ (B_i)
13	-0.32	(C_i)——(C_i) \ (B_i)
14	-0.01	(B_i)——(C_i) \ (C_i)
15	-2.32	(C_i)——(C_i) \ (C_i)

Here: $B_i = i/2$; $C_i = (3/2)i^2-1$; where $i = -1, 0, 1$ for Al, Ni, Ti respectively (15).

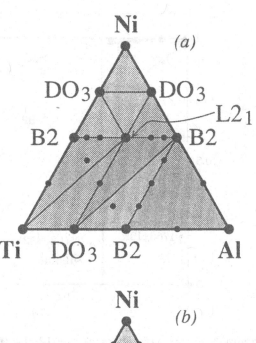

(a)

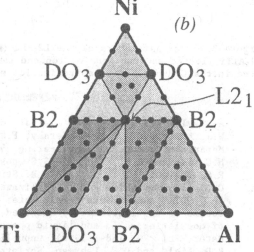

(b)

Figure 2: Projections of the BCC convex hull in which large circles are vertices and smaller circles lie above the convex hull;
a) Composition points at which LMTO calculations were performed;
b) All points considered in the BCC based ground-state analysis.

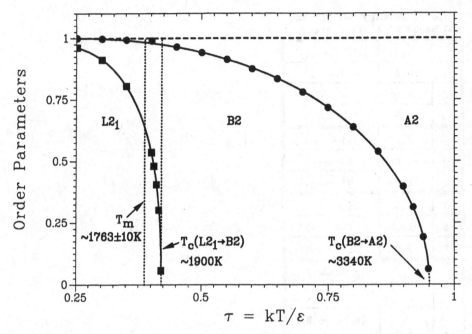

Figure 3: Order parameters of the $L2_1 \to B2$ (■) and $B2 \to A2$ (●) transitions in Ni_2AlTi plotted as functions of reduced temperature (τ). Calculations were done in the CVM octahedron approximation with the ECI in Table 2.

REFERENCES

1.   A. Taylor and R.W. Floyd, J. Inst. Metals, 81, 25- (1953).
2.   W.J. Boettinger, L.A. Bendersky, F.S. Biancaniello, and J.W. Cahn, Materials Science and Engineering, 98 273- (1988).
3.   N.C. Tso and J.M. Sanchez MRS Symposium Series V133, 63- (1989).
4.   R.S. Polvani, W. Tzeng, and P.R. Strutt, Met. Trans. 7A, 33- (1976).
5.   C.C. Jia, K. Ishida and T. Nishizawa CALPHAD XIX 82 (1990).
6.   In "QDTA" experiments samples are: 1) melted in a DTA apparatus; 2) cooled to the liquidus; 3) Quenched and analyzed for the compositions of coexisting solid and liquid phases. e.g. M. Durand-Charre J. Mater. Sci. 25, 168- (1990); unpublished NiAl-NiTi data in Figure 1a.
7.   R.D. Field and D.F. Lahrman, Scripta. Met. 23, 1469- (1989).
8.   B.P. Burton, J.E. Osburn, and A. Pasturel, in "High-Temperature Ordered Intermetallic Alloys IV", L.A. Johnson, D.P. Pope, and J.O. Stiegler Eds. MRS Symposium proceedings, 107- (1991).
9.   J.W.D. Connolly and A.R. Williams Phys. Rev. B27, 5169- (1983).
10. H. L. Skriver <u>The LMTO Method: Muffin-Tin Orbitals and Electronic Structure.</u> Springer-Verlag (1984).
11. O.K. Andersen Phys. Rev. B12, 3060- (1975).
12. M. Van Schilfgaard, A.T. Paxton, M. Methfenel, O. Jepsen and O.K. Andersen, private communication (A.P.).
13. U. Von Barth and L. Hedin, J. Phys. C. 5, 1629- (1972).
14. V.L. Moruzzi, J.F. Janak, and A.R. Williams, Calculated Electronic Properties of Metals, Pergamon, NY, (1978).
15. R. Kikuchi, J.M. Sanchez, D. defontaine, and H. Yamauchi Acta Met. 28, 651- (1980).

STATISTICAL THERMODYNAMICS OF ORDERING SYSTEMS:
AN IMPROVED QUASI CHEMICAL APPROXIMATION

P.L.CAVALLOTTI,M.ALBERTI,B.BOZZINI,A.IUDICA,L.NOBILI',P.OSSI=
'Dip.Chimica Fisica Applicata,Politecnico, P.za L.da Vinci 32,
 20133 Milano, Italy.
=CESNEF,Politecnico,Milano & Unita' cINFM,Trento, Italy

ABSTRACT

A statistical thermodynamic approximation for cooperative
phenomena in model systems is presented, to improve the quasichemical constant coupling approximation by introduc i n
a second pairs quasichemical reaction after averaging atomic
groups. By means of two suitable temperature functions related
to fluctuations, an analytical expression of the free energy
function is obtained. The first parameter takes into account
the change of the order parameter after averaging and the decreasing importance of the second reaction at increasing temperatures. The second one is a correction of the system temperature as a consequence of fluctuations' influence on the mean
thermodynamic behaviour. Relations to obtain the right critical exponents and well approximated thermodynamic critical quantities for ordering systems are given. The approximation is
applied to real systems, such as beta brass, taking into account next nearest neighbour interaction.

INTRODUCTION

The critical behaviour of systems with cooperative interactions depends on fluctuations over as many as possible
scales of length. Proposed statistical thermodynamic approximations do not give a suitable description of critical phenomena, because they start from a definition of low temperature ordered and high temperature well disordered systems.
We have previously shown [1-3], that it is possible, by modi-
fying the well known QuasiChemical (QC) approximation (4), to
obtain critical parameters for Ising systems:

$$K_o = -\Delta \mathcal{E}/(2k_B T_o) \qquad (1)$$

very close to exact or best known values, where $\Delta \mathcal{E}$ is the in-

teraction energy for neighbour pairs, k_B the Boltzmann constant and T_c the critical temperature.
We introduce a second QC equilibrium, taking into account interaction between atomic groups. By this additional interaction term we take into account structural connections in the lattice. In the QC approximation only direct atomic interactions are considered, supposing the system as formed by open chains; chain closures with polygon formation are encompassed in the new approximation by introducing the second QC reaction between averaged atomic groups.
To take into account fluctuations in the system, we have introduced two suitable temperature functions in the free energy analytical expression. We have tried to point out the constraints in the two functions, when they are applied in the critical range, to obtain the proper critical behaviour. We have also applied the new approximation to obtain an expression of the free energy function for the order-disorder transformation in β-brass.

THE IMPROVED QUASICHEMICAL APPROXIMATION

When the QC approximation is applied to an Ising system, a QC reaction between atomic pairs at $\tau = T/T_c$ is considered with a QC equilibrium constant:

$$AA + BB = 2 \cdot AB \qquad K_{eq} = \exp(-2 \cdot K_o/\tau) \qquad (2)$$

The pair number in the mean field approximation (MFA) is modified dividing it by the correction factor: $(1+\beta)$, with:

$$\beta = [1 + (1-\eta^2)(\exp 2K_o/\tau - 1)]^{1/2} \qquad (3)$$

where z is the coordination number, N_o the number of lattice sites and $\eta = 2N_A/N_B - 1$ the order parameter, obtaining:

$$N_{AB} = zN_o(1-\eta^2)/(4(1+\beta)) \qquad (4)$$

With reference to Kadanoff's construction of block atoms [5], even at T_c two types of interaction are present: direct interaction of neighbour atom pairs and interaction between atomic groups represented by fixed atoms.
After averaging n atom interactions we obtain a new lattice of fixed atoms, related to the original one according to the marginal dimensionality concept [6], that is, if the starting lattice is 2D the new lattice is its 2D dual, while if it is a 3D lattice the new one will be 1D with coordination number $z_I = 2$.
The new QC equation between atomic groups is:

$$(AA + BB = 2 \cdot AB)_I \qquad K_{eq}^I = \exp(-4 \cdot K_o/\tau) \qquad (5)$$

The number of AB couples becomes:

$$N_{AB}^I = (z_I/n) \cdot N_o \cdot (1-\eta_I^2)/\{4(1+\beta_I)\} \qquad (6)$$

with $\eta_I = \eta/\sqrt{\varphi}$ and:

$$\beta_x = \left[1 + (1 - \eta_x{}^2)(\exp 4K_o/\tau - 1)\right]^{1/2} \qquad (7)$$

$\varphi(\tau)$ is a correction factor taking into account the change of the order parameter after averaging and the decreasing importance of the second term at increasing temperature, if $\tau < 1$. Another temperature dependent function is introduced to take into account the influence of fluctuations on the system thermodynamic behaviour; replacing τ in K_{eq} by $\rho(\tau)$ we obtain:

$$K_{eq} = \exp(-2 \cdot K_o / \rho) \qquad (8)$$

The partition function for an Ising system, neglecting internal freedom degrees, becomes:

$$Q = \exp\left[-\frac{U_o}{k_s T} - \frac{N_{AB}K_o o}{k_s o} - \frac{2\varphi N_{AB}{}^x K_o o}{k_s o}\right] (W \cdot W_x{}^\varphi \cdot h)^\delta \qquad (9)$$

with:

$$W = \frac{(z\,N_o/2)!}{N_{AA}! N_{BB}! N_{AB}! N_{BA}!} \qquad W_x = \frac{(z_x N_o/2/n)!}{N_{AA}{}^x! N_{BB}{}^x! N_{AB}{}^x! N_{BA}{}^x!}$$

these are the number of pair configurations in the original lattice and after averaging, and with the normalizing factors:

$$h = \left[N_o! / (N_A! N_o!)\right]^{2-x-\delta} \qquad \delta = 1 + \varphi \cdot z_x/2n \qquad (10)$$

The configurational free energy function is obtained from:

$$F_o = -N_o k_s T \cdot \ln Q$$

and is given by:

$$\frac{F_o}{N_o k_s T_o} = \frac{\tau}{2o}\left\{(2-z)\left[\ln\frac{1-\eta^2}{4} + \eta \ln\frac{1+\eta}{1-\eta}\right] + z\left[\ln\frac{\beta^2-\eta^2}{(1+\beta)^2} + \eta \ln\frac{\beta+\eta}{\beta-\eta}\right] + \right.$$

$$\left. + (z_x/n)\varphi\left[\ln\frac{\beta_x{}^2-\eta_x{}^2}{(1+\beta_x)^2} + \eta_x \ln\frac{\beta_x+\eta_x}{\beta_x-\eta_x}\right]\right\} - \frac{zK_o}{4} \qquad (11)$$

In this function we take into account the ideal configurational free energy, the QC correction and the group correction, reflecting the presence of symmetry constraints in the lattice. The last correction is introduced to improve the QC approximation and gives the IQC approximation. As a consequence of long range correlations the configurational entropy results from the three contributions, with the further introduction of the temperature correction factor.
From the first derivative of F_o in eq.11 with respect to the order degree n, set equal to zero, we obtain:

$$(2-z)\ln\frac{1+\eta}{1-\eta} + z \cdot \ln\frac{\beta+\eta}{\beta-\eta} + \sqrt{\varphi}(z_x/n)\ln\frac{\beta_x+\eta_x}{\beta_x-\eta_x} = 0 \qquad (12)$$

The derivative of F_o/τ with respect to $1/\tau$ gives the internal energy, with two terms, one proportional to ϱ' and the other to ψ', of infinite value at $\tau = 1$ and opposite sign, giving an infinite specific heat in critical conditions:

$$\frac{U_o}{N_o k_B T_o} = -\frac{zK_o}{4} + \varrho' U_2 + \psi' U_3 \tag{13}$$

$$U_2 = \frac{\tau^2 K_o}{\varrho^2 \delta} \left[\frac{z(1-\eta^2)}{1+\beta} + \frac{(z_x/n)(\psi-\eta^2)}{1+\beta_x} \right] \tag{14}$$

$$U_3 = \frac{\tau^2 (z_x/n)}{8 \, \delta^2} \left[z\ln\frac{\beta^2-\eta^2}{(\beta+1)^2} + z\eta \ln\frac{\beta+\eta}{\beta-\eta} - 2\ln\frac{\beta_x^2-\eta_x^2}{(\beta_x+1)^2} + \right.$$

$$\left. +(2\eta-4\eta_x+\eta\delta)\ln\frac{\beta_x+\eta_x}{\beta_x-\eta_x} + (2-z)\ln\frac{1-\eta^2}{4} + (2-z)\ln\frac{1+\eta}{1-\eta} \right] \tag{15}$$

CRITICAL THERMODYNAMIC QUANTITIES

Critical thermodynamic quantities for Ising model systems are often used to prove the degree of approximation for the different theories in the critical range.
The critical parameter K_o is obtained from the second derivative of $F_o/(RT_o)$ with respect to η at $\tau=1$ and $\eta=0$:

$$z/\beta_o + (z_x/n)/\beta_o^2 + 2-z = 0 \tag{16}$$

with $\beta_o=\exp K_o$.
The critical free energy value is obtained from eq.11 assuming a critical value for ψ:

$$\psi_o = .44 \qquad \text{for 2D-lattices}$$
$$\psi_o = .25 \qquad \text{for 3D-lattices}$$

We obtain the expression:

$$\frac{F_{or}}{N_o k_B T_o} = -\frac{zK_o}{4} + \frac{1}{2\delta} \left[z \ln\frac{2\beta_o}{\beta_o+1} + \psi_o(z_x/n) \ln\frac{\beta_o^2}{\beta_o^2+1} - \ln 4 \right] \tag{17}$$

Critical values for the two coefficients in eq.13 are:

$$U_{2or} = \frac{K_o}{2\delta} \left(\frac{z}{\beta_o+1} + \frac{2\psi z_x/n}{\beta_o^2+1} \right) \qquad U_{3or} = \frac{z_x/n}{4 \, \delta^2} \left(z\ln\frac{2\beta_o}{\beta_o+1} - 2\ln\frac{2\beta_o^2}{\beta_o^2+1} \right) \tag{18}$$

The critical properties for the Ising model obtained from exact results or best approximations (series expansion) are reported in table 1, together with the values from IQC approximation for K_o and F_{or}/RT_o.

TABLE 1.
Thermodynamic quantities in the critical range
for Ising model systems.

2D-LATTICE	honeycomb	square	kagome'	diced	triangular
z	3	4	4	4	6
z_I/n	2.75	2	8/3	4/3	1.6

Critical point $K_o=-\Delta\mathcal{E}/(2k_BT_o)$

B.A.*	1.31696	.88137	.93313	.83140	.54931
IQC (eq.16)	1.31803	.88137	.92724	.82899	.54874

Free energy $-F_{or}/(k_BT_o)$

B.A.*	1.0250	.92969	.97479	-	.87959
IQC (eq.17)	1.0203	.92976	.96435	.89603	.88445

Internal energy $-U_{or}/(k_BT_o)$

B.A.*	.76035	.62322	.69427	-	.54931

Order parameter (magnetization) $\eta=C_\eta\cdot(1-\tau)^{1/8}$

B.A.*	1.253	1.222	1.2..	-	1.203

3D-LATTICE	diamond	simple cubic	body-centred cubic	face-centred cubic
z	4	6	8	12
z_I/n	2/5	2/6	2/6	2/6

Critical point $K_o=-\Delta\mathcal{E}/(2k_BT_o)$

B.A.*	.73964	.44342	.31492	.20420
IQC (eq.16)	.73977	.44060	.31756	.20470

Free energy $-F_{or}/(k_BT_o)$

B.A.*	.8335	.7779	.7540	.7418
IQC (eq.17)	.8326	.7751	.7560	.7405

Internal energy $-U_{or}/(k_BT_o)$

B.A.*	.32318	.21996	.17201	.15156

Order parameter (magnetization) $\eta=C_\eta\cdot(1-\tau)^{8/10}$

B.A.*	1.666	1.569	1.506	1.487

* Exact or best approximate values [7-8].

In the critical range it is possible to relate the critical exponents of the thermodynamic quantities:

$$\eta = C_\eta \, (1-\tau)^{\bar\beta} \qquad\qquad U = U_{cr} + C_U \, (1-\tau)^{1-\bar\alpha}$$

to the exponents of ϱ and ψ in the analytical expressions:

$$\varrho = \sum_i C_{\varrho i} \, (1-\tau)^{a_i} \qquad\qquad \psi = \sum_j C_{\psi j} \, (1-\tau)^{b_j}$$

From the condition: $d(\partial F_\sigma/\partial n)/d\tau = 0$:

$$(2-z)\,\frac{\eta'}{1-\eta^2} + z\,\frac{\beta\eta'-\eta\beta'}{\beta^2-\eta^2} + (z_x/n)\,\frac{\beta_x\eta'-\eta\beta_x'-\eta\beta_x\psi'/(2\psi)}{\beta_x^2-\eta_x^2} +$$

$$+ \frac{\psi'(z_x/n)}{4\sqrt\psi}\,\ln\frac{\beta_x+\eta_x}{\beta_x-\eta_x} = 0 \qquad\qquad (19)$$

$$\beta' = -\left[\eta\cdot\eta'(\beta^2-1)/(1-\eta^2)+K_\sigma\varrho'(\beta^2-\eta^2)/\varrho^2\right]/\beta \qquad (20a)$$

$$\beta_x' = \left[(\eta_x^2\psi'/\psi/2-\eta')(\beta_x^2-1)/(1-\eta_x^2)-2K_\sigma\varrho'(\beta_x^2-\eta_x^2)/\varrho^2\right]/\beta_x \quad (20b)$$

we obtain: $a_1 = b_1 = 2\bar\beta$ ($= 5/8$ for 3D and $1/4$ for 2D) Taking into account that:

$$U_2 = U_{2cr}+C_{U21}\,(1-\tau)^{2\bar\beta}+\cdots \qquad U_3 = U_{3cr}+C_{U31}\,(1-\tau)^{2\bar\beta}+\cdots$$

and of the values of the coefficients:

$$C_{U21} = -\frac{C_\eta^2 K_\sigma}{2\delta_\sigma\beta_\sigma}\left(\frac{z}{2}+\frac{z_x/n}{\beta_\sigma}\right) - \frac{C_{\varrho1}K_\sigma^2}{2\delta}\left[\frac{z\beta_\sigma}{(1+\beta_\sigma)^2}+\frac{4\psi_\sigma\beta_\sigma^2 z_x/n}{(1+\beta_\sigma^2)^2}\right] +$$

$$+ 2C_{\varrho1}U_{2cr} + C_{\psi1}\frac{z_x/n}{2\delta}\left(U_{2cr}-\frac{2K_\sigma}{1+\beta_\sigma^2}\right)$$

$$C_{U31} = \frac{K_\sigma C_{\varrho1}z_x/n}{4\,\delta^2}\left(\frac{z}{1+\beta_\sigma}-\frac{4}{1+\beta_\sigma^2}\right) + \frac{C_{\psi1}U_{3cr}z_x/n}{\delta} +$$

$$+ \frac{C_\eta^2 z_x/n}{4\,\delta^2}\left(\frac{z}{2}\frac{1-\beta_\sigma}{\beta_\sigma} - \frac{1-\beta_\sigma^2-(1-\sqrt w\beta_\sigma^2)^2}{\psi\,\beta_\sigma^4} + \frac{2\sqrt\psi-2+\delta}{\psi\,\beta_\sigma^2}\right)$$

in 3D systems, it is possible to obtain for the expressions:

$$\varrho = C_{\varrho1}(1-\tau)^{5/8}+C_{\varrho2}(1-\tau)+C_{\varrho3}(1-\tau)^{5/4}+C_{\varrho4}(1-\tau)^{7/8}+C_{\varrho5}(1-\tau)^{3/2}$$

$$\psi = C_{\psi1}(1-\tau)^{5/8}+C_{\psi2}(1-\tau)+C_{\psi3}(1-\tau)^{5/4}+C_{\psi4}(1-\tau)^{7/8}+C_{\psi5}(1-\tau)^{3/2}$$

the following constraints:

$$C_{\varrho1}U_{2cr}+C_{\psi1}U_{3cr}= 0 \qquad\qquad C_{\varrho2}U_{2cr}+C_{\psi2}U_{3cr}= U_{cr}+zK_\sigma/4$$

$$C_{\varrho3}U_{2cr}+C_{\psi3}U_{3cr}+1.6\,\bar\beta C_{\varrho1}C_{U21}+1.6\,\bar\beta C_{\psi1}C_{U31}= 0$$

$$C_{\psi4}U_{2cr}+C_{\psi4}U_{3cr}= 0$$

$$C_{\varrho5}U_{2cr}+C_{\psi5}U_{3cr}+7/12\,C_{\psi4}C_{U21}+7/12\,C_{\psi4}C_{U31}= 0$$

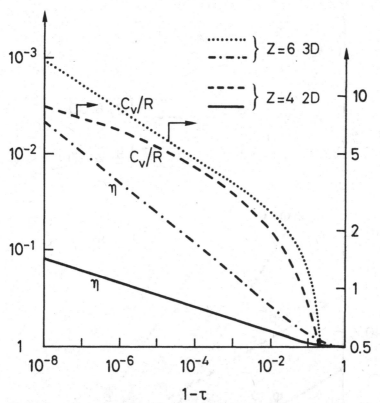

Fig.1. Long range order η and specific heat C_v/R for a simple cubic (z=6;3D) and a square (z=4;2D) Ising lattice at τ<1.

We have determined the two correction factors ϱ and φ in the low temperature field (τ<1) for several 2D and 3D lattices.
The values have been computed from eqs.12, 13-15 and taking eq.19 with eqs.20 into account.
These values are dependent on the long range order η; but they must be given with the highest precision to obtain acceptable values for the internal energy and moreover for the specific heat. In fig.1 the values of the two thermodynamic quantities η and C_v/R are reported for the Ising simple cubic (z=6; 3D) and square (z=4; 2D) lattices.
The values of the ϱ and φ functions for the same lattices are reported in fig.2, showing the critical range at (1-τ)<10⁻⁴ with almost linear behaviour in the double logarithmic plot.
However, short deviations from the linear behaviour determine the internal energy value.
Dimensionality determines the exponents of the ϱ and φ power expansion in (1-τ), being directly related to the critical exponents of η and U/RT_o.

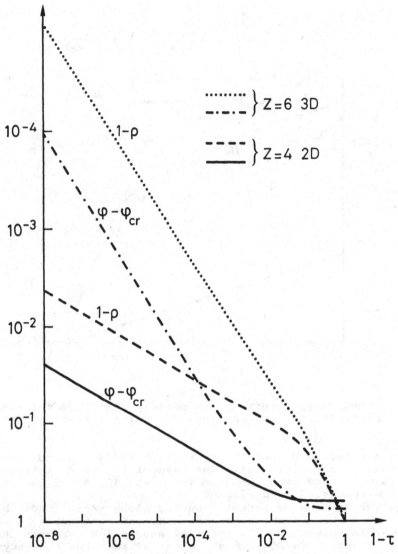

Fig.2. Values of the correction functions 1-ρ and φ-φ_cr for the two Ising systems of fig.1, the simple cubic (z=6;3D) and the square (z=4;2D)) Ising lattice at τ<1.

APPLICATION OF IQC APPROXIMATION TO ORDERING OF β-BRASS.

Cu-Zn alloys with bcc structure (CuSO-ZnSO) show ordering at T<741K with short relaxation times near T_*. It has been care-fully studied, because its behaviour is in surprising agree-ment with the prediction of the antiferromagnetic Ising model. Thermodynamic data show some disagreement with respect to sim-ple bcc Ising model.Corrections proposed take into account lattice compressibility [9] and strain energy influence [10].

By applying IQC approximation to interpret long range orderη [11-12] and specific heat[13-15] data for β-brass, we have pre- ferred to follow the Kajitani and Cook[10] suggestion, in- troducing an elastic energy contribution:

$$F_{el}/(N_o k_o T_o) = -.0299 \cdot \eta^4.$$

Depending on the value of this term we could distinguish when transition are second order(CuZn) or first order(CuAu,CuPt) [1]. Critical exponents suggested for β-brass are $\bar{\beta}=.293-.3$ [11-12] and $\bar{\alpha}=.11$. η values are reported until $(1-\tau)=10^{-2}$ and C_p val- ues (a better choice for application of Ising model according to Baker and Essam[9]) have been reanalyzed [16]. The results are presented in fig.3.

We have taken into account the limitation of specific heat data at low temperature, where it is difficult to reach equi- librium conditions [17], and that the entropy and internal en- ergy for the short range order above T_o are very low.

Thus, we have adopted a next nearest neighbour Ising model of the bcc lattice, increasing the z value to 14. The correction functions o and w are reported in fig.4, showing a behaviour very similar to the theoretical Ising model; main differences are a decrease of the first term constants for the correction factors and of ψ_{or}.

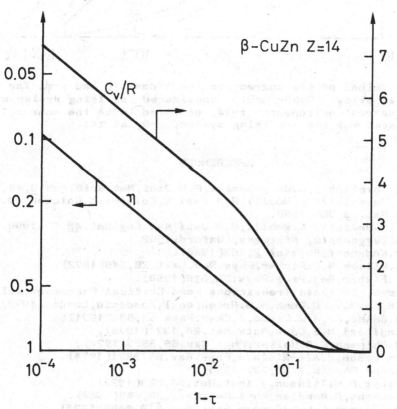

Fig.3. Long range order η and specific heat C_v/R for the b- brass alloy (CuSO-ZnSO) at $\tau<1$.

238

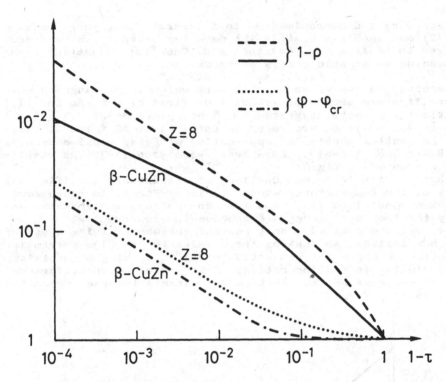

Fig.4. Values of the correction functions 1-ρ and φ-φ₀ for the
β-brass alloy (CuSO-Zn50), considered an Ising system with
next nearest neighbours z=14, compared with the same values
calculated for the bcc Ising system, z=8, at τ<1.

REFERENCES.

1. P.L.Cavallotti,A.Bergamaschi,P.M.Ossi,Met.Sci&Tech.$\underline{3}$,46,1985.
2. P.L.Cavallotti,L.Nobili,P.M.Ossi,B.Colombi,M.Colombo, J.Mag.&
   Mag.Mat.$\underline{83}$,300,1990.
3. P.L.Cavallotti,L.Nobili,P.M.Ossi,Key Eng.Mat.$\underline{48}$,47,1990.
4. E.A.Guggenheim, Mixtures, Oxford,1952.
5. L.P.Kadanoff,Physics $\underline{2}$,263(1966).
6. K.G.Wilson,M.E.Fisher,Phys.Rev.Lett.$\underline{28}$,240(1972).
7. M.E.Fisher,Rep.Prog.Phys.$\underline{30}$,615(1968).
8. C.Domb, in Phase Transitions and Critical Phenomena,vol.3,
   Ch.6, p.357, C.Domb,M.S.Green(eds),Academic,London(1973).
9. G.A.Baker,Jr,J.W.Essam,J.Chem.Phys.$\underline{55}$,861(1971).
10.T.Kajitani,H.E.Cook,Acta Met.$\underline{26}$,1371(1978).
11.D.R.Chipman,C.B.Walker,Phys.Rev.$\underline{B5}$,3823(1972).
12.D.Rathmann,J.Als-Nielsen,Phys.Rev.$\underline{B9}$,3921(1974).
13.H.Moser,Physik.Z.$\underline{37}$,737(1936).
14.C.Sykes,H.Wilkinson,J.Inst.Met.$\underline{61}$,223(1937).
15.J.Ashmann,P.Handler,Phys.Rev.Lett.23,642(1969).
16.M.B.Salamon.F.L.Lederman,Phys.Rev.$\underline{B10}$,4492(1974).
17.G.Inden,Z.Metallkunde $\underline{66}$,648(1975).

MARTENSITIC PHASE TRANSITION IN SYSTEMS
WITH FROZEN ATOMIC ORDER

A. PLANES, E. VIVES, T. CASTAN
Departament d'Estructura i Constituents de la Matèria.
Facultat de Física, Universitat de Barcelona, Diagonal 647,
08028 Barcelona, Catalonia (Spain).

ABSTRACT

We study systems that can be described in terms of two kind of degrees of freedom when the corresponding ordering modes couple one to the other. The primary ordering mode gives rise to a diffusionless first-order phase transition and the secondary mode, that we suppose can be externally controlled, is associated to a continuous phase transition. Our interest is to analyze how the thermodynamic properties of the primary phase transition change as a function of the secondary ordering-mode. This can be experimentally realized in metallic alloys undergoing, at low temperatures, a martensitic phase transition and an order-disorder transition at much higher temperatures. In this case we obtain that, because of the directional character of the martensitic transition, the entropy-change dependence on atomic order is very small.

INTRODUCTION

The phase diagrams of a wide variety of physical systems can be understood in terms of the interplay between different ordering modes. This is a problem of both fundamental and applied interest. Among others, this is the case of magnetic alloys [1,2]. The coupling between the degrees of freedom associated to the magnetic interaction and the configurational ordering is essential to explain the rich complexity of phases observed.

Equilibrium coupling effects appear when the phase transitions (PTs) associated to the different ordering modes are close one to the other and consequently have similar energies. Here we address to non-equilibrium coupling effects and consider two relevant degrees of freedom respectively associated with PTs mutually separated so that no direct effect is produced. The primary ordering mode gives rise to a first-order PT and the secondary mode, that we suppose can be externally controlled, are associated to a continuous PT. This situation is experimentally realized in metallic alloys undergoing a high-temperature order-disorder transition and a displacive, structural PT of martensitic type at much lower temperatures [3]. Examples are Cu-based bcc alloys as Cu-Zn and Cu-Zn-Al.

In this kind of system, a given atomic high-temperature ordering-state at T_i can be retained to lower temperatures by means of a quench. Therefore the system can undergo the martensitic PT with different internal ordering states. The absence of diffusion ensures that the atomic distribution is not modified by the structural PT. Our interest is to study how the properties of the martensitic PT are modified by the non-equilibrium coupling introduced by means of the quench. This unavoidable fast-cooling treatment forms part of the experimental standard routine of getting the phases suitable to undergo the structural PT and involves the freezing of the secondary ordering degrees of freedom. This, apart from being a problem of theoretical interest, has important technological implications. In the case of Cu-Zn-Al, it has been observed that both the martensitic temperature (T_M) and the enthalpy change (ΔH) are substantially modified by the quench while the entropy change seems not to be affected [4,5]. We shall show here that the large elastic anisotropy, commonly observed in martensitic materials [6], is the reason behind this experimental observation.

The paper is organized as follows. Firstly, the problem is framed in the context of a Landau free-energy. We then analyze the case of Cu-based martensitic alloys undergoing at high temperatures an order-disorder PT and compare the theoretical predictions with available experimental data. Finally, we discuss our conclusions.

GENERAL CONSIDERATIONS

The equilibrium condition for a first-order PT to occur is given by:

$$\Delta F = \Delta E - T_0 \Delta S = 0 \tag{1}$$

where ΔF, ΔE and ΔS are respectively the free-energy, energy and entropy changes between the high (H) and the low (L) temperature phases and T_0 is the equilibrium transition temperature. Equation (1) assumes that no significant volume change takes place at the PT, as it occurs in a wide variety of martensitic materials [3].

Let us characterize the state of the system associated with the externally controlled degrees of freedom (we call secondary ordering modes) by means of the state vector $\sigma = (\sigma_1, \sigma_2, \dots \sigma_N)$, where σ_i describes the single internal atomic-state. In our case σ_i will be an occupation variable.

Suppose we now change the internal state from σ to $\sigma + \delta\sigma$. Thus the transition temperature changes from T_0 to $T_0 + \delta T_0$, where δT_0 is a solution of:

$$\delta \Delta E = \delta (T_0 \Delta S) \tag{2}$$

being $\delta\Delta E = \Delta E(\sigma + \delta\sigma) - \Delta E(\sigma)$.

We consider only cases for which both kind of ordering modes can be described in terms of scalar magnitudes and define ζ and σ as the primary and secondary order parameters respectively. The general free-energy density contains three contributions:

$$f=f_1(\zeta,T)+f_2(\sigma,T)+f_{12}(\zeta,\sigma) \tag{3}$$

f_1 gives rise to the first-order PT and f_2 to the second-order PT. f_{12} accounts for the coupling between both kind of orderings. If h is an external field that couples to ζ then [7],

$$f=a_1(T-T_{c1})\zeta^2+\psi_1(\zeta)+a_2(T-T_{c2})\sigma^2+\psi_2(\sigma)+k\zeta^x\psi_{12}(\sigma)-h\zeta \tag{4}$$

a_1, a_2, T_{c1}, T_{c2} and k are constants. The integer x and the functions ψ_1, ψ_2 and ψ_{12} depend on particular symmetry properties of the system.

We further assume that σ remains constant. The ΔS is then given by:

$$\Delta S=-\left\{\left(\frac{\partial f}{\partial T}\right)_\zeta-\left(\frac{\partial f}{\partial T}\right)_{\zeta_0}\right\}-a\left(\zeta^2-\zeta_0^2\right) \tag{5}$$

where $\zeta - \zeta_0$ measures the order parameter jump at the first-order PT point, ζ and ζ_0 being solutions of:

$$\left.\frac{\partial f}{\partial \zeta}\right)_\zeta-\left.\frac{\partial f}{\partial \zeta}\right)_{\zeta_0}=0 \tag{6}$$

$$\Delta f=f(\zeta)-f(\zeta_0)=0$$

Different situations can be analyzed. In particular, it is straightforward to see that, only when h=0 and x=2, ΔS is independent of σ [7]. In this case, the coupling is merely energetic and it shows up as a shift in T_0 only:

$$\delta T_0=\frac{\delta\Delta E}{\Delta S} \tag{7}$$

MARTENSITIC TRANSITION IN BCC BINARY ALLOYS

A large number of bcc binary alloys, $A_x B_{1-x}$, transforming martensitically exhibit, at low and moderate temperatures, either DO_3 (x $\approx$ 0.75) or $L2_1$ (x $\approx$ 0.50) ordered structures. At higher temperatures, a more symmetric B_2-ordered structure is observed. Let T_{c1} and T_{c2} be the temperatures corresponding to the transitions A_2 (disordered bcc alloy) $\rightarrow$ B_2 and $B_2 \rightarrow DO_3$ (or $L2_1$) respectively.

Although the L-phases are usually rather complex here, for the sake of simplicity, we shall consider the simplest case and assume it is a fcc obtainable from the (bcc) H-phase through a Bain distortion [8].

Following standard procedures we divide the bcc lattice in four sublattices (see Fig. 1). If x > 0.5, the different occupation probabilities P_α^i (α= A, B and i = 1,2,3,4) of the $L2_1$ ordered structure verify $P_A^1 = P_A^2 \neq P_A^3 \neq P_A^4 \neq P_A^1$. The DO_3 ordered structure is defined by $P_A^1 = P_A^2 = P_A^3 \neq P_A^4$. As we mentioned above, due to the displacive character of the martensitic PT, the atomic distribution of both, the L- and the H-phases, will be the same.

Bain Mechanism

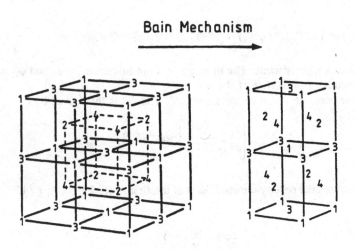

Figure 1. Schematic sublattice representation for the bcc and fcc structures.

The Hamiltonian of the system can be written as:

$$H=\sum_{<ij>} \{\varepsilon_{AA}(x_{ij})\delta(\sigma_i-1)\delta(\sigma_j-1)+\varepsilon_{AB}(x_{ij})(\delta(\sigma_i+1)\delta(\sigma_j-1)$$
$$+\delta(\sigma_i-1)\delta(\sigma_j+1))+\varepsilon_{BB}\delta(\sigma_i+1)\delta(\sigma_j+1)\} \tag{8}$$

We assume pair-wise interaction and extend the summation up to the minimum range required to ensure mechanical stability for both the bcc and the fcc phases. Thus for the H- and L- phases we write:

$$E_H=\sum_{\alpha\beta} N_{\alpha\beta}^{(1)}e_{\alpha\beta}^{(1)}+\sum_{\alpha\beta} N_{\alpha\beta}^{(2)}e_{\alpha\beta}^{(2)}, \quad E_L=\sum_{\alpha\beta} \hat{N}_{\alpha\beta}^{(1)}\hat{e}_{\alpha\beta}^{(1)} \tag{9}$$

$\varepsilon_{\alpha\beta}^{(i)}$ ($\hat{\varepsilon}_{\alpha\beta}^{(i)}$) is the interaction energy corresponding to the value of $\varepsilon_{\alpha\beta}^{(i)}(x)$ at the equilibrium distance between ith-neighbour pairs in the H (L) phase. $N_{\alpha\beta}^{(i)}$ ($\hat{N}_{\alpha\beta}^{(i)}$) is the number of ith-neighbour $\alpha\beta$-pairs in the H (L) phase, verifying the relationship:

$$\hat{N}_{\alpha\beta}^{(1)}=N_{\alpha\beta}^{(1)}+\frac{2}{3}N_{\alpha\beta}^{(2)} \tag{10}$$

Since the atomic distribution remains the same in both phases, the shift in the energy change, $\delta\Delta E$, corresponding to prepare the system in two different internal ordering states, is :

$$\delta\Delta E=\delta E_L-\delta E_H=(\hat{V}^{(1)}-V^{(1)})\delta N_{AB}^{(1)}+(\frac{2}{3}\hat{V}^{(1)}-V^{(2)})\delta N_{AB}^{(2)} \tag{11}$$

where $V^{(i)} = \epsilon_{AB}^{(i)} - \frac{1}{2}(\epsilon_{AA}^{(i)} + \epsilon_{BB}^{(i)})$ and $\hat{V}^{(i)} = \hat{\epsilon}_{AB}^{(i)} - \frac{1}{2}(\hat{\epsilon}_{AA}^{(i)} + \hat{\epsilon}_{BB}^{(i)})$ are the ordering energies in the H and L phases respectively.

In order to calculate the corresponding shift in the entropy change, $\delta\Delta S$, we assume that the lattice vibrations are the origin of ΔS. In the quasiharmonic approximation, for high enough temperatures:

$$\Delta S=3Nk_B\alpha\ln\frac{\omega_H}{\omega_L} \tag{12}$$

where ω_H and ω_L are the characteristic frequencies and α is the fraction of soft modes in the system [9]. The change $\omega_L-\omega_H$ is due to changes in both, lattice symmetry and pair interaction energies. If C_1 and C_2 are respectively the strength of n.n. and n.n.n. bonds, z_1 (=8) and z_2 (=6) are the number of n.n. and n.n.n. in the L-phase, and taking $C_1 >> C_2$, then [10]

$$\omega_H\sim(z_1C_1+z_2C_2)^{\frac{1}{2}} \tag{13}$$

Besides, we neglect the change in C_1 and write, for the L-phase:

$$\omega_L\sim(\hat{z}_1C_1)^{\frac{1}{2}} \tag{14}$$

$\hat{z}_1$ (=12) is the number of n.n. pairs in the (fcc) L-phase.
The elastic constants C_{44} and $C' = \frac{1}{2}(C_{11}-C_{12})$ depend, in the case of central forces, respectively on C_1 and C_2 only so that the elastic anisotropy A is given by $A = C_{44}/C' = C_2/C_1$. We finally get:

$$\Delta S=\frac{3}{2}Nk_B\alpha\ln\left\{\frac{z_1}{\hat{z}_1}+\frac{z_2}{\hat{z}_1}A^{-1}\right\} \tag{15}$$

The corresponding shift is:

$$\delta\Delta S\approx\frac{3}{2}Nk_B\alpha\frac{z_2}{z_1}\left\{1-\frac{\delta C_{44}C'}{\delta C'C_{44}}\right\}\frac{\delta C'}{C_{44}} \tag{16}$$

where $\delta C'$ and δC_{44} are the variations of the elastic constants associated to changes in the internal ordering state of the system. To proceed further we make use of the three following known results: a) Martensitic materials have large elastic anisotropy; typically $A \approx 15$ for Cu-based alloys [6]. b) In the central-force approximation, δC_{44} and $\delta C'$ are respectively proportional to $\delta N_{AB}^{(1)}$ and $\delta N_{AB}^{(2)}$ [11]. c) For the alloys of interest here, the number of n.n. AB-pairs remain nearly constant for quenches performed from temperatures higher or slightly higher that T_{c2} [12]. Then, the shift in the thermodynamic properties can be rewritten as:

$$\delta \Delta E = (\frac{2}{3}\hat{V}^{(1)} - V^{(2)})\delta N_{AB}^{(2)} \qquad (17)$$

$$\delta \Delta S \sim \frac{3}{2}Nk_B\alpha\frac{z_2}{z_1}\frac{\delta C'}{C_{44}} \sim 0 \qquad (18)$$

and

$$\delta T_0 = \frac{\delta \Delta E}{\Delta S} = \frac{1}{\Delta S}(\frac{2}{3}\hat{V}^{(1)} - V^{(2)})\delta N_{AB}^{(2)} \qquad (19)$$

where ΔS is taken as constant.

To compare later with experimental data, it is useful to introduce the next long-range order parameter:

$$\sigma = (p_A^{(2)} - p_B^{(2)}) - (4x - 3) \qquad (20)$$

The mean-field expression of $\delta N_{AB}^{(2)}$ in terms of (20) can be easily obtained and it yields the following expression for δT_0, measured with respect to the perfect ordered state:

$$\delta T_0 \sim \frac{1}{\Delta S}(\frac{2}{3}\hat{V}^{(1)} - V^{(2)})(\sigma^2 - (4x - 2)^2) \qquad (21)$$

COMPARISON WITH EXPERIMENTAL RESULTS AND DISCUSSION

In this section we shall analyze the previous results in terms of available experimental data existing for the alloy $Cu_x(ZnAl)_{1-x}$ which can be regarded as a quasi-binary alloy. This is because the CuAl-ordering energy is around 1.5 times greater than the ordering energy for CuZn pairs but 20 times greater than for ZnAl pairs [3]. Consequently, we assume:

$$\hat{V}^{(1)} = (x_{Zn}\hat{V}_{Cu-Zn}^{(1)} + x_{Al}\hat{V}_{Cu-Al}^{(1)})/(1-x) \qquad (22)$$

$$V^{(2)} = (x_{Zn}V_{Cu-Zn}^{(2)} + x_{Al}V_{Cu-Al}^{(2)})/(1-x) \qquad (23)$$

where, x_{Zn} and x_{Al} $(1-x = x_{Zn} + x_{Al})$ are the atomic fractions of Zn and Al respectively and $V_{\alpha\beta}^{(k)}$ $(\hat{V}_{\alpha\beta}^{(k)})$ are the ordering energies between kth- pairs in the H (L) phase.

For the composition of interest here ($x \approx 0.65$), this system exhibits B_2 and $L2_1$ order at high and low temperatures respectively. In particular, for $x_{Zn} = 0.2809$, $x_{Al} = 0.0995$ $(1-x = 0.3804)$, we evaluate (in units of R) $\hat{V}^{(1)} = 863$ K and $V^{(2)} = 610$ K. Taking $\Delta S = -1.30$ J/Kmol [13], equation (21) leads to a maximum shift $\delta T_0 = -38$ K while experimentally one obtains $(\delta T_0)_{exp} = -62$ K [14]. For Cu_3Al, $\hat{V}^{(1)} = 1250$ K and $V^{(2)} = 825$ K, and the maximum $\delta T_0 = 49$ K, of opposite sign to the previous case. This result is consistent with experiments performed on $Cu_x(AlBe)_{1-x}$ ($x = 0.74$) with only 2 at% of Be for which the shift in T_0 is positive [15].

Figure 2 shows δT_0 versus the relative change in the X-ray intensity ($\Delta I/I$) of 111 superlattice reflections arising from the $L2_1$ ordering, for the Cu; 16 at% Zn; 16 at% Al alloy. Although more accuracy is desirable, the dependence predicted by (21) seems to be confirmed, at least for small variations of the ordering degree.

Looking at eq. (18), we see that the result $\delta \Delta S \approx 0$ reflects the fact that martensitic materials have a large elastic anisotropy which in turn has to do with the directional character of the martensitic transition. This absence of sensitivity of the entropy change with ordering is actually predicted by the free energy (4) if the external field is $h = 0$ and the coupling between the respective order parameters is such that $x = 2$. In this sense, it is interesting to note that in a system where $x \neq 2$ we have obtained that ΔS is dependent on the ordering state [7]. We finally suggest that new experiments, performed under stress ($h \neq 0$), can be very interesting and contribute to the understanding of the problem.

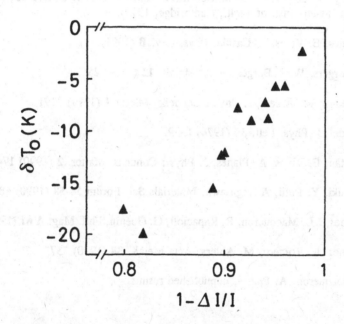

Figure 2. δT_0 (see eq. 2) as a function of $(1-\Delta I/I)$ for Cu-16 at%Zn-16 at%Al alloy. ΔI is the change of the X-ray intensity of 111 superlattice reflection.

ACKNOWLEDGEMENTS

This research was supported by CICyT (Spain) under project n° MAT89-0748.

REFERENCES

1. M. C. Cadeville, J.L. Morán-López, Phys. Rep. **153** (1987) 331.

2. B. Dunweg, K. Binder, Phys. Rev. **B 36** (1987) 6953.

3. M. Ahlers, Prog. Mater. Sci. **30** (1986) 135.

4. R. Rapacioli, M. Ahlers, Acta Metall. **27** (1979) 777.

5. A. Planes, J. Viñals, V. Torra, Phil. Mag. **A 48** (1983) 501; J. Viñals, V. Torra, A. Planes, J.L. Macqueron, Phil. Mag. **A 50** (1984) 653.

6. L. Delaey, P. F. Gobin, G. Guénin, H. Warlimont, *Proceedings of the International Conference on Martensitic transformations*, ICOMAT-79, (Massachussetts Inst. of Tech., Cambridge, 1979).

7. A. Planes, E. Vives, T. Castán, Phys. Rev. **B** (1991).

8. A.J. Borgers, W.G. Burgers, Acta Metall. **12** (1964) 255.

9. R. Romero, M. Ahlers, J. Phys.: Condens. Matter **1** (1989) 3191

10. J. Friedel, J. Phys. Lett. **35** (1974) L-59.

11. T. Castán, E. Vives, A. Planes, J. Phys.: Condens. Matter. **2** (1990) 1743.

12. T. Suzuki, Y. Fujii, A.Nagasawa, Materials Sci. Forum **56-58** (1990) 481.

13. A. Planes, J.L. Macqueron, R. Rapacioli, G. Guénin, Phil. Mag. **A 61** (1990)221.

14. A. Planes, R. Romero, M. Ahlers, Acta Metall. **38** (1990) 757.

15. J.L. Macqueron, A. Planes, unpublished results.

AN ISOTHERMAL CALORIMETRIC STUDY OF THE ORDERING PROCESS IN QUENCHED β-Cu-Zn-Al ALLOY

A. PLANES*, J. ELGUETA*, J.L. MACQUERON**

* Departament d'Estructura i Constituents de la Matèria. Facultat de Física. Universitat de Barcelona. Diagonal 647. 08028 Barcelona. Catalonia.

** Laboratoire de Traitement du Signal et Ultrasons. INSA, Bât 502. 20, Av. A.Einstein. 69621 Villeurbanne. France.

ABSTRACT

This paper presents the results obtained from an isothermal calorimetric study of the ordering process, at different temperatures T_a, in a Cu-Zn-Al alloy after quenches from several temperatures T_q. In all cases the ordering process involves the formation of the $L2_1$ superlattice through a vacancy assisted mechanism. The dissipated energy in the process was measured and the ordering kinetics studied by calorimetric tests.

The dissipated energy increases with the quench temperature, T_q, up to an intermediate value T_{max}, then decreasing for higher temperatures T_q. This behaviour is determined by reordering taking place during the quench, which is due to a large vacancy concentration frozen in from $T_q > T_{max}$.

The relaxation process does not follow a single exponential decay, but instead shows an initial fast decay followed by a progressive reduction of the relaxation rate. The Arrhenius dependence of the characteristic relaxation time of the process from T_q and T_a gives activation energies of 0.76 ± 0.03 eV and 0.43 ± 0.03 eV for the vacancy migration and formation respectively.

INTRODUCTION

The $Cu_x(Zn-Al)_{1-x}$ alloy system $(x \sim 0.70)$ has a bcc (β-phase) structure stable at high temperature, a B_2 superstructure ($Pm3m$ space group) below T_{B2} which turns to a $L2_1$ superstructure ($Fm3m$ space group) at a temperature T_{L21} ($< T_{B2}$) [1].

When such a system, in equilibrium (in a disordered or partially ordered state) at a temperature T_q, is rapidly quenched to a lower temperature T_a, it evolves towards the corresponding equilibrium state at this temperature. This relaxation is essentially associated with an ordering process involving exchanges of atoms among the different lattice sites through a vacancy assisted mechanism.

The present work deals with the study of this ordering process by measuring the dissipation of the excess energy as a function of time by means of a calorimetric technique. In the next section, sample preparation, experimental equipment and experimental procedure are described. The experimental results are presented in the second part and discussed in the last section.

EXPERIMENTAL

Measurements were performed using a high sensitivity, fast response, differential conduction calorimeter [2]. The calorimeter was immersed into a copper block, which operates as heat reservoir. The block was enclosed by a multilayer system of large thermal inertia and the temperature was externally controlled.

The data acquisition of the power output from the calorimeter was done every second and then averaged over N readings. N was taken equal to 30 for the early stages of the relaxation and to 300 for the late stages.

Measurements were performed on a Cu-Zn-Al single crystal (grown by the Bridgman method from elements of 99.99% purity). Its nominal composition is Cu; 14 at% Zn; 17 at% Al. Disk shaped specimens of diameter 13.5 mm and thickness 1.5 mm. were cut from the original crystal with a low-speed diamond saw. The order-disorder transition temperatures, obtained from neutron diffraction experiments, are $T_{B2} = 860$ K and $T_{L21} = 660$ K [3].

In order to avoid the influence of previous heat treatments, in all experiments the sample was first annealed at 1073 K for 360 s and then immersed into a oil-bath at the temperature T_q for a time long enough to ensure that the corresponding equilibrium degree of atomic order and vacancy concentration were reached. Each sample was then quenched into an ice-water mixture and after that immersed in an oil-bath at the annealing temperature T_a. Finally, the sample was placed, as quickly as possible, inside the calorimeter. The elapsed time between the quench and the first averaged acquisition was regularly around 120 s.

RESULTS

In fig. 1 we present typical curves giving the thermal power W as a function of time for a fixed annealing temperature T_a (= 302 K) and different quenching temperatures T_q. The origin of the time scale is taken at the moment when the sample was immersed in the oil-bath at T_a. The dashed lines in fig. 1 were extrapolated, assuming an exponential decay with a "time constant" varying linearly with time.

All the recorded curves show long tails. The time t_f at which the signal becomes masked by the (inevitable) noise, can be as long as several days. Since calorimetric measurements are performed at constant pressure, the enthalpy change of the process can be obtained from:

$$\Delta H - H(t_f) - H(0) = \int_o^{t_f} W dt$$

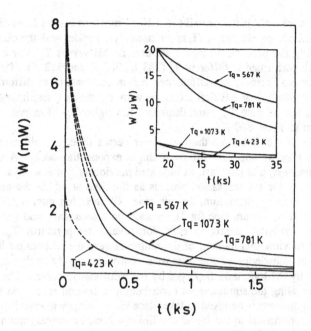

Figure 1. Typical graphs of the thermal power W as a function of time for different T_q and $T_a = 302$ K. The dashed part of the curves is obtained by extrapolation (see text). The inset shows the same curves for long times.

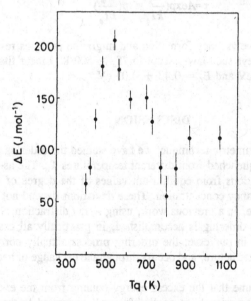

Figure 2. Excess energy dissipated during the relaxation process as a function of T_q. Points represent mean values obtained at different T_a temperatures (= 283K, 302K and 312K)

As the hydrostatic work $p\Delta V$ is very small ($< 3\ 10^{-6}$ J mol^{-1}, as results from dilatometric measurements), the enthalpy change, $-\Delta H$, approximately coincides with the excess energy ΔE dissipated during the process. In fig. 2 we present ΔE versus T_q. For a given T_q, measurements were performed at different T_a (283 K, 302 K and 312 K). We note that, within the experimental errors, no differences for measurements at different T_a are revealed. The results show that ΔE first increases with T_q, shows a maximum for $T_q \approx$ 550 K (we call this temperature T_{max}) and decreases for higher T_q. Moreover, ΔE shows a relative minimum at $T_q \approx 800$ K.

The analysis of the dissipated thermal power versus time, unambiguously shows that the relaxation process does not follow a single exponential decay. A continuous slowing down is observed, and may last, as indicated previously, for few days. We define a characteristic time τ for the relaxation process as the time at which the product t.W shows its most pronounced maximum. In our case, all recorded curves t.W versus t present a well defined maximum (see fig.3). In addition a secondary and generally less noticeable relative maximum seems to exist for several temperatures T_q. From an experimental point of view this method of defining τ is very useful because it is barely affected by the poor knowledge of the dissipated power during the early stages of the process, as well as by the long tails displayed by the calorimetric curves. The method is also effective in revealing the simultaneous contribution of several relaxation times.

The ordering process is believed to take place via a vacancy mechanism. Thus we assume that τ^{-1} is proportional to the initial equilibrium vacancy concentration at T_q and to the vacancy mobility at T_a. Consequently, we consider that τ depends on T_a and T_q through an Arrhenius law:

$$\tau = A \exp(\frac{E_f}{kT_q}) \exp(\frac{E_m}{kT_a})$$

E_f and E_m are the effective vacancy formation and migration energies respectively. Fig. 4 shows that τ indeed obeys such law, except for $T_q \geq 800$ K. Linear fits ($T_q < 800$ K) give $E_m = 0.76 \pm 0.03$ eV and $E_f = 0.43 \pm 0.03$ eV.

DISCUSSION

Using an isothermal calorimetric technique, we have studied the ordering process in a β-Cu-Zn-Al single crystal quenched from different temperatures T_q. The as-quenched state is characterized by deviations from equilibrium values of the degree of configurational long-range order and vacancy concentration. These deviations depend not only on T_q but also on the quenching rate. In a previous work, using X-ray diffraction, Suzuki et al. [4] have shown that, the B2 ordering is accomplished, in practically all cases, during the quench. We believe that in our case, the ordering process actually corresponds to the formation of the L2$_1$ superstructure, and occurs through the exchange of nearest neighbour atoms.

As usual, we assume that the excess energy coming from the excess of thermal vacancies existing after the quench, is negligibly small compared to the excess energy associated with the purely configurational disorder. Thus, the total measured dissipated energy gives the difference between the configurational energy soon after the quench from T_q and the configurational energy of the state reached after the relaxation at T_a. In another work we have shown that these final states are still dependent on the quenching

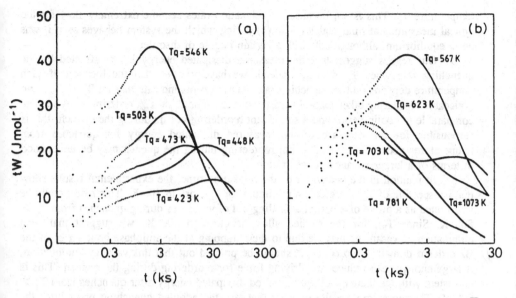

Figure 3. Curves giving t.W versus t (in logarithmic scale) for different temperatures T_q. T_a = 302 K. (a) T_q up to 546 K (b) T_q from 546 K to 1073 K.

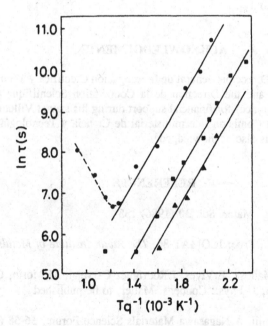

Figure 4. lnτ versus T_q^{-1} for T_a = 312 K (▲), 302 K (■) and 283 K (●). Continuous lines are linear fits to the experimental points.

temperature T_q. This is so, because the relaxation rates become extremely slow for the typical measurement time scales ($\sim$ days) during which the system behaves as if it was be in equilibrium, although still with a certain degree of disorder.

Our results suggest that the measured dissipated energy is not affected by the quenching rate when $T_q < T_{max}$. Indeed, we have verified that the location of such temperature depends on the quenching rate. These results indicate that for $T_q < T_{max}$, no noticeable reordering takes place during the quench. For $T_q > T_{max}$ ordering rates become comparable to cooling rates and a significant reordering occurs during the quench: this is responsible for the decrease of the measured dissipated energy for quenches from temperatures above T_{max}. Such partial reordering during the quench may be understood in terms of a larger vacancy concentration.

In relation to the behaviour of the relaxation time, the experimental results show that irrespective of T_a, τ presents a minimum for T_q close to 800 K. This result could be interpreted as a prove of a partial annealing-out of vacancies during quenches from $T_q > 800$ K. Since T_{B2}, for the studied alloy, is close to 800 K, we suggest that such elimination of vacancies may be due to their trapping at the antiphase boundaries of the B2 ordered domains. Moreover, it should be pointed out that this vacancy annihilation, if large enough, can reduce the driving force for reordering during the quench. This is consistent with the increase of the measured dissipated energy after quenches from $T_q > 800$ K. Therefore, our results indicate that, for the adopted quenching procedure, the configurational ordering state at T_q is effectively frozen by the quench only when $T_q < T_{max}$, while the temperature range enabling the freezing-in of high concentration of excess vacancies extends up to T_{B2}.

ACKNOWLEDGEMENTS

AP acknowledges the Dirección General de Investigación Científica y Técnica (Ministerio de Educación, Spain) and the Direction de la Coopération Scientifique (Ministère des Affaires Etrangères, France) for financial support during his stay at Villeurbanne. Partial financial support from Comisión Interministerial de Ciencia y Tecnología (Spain) under project MAT89-0748 is also acknowledged.

REFERENCES

1.   M.Ahlers, Prog. Mater. Sci. **30** (1986) 135

2.   G.Guénin et al., Proc. ICOMAT-86, *The Japan Institute of Metals*, p.794

3.   A.Planes, Ll.Mañosa, E.Vives, J.Rodríguez-Carvajal, M.Morin, G.Guénin, J.L.Macqueron, J. Phys.: Condens. Matter, to be published

4.   T.Suzuki, Y.Fujii, A.Nagasawa Materials Science Forum, **56-58** (1990) 481

PART II

MECHANISMS OF RADIATION INDUCED AMORPHIZATION

ARTHUR T.MOTTA and CLEMENT LEMAIGNAN,

CEA/CEN-Grenoble, DTP/SECC, 85x, 38041 Grenoble CEDEX,

France

ABSTRACT

A brief review is presented of the mechanisms of energy storage leading to irradiation induced amorphization, with special attention to the comparison between chemical disordering and point defects. It is shown that in addition to chemical disordering, there is considerable experimental evidence for the point defect mechanism. A summary is presented of experimental results of amorphization induced by neutron, electron and ion irradiation of $Zr(Cr,Fe)_2$ precipitates in Zircaloy, and of their theoretical interpretation. In that system, there is evidence for forms of energy storage other than chemical disordering, notably amorphization by departure from stoichiometry under neutron irradiation.

INTRODUCTION

The crystalline-to-amorphous transformation (amorphization) of intermetallic compounds under irradiation has received considerable attention lately. It was first observed under neutron irradiation in U_6Fe by Bloch [1], and more recently during charged particle irradiation (electron and ion) [2-5].

It was noticed from the start that some compounds underwent amorphization while others remained crystalline under similar irradiation conditions. The explanation for this difference in amorphization susceptibility was investigated by several workers, who developed various empirical criteria to predict which compounds would amorphize under irradiation. Those criteria included: ionicity of the bonds and melting temperature [6], width of phase field [5], position of the compound in the phase diagram [7], or a combination of separation in the periodic table, complexity of structure and stoichiometry > 0.33 [8]. Although useful, the empirical approach fails to explicitly take into account the interplay between the nature of the damage and the annealing mechanisms. For example, a material that does not become amorphous under electron irradiation at a given temperature might do so under cascade-producing irradiation or at a lower temperature.

Another approach, emphasized in other works [9-14] as well as in the present one, is to recognize that a certain amount of energy must be stored in the material during irradiation in order for amorphization to occur. By inferring what form of energy is stored, and studying the kinetics of that energy accumulation, a deeper insight into the amorphization process can be obtained.

In the first part of this work we discuss the two main models of energy storage that have been proposed in the literature, namely chemical disordering and increase in point defect concentration. It is our intent to show that forms of energy storage other than chemical disordering also contribute to irradiation induced amorphization of intermetallics.

The results obtained for irradiation induced amorphization of the system $Zr(Cr,Fe)_2$ precipitates in Zircaloy are then presented. These intermetallic precipitates have been made amorphous by the three different types of irradiation (neutron, ion and electron), at various temperatures. These results support the thesis above that several different mechanisms of energy storage contribute to the occurrence of irradiation induced amorphization.

Secondarily, our aim is to show that by studying amorphization in a single system, and cross checking the information from different types of irradiation in a range of temperatures, a view of ensemble can be gleaned that gives insight into the mechanisms of irradiation induced amorphization.

ENERGY STORAGE DURING IRRADIATION

Since the amorphous state is metastable with respect to the unirradiated crystalline state, a rationale must be found for the occurrence of the crystalline to amorphous transformation under irradiation. A necessary condition for amorphization is [15]:

$$G_{irr} > G_{ac} \tag{1}$$

where G_{ac} is the difference in free energy between the unirradiated crystalline and amorphous phases, and G_{irr} is the increase in free energy brought about by irradiation. Some of the energy input from irradiation must therefore be stored permanently (or for a time comparable to the amorphization time) in the material. Since most of the irradiation energy dissipates as heat, the question is then posed: what mechanisms can accumulate enough energy in the lattice during irradiation to satisfy the criterion above and permit amorphization to occur?

The two most commonly proposed mechanisms are an increase in the concentration of point defects (topological disorder) or of anti-site defects (chemical disorder). There has been a great deal of controversy as to whether point defects or anti-site defects are responsible for amorphization under irradiation. We detail some of this discussion below.

Chemical Disordering

As has been pointed out in the literature [16-20], there is strong evidence linking the occurrence of chemical disorder and amorphization. The main experimental evidence for the chemical disordering model comes from the following facts:

1) No pure metals (with the exception of Ga), and no solid solutions have been amorphized under irradiation [19].

2) Luzzi et al [20] found that the critical temperatures for electron irradiation induced chemical disordering and electron irradiation induced amorphization in intermetallic compounds of the Cu-Ti, system were identical.

In addition, calculations show that the amount of chemical disordering attainable under most irradiation conditions, can store enough energy in the lattice to drive the crystalline-amorphous transformation [16,17].

Furthermore, most of the empirical criteria cited in the introduction can be related to either the ease of occurrence or the consequences of chemical disordering. For example, random atom exchanges, of the type brought about by irradiation, have a larger probability of producing anti-site defects in complex structures than in high symmetry structures, so the complex structure criterion could be related to the ease of occurrence of chemical disordering. The same argument is valid for A_xB_{1-x} compounds where x is either very small or close to 1 (the stoichiometry criterion). In that case a random atom exchange has a higher probability of occurring within the majority sublattice, causing no disorder, compared to the case when x = 0.5. The ionicity, solubility and periodic table separation criteria can all be related to the strength of the chemical bond, and are therefore related to the consequences of creating anti-site defects.

From the above, it appears that the role of chemical disordering in the amorphization of intermetallic compounds has been established. It can thus be said to be a necessary condition for the amorphization of intermetallic compounds [17,21].

Point Defect Accumulation

Given the role of chemical disordering in the amorphization process explained above, it remains to be determined whether point defects contribute to irradiation induced amorphization as well. In contrast to chemical disordering, the role of point defects is less clearly established [17-19,21]. We will employ a two-step approach to this discussion: first we will discuss the kinetic feasibility of obtaining high point defect concentrations and then discuss the experimental evidence supporting it.

The Stability Limit of Point Defect Accumulation: It was recognized early on that defect concentrations of the order of 0.01 to 0.02 would be necessary to bring about amorphization [5,15]. If both vacancies and interstitials accumulate in the solid, with a spatially even concentration distribution, then the concentrations above are very close to the athermal recombination limit, where each newly created defect would instantaneously recombine with an existing point defect. This limit is roughly equal to the inverse of the number of atoms in the athermal recombination volume, so a number of atoms equal to 50 would give a maximum concentration of 0.02.

This stability limit has been invoked by several authors [17,19,20] as an argument against the point defect model. Indeed, the burden of proof is on the proponents of point defects as a driving force for amorphization to come up with a mechanism for obtaining high point defect concentrations.

There are, in fact, several good answers to that limitation. Pedraza et al.[11,12] propose that the recombination limit can be circumvented if a locally favorable chemical environment stabilizes Frenkel pairs (called defect complexes) against recombination. High enough concentrations of point defects to cause amorphization can be achieved via that mechanism for a complex binding energy of 0.7 eV. The author also notes that this model can explain the inhibition of amorphization observed in simultaneous electron and ion irradiation of Zr_3Al, compared to pure ion irradiation [22].

Equally, if only one type of point defect accumulates in the lattice, the recombination limit does not apply. This can happen whenever there is preferential elimination of one type of point defect at sinks, either because of a bias or because only one point defect is mobile. Simonen [9] calculated the accumulation rate of vacancies when mobile interstitials annihilate at dislocations, and found that a vacancy concentration of 0.01 could be reached in the observed amorphization time for an interstitial migration energy of 1.0 eV. For lower values of the interstitial migration energy, (or for higher temperatures) the vacancy concentration is smaller than the critical value. The proposed model for electron irradiation induced amorphization of $Zr(Cr,Fe)_2$ precipitates in Zircaloy mentioned later also falls into the category of models in which the concentration of one defect increases due to preferential absorption of the other defect at sinks , in that case the free surface.

Therefore, there are instances when point defect buildup to the level required for amorphization is kinetically feasible.

The discussion above applies mostly to electron irradiation. In the case of cascade producing irradiation, the recombination limit does not apply because the assumption of an evenly distributed production of point defects is not valid. Very high defect concentrations can be achieved locally, in the core of a collapsed cascade or after cascade superposition [9].

Experimental Evidence: There is also experimental evidence that chemical disordering is not always enough to cause amorphization:

a) Koike et al [23] have found different critical temperatures for amorphization of CuTi when induced by Kr ion, Ne ion and electron irradiation. This mirrors the observation presented below of different critical temperatures for amorphization of $Zr(Cr,Fe)_2$ precipitates when induced by neutron, electron and ion irradiation. Such a diversity of critical temperatures is more easily explained by a model that considers other contributions to the free energy in addition to chemical disordering, than by a pure chemical disordering model.

b) The correspondence between the critical temperatures for disordering and amorphization observed by Luzzi [20], is not always present:

Howe and Rainville [4] observed complete disordering of Zr_3Al when subjected to Ar ion irradiation, up to a temperature of 693 K, but only saw complete amorphization below 300 K. Between those two temperatures a mixture of disordered crystalline and amorphous phases was observed, showing that the disordered crystalline phase is stable against amorphization. Also, Koike et al.[24] suggest that the abnormally high dose-to-amorphization observed during electron irradiation of Zr_3Al is due to accumulation of point defects, in addition to chemical disordering.

During proton irradiation of NiTi, Cheng et al [25,26], reported the occurrence of amorphization without any significant chemical disordering, as measured by the intensity of the superlattice reflections. They conclude that chemical disordering is not enough and that an increase in the concentration of point defects must be the main energy source for amorphization.

Nastasi [27] observed the appearance of a completely chemically disordered structure in Ni_2Al_3, when irradiated with Kr ions to a fluence of 7.5×10^{13} ion.cm^{-2}, whereas only partial amorphization was seen at 1.5×10^{14} ion.cm^{-2} and complete amorphization appeared at 3×10^{14} ion.cm^{-2}. The author concludes that the free energy of the completely disordered state is lower than that of the amorphous state, and that additional forms of damage besides chemical disordering are required for amorphization. This behavior has a parallel in one of Luzzi's experiments (fig.7 in ref. [20]), where during electron irradiation of Cu_4Ti_3 at 265 K, the degree of long range order S decreases from 1 to 0.3 at a dose of 2 dpa, then remains constant. Amorphization occurs only at 3.3 dpa, at the same level of chemical disorder. This could mean that another type of damage goes on accumulating in the lattice and finally reaches a critical level to cause the transformation [21].

Molecular dynamics simulations provide yet another window with which to study amorphization. In addition to an early attempt by Limoge et al on a pure metal [28], two molecular dynamics simulations of amorphization have been done so far to investigate the role of point defects and chemical disordering in amorphization, now in intermetallics. Massobrio et al [29] found that Zr_2Ni amorphized upon chemical disordering, after a reduction in S from 1 to 0.6. This shows that, for their potential, which correctly predicts several macroscopic properties of Zr_2Ni, chemical disordering can store enough energy for amorphization. No similar attempt was made with point defects. The opposite conclusion was arrived at in Sabochik and Lam's study of amorphization in CuTi [30], where amorphization was not achieved even for $S=0$. Only when point defects were introduced did the lattice amorphize.

Summary

From the foregoing, it should be clear that chemical disordering plays a very important role in irradiation induced amorphization, but that there is evidence for contributions of other forms of energy storage leading to amorphization, notably point defects.

In fact, although this discussion centered on the role of defects on amorphization, there is extensive evidence of preferential amorphization at lattice imperfections such as grain boundaries [20], dislocations [31], free surfaces [32] and anti-phase boundaries [33]. This is reasonable to expect since these are high energy regions, where the local distortion of the lattice or the high chemical energy gives an additional contribution to the free energy rise due to irradiation. Yet another form of stored energy, present in two-phase systems, a departure from stoichiometry, is presented in this paper.

It still must be assumed that chemical disordering is essential to amorphization, since no combination of the other forms of energy storage under irradiation has been shown to cause amorphization in pure metals. What is the specific role of chemical disordering in cases where amorphization occurs without any chemical disordering [26] is not clear. It is possible that the existence of chemical order either enhances the accumulation kinetics [11] or increases the formation energy of the other forms of defects (point defects, line defects, grain boundaries).

The question of which mechanisms are important for amorphization under irradiation is, therefore, not completely resolved. A three fold approach, combining several types of irradiations that have different replacement to displacement ratios, with modeling of the irradiation induced microstructural evolution with chemical rate equations, and with the investigation of the point defect structure with molecular dynamics [34,35], should prove the most fruitful.

A case in point is CuTi, where the observation of a greater amorphization susceptibility of $Cu_{0.485}Ti_{0.515}$ compared to $Cu_{0.525}Ti_{0.475}$ [36], was explained by Shoemaker et al [34], by the greater ability of the Cu vacancy to cross Ti planes when some Cu atoms are present in those planes.

It must be noted, finally, that this whole discussion is simplified, in that the very notion of a localized point defect may not be applicable to intermetallics, where the equlibrium defect configuration can be quite complex and delocalized [34], as is the case in concentrated solid solutions [37].

AMORPHIZATION OF Zr(Cr,Fe)$_2$ PRECIPITATES IN ZIRCALOY

We now show some results of irradiation induced amorphization of intermetallic precipitates in Zircaloy. This is an interesting system to study because there is a large amount of neutron irradiation data from power reactors, at 350 K (Candu reactors) and 550 to 650 K (Light Water Reactors). In addition to neutron irradiation, those precipitates have been extensively irradiated with 127 MeV Ar ions from 260 to 600 K [38] and with 1.5 MeV electrons from 7 to 320 K [39]. Moreover, the peculiarity of studying the amorphization of precipitates in a matrix can induce interesting interface effects, as shown below.

Laves phase Zr(Cr,Fe)$_2$ precipitates are present in as-fabricated Zircaloy. Their structure and composition has been extensively described elsewhere [40,41]. Those precipitates have been made amorphous by neutron [42,43], ion [38], and electron [13] irradiation.

The experimental results are as follows:

a) Amorphization occurs under the three types of irradiation, when the irradiation temperature is below a critical temperature T_c. At that temperature the dose-to-amorphization increases exponentially, and above T_c it is not possible to amorphize the material.

b) The critical temperature T_c is specific to each type of irradiation. The lowest critical temperature occurs under electron irradiation (300 K), the next higher occurs under ion irradiation (400 K) and the highest occurs under neutron irradiation (580 K), as can be seen in figure 1.

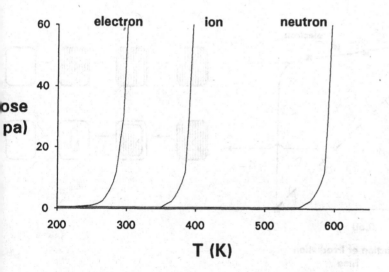

Figure 1. Dose to amorphization of Zr (Cr,Fe)$_2$ precipitates versus irradiation temperature for electron, ion and neutron irradiation.

c) Under neutron irradiation, the transformation starts at the precipitate/Zr matrix interface, and gradually moves into the precipitate until it is completely amorphous. There is a depletion of Fe from 40 at.% to 10 at.% in the amorphous layer. The increase in the thickness of the amorphous layer is linear with irradiation dose.

d) Under electron irradiation amorphization occurs homogeneously and abruptly, towards the end of the irradiation time. There is no variation in composition either during or after amorphization.

e) Under ion irradiation the transformation morphology could not be determined, since partially amorphous precipitates were not found in the samples examined. It could be ascertained, however, that there is no variation in composition associated with amorphization.

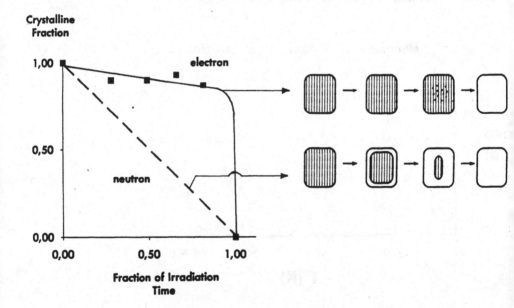

Figure 2. Measured crystalline fraction as a function of fraction of total irradiation time under electron and neutron irradiation of Zr (Cr,Fe)$_2$ precipitates in Zircaloy.

Figure 2 compares the amorphization kinetics under neutron and electron irradiation, by plotting the crystalline fraction as a function of the fraction of the total irradiation time. It can be seen that under electron irradiation, the crystal retains its crystallinity until close to the end of the irradiation time and then transforms abruptly to amorphous, whereas under neutron irradiation the amount transformed is directly proportional to the dose. A qualitative scheme of the two transformations is also shown in fig.2, illustrating the two different transformation morphologies, explained in c) and d) above.

ENERGY STORAGE MODELS FOR Zr(Cr,Fe)$_2$

Electron Irradiation

Under electron irradiation it was found that the kinetics of Zr(Cr,Fe)$_2$ precipitate amorphization could be accounted for by a model incorporating the contributions of both point defects and anti-site defects to the free energy rise under irradiation [21].

The chemical disordering contribution was calculated using the Bragg-Williams model to find the change in the number of unlike atom pairs when the long range order parameter S decreases due to random recombination. The increase in free energy due to disordering is then proportional to the decrease in the number of unlike atom pairs.

The mechanism of point defect accumulation was proposed to be a thin foil artefact, related to the elimination of a "fast" (interstitial) defect at the foil surface, while the "slow" defect (vacancy) accumulates in the lattice. There is then a competition between recombination in the bulk and elimination at the surface. The results of the calculation show that at the critical temperature, there is a drastic reduction in the fraction of defects eliminated at the surface, which reduces the accumulation of point defects in the lattice and impedes amorphization. The defect energies required for this model to account for the temperature dependence of the dose-to-amorphization are a fast defect with a migration energy of 0.45 eV and a "slow" defect of the opposite polarity with a migration energy bigger than 0.95 eV.

Neutron Irradiation

In contrast, neutron irradiation induced amorphization of $Zr(Cr,Fe)_2$ precipitates is proposed to be controlled by a departure from stoichiometry caused by cascade (or ballistic) mixing in a thin layer close to the crystalline/amorphous precipitate interface [44]. The departure from stoichiometry in line compounds causes a large increase in its free energy . This explains why there is preferential amorphization in the affected zone close to the interface. The Fe concentration in the amorphous phase of 0.33 no longer being in equibrium with that in the matrix, amorphization is followed by a discharge of iron from the amorphous precipitate into the matrix, which allows amorphization to advance into the precipitate. The fact that the amorphous layer thickness has a linear, rather than square root, dependence on the dose, indicates a ballistic rather than a thermal mixing process. The temperature of 0.3 T_m has been associated with a decrease in cascade mixing due to the onset of the thermally assisted mixing regime [6,45]. Since the critical temperature is in this case equal to 0.3 T_m, it is likely that a reduction in cascade mixing efficiency is responsible for the absence of amorphization above that temperature.

Amorphization and Annealing Mechanisms

In table 1 a tentative assignment is made of radiation damage mechanisms that cause amorphization in $Zr(Cr,Fe)_2$ precipitates with their respective annealing mechanisms operative at the critical temperatures.

For each type of irradiation, the proposed damage mechanism is shown in the first line. At the critical temperature, listed in the second line, this damage mechanism is swamped by the annealing process shown. The defect that activates such process is also shown along with its activation energy. Lastly a comparison is made of the irradiation induced amorphization of $Zr(Cr,Fe)_2$ with that of $Zr_2(Ni,Fe)$.

The annealing of <u>electron</u> irradiation damage is controlled by the motion of the interstitial (or "fast" defect) with a migration energy of 0.4 eV, found by fitting to the critical temperature. In the case of ion irradiation, the amorphization mechanism is largely unknown, but the hypothesis of a long range reordering process taking place around 400 K is not incompatible with the requirements of a slow defect with a migration energy of 0.9 eV, originating from the electron irradiation model. Given a disordering rate dS/dt of 10^{-3} to $10^{-2}s^{-1}$, a calculation using any one of the standard expressions for vacancy reordering [21,46-49] yields a migration energy of 1 eV for the reordering defect, in order for the equality between disordering and reordering to happen at 400 K.

Above the critical temperature, in the case of <u>ions</u>, although cascades are still present and still cause disorder, long range reordering by free vacancies can offset that disordering and prevent amorphization.

At the higher temperatures and longer irradiation times present under <u>neutron</u> irradiation, the storage mechanism based on a departure from stoichiometry becomes available, because now the product $(Dt)^{1/2}$ is large enough for the amorphous layer to discharge its iron into the matrix, which

TABLE 1 Mechanisms of Irradiation Induced Amorphization in $Zr(Cr,Fe)_2$

	Electron	Ion	Neutron
Mechanism of Free Energy Rise Under irradiation	Accumulation of vacancies in the lattice because of interstitial migration to surface, supplemented by chemical disordering	Cascade Damage Point Defects + Chemical Disorder (?)	Departure from stoichiometry caused by ballistic mixing and cascade damage
Tc (K)	300	~ 400	580
Annealing Mechanism at Tc	Sustainable vacancy supersaturation is lower near Tc because of increased recombination in the bulk	(?) Long range vacancy reordering activated at critical temperature	Increased annealing in cascades close to critical temperature 0.3 Tm .=> less mixing and disorder
Defect & Activation Energy	"Fast" defect with Em = 0.4 eV and Em fast - Em slow > 0.5 eV	Migration Energy of Reordering Defect ~ 1 eV	N/A (Mobility in Cascades)
Comparison with $Zr_2(Ni,Fe)$	$Zr_2(Ni,Fe)$ amorphizes easier Amorphization easier at higher dose rates	Low temperature irradiations: $Zr_2(Ni,Fe)$ amorphizes easier High temperature irradiations: $Zr(Cr,Fe)_2$ amorphizes easier	$Zr(Cr,Fe)_2$ amorphizes easier

allows it to advance into the precipitate. Given that the disappearance of amorphization at the critical temperature in the case of neutron irradiation is thought to be due to increased annealing in the cascades, no defect is assigned to that temperature. At the critical temperature, the cascades become ineffectual at producing mixing and disorder, so that the damage is eliminated at the source, and amorphization is again not possible.

In comparing the amorphization susceptibility of the two types of precipitates present in Zircaloy, $Zr(Cr,Fe)_2$ and $Zr_2(Ni,Fe)$, it is interesting to note that for the low temperature irradiations (electron and low temperature ion), $Zr_2(Ni,Fe)$ precipitates are less stable against amorphization than $Zr(Cr,Fe)_2$ precipitates are, whereas the reverse is true at higher temperature irradiations (neutron and high temperature ion) [50]. This could indicate that different amorphization mechanisms are present at low and high temperatures.

Another noteworthy observation is that $Zr_2(Ni,Fe)$ precipitates are found completely amorphous after 1 dpa neutron irradiation at 350 K, and are completely crystalline after 10 dpa irradiation at 523 K. This is not inconsistent with a critical temperature of for neutron irradiation induced amorphization at 0.3 T_m, equal to 440 K for $Zr_2(Ni,Fe)$. Although it would be interesting to observe the $Zr_2(Ni,Fe)$ precipitates in Zircaloy 2 irradiated to 1-10 dpa, at 400-440 K such samples are not available since this is, unfortunately, not a common power reactor temperature.

The dose-to-amorphization under electron irradiation in $Zr_2(Ni,Fe)$ was also found to be lower for higher dose rates, in accord with the observations of Xu et al. in CuTi [51], Basu et al. in YBaCuO [52] and Linnros et al. in Si [53].

CONCLUSIONS

A brief review was made of the models in the literature for energy storage in the lattice during irradiation-induced amorphization.

Chemical disordering is probably the main contribution to energy storage during irradiation, but several other mechanisms contribute as well. Those are point defect buildup, departure from stoichiometry, and all existing lattice imperfections (grain boundaries, dislocations, and anti phase boundaries). Although they are marginal mechanisms, they are often the controlling mechanism for amorphization, providing the extra energy needed to amorphize the material.

The system $Zr(Cr,Fe)_2$ precipitates in Zircaloy provides interesting examples of alternative forms of energy storage, notably in the case of neutron irradiation, where amorphization is caused by a departure from stoichiometry at the precipitate/matrix interface.

Acknowledgments:

The authors would like to thank D.Pêcheur for the use of his unpublished results. Helpful discussions with P.Desré, J.Hillairet, Y.Limoge and F.Rossi are also acknowledged.

REFERENCES

1. J.Bloch, J.Nuc.Mat., 6 (1962) 203.
2. G.Thomas, H.Mori, H.Fujita and R.Sinclair, Scripta Met., 16

(1982), pp. 589-592.

3. A.Mogro-Campero, E.L.Hall, J.L.Walter and A.J.Ratkowski, in Metastable Phase Formation by Ion Implantation, S.T.Picraux, W.J.Choike eds., Lausanne, Elsevier Science 1982, pp.203-210.

4. L.M.Howe and M.H.Rainville, J.Nuc.Mat., 68 (1977), pp. 215-234.

5. J.L. Brimhall, H.E. Kissinger and L.A.Charlot, Rad.Effects, 77 (1983), pp. 237-293.

6. H.M.Naguib and R.Kelly, Rad.Effects,25 (1975), pp. 1-12.

7. H.Mori, H.Fujita, this Workshop.

8. D.E.Luzzi and M.Meshii, Scripta Met., 20 (1986), pp. 943-948.

9. E.P.Simonen, Nuc.Inst. and Meth. in Phys.Res. B16 (1986), pp. 198-202.

10. Y.Limoge and A.Barbu, Phys.Rev. B, 30, No.4 (1984), pp. 2212-2215.

11. D.F.Pedraza and L.K.Mansur, Nucl.Inst. and Meth. in Phys.Res., B16 (1986) pp.203-211.

12. D.F.Pedraza, J.Less Common Metals, 140 (1988) pp.219-230.

13. A.T.Motta, D.R.Olander and A.J.Machiels, 14th Int.Symp. on the Effects of Irradiation on Materials, Andover 1988, ASTM STP 1046, 457.

14. D.E.Luzzi, H.Mori, H.Fujita and M.Meshii, Proc. of the Int. Symp. on In-Situ Exp. in the HVEM, Osaka University, 1985, pp.472-476.

15. M.L.Swanson, J.R.Parsons and C.W.Hoelke, Rad.Effects, 9 (1971), pp. 249-256.

16. D.E.Luzzi and M.Meshii, Res. Mechanica, 21 (1987), pp. 207-247.

17. P.Okamoto, M.Meshii, to appear in the ASM publication Science

of Advanced Materials, eds.H.Wiedersich and M.Meshii.

18. W.L.Johnson, Prog.Mat.Sci. 30 (1986) 81-134.

19. R.W.Cahn and W.L.Johnson, J.Mat.Res., 1 (5) (1986), pp. 724-732.

20. D.E.Luzzi, H.Mori, H.Fujita and M.Meshii, Acta Met. 34, No. 4 (1986), pp. 629-639.

21. A.Motta. and D.Olander, Acta Met.&Mat., vol.38 (11), (1990) 2175-2185.

22. D.F.Pedraza, Phys.Rev.B 38 (7) (1988) 4803-4809, and ref [13] in it.

23. J.Koike, P.R.Okamoto, L.E.Rehn and M.Meshii, J.Mat.Res. 4(5) (1989), pp.1143-1150.

24. J.Koike, P.R.Okamoto, L.E.Rehn and M.Meshii, Met.Trans.A 21A (1990), pp.1799-1808.

25. J.Cheng, M.Yuan, C.N.J.Wagner and A.Ardell, J.Mat.Res. 4(3) (1989) pp.563-578.

26. J.Cheng and A.J.Ardell, Nuc.Inst.Meth.Phys.Res. B44 (1990) pp. 336-343.

27. M.Nastasi, J.Less Common Metals, 168 (1991) pp.91-102.

28. Y.Limoge, A.Rahman and S.Yip, Mat.Sci.Forum, 15-18 (1987) pp.1421-1426.

29. C.Massobrio, V.Pontikis, and G.Martin, Phys.Rev.B, vol.41 no.15, pp.10486-10497.

30. M.J.Sabochik and N.Q.Lam, Scripta Met.&Mat., 24 (1990) pp.565-570.

31. H.Mori, H.Fujita and M.Fujita, Jap.Journal of Appl.Phys 22 no.2 (1983 pp.L94-L96.

32. H.Mori, H.Fujita and M.Fujita, Proc.of the 7th Int.Conf. on

HVEM, Berkeley, 1983, pp.233-238.

33. H.Mori and H.Fujita, Proc. of the Yamada Conf. on Disl. in Solids, ed.H.Suzuki, (University of Tokyo, Tokyo 1985), p.563.

34. J.R.Shoemaker, R.T.Lutton, D.Wesley, W.R.Wharton, M.L.Oehrli, M.S.Herte, M.J.Sabochik, and N.Q.Lam, J.Mat.Res. 6(3) (1991), pp.473-482.

35. A.Caro, M.Victoria and R.S.Averback, J.Mat.Res. 5(7) (1990), pp.1409-1413.

36. D.E.Luzzi, H.Mori, H.Fujita and M.Meshii, in Beam-Solid Interactions and Phase Transformations, eds. H.Kurz, G.L.Olson, and J.M.Poate, Mat.Res.Soc.Symp.Proc. 51, Pittsburgh, 1986, P.479.

37. M.Halbwachs, J.T.Stanley and J.Hillairet, Phys.Rev.B 18, no.9 (1978), pp.4938-4944.

38. F.Lefebvre and C.Lemaignan, J.Nuc.Mat., 165 (1989), pp.122-127.

39. A.Motta, F.Lefebvre and C.Lemaignan, to be published in the Proc. of the 9th Conference on Zirconium in the Nuclear Industry, Kobe, Japan, November 1990.

40. P.Chemelle, D.B.Knorr, J.B.Vander Sande and R.M.Pelloux, J.Nuc.Mat. 113 (1983), pp. 58-64.

41. D.Charquet and E. Alheritiere, Workshop on Second-Phase Particles and Matrix Properties in Zircaloys, Erlangen, W.Germany, July 1-2, 1985.

42. M.Griffiths, R.W.Gilbert and G.J.C.Carpenter, J.Nuc.Mat. 150 (1987), pp. 53-66.

43. W.J.S.Yang, R.P.Tucker, B.Cheng and R.B.Adamson, J.Nuc.Mat.138 (1986), pp. 185-195.

44. A.T.Motta and C.Lemaignan, to be submitted to the Journal of

Nuclear Materials.

45. F.Rossi and M.Nastasi, J.Appl.Phys. 69(3) (1991) pp.1310-1319.

46. G.H.Vineyard, Phys.Rev. 102 (4) (1956), pp. 981-992.

47. G.J.Dienes, Acta Met., 3 (1955), pp. 549-557.

48. R.Zee and P.Wilkes, Phil.Mag.A, 42 (1980), pp. 463-482.

49. N.Njah, J.Nuc.Mat. 170 (1990) pp.232-235.

50. D.Pêcheur, C.E.N.G., unpublished research.

51. G.Xu, J.Koike, P.R.Okamoto, and M.Meshii, Proc. of the 47th EMSA Meeting (1989) p.658.

52. S.N.Basu, T.E.Mitchell, and M.Nastasi, J.Appl.Phys. 69(5) (1991) pp.3167-3171.

53. J.Linnros, R.G.Elliman, and W.L.Brown, J.Mat.Res. 3(1988) p.1208.

ELECTRON-IRRADIATION-INDUCED CRYSTALLINE-TO-AMORPHOUS TRANSITION IN METALLIC AND NON-METALLIC COMPOUNDS

H. MORI and H. FUJITA
Research Center for Ultra-High Voltage Electron Microscopy,
Osaka University, Suita, Osaka, 565 Japan

ABSTRACT

In order to clarify the factor controlling the electron-irradiation-induced crystalline-to-amorphous(C-A) transition in crystals, a series of metallic and non-metallic compounds were irradiated with 2 MeV electrons in a HVEM, and the presence (or absence) of the C-A transition was examined. Analysis of the results reveals that the tendency toward the C-A transition under electron irradiation is best correlated with the position of the compounds in the corresponding T-C phase diagram. Namely, those compounds whose position in the phase diagram is close to the bottom of a deep valley of liquidus in the diagram have a strong tendency toward the C-A transition, while those away from such a valley show little tendency toward the C-A transition.

INTRODUCTION

In the early years of the 1980's it was discovered that high energy (mega electronvolts) electron irradiation can induce a crystalline-to-amorphous (C-A) transition in some metallic compounds such as NiTi[1,2], Fe_3B[3], and others [4]. The C-A transition is caused not by ionization damage but by displacement damage[1,2]. This C-A transition is of interest due to the simplicity of electron irradiation damage. The simplicity firstly comes from the fact that the energy transferred to the primary knock-on atoms from a few MeV incident electrons is sufficient to produce only single or at most double atom displacements. This is in sharp contrast to the massive damage in the form of displacement cascade produced by ion and neutron irradiation, where a complex process analogous to liquid phase quenching or vapour phase quenching might be operative in the damage core region. Secondly, the chemical composition of target materials remains constant during the electron irradiation since impurity atoms are neither introduced nor created in the materials, which is not the case with ion and neutron irradiation. Such simplicity in experimental conditions that neither any quenching process nor any changes in chemical composition is involved in the irradiation makes the C-A transition to be a perfect candidate for an investigation of the solid-state amorphization process and amorphous states themselves. Based upon these premises, a systematic research has been made on the C-A transition by the authors' group, with the use of a 2000 kV high voltage electron microscope (HVEM). As a part of the research a series of metallic and non-metallic compounds have been irradiated with 2 MeV electrons in the HVEM, and the presence (or absence) of the C-A transition has been examined, in an attempt to establish the generality of the C-A transition. In this note the experimental results obtained by the authors' group are summarized with emphasis on the results of metallic compounds, and based upon the results the factor controlling the amorphization tendency will be discussed.

EXPERIMENTAL PROCEDURES

The 67 metallic compounds examined in the authors' laboratory are listed in the first column of Table 1. They are divided into three groups, i.e. intermetallics in transition metal(T)-transition metal(T) systems, in aluminum-T systems and in T-boron systems. Details of alloy preparation are given elsewhere[5,6]. All the specimens were irradiated with 2 MeV electrons in a Hitachi HU-3000 type HVEM, and changes in the bright-field images as well as in the selected area diffraction patterns (SAD's) were monitored in-situ during irradiation. The criterion used to define the

TABLE 1
Crystal stability under electron irradiation of intermetallic compounds

Compound	G→A transition	Crystal structure or Crystal system	Transformation temp. (K)	Composition phase field(at%)
(T-T systems)				
$TiCr_2$	no	C15	1423	5
$TiMn_2$	yes	C14	1598	7
TiFe	no	B2	1590	2.8
$TiFe_2$	yes	C14	1700	7.6
TiCo	no	B2	1573	4
$TiCo_2$	yes	C15	1523	0
Ti_2Ni	yes	$E9_3$	1536	1
TiNi	yes	B2	1583	<1
$TiNi_3$	no	DO_{24}	1926	0
Ti_2Cu	yes	C11b	1288	2
TiCu	yes	B11	1177	4
Ti_3Cu_4	yes	Ti_3Cu_4 type	1191	0
Ti_2Cu_3	yes	Os_2Al_3 type	1143	0
$TiCu_4$	no	Au_4Zr type	1143	2
Zr_2Ni	yes	C16	1413	0
ZrCu	yes	ZrCu type	713	0
Zr_7Cu_{10}	yes	Zr_7Ni_{10} type	1168	0
Nb_7Ni_6	yes	$D8_5$	1568	4
MoNi	yes	tetragonal	1635	3
(Al-T systems)				
Al_3Ti	no	DO_{22}	1613	1
$Al_{10}V$	yes	cubic	943	0
$Al_{45}V_7$	yes	monoclinic	961	0
$Al_{23}V_4$	yes	monoclinic	1009	0
Al_3V	no	DO_{22}	1633	0
Al_8V_5	no	cubic	1943	1
Al_7Cr	yes	monoclinic	1063	1
Al_5Cr	yes	monoclinic	1213	2
Al_4Cr	yes	monoclinic	1303	1
Al_9Cr_4	no	cubic	1443	3
Al_8Cr_5	no	rhombohedral	1398	4
Al_6Mn	yes	D2h	983	0
Al_4Mn	yes	orthorhombic	1303	1
Al_3Mn	no	orthorhombic	1275	3
$Al_{11}Mn_4$	no	triclinic	1168	0
AlMn (γ_2)	no	hexagonal	1261	15
Al_3Fe	yes	monoclinic	1433	1.5
Al_5Fe_2	no	orthorhombic	1444	2
Al_2Fe	no	triclinic	1426	1
Al_9Co_2	yes	monoclinic	1216	0
Al_5Co_2	no	$D8_{11}$	1443	1
Al_3Ni	no	DO_{20}	1127	0
Al_3Ni_2	no	$D5_{13}$	1406	4.5

TABLE 1 (continued)
Crystal stability under electron irradiation of intermetallic compounds

Compound	G→A transition	Crystal structure or Crystal system	Transformation temp. (K)	Composition phase field(at%)
AlNi	no	B2	1911	13
AlNi$_3$	no	L1$_2$	1668	4
Al$_2$ Cu	no	C16	1137	1
AlCu (η_2)	no	orthorhombic	833	1
Al$_3$ Zr	no	DO$_{2\,3}$	1853	0
Al$_2$ Zr	yes	C14	1918	0
Al$_3$ Zr$_2$	yes	orthorhombic	1868	0
AlZr	yes	Bf	1523	0
Al$_4$ Zr$_5$	yes	hexagonal	1800	
Al$_3$ Zr$_4$	yes	hexagonal	1300	0
Al$_2$ Zr$_3$	yes	tetragonal	1753	0
AlZr$_2$	yes	C16	1523	0
AlZr$_3$	no	L1$_2$	1248	0
Al$_{12}$ Mo	no	cubic	979	0
Al$_8$ Mo$_3$	no	monoclinic	1843	1
Al$_{12}$ W	no	cubic	970	0
Al$_2$ Au	no	Cl	1333	0
(T-B system)				
Co$_3$ B	yes	orthorhombic	1398	0
Co$_2$ B	yes	C16	1553	0
CoB	no	B27	1733	0
Ni$_3$ B	no	DO$_{11}$	1429	0
Ni$_2$ B	yes	C16	1398	0
o-Ni$_4$ B$_3$	yes	orthorhombic	1298	0
m-Ni$_4$ B$_3$	no	monoclinic	1304	0
NiB	no	orthorhombic	1291	0

occurrence of the amorphization was the complete replacement
of crystalline spot patterns by diffuse ring patterns in the
SAD's. The electron irradiation of the intermetallics in T-T
systems[5] and in Al-T systems[7] was carried out at around
160K. The irradiation of compounds in the Co-B and Ni-B
systems was carried out at approximately 100K and 4.2K,
respectively. The electron flux was fixed at around 1×10^{24}
e/m^2 s, throughout the experiment.

RESULTS

Of the 67 intermetallic compounds investigated in this
laboratory, 35 compounds proved to undergo a C-A transition
during irradiation, while the other 32 remained crystalline.
This fact indicates the C-A transition to be a phenomenon of

wide generality in metallic compounds. The results are summarized in Table 1. The presence (or absence) of the C-A transition is given in the second column of the table, where the words yes and no denote the presence and absence of the C-A transition, respectively.

DISCUSSION

The third, fourth and fifth columns of Table 1 list crystal structure(or crystal system), phase transformation temperature(T_{trans}), and composition phase field of each compound, respectively. Here, phase transformation means either melting, peritectic decomposition or peritectoid decomposition, depending on the particular compound. Of these three parameters crystal structure and composition phase field are essential parameters in the structure-type criterion[8] and solubility criterion[9] respectively, which have been proposed for amorphization of compounds by ion irradiation. Namely, Matzke and Whitton[8] have proposed a structure-type criterion for the case of ion irradiation of non-metallic compounds, which states that non-cubic materials tend to undergo the C-A transition whereas cubic materials never undergo the C-A transition. Brimhall et al.[9] have made a study of ion-irradiation-induced C-A transition in intermetallic compounds and found that those compounds with narrow solubility ranges in the phase diagram tend to become amorphous under irradiation, while those with wide solubility ranges tend to remain crystalline. The parameter of phase transformation temperature is, on the other hand, listed here because of the following two reasons. Firstly, the temperature is a parameter relating to the structural stability of the compounds against heating and is therefore expected to have any relation with the amorphization tendency. Secondly, compounds with high transformation temperatures might be anticipated to have low mobility of point defects at the fixed irradiation temperature, T_{irr}, according to the low homologous temperature, T_{irr}/T_{trans}, and therefore in such compounds the defect would be rapidly accumulated to a high level, as compared with compounds with

low transformation temperatures. However these three
parameters do not appear to be prime factors controlling the
tendency toward amorphization under electron irradiation,
because it is difficult to make a consistent arrangement of
the experimental results based upon these parameters.
Examples of violations to the criteria are as follows. TiNi
and $Al_{1.0}V$ which have the cubic symmetry undergo the C-A
transition, in disagreement with the structure-type
criterion. $TiFe_2$ can be rendered amorphous while TiFe
remains crystalline although the former has a wider
solubility range (i.e. 7.6 at%) than the latter (i.e. 2.8
at%). This result is in contradiction to the solubility
criterion. Furthermore the transformation temperatures of
the compounds that undergo the C-A transition range from 713
to 1918 K, which is not so widely different from the range of
T_{trans} of the compounds that remain crystalline.

It is now well known that the form of the equilibrium
phase diagram provides a useful guide to the ease of
amorphous material formation by liquid phase quenching[10].
Namely, the amorphous forming ability of alloys is enhanced
in the deep eutectic region of a phase diagram. In the
following, the observed amorphization tendency under electron
irradiation is discussed in terms of the position of the
compounds in the corresponding phase diagram.

In the cobalt-boron binary system, there exist in total
three equilibrium intermediate-phases, i.e., Co_3B, Co_2B,
and CoB, as shown in Fig.1. All these three compounds were
examined, and it was revealed that Co_3B and Co_2B can be
rendered amorphous while CoB remains crystalline under
electron irradiation, as shown in the figure by symbols O
(amorphization) and X(no amorphization), respectively. Also
given in the diagram is the composition range, L.Q., over
which amorphous materials can successfully be produced by the
liquid quenching method[11]. It is evident from the diagram
that Co_3B and Co_2B which become amorphous are located near
the bottom of a deep valley of liquid phase region in the
diagram and are situated within the composition range L.Q.,
or immediately on the boundary of the composition range,

whereas CoB which remains crystalline is out of the L.Q..
This fact suggests that the amorphization tendency under
electron irradiation is similar to that obtained by the
liquid quenching method.

Another example of such a close correlation between the
amorphization tendency and the position of the compounds in
the phase diagram is found in the aluminum-vanadium alloy
system. In this system, there exist totally five
intermediate-phases, i.e. $Al_{10}V$, $Al_{45}V_7$, $Al_{23}V_4$, Al_3V and
Al_8V_5, as shown in Fig.2. Of the five intermetallic

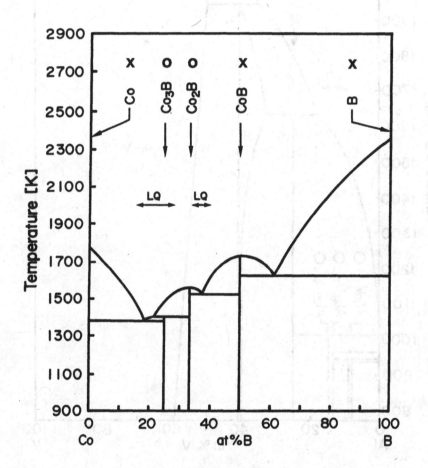

Figure 1. Equilibrium phase diagram for the Co-B system. The symbols O
and X on the compounds represent presence and absence of the electron-
irradiation-induced amorphization in the compounds, respectively. The
composition range over which amorphous materials are successfully
obtained by the liquid quenching technique[11] is marked as L.Q..

compounds only the first three can be rendered amorphous
while the last two remain crystalline. This is shown in the
figure by symbols O and X which denote the presence and

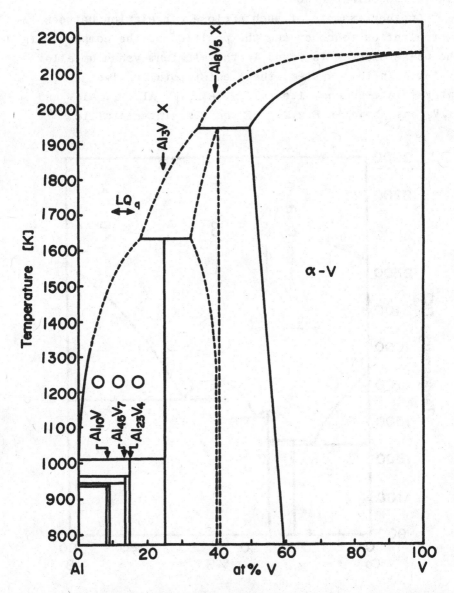

Figure 2. Equilibrium phase diagram for the Al-V system. The symbols O
and X on the compounds represent presence and absence of the electron-
irradiation-induced amorphization in the compounds, respectively. The
composition range over which quasicrystals are successfully obtained by
the liquid quenching technique[12] is marked as LQq.

absence of amorphization, respectively. Also given in the diagram is the composition range, LQq, over which quasicrystals can successfully be produced by liquid phase quenching[12]. It is evident that the three compounds which can be rendered amorphous, are all located within a composition range where the liquidus shows a large drop to form a deep valley in the phase diagram, and that the compositions of the individual compounds are very close to the composition range LQq.

Similar linkage between the amorphization tendency and the position of the compounds in the phase diagram is observed in almost all the alloy systems examined (Here both equilibrium and non-equilibrium liquidus are taken into account).

Based on these observations, it can be said that the tendency toward the C-A transition in metallic compounds under electron irradiation is best correlated with the position of the compounds in the T-C phase diagram. Namely, those compounds whose position in the phase diagram is close to the bottom of a deep valley of liquidus in the diagram have a strong tendency toward the C-A transition, while those away from such a valley show little tendency toward the C-A transition. This conclusion is valid also in non-metallic compounds[13]. The reason why such parameters as the form of the phase diagram and the relative position of the compounds in the diagram are a good guide to the amorphization tendency , in contrast with such single parameters as listed in Table 1, may be that the former parameters involve more complete information as to the interaction among the constituent atoms. All these facts suggest that the main factor controlling the amorphization tendency of compounds under electron irradiation is chemical in its nature and such parameters as the manifold in coordination, or in other words the extent to which a variation is admissible in the atomic coordination in the material, play an essential role in determining the ease with which the amorphization takes place.

ACKNOWLEDGEMENTS

The authors are grateful to Dr. M. Komatsu and to Messrs. K. Yoshida and T. Sakata for their help in the maintenance of the 3 MV HVEM. This work was supported in part by the Ministry of Education, Science and Culture, Japan, through a grant-in-aid for scientific research on priority areas (grant number 03205084).

REFERENCES

1. Thomas, G., Mori, H., Fujita, H. and Sinclair, R., Scripta Metall., 1982, 16, 589-92.
2. Mori, H. and Fujita, H., Japan. J. Appl. Phys., 1982, 21, L494-96.
3. Mogro-Campero, A., Hall, E. L., Walter, J. L. and Ratkowski, A. J., Metastable Materials Formation by Ion Implantation, ed. Picraux, S.T. and Choyke, W.J., North-Holland, Amsterdam, 1982, pp.203-10.
4. Mori, H., Fujita, H., Tendo, M. and Fujita, M., Scripta Metall., 1984, 18, 783-88.
5. Tendo, M., Diploma Thesis, Osaka University, Osaka, March 1983.
6. Nakajima, M., Diploma Thesis, Osaka University, Osaka, March 1987.
7. Mori, H. and Fujita, H., Non-Equilibrium Solid Phases of Metals and Alloys, Japan Inst. Metals, Sendai, 1988, pp.93-96.
8. Matzke, H. and Whitton, J. L., Can. J. Phys., 1966, 44, 995-1010.
9. Brimhall, J. L., Kissinger, H. E. and Charlot, L. E., Radiat. Eff., 1983, 77, 273-93.
10. Massalski, T. B., Proc. 4th Int. Conf. on Rapidly Quenched Metals, ed. Masumoto, T. and Suzuki, K., Japan Inst. Metals, Sendai, 1981, pp.203-08.
11. Inoue, A., Kitamura, A. and Masumoto, T., Trans. Japan Inst. Metals, 1979, 20, 404-06.
12. Inoue, A., Arnberg, L., Lehtinen, B., Oguchi, M. and Masumoto, T., Met. Trans., 1986, 17A, 1657-64.
13. Inui, H., Mori, H. and Fujita, H., Acta Metall., 1989, 37, 1337-42.

DISORDERING AND AMORPHIZATION OF Zr_3Al
BY 3.8 MeV Zr^{3+} ION BOMBARDMENT

F. C. CHEN[*], D. F. PEDRAZA[†] AND A. J. ARDELL[*]
[*]Department of Materials Science and Engineering, University of California,
Los Angeles, CA 90024
[†]Metals and Ceramics Division, P. O. Box 2008, Oak Ridge National Laboratory,
Oak Ridge, TN 37831

ABSTRACT

The ordered intermetallic compound Zr_3Al was irradiated with 3.8 MeV Zr^{3+} ions at various fluences up to 5×10^{12} ions/mm^2 at a temperature of 250 °C and the irradiation-induced microstructures were investigated by transmission electron microscopy. Disordering began at the lowest dose, 0.0033 dpa, and complete loss of chemical long-range order occurred at a dose of 0.33 dpa. The onset of amorphization was also observed at this dose. Electron diffraction patterns from irradiated samples showed satellite reflections along $\langle 011 \rangle$ in thin foils in [100] orientation and streaking along $\langle 11\bar{1} \rangle$ in foils oriented [011]. These diffraction effects are attributed to the presence of irradiation-induced microstructural defects that, when imaged in dark field, resemble rows of dislocation loops. A model is proposed for these arrays of loops, which are suggested to have Burgers vectors of the Frank type. The model accounts for the contrast effects observed in the images and the streaking and satellites seen in the diffraction patterns. At the highest dose, 1.6 dpa, a new phase, Zr_5Al_3, appeared unexpectedly, most likely as a consequence of irradiation-induced solute segregation.

INTRODUCTION

It is now well known that disordering and, in many cases, subsequent amorphization of the ordered $L1_2$ alloy Zr_3Al can be induced by irradiation with charged particles. This has been demonstrated using 0.5 to 2.0 MeV C^+, N^+ and Ar^+ ions at temperatures ranging from -243 to 477 °C [1-3], 1 MeV Kr^+ ions at 22 °C [4], 2 MeV protons at -192 °C [5], and 1 MeV electrons at -263 to 502 °C [4,6]. However, there apparently have been no studies concerning its behavior when irradiated by self-ions, wherein the bombarding species represents one of the components of the irradiated alloy. Apart from the possibility of shifting the phase equilibria of the material because of compositional changes brought about by the implantation of the self ions at large fluences, other potential alloying effects are eliminated entirely while retaining the advantages of producing radiation damage at relatively high rates. At small fluences the alteration of the composition is negligible, and the only effect of irradiation is caused by displacement damage. In this paper we report the microstructural changes that occur in Zr_3Al under irradiation with energetic 3.8 MeV Zr^{3+} ions at a temperature of 250 °C.

EXPERIMENTAL PROCEDURES

Details of the preparation of the Zr_3Al alloy are described elsewhere [7]. Samples of the material in the form of 3 mm disks were irradiated at the Triple Ion Beam Facility at the Oak Ridge National Laboratory using 3.8 MeV Zr^{3+} ions at a temperature of 250 °C. Seven different fluences were used, ranging from 10^{10} ions/mm^2 to 5×10^{12} ions/mm^2. The damage and ion deposition curves, calculated with the PC version of TRIM 89, assuming a threshold displacement energy, E_d, of 25 eV, are shown in Fig. 1, where it is seen that the range of the 3.8 MeV Zr^{3+} ions is approximately 2 µm (Fig. 1a). The doses at the depths corresponding to the maximum damage for each fluence are shown in Table 1 in units of displacements per atom, dpa. The average concentrations of implanted Zr ions are also shown in Table 1. The doses were calculated from the damage curve by multiplying the ordinate by $øt/N_v$, where ø is the ion flux, t is the irradiation time and N_v is the number of atoms per unit volume. These doses are lower than those reported previously [7] because of the higher value of E_d used (25 cf. 20 eV), consistent with previous work [1-4].

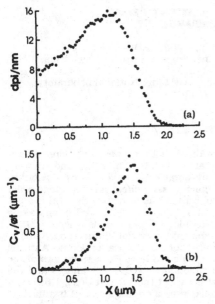

FIGURE 1. TRIM code calculations of (a) the damage displacement curve and (b) the ion deposition curve for 3.8 MeV Zr^{3+} ions incident on Zr_3Al. In (a) dpi represents displacements per ion. In (b) C_v represents the number of Zr ions deposited per unit volume; øt is the fluence.

TABLE 1

Relevant values of the parameters characteristic of the 3.8 MeV Zr^{3+} ion irradiations of Zr_3Al used in this investigation.

Fluence (ions/mm^2)	max. dose (dpa)	C_v (ions/mm^3)
1×10^{10}	0.0033	5.26×10^{12}
5×10^{10}	0.0167	2.63×10^{13}
1×10^{11}	0.0330	5.26×10^{13}
3×10^{11}	0.1000	1.58×10^{14}
1×10^{12}	0.3300	5.26×10^{14}
5×10^{12}	1.6700	2.63×10^{15}

For transmission electron microscopy (TEM) analysis, the irradiated sides of the disk samples were thinned to the peak damage region, then back-thinned to perforation using a South Bay Technology, Inc. Model 550C Jet Thinning Instrument. For the irradiated sides of the samples, two electrolytes were used: a solution of 20% perchloric acid in acetic acid at room temperature that yielded a polishing rate of 2.09 µm/s; a solution of 5% perchloric acid in ethanol at -50 °C that yielded a polishing rate of 0.13 µm/s. The polishing time required to reach the peak dose was determined by dividing the total damage depth by the polishing rate. To avoid overshooting the peak damage depth, the actual polishing times used were slightly less. For the perchloric acid in acetic acid solution the actual polishing time was reduced by 25%. With the perchloric acid in ethanol electrolyte greater control was possible due to its much lower rate of attack; this enabled the peak damage region to be more closely approached. Thus, the actual polishing time was curtailed by just 5%. In either case, the error introduced into the calculated peak doses by thinning just short of the maximum damage depth is not greater than approximately 10%. After electropolishing, the samples were examined in a JEOL model 100CX TEMSCAN transmission electron microscope (TEM) operating at 100 keV.

Values of the long range order parameter S were determined from the formula [2]

$$S = \left[\frac{(I_s/I_f)_{irr}}{(I_s/I_f)_{unirr}} \right]^{1/2} .$$

where I_s and I_f refer to the intensities of the superlattice and fundamental reflections, respectively, taken from an electron diffraction pattern. These intensities were assessed by scanning the negatives of the diffraction patterns with an Optronics International P-1000 microdensitometer. To ensure consistent intensity readings, the samples were tilted so that the intensities of the low order reflections were symmetric about the transmitted spot. The {200}/{100} and {220}/{110} pairs of reflections in the ⟨001⟩ zone orientation were used in the calculations of S. Between 2 and 5 pairs of intensities were measured for each determination of S. The intensities were measured from films that would normally be regarded as underexposed to ensure that the much more intense fundamental reflections did not saturate the film.

RESULTS

TEM analysis of the irradiated Zr_3Al shows that the Zr^{3+} ions are disordering the alloy. A series of diffraction patterns illustrating the gradual loss of order from the unirradiated state to a dose of 0.33 dpa is shown in Fig. 2. It can be seen that at the very lowest dose, 0.0033 dpa, disordering has already commenced, and at a dose of 0.33 dpa the Zr_3Al is essentially completely disordered. Residual intensity is visible at the positions of the superlattice reflections in the samples irradiated to 0.33 dpa (Fig. 2) in diffraction patterns exposed for longer times than those normally used for measuring the values of S. Additionally, at 0.33 dpa amorphization has begun, as evidenced by the faint amorphous ring.

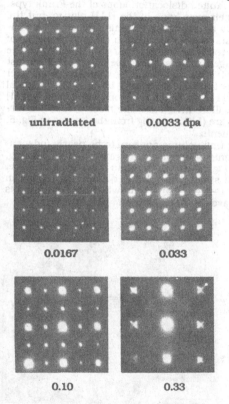

FIGURE 2. Illustrating the progression of the loss of long range order with increasing dose in 3.8 MeV Zr^{3+}-ion irradiated Zr_3Al. All zone axes are [001].

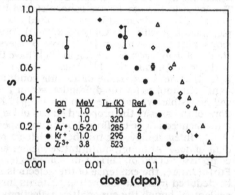

FIGURE 3. The variation of long-range-order parameter, S, with dose in Zr_3Al under a variety of irradiation conditions

The variation of S with dose is shown in Fig. 3, in which the data of Howe and Rainville [2], Koike et al. [4], Carpenter and Schulson [6] and Rehn et al. [8] are included for comparison. Except for the unexpectedly large reduction of S at the smallest dose, 0.0033 dpa, the agreement with the previous work is very good.

It was noted previously [7] that the Zr^{3+} irradiations cause mottled contrast visible in the Zr_3Al grains under bright field imaging conditions, and that this contrast is absent in unirradiated samples. The possibility that the defects might somehow be electropolishing artifacts was investigated by preparing thin foils of unirradiated material using the same procedure as that for the irradiated samples, so that a region near the free surface of unirradiated electropolished Zr_3Al was examined. These specimens showed no evidence of the defects observed in the irradiated samples, providing assurance that the defects observed are irradiation induced. Figure 4 shows that the mottled contrast is apparent even at the lowest dose, 0.0033 dpa. The arrows indicate some of the defects, which are readily visible in bright field. The diffraction pattern inset in Fig. 4 shows that streaking occurs along $\langle 11\bar{1}\rangle$ directions in [011] diffraction patterns, and provides a clue into the crystallography and possible morphology of the defects. The defects lie on $(11\bar{1})$ and $(\bar{1}1\bar{1})$ and can be either planar or rod-shaped, based on their projected images. Figure 5 is a centered dark-field image of a sample irradiated to 0.33 dpa taken using a portion of a streak close to the $[11\bar{1}]$ reflection, shown inset in the diffraction pattern. Despite the difficult imaging conditions, the images suggest that the defects are not planar, but resemble rows of circular loops, approximately 6 nm in diameter, sharing a common axis.

To examine the possibility that the defects seen in Fig. 5 are rows of dislocation loops, which have not been reported previously in irradiated Zr$_3$Al, a diffraction contrast experiment was performed on a large grain in a sample irradiated to 0.033 dpa. Figure 6 shows the same area of a foil in [011] orientation, imaged using the $(11\bar{1})$ (Fig. 6a) and $(\bar{1}1\bar{1})$ (Fig. 6b) reflections. It is evident that a different set of defects goes out of contrast in each of the figures, suggesting that they might be rows of dislocation loops. Also, based on what is observed in the diffraction patterns, particularly the satellite reflections along $\langle 110\rangle$ in thin foils oriented [001] [7], the streaks in the diffraction patterns inset in Figs. 4 and 5 are relrods rather than relplanes.

A model of the defect structure consistent with all these observations is shown schematically in Fig. 7. It consists of stacks of faulted dislocation loops of the Frank type. The loops in the stacks are concentric, with their axes lying in the $\{11\bar{1}\}$ planes, but the planes of the individual loops are tilted towards another plane in this family. For example, the stack with its axis lying in $(11\bar{1})$ consists of loops parallel to $(\bar{1}1\bar{1})$. The stacks of loops with axes lying in $(\bar{1}1\bar{1})$ have Burgers vectors of the type $\pm^a/_3[11\bar{1}]$, whereas the stacks of loops with axes lying in $(11\bar{1})$ have Burgers vectors of the type $\pm^a/_3[\bar{1}1\bar{1}]$. Contrast at the loops parallel to $(\bar{1}1\bar{1})$ vanishes for $\mathbf{g} = (11\bar{1})$ because $\mathbf{g}\cdot\mathbf{b} = \pm^1/_3$, while the set parallel to $(11\bar{1})$ under these same diffracting conditions is visible because $\mathbf{g}\cdot\mathbf{b} = \pm1$. The identical argument applies for the diffraction vector $\mathbf{g} = (\bar{1}1\bar{1})$, for which the loops with $\mathbf{b} = \pm^a/_3[\bar{1}1\bar{1}]$ will be visible, while those with $\mathbf{b} = \pm^a/_3[11\bar{1}]$ will be invisible. The most probable directions of the axes of the proposed stacks of loops are $\langle\bar{1}12\rangle$, judging from the images in Fig. 5, but this needs to be confirmed by future experiments.

Interestingly, the number of defects per unit volume appears to be nearly independent of dose. Although quantitative measurements of the defect concentration are lacking, comparison of various bright-field images (e. g. Figs. 4 and 6) suggests that this is the case. Furthermore, the existence of the defects is not strongly coupled to disordering. This can be deduced from the diffraction patterns inset in Figs. 4 and 5, in which the streaking is evident, although the superlattice reflections have disappeared in the latter figure.

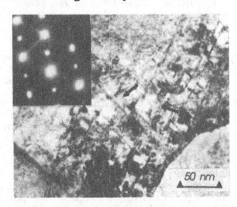

FIGURE 4. Bright-field TEM image of a sample of Zr$_3$Al irradiated to the lowest dose, 0.0033 dpa. The arrows indicate the defects referred to in the text. These produce the streaking in the diffraction pattern, correctly oriented, inset.

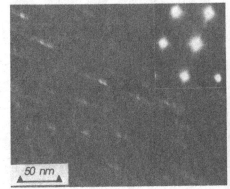

FIGURE 5. Dark-field TEM image of a sample of Zr$_3$Al irradiated to 0.33 dpa. The circle in the diffraction pattern inset, which is correctly oriented, shows the part of the streak used to take the dark field image.

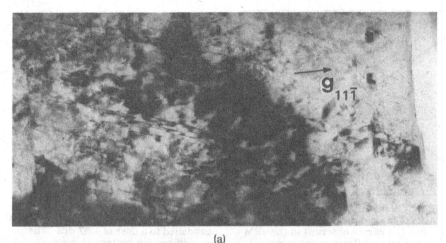

(a)

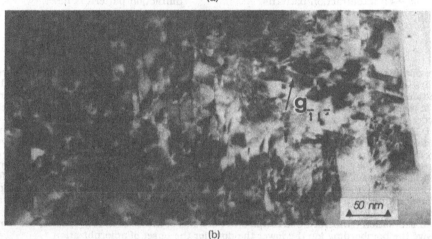

(b)

FIGURE 6. Bright field images of the same region of a sample of Zr₃Al irradiated to a dose of 0.33 dpa, taken under the diffraction conditions indicated. A different set of defects becomes invisible in each image. This is most apparent near the center of each photograph, where two-beam diffracting conditions are closer to being realized.

At the very highest dose, 1.67 dpa, the presence of the phase Zr₅Al₃ was noted. This has the tetragonal D8ₘ structure, with a = 1.1049 nm and c = 0.5396 nm. The appearance of this phase is seen in Fig. 8, which also shows a [001] diffraction pattern. The Zr₅Al₃ phase is quite featureless, but the diffraction pattern indicates the coexistence of an amorphous phase. These observations are preliminary, and more work is required to sort out the details of the microstructures of the specimens irradiated to the larger doses. Nevertheless, Zr₅Al₃ was not found in the unirradiated state, despite extensive examination, whereas several strongly diffracting grains of Zr₅Al₃ were readily observed in the Zr₃Al irradiated to 1.67 dpa. The Zr₅Al₃ phase is clearly irradiation-induced and attributable to displacement damage, since the concentration of implanted Zr atoms is negligible, and in any event would push the phase equilibria in the direction of α-Zr from Zr₃Al or to Zr₃Al from Zr₂Al (which is always present in minor amounts [10]).

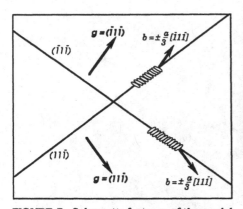

FIGURE 7. Schematic features of the model of stacks of faulted dislocation loops used to explain the effects observed in the TEM images and diffraction patterns.

FIGURE 8. A bright-field TEM image of the Zr_5Al_3 phase observed in a sample of Zr_3Al irradiated to a dose of 1.67 dpa. The diffraction pattern is inset.

DISCUSSION

There are interesting differences between the results of our work and those of other investigators. For example, the work of Howe and Rainville [1-3] closely approximates the conditions of the present study. They bombarded Zr_3Al with 0.5 to 2 MeV C^+, N^+ and Ar^+ ions at temperatures as high as 477 °C. At temperatures between 12 °C and 420 °C complete disorder was attained at doses ranging from 1.4 to 2.8 dpa. This compares with a dose of only 0.33 dpa to induce total disorder using 3.8 MeV Zr^{3+} ions at 250 °C. The C^+, N^+ and Ar^+ ions are apparently less effective in producing damage than the Zr^{3+} ions, at least over the temperature range studied by Howe and Rainville. This perhaps can be attributed to the greater mass of the Zr^{3+} ions. Among the C^+, N^+ and Ar^+ ions the most massive is Ar^+, which has a mass of 39.9 atomic mass units (amu), compared with a value of 91.2 amu for Zr^{3+} ions. Rehn et al. [8] observed the onset of amorphization in Zr_3Al at about 0.15 dpa on room-temperature bombardment with 1 MeV Kr^+ ions, which have nearly the same mass as Zr^+ (83.8 amu). This result adds to the evidence that the onset of amorphization at a fixed irradiation temperature is coupled to the mass of the bombarding species, the more massive the bombarding ion the lower the dose for the onset of amorphization.

Howe and Rainville also noted the presence of irradiation-induced defects, though theirs differed significantly from those observed in this study. In their investigation the defects appeared as dark, roughly circular spots in dark-field images. They postulated the defects to be three-dimensional vacancy or interstitial clusters with spherically symmetric strain fields. We attempted to image the irradiated samples in dark field using the same reflections as those used by Howe and Rainville, but did not find any features resembling the dark spots they observed. There can be no doubt that the defects seen in our specimens (Figs. 4 to 6) have not been observed under any other conditions of irradiation. Since they produce such distinctive diffraction effects it is highly unlikely that they could have been overlooked by other investigators. They seem to be unique to material irradiated by Zr^{3+} ions, for reasons which are not known.

It is necessary to consider whether the proposed model of these defects is consistent with our other observations. If the defects were truly rod shaped they would produce relplanes, which could not account for the satellites consistently observed at reflections in the [001] diffraction pattern. This supports the conclusion that the essential components of the defects must be planar in nature. Faulted dislocation loops produce intense streaking in the form of relrods normal to their own planes, i. e. planes of the type {$11\bar{1}$}, which readily accounts for the satellite reflections. However, the planes of the individual dislocation loops in the stack should be parallel to the [011] zone axis, and should therefore appear as short segments in projected images, such as the dark-field image in Fig. 5.

We believe that the circular appearance of the features seen in this figure is due to a combination of tilting of the foil from the exact [011] zone axis, the effect of spherical aberration in the image and the contribution of specimen drift during the relatively long exposure time (64 s) used to photograph the image. Whereas the model successfully explains the features seen in TEM images and diffraction patterns, we have as yet made no attempt to determine whether the postulated stacks of faulted dislocation loops are energetically stable. The model must therefore be viewed with this qualification in mind.

In a study involving electron bombardment of Zr_3Al, Koike et al. [4] observed diffuse intensity along the $\langle 011 \rangle$ directions in foils oriented [100]. The diffuse intensity observed by them can be easily duplicated in our diffraction patterns recorded at moderate exposure times; longer exposures produce the intense satellites seen in Fig. 2, 0.33 dpa. Koike et al. [4], following the suggestion of Rehn et al. [8] and Okamoto et al. [11], attribute the diffuse intensity to softening of the shear elastic constant $C' = (C_{11} - C_{12})/2$, whereas in our case similar diffraction effects are definitely caused by the irradiation-induced microstructural defects visible in Figs. 4 to 6. Koike et al. observed what they call black-dotted contrast in their samples electron irradiated at 10 and 57 K, and attributed it to the presence of lattice defects of an unspecified nature. Unfortunately, they did not publish images of their samples irradiated at 295 K, hence it is not possible to determine whether these lattice defects would condense at higher irradiation temperatures into arrays of loops similar to those observed by us.

An implication of the existence of dislocation loops is that point defects (probably, though not necessarily, interstitials) are removed from the irradiated microstructure, thereby reducing the driving force for amorphization. Since current theories of amorphization [12-17] invoke a significant buildup of point defects as a mechanism of raising the free energy of the crystalline state, the presence of the defects we have observed would appear to be inconsistent with the observation that the onset of amorphization occurs at a relatively low dose for heavy ions (~0.33 dpa). We believe, however, that since the concentration of the loop-type defects saturates at low doses, the new point defects created on continued irradiation enable their continued buildup. This buildup, combined with the destruction of long-range order, continues until the concentration of point defects necessary for amorphization is reached.

At the highest dose, 1.67 dpa, the phase Zr_5Al_3 was found (Fig. 8). This is probably a result of irradiation-induced solute segregation, and a possible mechanism that explains this phenomenon has been proposed [9]. It states that the fraction of atoms among the interstitials created during irradiation by the production of Frenkel pairs will not be the same as the proportion of atoms in the alloy. Instead, the interstitials will be populated preferentially by the undersized atoms, and interstitial fluxes to point defect sinks will drag the undersized solute atoms with them; Al is the undersized atom in Zr-Al alloys. The formation of Zr_5Al_3 requires that Al be added to the existing phases in the material. Since Zr_2Al is commonly present in Zr_3Al alloys, segregation of Al to particles of Zr_2Al could easily account for the formation of Zr_5Al_3.

SUMMARY

1. Under bombardment with 3.8 MeV Zr^{3+} ions at 250 °C, Zr_3Al begins to disorder at a dose as low as 0.0033 dpa and is very nearly completely disordered at a dose of 0.33 dpa. At that dose, the onset of amorphization also occurs, as evidenced by the amorphous ring in the $\langle 100 \rangle$ electron diffraction pattern.

2. Defects form as a result of irradiation, causing satellite reflections along $\langle 011 \rangle$ in [100] electron diffraction patterns, and producing streaks parallel to $\langle 11\bar{1} \rangle$ in [011] electron diffraction patterns. When imaged in dark field, they resemble rows of dislocation loops. It is proposed that the defects consist of concentric stacks of faulted dislocation loops of the Frank type. These proposed arrays are capable of producing the diffraction contrast effects seen in the TEM images and the streaking seen in the diffraction patterns.

3. The phase Zr_5Al_3 is seen at a dose of 1.67 dpa, and most likely forms as a result of irradiation-induced solute segregation.

ACKNOWLEDGEMENTS

Financial support for this research was provided by the U. S. Department of Energy, Division of Materials Science, under grant DE-FGO3-86ER45264 (F. C. Chen and A. J. Ardell) and under contract DE-AC05-84OR21400 with Martin Marietta Energy Systems, Inc. (D. F. Pedraza). We are grateful to Professor D. Eisenberg and Mr. D. Cascio of the Molecular Biology Institute at UCLA for assistance with the microdensitometer scans, to Mr. R. Lysse of UCLA for printing the micrographs, and to Mr. S. Cook of ORNL for performing the irradiations.

REFERENCES

1. Howe, L. M. and Rainville, M. H., A study of the irradiation behavior of Zr$_3$Al. J. Nucl. Mater., 1977, **68**, 215-34.
2. Howe, L. M. and Rainville, M. H., The nature of irradiation-produced damaged regions in ordered Zr$_3$Al. Phil. Mag. A, 1979, **39**, 195-212.
3. Howe, L. M. and Rainville, M. H., Ion bombardment of ordered Zr$_3$Al. Rad. Eff., 1980, **48**, 151-6.
4. Koike, J., Okamoto, P. R., Rehn, L. E. and Meshii, M., Amorphization in Zr$_3$Al with 1-MeV e$^-$ and Kr$^+$. Metall. Trans. A, 1990, **21**, 1799-1808.
5. Cheng, J., Yuan, M., Wagner, C. N. J. and Ardell, A. J., Fractographic fingerprinting of proton-irradiation-induced disordering and amorphization of intermetallic compounds. J. Mater. Res., 1989, **4**, 565-78.
6. Carpenter, G. J. C. and Schulson, E. M., The disordering of Zr$_3$Al by 1 MeV electron irradiation. J. Nucl. Mat., 1978, **73**, 180-9.
7. Chen, F. C., Ardell, A. J. and Pedraza, D. F., Miniaturized disk-bend testing and microstructure of 3.8 MeV Zr^{3+} irradiated Zr$_3$Al. In High Temperature Ordered Intermetallic Alloys IV, eds. L. A. Johnson, D. P. Pope and J. O. Stiegler, MRS Symp. Proc. Vol. **213**, Materials Research Society, Pittsburgh, PA, 1991, pp. 763-68.
8. Rehn, L. E., Okamoto, P. R., Person, J., Bhadra, R., and Grimsditch, M., Solid-state amorphization of Zr$_3$Al: evidence of an elastic instability and first-order phase transformation. Phys. Rev. Lett., 1987, **59**, 2987-90.
9. Okamoto, P. R. and Wiedersich, H., Segregation of alloying elements to free surfaces during irradiation. J. Nucl. Mat., 1974, **53**, 336-45.
10. Schulson, E. M., Flow and fracture of Zr$_3$Al. Int. Met. Rev., 1984, **29**, 195-209.
11. Okamoto, P. R., Rehn, L. E., Pearson, J., Bhadra, R. and Grimsditch, M., Brillouin scattering and transmission electron microscopy studies of radiation-induced elastic softening, disordering and amorphization of intermetallic compounds. J. Less-Comm. Metals, 1988, **140**, 231-44.
12. Limoge, Y. and Barbu, A., Amorphization mechanism in metallic crystalline solids under irradiation. Phys. Rev. B, 1984, **30**, 2212-5.
13. Pedraza, D. F., Mechanisms of the electron irradiation-induced amorphous transition in intermetallic compounds. J. Mater. Res., 1986, **1**, 425-41.
14. Simonen, E. P., Theory of amorphization kinetics in intermetallics. Nucl. Instr. Meth. Phys. Res., 1986, **B16**, 198-202.
15. Pedraza, D. F. and Mansur, L. K., The effect of point defects on the amorphization of metallic alloys during ion implantation. Nucl. Instr. Meth. Phys. Res., 1986, **B16**, 203-11.
16. Pedraza, D. F., Radiation-induced collapse of the crystalline structure. J. Less-Comm. Metals, 1988, **140**, 219-30.
17. Motta, A. T. and Olander, D. R., Theory of electron-irradiation-induced amorphization. Acta Metall. et Mater., 1990, **38**, 2175-85.

DISORDERING INDUCED BY VERY HIGH ENERGY HEAVY IONS IN METALLIC COMPOUNDS.

A. BARBU, A. DUNLOP, D. LESUEUR and G. JASKIEROWICZ
CEA/DTA/CEREM/DTM/Laboratoire des Solides Irradiés
Ecole Polytechnique, 91128 PALAISEAU cedex, FRANCE

ABSTRACT

It is now well established that swift ions are able to induce disordering by electronic excitations in metallic compounds at low temperature. It is shown here that at least in Ni Zr$_2$ this effect is also observed at room temperature.

INTRODUCTION

As an energetic ion penetrates into a solid, it loses its energy through two nearly independent processes: (i) collisions with the target electrons leading to ionization and electronic excitation and (ii) collisions with the target nuclei leading to atomic displacements.

It was recognized a long time ago that elastic collisions were able to induce structural modifications both in insulators and metallic materials. In contrast, it was known that electronic excitations could result in structural modifications in insulators along the path of swift ions, but were completely unable to damage metallic materials.

In this short paper we will show that this point of view is totally false. We will summarize our previously published articles [1,2] and give some new results.

MATERIALS AND METHODS

Several kinds of metallic materials with a wide range of ordering energies were selected for these studies:

a) Cu3Au with a low ordering energy which is known to be impossible to amorphize whatever the method used.

b) Zr3Al, NiTi, Ni-Zr compounds with high ordering energies which are known to be more or less amorphizable by elastic collisions during low energy ion or high energy electron irradiations [3,4].

The irradiations were performed at GANIL Caen and GSI Darmstadt. Ions ranging from Argon to Uranium with energies near 1 GeV (for which the ratio of electronic to nuclear energy loss is typically $(dE/dx)_e/(dE/dx)_n \approx 2000$) were used.

The microstructure of the regions located near the entrance surface of the ions (i.e. in which the ions are mainly slowed down by electronic excitations) was studied by means of electron transmission microscopy and X rays diffraction.

RESULTS AND DISCUSSION

After irradiation at 80 K of Cu3Au [5] and Zr3Al, nothing but rare contrasts that can be attributed to displacement cascades induced by elastic collisions has been observed.

In Ni-Zr compounds (tetragonal NiZr2 and orthorhombic NiZr) irradiated at 80K with 0.7 GeV Pb ions ($(dE/dx)_e$=4.8 keVÅ^{-1}) and 2.8 GeV U ions ($(dE/dx)_e$=5.7 keVÅ^{-1}) more or less continuous amorphous tracks have been observed [1]: in the case of NiZr2 they were quasi continuous and consisted of 8 nm diameter strongly overlapping amorphous spheres, whereas in NiZr they were extremely discontinuous. At high fluences, when the tracks overlap, the sample is totally amorphized and anisotropic growth, previously discovered in metallic glasses, appears [6]: it consists in an increase of the two dimensions of the sample perpendicular to the beam direction and in a reduction of the dimension parallel to the beam.

In order to check the effect of the irradiation temperature, we irradiated NiZr2 at room temperature with 1.6 GeV Pb ions ($(dE/dx)_e$=4.8 keVÅ^{-1}). As shown in figure1, it is clear that latent tracks are also induced at room temperature. They seem to be slightly more discontinuous than in the experiment performed at 80 K with 0.7 GeV Pb ions but it is not possible to know whether it is an effect of the temperature or of the velocity of the ions: $(dE/dx)_e$ has the same value for both energies, but the velocity is 1.5 times higher for the room temperature experiment.

In NiTi irradiated at 80K with 0.7 GeV Pb ions, no tracks were observed [2], but total amorphization appears for a fluence very close to that needed for total amorphization in NiZr2 (10^{13}ions cm^{-2}). As during high energy electron irradiations, X ray diffraction spectra show that the initial martensitic structure transforms into cubic B2 structure before amorphization. Anisotropic growth is again observed at high fluences (Figure 2).

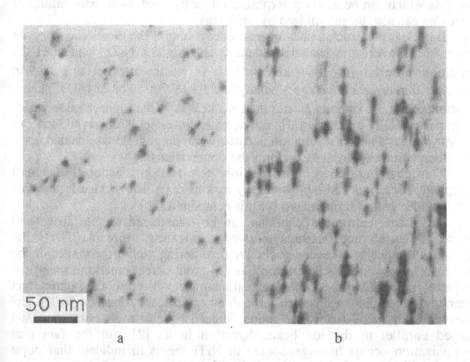

Figure1: NiZr$_2$ irradiated with 1.6 GeV Pb ions at room temperature at a fluence of 10^{11}ions cm^{-2}. Views under two different orientations: a) in the ion direction; b)tilted of 20°. The number of tracks and incoming ions are identical.

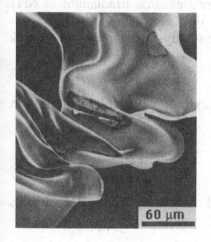

Figure 2: SEM observation of the ripples due to anisotropic growth induced in a thin area of a NiTi sample prepared for transmission electron microscopy observations (7. 10^{13} 0.7GeV Pb cm^{-2}).

It is probably too early to derive a general criterion for disordering induced by swift ions in metallic compounds. However it seems that only materials which can be easily amorphized either by irradiation, ball-milling or quenching can also be amorphized by swift ions.

It is clear than such a phenomenon occurs only if the electronic stopping power lies above a material dependent threshold. For $NiZr_2$ and NiTi the threshold value is close to 4 keV $Å^{-1}$. It is worthwhile noticing that in insulators this threshold can be smaller than 1 keV $Å^{-1}$. For Zr_3Al, if such a threshold exists, its value is greater than 4.3 keV $Å^{-1}$, the highest value which can experimentally be reached. However, when no effect is observed, one can always wonder, whether restoration occured when the sample was heated up to room temperature in order to perform the observations.

The observation by transmission electron microscopy of small amorphous zones induced by low energy ions in Zr_3Al shows clearly that this explanation is not relevant at least for this compound [10].

How the electronic excitations can be transferred to the lattice and induce structural modifications remains a puzzling problem [7]. Is the amorphization due to a sudden collective disordering within the wakes of the ions or to a more local phenomenon like (1) point defect creation around the ion paths and/or (2) to the local chemical disorder induced by the point defect generation when they are created through a replacement sequence [8,9]? The observation of small dislocation loops induced by electronic excitation and aligned parallel to the ion beam direction in Ti [2] and the fact that amorphization occurs homogeneously in NiTi seems to indicate that point defects could play a role in the disordering phenomena.

Within this frame, the behaviour of Zr_3Al can be explained by assuming that the point defect concentration created in each individual track is not sufficient to trigger the crystalline to amorphous transition and that the point defects are not stable enough to be accumulated by track overlap.

Similar discrepancies are observed during electron irradiations of NiTi and Zr_3Al: NiTi is easily amorphized by 1 MeV electrons at low and room temperature [4], whereas Zr_3Al can only be totally amorphized below 60 K [11].

The authors are very grateful for the elaboration of $NiZr_2$ samples in C.E.C.M. Vitry by J. Bigot and E. Gaffet.

REFERENCES

1) Barbu A., Dunlop A., Lesueur D. and Averback R.S. : Europhysics Letters,1991, 15, 37

2) Barbu A, Dunlop A, Henry J., Lesueur D., Lorenzelli N. : in "PM'91",Materials Science Forum. In press.
3) Moine P. Rivière J.P.Junqua N. and Delafond J. :Nucl. Instr. and Meth. 1983,209-210, 267
4) Mori H., Fujita H., Tendo M., Fujita M.: Scripta. Met. 1984, 18, 783
5) Barbu A., Martin G., Toulemonde M., Jousset J.C.: C. R. Acad Sci. Paris, 1984, tome 299, série II, n° 8, 409
6) Klaumünzer S., Schumacher G. : Phys. Rev Lett. 1983, 51, 1987
7) Dunlop A. and Lesueur D. :PM'91, Materials Science Forum. In press.
8) Limoge Y. and Barbu A. : Phys. Rev. B, 1984, 30, (4), 2212
9) Pedraza D.F. : J. Mat. Res. , 1986, 1, (3), 425
10) Howe L.M. and Rainville M.H. J. Nucl. Mat. vol. 68, 1977, 215
11) Koike J., Okamoto P.R., Rehn L.E. and Meshii M. Met. Trans. 21A, 1990, 1799

STABILITY OF CRYSTALS AGAINST MELTING AND VITRIFICATION

Hans J. FECHT
Universität Augsburg, Institut für Physik
Memmingerstr. 6, W-8900 Augsburg, F.R.G.

ABSTRACT

The long-range atomic order of crystalline alloys can be destroyed by a variation of temperature resulting in regular melting or by non-equilibrium solid state processing resulting in glass formation. The destabilization of the crystalline lattice and evolution of the amorphous phase is triggered by a variety of methods including, for example, alloying beyond the equilibrium limit, mechanical deformation, irradiation, thermal annealing of thin film superlattices and application of high pressures. Under local equilibrium conditions, the crystal-to-glass transition exhibits similar features as equilibrium melting, whereas under non-equilibrium conditions catastrophic vitrification can occur at the T_0-temperature extended into the glass forming regime. It is shown that randomly frozen-in impurities or lattice defects cause rounding of the first-order melting transition to an isentropic transition resulting in a triple point between supersaturated crystal, undercooled liquid and glass at the crossover of the ideal glass transition and melting lines.

INTRODUCTION

The interest in melting has recently been revitalized with the discovery of the destabilization and amorphization of highly metastable crystalline alloys. In practice, solid state amorphization can be induced by several methods, such as hydriding of intermetallic compounds [1], thermal annealing of multilayered thin film specimen [2] or metastable crystalline bulk samples [3], mechanical deformation [4] and mechanical alloying of powder particles [5] or thin metal sheets [6], ion beam [7], electron beam [8] or ultraviolet irradiation [9], compression [10], de-compression [11] etc. For this process to occur both thermodynamic conditions and kinetic constraints must be satisfied with (i) the crystalline phase being prepared in or driven to a non-equilibrium state energetically above that of the amorphous phase, and (ii) the formation of crystalline equilibrium phases or phase mixtures - often stoichiometric intermetallic compounds - being frustrated by the kinetic constraints imposed at the low reaction temperatures. Then the amorphous phase is found to be formed instead of the stable phase or phase mixture. Such a transformation to a non-equilibrium state of matter can be described by standard thermodynamic methods if the kinetic constraints are defined appropriately [12]. Within such a framework, a thermodynamic scenario is developed and described here relating the crystal-to-glass transition to regular melting (crystal to liquid) and to the glass transition itself (liquid to glass).

MELTING AND GLASS FORMATION

Glasses are conventionally produced from the highly undercooled liquid state. The levels of undercooling required for glass formation can be achieved either kinetically by rapid quenching methods or during slow cooling of liquid droplet emulsion samples by the effective control of potent heterogeneous nucleation sites [13]. The nature of the glass transition is still a matter of controversy. Experimental evidence and theoretical models suggest the glass transition to be either first order (free volume approach [14]), second order [15], third order [16] or no phase transition at all, e.g. kinetic freezing [17]. However, there is general agreement that the maximum undercooling of a liquid is limited to the isentropic temperature in order to avoid the paradoxical situation (Kauzmann paradox [18]) where the configurational entropy of the disordered state (metastable liquid) becomes smaller than the configurational entropy of the ordered equilibrium state (stable crystal). Consequently, as long as crystallization can be prevented, the undercooled liquid will freeze to a glass close to the ideal glass transition temperature, T_{go}, where the entropy difference ΔS between liquid and crystal would vanish. In practice, the glass transition sets in somewhat above T_{go} with the deviation from internal equilibrium and related relaxation rates being determined by the applied cooling rate. Therefore, only at infinitely slow cooling, would ergodic conditions prevail and all configurations in phase space be sampled, resulting in a second order phase transition at T_{go}. On realistic time scales the system exhibits non-ergodic behavior with the glass having excess entropy in comparison to the crystal.

However, taking into account experimental values for the specific heat of the stable and highly undercooled liquid and measured values for the enthalpy of crystallization of amorphous alloys, the undercooled liquid at the glass transition temperature of, for example Pd-Ni-P [19] and Au-Pb-Sb alloys [20] is found to exhibit only a very small excess entropy in comparison with the stable crystalline phase. CALPHAD calculations by Bormann [21] support such a trend and give an estimated excess entropy at the glass transition temperature of about 0.1 of the entropy of fusion for Zr-50a/o Ni alloys. In addition, specific heat, thermal expansion and Mössbauer spectroscopy data on $Pd_{40}Ni_{40}P_{20}$ glasses indicate that the glass transition can be approached under internal equilibrium conditions, i.e. that T_g (at sufficiently small rates) becomes independent of the rate of temperature change [22]. These observations suggest that for fully relaxed metallic glasses the glass transition can be treated as an equilibrium phenomena. In addition, the specific heat measurements and CALPHAD estimates suggest that the actual glass transition temperature can be well approximated (within less than thirty degrees) by the Kauzmann temperature [19-21].

Consequently, the glass can be considered thermodynamically as a frozen highly undercooled liquid and therefore, the transition from crystal to glass can be compared with the melting process. Despite the strong interest in melting over the past hundred years, a generally accepted theory for the crystal-liquid transition has not been developed [23]. Thermodynamically, the melting point T_m is defined as the temperature where the Gibbs free energies or corresponding chemical potentials of liquid and crystal are equal. All melting theories are based on some form of instability or catastrophe which are either vibrational [24], elastic [25], isochoric [26,27], defective [28] or entropic [29] in nature. Furthermore, these melting theories rely on a homogeneous melting process throughout the entire crystal, even though it is a well known experimental fact that melting starts at lattice defects, such as surfaces and internal interfaces. Thus, it is observed experimentally that regular melting occurs at T_m, i.e. before an instability develops and, as such, represents a first order phase transition as indicated by a well-defined crystal / liquid interface and an entropy of fusion on the order of 1 k_B. If the effective heterogeneous nucleation sites for the formation of a liquid nucleus can be eliminated, considerable crystalline superheating can be achieved as shown for a variety of materials including oxides [30], pure metals [31] and solid Helium [32].

It is generally agreed that thermally induced vibrations of atoms in solids play a major role in melting [33]. A simple vibrational model of Lindemann predicts a lattice instability when the root-mean-square amplitude of the thermal vibrations reaches a certain fraction f of the next neighbor distances [24]. However, the Lindemann constant f varies considerably for different substances because lattice anharmonicity and soft modes are not considered, thus limiting the predictive power of such a law considerably. Furthermore, Born proposed the collapse of the crystal lattice to occur when one of the effective elastic shear moduli vanishes [25]. Experimentally, it is found instead that the shear modulus as a function of dilatation is not reduced to zero at T_m and would vanish at temperatures far above T_m for a range of different substances [34]. In addition, the role of equilibrium lattice defects is not negligible close to the melting point. Due to the entropy stabilization of vacancies close to the melting point these equilibrium vacancies contribute to the specific heat at constant pressure exhibiting an increase in addition to the linear term by [29]

$$c_p^v = \frac{A (\Delta H^v)^2}{k_B T^2} \cdot \exp\left(-\frac{\Delta H^v}{k_B T}\right) \tag{1}$$

where ΔH^v represents the heat of formation of a vacancy in the dilute limit and the prefactor A equals $\exp(\Delta S^v / k_B)$, ΔS^v being the activation entropy of a defect (about 2k_B [35]). Vacancies in metals are strongly related with the cohesive energy and as such, the heat of formation of a vacancy ΔH^v is related to the equilibrium melting point T_m by the following universal relationship [36]:

$$\Delta H^v = 8 \cdot 10^{-4} \cdot T_m \text{ (eV/K-atom)} \tag{2}$$

Extrapolations of measured heat capacity data into the superheated crystal regime allow to determine the entropy of a superheated crystal as function of vacancy concentration. Such data are available for Al [37], W [38] and Nb [39]. From the experimental values for the crystalline phase and an approximation for the excess specific heat of the undercooled liquid Δc_p^L for pure metals as [40]

$$\Delta c_p^L = 9 - 9\, T / T_m \quad (J\,mol^{-1}K^{-1}) \tag{3}$$

the enthalpy $\Delta H\ (= \int c_p\, dT)$, entropy $\Delta S\ (= \int c_p/T\, dT)$ and free energy $\Delta G\ (= \Delta H - T\Delta S)$ of liquid in comparison with crystalline Al (reference, $G = 0$) can be evaluated. The results are shown in Fig. 1 indicating the specific heat (a) and the thermodynamic stability of liquid versus crystal indicated by the Gibbs free energy ΔG in (b). The Kauzmann temperature, T_{go}, where the excess entropy ΔS of the undercooled liquid vanishes is found at about 0.25 T_m in agreement with Turnbull's approach [41]. In analogy to the isentropic temperature delimiting liquid undercooling an isentropic temperature is found above the melting point delimiting crystalline superheating. Thus, the Kauzmann paradox can be inverted and applied to melting, suggesting that the stability of a superheated crystal (including equilibrium lattice defects) is ultimately limited by an isentropic condition, where the entropy of the superheated crystal would become larger than that of the liquid. At this temperature T^i_s the interfacial tension between crystal and liquid would also vanish [42] and melting would become catastrophic. This temperature is located for Al at 1.38 T_m and can vary for different substances, for example between 1.18 T_m for W and 1.43 T_m for Nb. In a similar approach the isentropic melting temperatures have been estimated for alkali-metals at about double the equilibrium melting point [43].

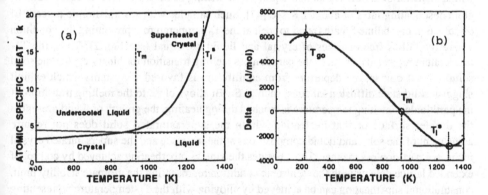

Figure 1. The specific heat (a) and Gibbs free energy (b) of an undercooled liquid and superheated crystal as function of temperature based on measured and extrapolated data for Aluminum.

Furthermore, a hierarchy of inner instabilities has been developed by Tallon based on the entropic instability criterion as an outer bound for undercooling and superheating [26]. This includes an elastic compressibility, elastic rigidity (Born) and isochoric instability before the entropic instability is reached. Thus, for melting as well as liquid undercooling a succession of catastrophe barriers exists including a thermodynamic barrier (equilibrium melting and crystallization), a kinetic barrier (glass formation and shear instability for melting) and finally an entropic barrier (Kauzmann points). The locations of these barriers can be estimated based on specific heat and thermal expansion data in addition to the thermodynamic functions measured in the equilibrium state. However, for an accurate estimate of the different barriers good experimental data, in particular, in the undercooled liquid and superheated crystalline regime are generally not available, and estimates are based on many assumptions which do not always agree with experimental findings, for example about the isentropic [44] and isochoric temperatures [45]. In addition, a pronounced asymmetry is generally observed between the ease of liquid undercooling and crystalline superheating which makes it very difficult to observe superheating above the equilibrium melting point [31] in comparison with liquid undercooling (glass formation) below the melting point [13].

Furthermore, this description of melting is strictly applicable only for pure metals. But pure metals are rather difficult to produce and maintain in the amorphous state because they exhibit pronounced relaxation and recovery processes even close to cryogenic temperatures. The glassy state is found to be stabilized by the addition of certain impurities with a negative enthalpy of mixing. The ease of glass formation increases with impurity content and thus, we consider binary or pseudo-binary alloys in the following.

Unlike pure metals, melting of a binary alloy is accompanied by the redistribution of solute between the crystalline and liquid phases. Therefore the kinetics of melting for crystalline solid solutions depends on the applied heating rate. Only at low heating rates equilibrium melting can occur whereas relatively high levels of superheating above the equilibrium temperature can be achieved for bulk samples through the application of rapid heating rates in the range of 10^3 K / sec [46]. Using fine single-crystalline powder samples with a diameter of about 10 μm, non-equilibrium melting of Sn-Bi droplets is observed even at modest heating rates of about 1 K/sec [47]. Such samples with several atomic percent Bi dissolved in crystalline Sn are found to melt at the T_0-temperature representing the condition where the Gibbs free energies of crystal and liquid are equal (see Fig. 2). Thus, the T_0-temperature represents a limit to the partitionless melting transition for binary crystalline solid solutions. Of course, the departure from equilibrium is favored in systems which exhibit sluggish solid state diffusion kinetics, such as Sn-Bi alloys close to the melting transition. If the partitionless melting temperature is reached during heating, the growth of liquid present at the droplet surface or that nucleating within the supersaturated solid does not require adjustment of the solid and liquid compositions at the interface and the supersaturated crystal melts in a partitionless manner. Thus, besides the kinetic superheating achieved by control of extended lattice defects with the isentropic condition representing the ultimate stability limit, constitutional superheating can be achieved by alloying with the T_0-temperature representing the ultimate limit against partitionless (polymorphous) melting. These ideas describing the thermodynamics and kinetics of melting have direct relevance for the solid-state amorphization reactions described below.

SOLID-STATE-AMORPHIZATION: THERMODYNAMICS AND KINETICS

Experimentally, it is often observed that the amorphous phase is nucleated at extended lattice defects - predominantly high-energy grain boundaries - in analogy to nucleation of the liquid when melting starts [48]. In this case, the amorphous phase grows into the crystalline phase under internal metastable equilibrium conditions resulting in a microstructure with a crystal / amorphous mixture exhibiting a well defined interface.

The nucleation of the amorphous phase can be treated thermodynamically similar to the crystallization from an undercooled liquid eutectic alloy where the phase selection during crystallization plays a significant role [49]. As indicated schematically in Fig. 2 a driving force for amorphous phase formation develops when the α-solid solution is supersaturated beyond c_B', i.e. the composition of the α-crystal corresponding to the two-phase metastable equilibrium with the amorphous phase of composition c_B''. For concentrations larger than c_B' the driving force for amorphous phase formation can be obtained by the conventional parallel tangent construction [50,51] and can be approximated by [52]

$$\Delta G_m = \frac{d^2 G_m{}^\alpha}{d c_B{}^2} \cdot (c_B{}^a - c_B{}^\alpha) \cdot \Delta c_B{}^\alpha \tag{4}$$

where $G_m{}^\alpha$ represents the Gibbs free energy of the α-solid solution and $c_B{}^a$ and $c_B{}^\alpha$ the composition of the amorphous phase and α-solid solution, respectively. Thus, the driving force for the solid-state-amorphization process ΔG_m increases with increasing curvature of the α-solid solution and increasing supersaturation $\Delta c_B{}^\alpha$ of B-atoms and depends on the composition of the amorphous nucleus $c_B{}^a$ to be formed. According to classical nucleation theory, the nucleus of the new phase most likely to form is the one with the lowest free energy barrier ΔG^*. This barrier is a function of the driving force ΔG_m, the interfacial tension σ_{xa} between crystal and amorphous phase to be formed and the nucleus' radius r. Because the interfacial tension and the radius are only weakly dependent on composition it is expected that a nucleus forms with a composition corresponding to the maximum driving force. This is indicated in Fig. 2. Thus, it is seen that in order to form an amorphous nucleus with the most favorable composition large concentrations fluctuations of B-atoms in the α-crystal are essential. Such fluctuations are small in the crystalline state but are greatly enhanced by the presence of large angle grain boundaries [53]. However, if the composition of the solid solution reaches the composition c_0 the Gibbs free energies of crystalline and amorphous phases become equal, and a massive or polymorphous transformation to the glassy phase becomes possible. For this transition large scale concentration fluctuations of B-atoms are not necessary anymore, thus lowering the nucleation barrier significantly, and "melting" to a glass can occur in analogy to melting of a constitutionally superheated crystal at the T_0-temperature as described above.

Thus, in analogy to the melting transition of a crystalline solution two different nucleation modes can be distinguished for solid-state-amorphization. In the case of, e.g., thin-film diffusion couples the amorphous phase is found to be nucleated at special high-energy grain boundaries or triple points. These grain boundaries act as regions where concentration fluctuations through segregational effects can become sufficiently large in order to form a critical nucleus of the amorphous phase [53] with the composition given by the maximum driving force condition. Thus, we observe a first order phase transition where the amorphous phase adopts a composition given by the common tangent construction of two-phase metastable equilibrium conditions.

If high-energy grain boundaries as effective heterogeneous nucleation sites for the amorphous phase are absent and the crystalline phase can be supersaturated up to the composition c_0, solid-state amorphization occurs in analogy with the polymorphous melting condition at the T_0-temperature. Thus, for high supersaturation in the absence of grain boundaries the critical composition fluctuations for amorphous nucleus formation can be reached at the T_0-condition, resulting in a massive, i.e. partitionless transformation to a glass. This second mode representing homogeneous nucleation can be described thermodynamically in a polymorphous phase diagram.

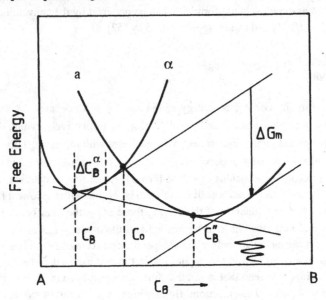

Figure 2. Schematic representation of the Gibbs free energy curves of a crystalline solid solution (α) and the amorphous phase (a) at a temperature below the glass transition temperature as function of solute concentration c_B.

The phase diagrams of typical glass-forming alloys are characterized by steeply plunging T_0-lines, e.g. for Zr-Ni, Zr-Co alloys etc. [54]. For such alloys a generic non-equilibrium phase diagram can be developed neglecting the kinetically excluded intermetallic compounds and illustrating the possible thermodynamic states of a metastable system constrained to be a single phase. For alloys with large negative slopes for the T_0-lines, the T_0-line ($\Delta G = 0$) must cross the ideal glass transition line T_{go} ($\Delta S = 0$) at a certain composition c^* since T_{go} is located close to the actual glass transition temperature. Under this condition, a triple point (c^*, T^*) is defined between supersaturated crystal, undercooled liquid and ideal glass with $\Delta G = \Delta H - T \Delta S = 0$ (due to the melting condition) and $\Delta S = 0$ (due to the Kauzmann condition) as depicted in Fig. 3. Thus, the compositional induced disorder reduces the melting point of the crystalline solid solution to the ideal glass transition temperature with the slope of the T_0-line, which corresponds to $dT_0/dc = - (\partial \Delta G/\partial c) / \Delta S$, approaching infinity at (c^*, T^*). Using realistic fitting parameters for the CALPHAD method [55] and including appropriate corrections for the heat capacities of the undercooled liquid such as given in eq. (3) [40], reasonable extrapolations for the location of this triple point can be made. For example, the triple point for hexagonal Zirconium supersaturated with Ni is predicted at 638 K and 11.5 a/o Ni [40]. In addition, the entropic instability line against melting T^i_s should pass through the triple point because of the vanishing entropy difference between crystal and liquid.

Since the interfacial energy σ_{xL} between crystal and liquid (or glass) is mainly entropic in origin according to Spaepen [42] with $\sigma_{xL} \approx \Delta S \ (T_m)$, the nucleation barrier for melting should decrease along the melting line and eventually vanish at the triple point. Thus, the crystalline phase is strictly unstable beyond the composition c^* and melting becomes catastrophic at the triple point.

Below the triple point, the T_0-line extends to lower temperatures as a straight line with infinite slope as long as non-ergodicity prevails and the Kauzmann argument holds. Thus, at or below the triple point the transition from the ordered phase (crystal exhibiting long range order) to the disordered phase (liquid or ideal glass characterized by the loss of long range order) would be continuous in entropy. Furthermore, through variation of pressure the appropriate Clapeyron equations can be defined [40]. In principle, this isentropic transition could become continuous in volume as well at a critical pressure representing a truly second-order phase transition for melting as long as the metastable constraints can be maintained.

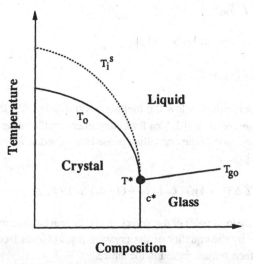

Figure 3. Schematic metastable polymorphous phase diagram which includes the instability temperature, T^i_s, the melting temperature, T_0, and the ideal glass transition temperature, T_{go}, and defines the stability range of the crystal, liquid and glass.

ROLE OF NON-EQUILIBRIUM LATTICE POINT DEFECTS

As mentioned above, equilibrium lattice defects - predominantly vacancies - develop at temperatures close to the melting point. This would result in an entropic lattice instability at a critical vacancy concentration of about 10%. However, under thermal equilibrium conditions this temperature is located in the superheated regime of the crystal.

On the other hand, irradiation and mechanical deformation experiments demonstrate that solid state amorphization can be achieved by creating sufficiently large concentrations of non-equilibrium lattice defects, such as vacancies and anti-site defects. If these defects which in general raise the energy of the crystal, remain frozen in the lattice due to the relatively low processing temperatures, the melting point of the crystal can be reduced considerably. Thus, the role of defects can be compared with the effect of an alloying element resulting in a similar phase diagram as shown in Fig. 3.

In particular, we want to consider the destabilization of the crystalline lattice by the incorporation of non-equilibrium vacancies, typical for externally driven amorphization reactions. By comparing the Gibbs free energies of the undercooled liquid with that of the crystal containing point defects (non-equilibrium melting) the relative stability of such non-equilibrium states can be evaluated [56]. The Gibbs free energy difference $\Delta G = \Delta H - T \Delta S$ between a liquid below its equilibrium melting point and a single crystal can be realistically approximated for pure metals by [57,58]

$$\Delta G = 7 \Delta S_f \cdot \Delta T \cdot T / (T_m + 6T) \tag{5}$$

and for glass forming metallic alloys by [20,59]

$$\Delta G = 2 \Delta S_f \cdot \Delta T \cdot T / (T_m + T) \tag{6}$$

with ΔT being the undercooling below the melting point T_m and ΔS_f the entropy of fusion (typically 1.1 k_B for pure metals and 1.5 k_B for binary intermetallic compounds [60]).

The free energy increase of the crystalline phase is obtained as a function of point defect concentration c_V as [35]

$$\Delta G^V = c_V (\Delta H^V - T \Delta S^V) + k_B T (c_V \ln c_V + (1 - c_V) \ln(1 - c_V)) \tag{7}$$

Thus, the virtual melting point of the defective crystal can be determined as a function of vacancy concentration by the equality of the crystal / liquid Gibbs free energies under the assumption that the defects remain frozen in the lattice.

As a result, the free energy relationships are summarized in the non-equilibrium phase diagram shown in Fig. 4 indicating a decrease of the virtual melting point as a function of vacancy concentration. This melting curve corresponds to the T_0-lines in binary phase diagrams. In addition, the isentropic and isenthalpic temperatures are included defining a triple point at (c_V^*, T^*) in analogy to the one for alloys. At this condition, the three different non-equilibrium states, i.e. the undercooled liquid, the crystal supersaturated with vacancies and the glass, become essentially indistinguishable ($\Delta G = \Delta S = \Delta H = 0$). Within our simple model, this triple point is predicted at $c_V^* = 0.077$ and $T^* = 0.52\ T_m$ for a typical glass forming alloy and at $c_V^* = 0.077$ and $T^* = 0.35\ T_m$ for pure metals, respectively. Below T^*, the liquid is replaced by the glass with the crystal-to-glass transition being non-ergodic by nature, thus ill-defining the actual melting curve.

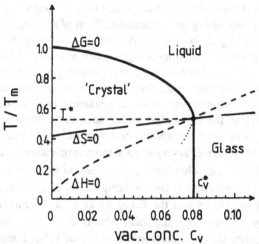

Figure 4. A universal phase diagram for the crystal supersaturated with vacancies, the undercooled liquid and the glass. Additionally included are the isentropic ($\Delta S = 0$) and isenthalpic temperatures ($\Delta H = 0$) as function of defect concentration crossing the melting line at the melting instability at (T^*, c_v^*).

These vacancy concentrations are rather large in comparison with the equilibrium values obtained close to the melting point. For certain intermetallic compounds such large vacancy concentrations have been measured close to T_m [61,62], whereas for pure metals such values are far beyond equilibrium data (10^{-3} close to T_m). However, for pure metals the vacancies (broken bonds) can condense to free surfaces, thus forming small metal clusters with the surface energy γ being proportional to the heat of formation of a vacancy [63,64] or to high-angle grain boundaries, thus forming nano-crystalline material [65,66]. Hence, the phase diagram of Fig. 4 is of universal character and applies to small particle and grain size as well. Due to the large heterophase fluctuations developing close to the triple point (liquid-like clusters in the crystalline phase [67]) the crystal becomes intrinsically unstable and collapses to an isentropic liquid or ideal glass in analogy with the entropy instability described above for binary alloys.

EXPERIMENTAL EVIDENCE

For regular melting in 3-D ergodic conditions prevail (i.e., the system can sample *all* configurations in phase space) and the kinetics are very fast, which makes it difficult to observe crystalline superheating experimentally. In comparison, the "melting" transition from a highly metastable crystal to a glass is very sluggish and the dissociation of the lattice can be followed by standard techniques. However, the inhomogeneous nature of most sample configurations which exhibit solid-state amorphization makes it often difficult to analyze the process. For example, steep concentration gradients for mechanically alloyed and thin-film samples, large defect concentrations during plastic deformation in a ball mill and / or localization of the disordered phase only in small sample regions during ion beam irradiation and at internal interfaces in thin-film superlattices add to the complexity of the situation. In principle the solid-state amorphization reaction can be described thermodynamically by the approach as outlined above taking into account the effects of alloying, non-equilibrium lattice defects and segregation at grain boundaries. Accordingly, most experimental results suggest that the amorphous phase is formed inhomogeneously at extended lattice defects, such as grain boundaries [68,69] or dislocation networks [8] resulting in a first order phase transition.

On the other hand, in some cases there is clear evidence supporting the ideas of an instability mode for solid-state amorphization. For example, by ball milling of Zr-Al powder samples it is found that the crystalline phase, here the hexagonal Zr-crystal, can be supersaturated with Al up to the instability condition before glass formation sets in. Even though the solubility of Al in α-Zr is very small at room temperature for the stable equilibrium between α-Zr and Zr_3Al, the solubility can be enhanced to approximately 15 a/o Al by mechanical alloying with the impurity content being limited to less than 300 ppm oxygen and 1 a/o iron [70]. During this process the Bragg peaks of α-Zr shift to higher angles and broaden considerably whereas the Bragg peaks relating to fcc Al completely disappear. For Al-concentrations of more than 17% glass formation is found after extended milling. The fact that the lattice parameters of α-Zr are reduced continuously with composition up to 15% and that the specific heat and stored enthalpies exhibit a a pronounced cusp at $c^* \approx 17\%$, suggests a polymorphous transition during crystal-to-glass transformation rather than a two-phase metastable equilibrium [70,71]. However, due to the complexity of the milling process, in particular with respect to the large plastic deformation and the formation of a large interfacial area resulting in an average grain size for the solid solution of about 10 nm, further work is required to analyze the mechanism at the atomic level.

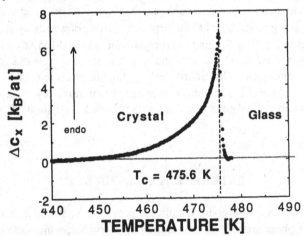

Figure 5. The atomic specific heat Δc_x at constant concentration in excess of the background of 2.9 k_B during the transition from crystalline $Fe_2ErH_{3.4}$ to amorphous $Fe_2ErH_{3.4}$.

To explore the possibility of a crystal instability in the vicinity of the assumed triple point, the most promising sample configuration seems to be represented by a stable intermetallic compound which can be homogeneously supersaturated through the incorporation of a fast moving atomic species, such as hydrogen, at temperatures where the metals atoms are still relatively immobile. For example, powder particles of the intermetallic compound Fe_2Er (C-15 Laves type structure) can be prepared containing a negligibly small fraction ($<10^{-4}$) of grain boundaries acting as heterogeneous nucleation sites. The Fe_2ErH_x specimens were supersaturated with hydrogen far above the equilibrium solubility (x = 2) and then "melted" to a glass as described in more detail elsewhere [72,73]. An expansion of the cubic lattice of the Laves phase is observed for x < 3.2 with an increase of the lattice parameter from a = 7.29 Å to a = 7.82 Å (measured at room temperature) corresponding to a volume increase of 23 %. For higher H concentrations, an anisotropic expansion of the cubic lattice is found with a trigonal distortion along the <111> direction resulting in the formation of a rhombohedral phase (a = 7.81 Å and α = 91°10'). Heating this crystalline sample at 5 atm hydrogen pressure to 510 K results in vitrification of the rhombohedral phase as indicated by the loss of the Bragg peak intensities (long-range order) in the X-ray diffractograms [71]. Thermal analysis performed on the hydrided crystalline $Fe_2ErH_{3.4}$ powder samples allow the "melting process" to be followed further. Fig. 5 represents the excess specific heat Δc_x in units of k_B / atom above the background of 2.9 k_B between 440 K and 478 K for a sample with constant composition. The measured data correspond to a lambda type transition which is unambiguously related to the solid state amorphization process.

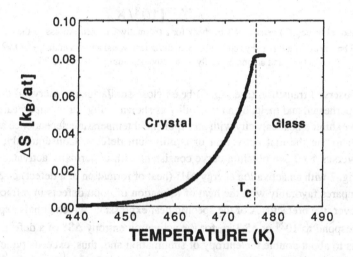

Figure 6. The excess entropy obtained through integration of Δc_x / T dT of Fig. 5.

These calorimetric data differ completely from ordinary first order melting and from calorimetric measurements during vitrification of multilayered thin film diffusion couples. Melting is characterized by a narrow (few degrees) and large signal (in comparison to the specific heat), the area of which is proportional to the heat of fusion. *In situ* measurements of the vitrification of Ni-Zr diffusion couples exhibit an exothermic signal corresponding to the (negative) enthalpy of mixing during diffusion-controlled growth of the amorphous phase [74]. In this case, the transition occurs via nucleation of the amorphous phase at grain boundaries followed by growth of the amorphous phase into the crystal. Whereas regular melting is characterized by a discontinuity in entropy (on the order of 1 k_B), the observed transition from the highly metastable rhombohedral hydride phase to the amorphous structure is continuous in entropy. If we calculate the excess entropy ΔS through integration over $\Delta c_x/T$ dT, a continuous increase to a maximum excess entropy of fusion of 0.08 k_B / atom would be obtained as shown in Fig. 6, which is small in comparison with regular melting ($\sim$ 1 k_B).

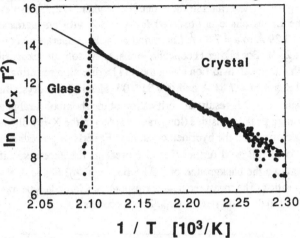

Figure 7. Plot of $\ln(\Delta c_x\ T^2)$ versus 1 / T to check for a thermally activated process. If the excess specific heat would be caused by a thermally activated defect, the defect would have an enthalpy of formation ΔH^v of 2.6 eV and an unreasonably high activation entropy ΔS^v of 55.3 k_B.

The observed transition (see Fig. 5) bears close similarities to a crystal that has been severely superheated and melts by an instability as shown in Fig. 1a. A superheated crystal is expected to exhibit a heat capacity with an exponential temperature dependence according to eq. (1) due to the thermal activation of equilibrium defects. Consequently, if we plot $\ln(\Delta c_x T^2)$ versus 1 / T we obtain a curve consistent with a thermally activated process as shown in Fig. 7 with an activation energy ΔH^v (heat of formation of a defect) of 2.6 eV. This energy compares favorably with the heat of formation of point defects in refractory metals [36]. However, the prefactor A of the specific heat expression, which equals $\exp(\Delta S^v / k_B)$, would correspond to 10^{24} or 55.3 k_B for the activation entropy ΔS^v of a defect. This value corresponds to about double the entropy of sublimation and, thus, exceeds typical expected values (1-3 k_B) by far. Therefore, a phase transformation induced through the formation of equilibrium lattice defects can be ruled out as a trigger for solid-state amorphization in the case of hydrides.

The shape of the heat capacity curve shown in Fig. 5 is rather suggestive of an underlying second order phase transition. The classic signatures of second-order phase transitions are singularities in thermodynamic functions. Although such a transition has not been observed when the glass is approached from the liquid state, it is clear evidence for critical fluctuations during the "melting" of a crystal to a glass. As the transition is approached, heterophase fluctuations develop, that are connected with the local dissociation of the lattice. Such a theory for melting based on heterophase fluctuations has been put forward by Yukalov [67].

In fact, if the excess specific heat is plotted against log ε ($\varepsilon = |T/T_c - 1|$) the λ-shaped cusp can be fitted over two decades to a logarithmic specific heat dependence as $\Delta c_x = A$ log ε

with $A = -3.405$ k_B for $\epsilon < 0.05$. Such a logarithmic ϵ-dependence is typical for the fluctuations accompanying second-order phase transitions and further supports theories for the glass transition based on scaling theories about an underlying critical point [75].

DISCUSSION

Based on the approach presented here a first-order phase transition can be reduced to an isentropic transition and possibly to a second-order phase transition by the variation of an additional external parameter. In a similar model such a possibility has been investigated by Imry describing the effects of microscopic random quenched impurities [76]. There it is shown that pronounced local fluctuations in impurity density develop when the free-energy gain more than offsets the energy cost of the interface produced, thus smearing out the otherwise pure-system-like transition. Similar arguments apply to our case of melting if one is close to the triple point, where the interfacial energy between crystal and liquid or glass becomes very small or even vanishes.

Furthermore, the metastable phase diagram proposed for binary alloys distinguishing the crystal, liquid and glass phases is of generic character. The phase diagram indicates the structural stability of non-equilibrium phases with strong similarities to the magnetic spin-glass transition including the corresponding Nishimori-lines [77]. The horizontal axes in Figures 3 to 4 represent a general, tunable parameter, characteristic for the degree of frustration of the system which can correspond to alloy composition, defect concentration, pressure, excess free volume, reciprocal particle or grain size etc. Above a certain degree of frustration, the ordered crystalline state becomes unstable against liquid-like heterophase fluctuations and the system collapses to a disordered glassy or liquid-like state as indicated in a universal phase diagram by the melting instability with intersecting isentropic, isenthalpic and isenergic lines.

Thus, it is possible experimentally to reduce the regular discontinuous melting transition to continuous-like behavior by the static disorder d built into the crystal lattice. With such an approach the melting point is reduced to the glass transition characterized by the triple point (c^*, T^*) with a vanishing entropy of fusion. This process is further elucidated in Fig. 8 by a family of free energy curves. If we consider the free energy difference ΔG between crystal and liquid as function of temperature at different degrees of static disorder d we obtain the following situation. For the pure metal $\Delta G = G^l - G^x$ is negative above T_m, positive below T_m and zero at T_m. The entropy of fusion at T_m is given by the slope $\partial \Delta G / \partial T = -\Delta S_f$ which is large and negative at T_m and zero at T^i_s and T_{go} (Fig. 8a) similar to the curve displayed in Fig. 1b. Below T_{go} the undercooled liquid becomes unstable with respect to the crystal and above T^i_s the superheated crystal becomes unstable with respect to the liquid based on the Kauzmann and inverse Kauzmann paradox. This regime is indicated by the dashed line of the free energy curves of Fig. 8. With increasing static disorder the melting point T_0 is lowered and ΔS_f at T_0 decreases as shown in Fig. 8b. In general, T_{go} is comparatively weakly dependent on the degree of disorder as shown for alloy and vacancy concentration. With further increase of disorder the melting point T_0 is reduced further until it becomes identical with the glass transition temperature T_{go} as shown in Fig. 8c. This point in T-d space is

314

identical with the triple point at a critical concentration c* and temperature T*. Furthermore, it is characterized by $\partial \Delta G / \partial T = 0$ and $\partial \Delta^2 G / \partial T^2 = 0$. At this point ΔG, ΔS and, consequently, ΔH become zero. This triple point then corresponds to an instability similar to the inflection point in P-V space for the Van der Waals equation of state. Such a triple point has also been predicted based on catastrophe theory by investigating the perturbations of catastrophe germs [78]. For larger degrees of disorder the entropy difference ΔS would be negative for all temperatures. Therefore, the crystalline phase becomes intrinsically unstable and ceases to exist. This generic approach results in the phase diagrams shown in Fig. 3 and Fig. 4 and allows now a generic approach to describe non-equilibrium crystals and their stability against melting and glass formation.

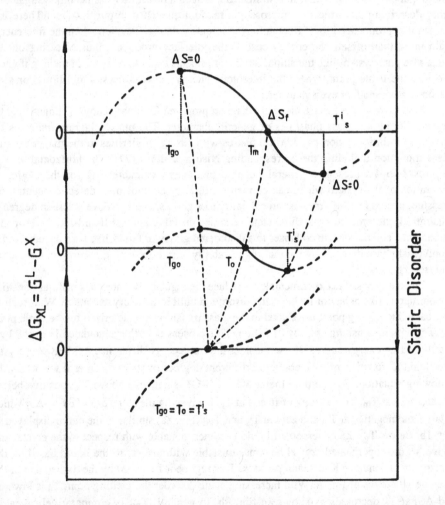

Figure 8. The Gibbs free energy difference ΔG between liquid and crystal (G = 0) as function of temperature with varying static disorder built into the crystal lattice and indicating the reduction of the equilibrium melting point to an isentropic transformation.

ACKNOWLEDGEMENTS

Stimulating discussions with P. Desré, W.L. Johnson, K. Samwer, and R. Yavari and the financial support through DARA (grant number 50 WM 9110-5) are gratefully acknowledged.

REFERENCES

1. Yeh, X.L., Samwer, K., and Johnson, W.L., Appl. Phys. Lett. 1983, **42**, 242
2. Schwarz, R.B., and Johnson, W.L., Phys. Rev. Lett. 1983, **51**, 415
3. Blatter, A., and Von Allmen, M., Phys. Rev. Lett. 1985, **54**, 2103
4. Schwarz, R.B., Petrich, R.R., and Saw, C.K., J. Non-Cryst. Solids 1985, **76**, 281
5. Koch, C.C., Cavin, O.B., McKamey, C.G., and Scarbrough, J.O., Appl. Phys. Lett. 1983, **43**, 1017
6. Atzmon, M., Veerhoven, J.R., Gibson, E.R. and Johnson, W.L., Appl. Phys. Lett. 1984, **45**, 1052
7. Okamoto, P.R., Rehn, L.E., Pearson, J., Bhadra, R., and Grimsditch, M., J Less-Common Met. 1988, **140**, 231
8. Mori, H., Fujita, H., Tendo, M., and Fujita, M., Scripta Metall. 1984, **18**, 783
9. Kouchi, A. and Kuroda, T., Nature 1990, **344**, 134
10. Mishima, O., Calvert, L.D., and Whalley, E., Nature 1984, **310**, 393
11. Ponyatovsky, E.G., Belash, I.T., and Barakalov, O.I., J. Non-Cryst. Sol. 1990, **117 / 118**, 679
12. Perepezko, J.H., Chang, Y.A., Fecht, H.J., and Zhang, M.X., J. Less-Comm. Met. 1988, **140**, 287
13. Perepezko, J.H. and Smith, J.S., J. Non-Cryst. Sol. 1981, **44**, 65
14. Cohen, M.H. and Crest, G.S., Phys. Rev. B 1979, **20**, 1077
15. Adam, G. and Gibbs, J.H., J. Chem. Phys. 1965, **43**, 139
16. Woodcock, L.V., J. Chem. Soc. Faraday II 1976, **72**, 1667
17. Birge, N.O. and Nagel, S.R., Phys. Rev. Lett. 1985, **54**, 2674
18. Kauzmann, W., Chem. Rev. 1948, **43**, 219
19. Kui, Y. and Turnbull, D., J. Non.Cryst. Sol. 1987, **94**, 62
20. Fecht, H.J., Perepezko, J.H., Lee, M.C., and Johnson, W.L., J. Appl. Phys. 1990, **68**, 4494
21. Bormann, R., Gärtner, F., and Zöltzer, K., J. Less.-Comm. Met. 1988, **145**, 19
22. Ström-Olsen, J.O., Ryan, D.H., Altounian, Z., and Brüning, R., J. Mat. Sci. Engg. 1991, **A133**, 403
23. Cahn, R.W., Nature 1986, **323**, 668
24. Lindemann, F.A., Z. Phys. 1910, **11**, 609
25. Born, M., J. Chem. Phys. 1939, **7**, 591
26. Tallon, J.L., Nature 1989, **342**, 658
27. Wolf, D., Okamoto, P.R., Yip. S., Lutsko, J.F., and Kluge, M., J. Mat. Res. 1990, **5**, 286
28. Gorecki, T., Scr. Metall. 1977, **11**, 1051
29. Fecht, H.J. and Johnson, W.L., Nature 1988, **334**, 50
30. Cormia, R.L., MacKenzie, J.D. and Turnbull, D., J. Appl. Phys. 1963, **34**, 2245
31. Däges, J., Perepezko, J.H. and Gleiter, H., Phys. Lett. 1986, **A119**, 79
32. Jung, J. and Franck, J.P., Jap. J. Appl. Phys. 1987, **26-3**, 399
33. Bilgram, J., Phys. Rep. 1987, **153**, 1

316

34. Tallon, J.L., Phil. Mag. 1979, A39, 151
35. Wollenberger, H.J. in Physical Metallurgy, Cahn, R.W. and Haasen, P. eds., Elsevier Science Publishers, 1983, p. 1139
36. Doyama, M. and Koehler, J.S., Acta Metall. 1976, 24, 871
37. Pochapsky, T.E., Acta Metall. 1953, 1, 747
38. Kraftmaker, Ya. A., Soviet Phys. Solid St. 1963, 5, 696
39. Kraftmaker, Ya. A. and Strelkov, P.G., Soviet Phys. Solid St. 1963, 4, 1662
40. Fecht, H.J., Desré, P. and Johnson, W.L., Phil. Mag. 1989, B59, 577
41. Turnbull, D., Contemp. Phys. 1969, 10, 473
42. Spaepen, F., Acta Metall. 1975, 23, 729
43. Lele, S., Ramachandrarao, P. and Dubey, K.S., Nature 1988, 336, 567
44. Sommer, F., J. Mat. Sci. Engg. 1991, A133, 434
45. Shiraichi, S.Y. and Ward, R.G., Can. Metall. Quart. 1964, 3, 117
46. Woodruff, D.P. and Forty, A.J., Phil Mag. 1967, 15, 985
47. Allen, W.P., Fecht, H.J. and Perepezko, J.H., Scr. Metall. 1989, 23, 647
48. Cahn, R.W. and Johnson, W.L., J. Mat. Res. 1986, 1, 724
49. Fecht, H.J., Acta Metall. Mater. 1991, 39, 1003
50. Hillert, M., Acta Metall. 1953, 1, 764
51. Thompson, C.V. and Spaepen, F., Acta Metall. 1983, 31, 202
52. Hillert, M., in Lectures on the Theory of Phase Transformations, Aaronson, H.I., ed., The Metallurgical Society, 1986
53. Busch, R., Schneider, S. and Samwer, K., Nachr. Akad. Wissensch. Göttingen 1991, 1, 1
54. Massalski, T.B. (ed.), Binary Alloy Phase Diagrams, American Society of Metals 1986
55. Saunders, N., Calphad 1985, 9, 297
56. Fecht, H.J., Nature submitted (1991)
57. Singh, H.B. and Holz, A., Solid State Comm. 1983, 45, 985
58. Perepezko, J.H. and Paik, J.S., J. Non.-Cryst. Sol. 1984, 61/62, 113
59. Thompson, C.V. & Spaepen, F. Acta Metall. 1979, 27, 1855
60. Goodman, D., Cahn, J.W., and Bennett, L.H., Bull. All. Phase Diag. 1981, 2, 29
61. Van Ommen, Waegemaekers, A.A., Moleman, A.C., Schlatter, H. and Bakker, H., Acta Metall. 1981, 29, 123
62. Bakker, H., these proceedings
63. Gorecki, T. Z. Metallkde. 1979, 70, 121
64. Buffat, P. & Borel, J.P. Phys. Rev. A 1976, 13, 2287
65. Gleiter, H., Prog. Mat. Sci. 1989, 33, 223
66. Fecht, H.J., Phys. Rev. Lett. 1990, 65, 610
67. Yukalov, V.I., Phys. Rev. B. 1985, 32, 436
68. Johnson, W.L., Prog. Mat. Sci. 1986, 30, 81
69. Johnson, W.L. and Fecht, H.J., J. Less-Comm. Met. 1988, 145, 63
70. Fecht, H.J., Han, G., Fu, Z., and Johnson, W.L., J. Appl. Phys. 1990, 67, 1744
71. Ma, E. and Atzmon, M., Int. Symp. Mech. Alloying Kyoto, Japan, May 1991
72. Fecht, H.J., Fu, Z. and Johnson, W.L., Phys. Rev. Lett. 1990, 64, 1753
73. Fecht, H.J. and Fu, Z., Mod. Phys. Lett. 1991, 5, 1
74. Cotts, E.J., Meng, W.J. and Johnson, W.L., Phys. Rev. Lett. 1986, 57, 2295
75. Sethna, J.P., Shore, J.D., and Huang, M., Phys. Rev. B 1991, in print
76. Imry, Y. and Wortis, M., Phys. Rev. B 1979, 19, 3580
77. LeDoussal, P. and Harris, A.B., Phys. Rev. Lett. 1988, 61, 625
78. Gilmore, R., Catastrophe theory, J. Wiley & Sons, New York (1976)

ON THE AMORPHISATION OF NiGe ALLOYS BY MECHANICAL ALLOYING

T. BENAMEUR and A.R. YAVARI

LTPCM -CNRS UA 29 Domaine Universitaire, B75, 38402 St Martin d'Hères France

J. MALAGELADA and M.D. BARO

Dept. de fisica, Universitat Autonoma de Barcelona, 08193 Bellaterra, Spain

ABSTRACT

We have studied the phase transformations induced in the Ni-Ge binary system by mechanical alloying of Ni and Ge powder mixtures and by milling of NiGe melt-spun ribbons. We observed that mechanical alloying is preceded by significant x-ray line broadening, consistent with an acceleration of the mixing kinetics with grain refinement. During mechanical alloying, crystalline compounds form first and are subsequently amorphised at least partially. The prealloyed melt-spun samples also become nanocrystalline after deformation. Heavy Fe contamination from the milling medium occurs before amorphisation at long milling times. Although some unalloyed iron-chromium stainless steel fragments are present, most of the iron is in solution and can play an important role in the observed amorphisation. Some amorphisation occurs in nearly all our samples. Maximum amorphisation is obtained for the starting composition $Ni_{60}Ge_{40}$ which yields a nearly fully amorphous $(NiFeCr)_{70}Ge_{30}$ alloy. Calorimetric measurements indicate a heat of crystallisation of about 3 kJ/mole for this composition. The amorphous phase is paramagnetic while the crystalline alloy shows a Curie temperature $T \approx 660$ K.

INTRODUCTION

Amorphisation by solid state reaction in multilayers, by mechanical alloying of metal powder mixtures or by mechanical grinding of certain alloys has been studied by many authors [1,2] but not yet fully understood. Amorphisation by mechanical alloying is of particular interest because it often allows formation of amorphous compositions inaccessible by solid-state reaction [3] or by rapid quenching from the liquid state [4]. For example we have previously obtained by mechanical alloying [5,6] amorphous Al_2Pt and Mg_2Ni which cannot be obtained in macroscopic quantities by the other cited techniques. Amorphisation of alloys by mechanical grinding is also of interest especially because it is related to the amorphisation of pure powder mixtures by mechanical alloying. As we pointed out elsewhere [7], if an alloy such as A_nB, can be amorphised by grinding, then any crystalline nuclii of the A_nB phase forming in A/B powder mixtures during mechanical alloying would be amorphised and crystalline A_nB would not be formed by mechanical alloying. Furthermore some authors have suggested that alloys with $L1_2$ superstructures should be easily amorphisable by grinding. We have therefore studied the behavior of Ni/Ge powder mixtures and Ni_xGe_{1-x} melt-spun alloys during ball milling. The system is of particular interest because ordered Ni_3Ge has the $L1_2$ stucture and the heats of formation of NiGe alloys are generally small.

EXPERIMENTAL

The NiGe powder samples were prepared from two initally different states using the ball milling process. First, elemental crystalline powders of pure Ni and Ge with particle sizes of about 20 μm were mechanically alloyed (m.a.) in a commercial milling device (Pulverisette 0) modified to allow a vacuum of 10^{-5} torr before introduction of a working pressure of 1.5 bars of innert gas. Its iron-chromium stainless steel vial contains a single ball of the same material with a typical weight of 0.5 kg, corresponding to a ball diameter of 5 cm. The vibration amplitude was remained constant for the investigated compositions. The second type of NiGe samples were alloy ribbons (intermetallics) of compositions $Ni_{60}Ge_{40}$ and $Ni_{75}Ge_{25}$ (at %) prepared by melt-spinning (m.s.) under controlled atmosphere. Without annealing, the melt-spun ribbons were introduced in the container for the grinding process. Milled materials were analysed by x-ray diffraction using Cu-K_α radiation. The particle morphologies were observed by scanning electron microscopy (SEM). The unavoidable contamination from the stainless steel container, which occurs during milling especially after long milling times, was determined by EDX (Energy Dispersive x-ray) analysis in the SEM using a semi-quantitative program with internal references. The average chemical analyses are listed in table 1. While the contamination occurs by the steel fragments that enter the milled agglomerate (see figure 1) it is subsequently dissolved into the alloy as confirmed by magnetic and x-ray results. Specimen for transmission electron microscopy (TEM) were prepared by embedding milled powders in Ni ribbons, dimpling and ion milling The specimens were examined in a Jeol 200 CX operating at 200 kV. The crystallisation behaviour of nearly fully amorphous samples obtained after 240 hours of milling of a powder of initial composition $Ni_{60}Ge_{40}$, were monitored by differential scanning calorimetry (DSC) using a DSC-7 in Barcelona and by magnetic measurements in the furnace-equipped magnetometer at the Laboratoire Louis Néel in Grenoble.

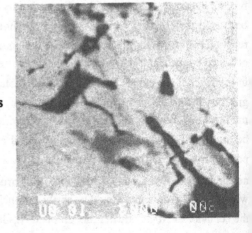

Figure 1. SEM micrographs showing back-scattering contrast (grey) from steel fragments in mechanically alloyed $Ni_{60}Ge_{40}$

Table I : Experimental conditions and the end product compositions (at %)

Initial composition	Initial state	B.M. duration	End product
$Ni_{50}Ge_{50}$	Powders	60 hrs	$Ni_{50}Ge_{49.7}$ Fe = 0.3
$Ni_{60}Ge_{40}$	P.	130 hrs	$Ni_{57}Ge_{38}$ Fe=4 Cr=1
$Ni_{75}Ge_{25}$	P.	90 hrs	$Ni_{73.5}Ge_{24.5}$ Fe=2 Cr=0.5
$Ni_{50}Ge_{50}$	P.	240 hrs	$Ni_{39.5}Ge_{39.5}$ Fe=17 Cr=4
$Ni_{60}Ge_{40}$	P.	240 hrs	$Ni_{46}Ge_{30}$ Fe=19 Cr=5
$Ni_{75}Ge_{25}$	P.	300 hrs	$Ni_{67}Ge_{22}$ Fe=9 Cr=2
$Ni_{60}Ge_{40}$	m.s	180 hrs	$Ni_{56}Ge_{37}$ Fe=6 Cr=1.5
$Ni_{75}Ge_{25}$	m.s	160 hrs	$Ni_{72}Ge_{24}$ Fe=3.2 Cr=0.8

* Chromium compositions under 0.5 at % not listed

RESULTS and DISCUSSION

Figure 2 shows the cross-section of m.a. $Ni_{75}Ge_{25}$ after 13 hrs of milling. It reveals a wide distribution of particle size particularly for the Ni component. The dark and bright areas correspond, according to EDX analysis, to Ni and Ge respectively. The grey regions were identified as containing both Ni and Ge. The mechanically alloyed $Ni_{75}Ge_{25}$ exhibits a few amorphous regions after 24 hrs of milling as observed in the TEM image of figure 3, but the compositions of these regions have not been determined.

Figure 4 shows the evolution of the x-ray diffraction pattern of a powder mixture of initial composition $Ni_{75}Ge_{25}$ with increasing milling time. During milling, the germanium Bragg peaks become slightly broader and their intensities decrease continuously and disappear after 130 hours of m.a. This is consistent with the large solid solubility of Ge in Ni (more than 10 at% at 473 K [8]). In contrast we observe a continuous broadening of the Ni Bragg peaks with a shift in the peak positions as shown in figure 4.

Figure 5 shows the crystallite size estimated from the Scherrer formula [9], $L = 0.9\lambda/\beta\cos\phi$, where L is the crystallite size, ϕ is the scattering angle, λ is the wavelength and β is the line width at one half the maximum intensity (determined by fitting a Lorentzian line shape and correcting for instrumental broadening).

The kinetics of transformation of m.s. $Ni_{75}Ge_{25}$ are only slightly different than those of mechanically alloyed $Ni_{75}Ge_{25}$ (see figure 6 as compared to figure 4). After 300 hrs, the mechanically alloyed powder contains an important volume fraction of amorphous phase but its composition has now become $(NiFeCr)_{78}Ge_{22}$ (see table I).

Figure 7 shows x-ray diffraction patterns with increasing milling time for a Ni/Ge powder mixture of initial composition $Ni_{60}Ge_{40}$. The decrease of the peak intensities and continuous broadening of the Bragg peaks are followed by the formation of crystalline phases and their subsequent transformation into an amorphous alloy but with a composition now near $(NiFeCr)_{70}Ge_{30}$ (see table I). The amorphisation therefore occurs with increasing iron content. When this alloy is milled in a silicon oxide vial with a zirconia ball, no evidence of an amorphous product is found even after long milling times (see bottom of figure 7).

Figure 8 shows the evolution of the x-ray pattern of melt-spun $Ni_{60}Ge_{40}$ with increasing milling time which is similar to the mechanically alloyed powder. We note that the spectrum after 70 hrs for m.s. $Ni_{60}Ge_{40}$ is nearly similar to that of m.a. $Ni_{60}Ge_{40}$ after 130 hours (see figure 7).

For the mechanical alloying of a powder mixture of composition $Ni_{50}Ge_{50}$, increasing the milling time results in the nucleation and growth of crystalline NiGe up to 110 hrs of milling as shown in figure 9. After 240 hours of milling, elemental Bragg peaks disappear together with those of intermetallic NiGe while a significant amorphous phase fraction is formed.

In all cases large atom fractions of Fe and some Cr were introduced in the alloys. $Ni_{75}Ge_{25}$ (m.s. ribbon or elemental powder mixture) shows less iron contamination probably due to the low hardness of the Ni_3Ge compound compared to compositions containing 37 % to 40 at % Ge [10] as listed in Table 1.

In all cases some amorphisation was observed after heavy iron contamination (see figure 10)

Calorimetric measurements on partially amorphous $(NiFeCr)_{78}Ge_{22}$ (obtained from a starting composition $Ni_{75}Ge_{25}$) indicate onset of crystallisation near $T = 670$ K as shown in figure 11. The total heat release upon crystallisation as well as that due to recrystallisation of the nanocrystalline fraction is measured to be about 4 kJ/mole.The activation enthalpy of the crystallisation process can be determined by plotting, for a number of scans at different rates b, $\ln b/T_p^2$ versus $1/T_p$, where T_p is the peak temperature (maximum rate) of the transformation. We found an activation enthalpy of $Q \approx 186$ kJ/mole.

For nearly fully amorphous $(NiFeCr)_{70}Ge_{30}$ obtained from an initial composition of $Ni_{60}Ge_{40}$, the DSC scan of figure 12 obtained at 40 K/mn, shows a sharp peak in the range of 570 K to 605 K. The crystallisation enthalpy ΔH_c is about 3 kJ /mole with an activation enthalpy $Q \approx$ 154 kJ/mole.

Figure 13 shows magnetisation measurements at magnetic field H=10 kOe as a function of temperature for the $(NiFeCr)_{70}Ge_{30}$ amorphous alloy. The amorphous phase is seen to be paramagnetic with negligible magnetization before the onset of crystallisation at T = 550 K after which the magnetization increases to reach another curve charateristic of the emerging crystal structures.

The paramagnetic behavior of the mechanically alloyed powder indicates that the FeCrNi steel contaminant which initially enters as fragments (see figure 1) is fully dissolved at this stage. The cooling curve after crystallisation is typical of a ferromagnetic intermetallic with $T_C = 660$ K. The small but significant difference between the heating and cooling curves near T =700 K is not fully understood but indicates that crystalline phases in this range are still evolving. For comparison with DSC data, we plot in figure 13, the crystallisation rate is defined as dX/dt = (dX/dT)·(dT/dt) and X(T)= $M_H(T)$ / $M_C(T)$ where X is the crystallised fraction, M_H the magnetization during heating and M_C the magnetization during cooling at the same temperature T in the range of 470 K to 610 K in figure 14. The results show a maximum rate T_p= 540 K for dT/dt = 5 K/mn compared to T_p=554 K obtained from DSC data projected to the same heating rate. As such, magnetisation measurements and their analysis are seen to be fully complementary to DSC results.

CONCLUSION

As with many other materials, Ni/Ge powder mixtures and NiGe (melt-spun) alloys become nanocrystalline after severe deformation during ball-milling.

While the extent of impurity fragments from the mill container and ball depends on the mechanical properties of the starting powder, there are no major differences between the phases obtained by mechanical alloying and by grinding of the prealloyed material. For all the investigated compositions, mechanical alloying resulted in the formation of the alloy phases present in the prealloyed samples.

DSC, x-ray diffraction, SEM and TEM observations and magnetic measurements indicate that long time milling with stainless steel equipment results in some amorphous phase formation in all samples. However selected experiments with a silicon oxide vial and a zirconia ball with the same weight as the steel one show no evidence of amorphisation.

The presence of iron, sometimes up to nearly 20 at% is therefore essential for significant amorphisation in the NiGe system. The reasons for this are not well understood because of the nearly equal atomic sizes of Ni and Fe and the fact that the heat of mixing ΔH_{mix} of Ge with Fe is even less important than that of Ge with Ni. Furthermore the two phase diagrams are quite similar and Fe is known to substitute for Ni. The behavior of the FeGe system is currently being investigated and may shed some light on the role of Fe in the amorphisation of NiGe alloys. One possibility is that in view of the similarities between amorphisation by solid state reaction and by m.a., the sharp concentration gradients at interlayer interfaces and the role of a third element as described by Bouanha et al [12] may favor amorphous phase formation during m.a. in presence of iron but this requires further studies for conformation.

The composition $Ni_{75}Ge_{25}$ was investigated because it forms an ordered phase with the $L1_2$ superlattice like Zr_3Al which has been found to be easily amorphisable. After long milling times partial amorphisation was observed but th composition had evolved and become $Ni_{67}Ge_{22}Fe_9Cr_2$. Since iron and chromium substitute for Ni[11], the ordered state of this alloy $(NiFeCr)_{78}Ge_{22}$, would most likely still be $L1_2$. However the highest amorphous phase volume fraction was obtained for a composition $(NiFeCr)_{70}Ge_{30}$ which was also found to be paramagnetic and Ni_3Ge itself had little aptitude towards amorphisation by heavy deformation.

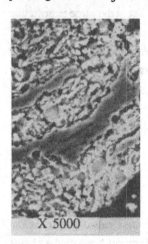 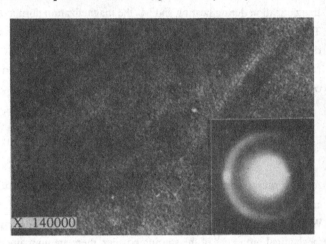

Figure 2. Secondary electron SEM image of cross section of Ni/Ge powder mixture of global composition $Ni_{75}Ge_{25}$ after 13 hours of milling (left)

Figure 3. TEM image of the $Ni_{75}Ge_{25}$ powder after 24 hours of milling: bright field image and corresponding electron diffraction patterns (right)

ACKNOWLEDGEMENT

This work was supported by the European Community's SCI-0262-C project.

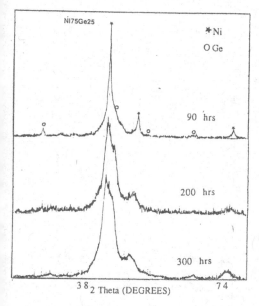

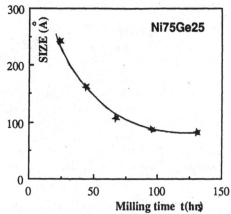

Figure 5: (right) Effective Ni crystallite size versus milling time t during mechanical alloying of $Ni_{75}Ge_{25}$ powder mixture

Figure 4. (left) x-ray diffraction pattern of Ni/Ge powder of composition $Ni_{75}Ge_{25}$ as a function of milling time t.

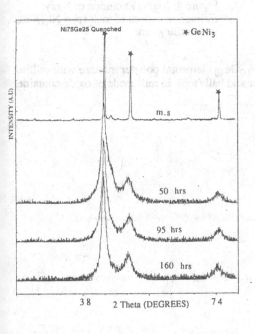

Figure 6: (left) Evolution of x-ray diffraction pattern of melt spun $Ni_{75}Ge_{25}$ with milling time t.

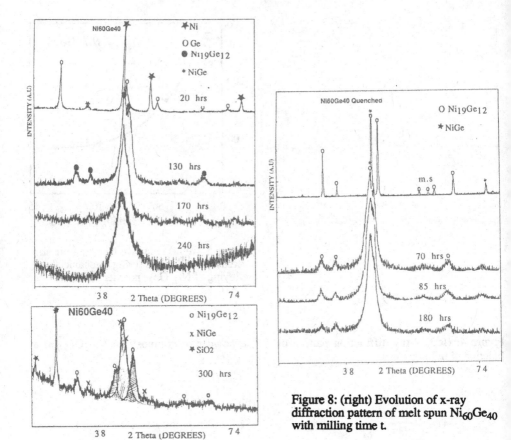

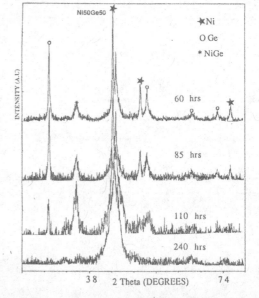

Figure 8: (right) Evolution of x-ray diffraction pattern of melt spun $Ni_{60}Ge_{40}$ with milling time t.

Figure 7.(left) Evolution of x-ray pattern of $Ni_{60}Ge_{40}$ elemental powder mixture with milling time t. In the usual mill made of steel container and ball (top). In mill made of oxide container and ball (bottom)

Figure 9. (right) Evolution of x-ray diffraction pattern of $Ni_{50}Ge_{50}$ powder mixture with milling time t

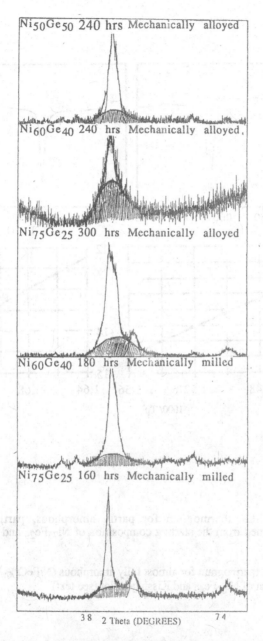

Figure 10. Schematic deconvolution of x-ray diffraction patterns of Ni Ge alloys With final compositions for powder mixtures: $Ni_{50}Ge_{50} \rightarrow Ni_{39.5}Ge_{39.5}Fe_{17}Cr_4$ $Ni_{60}Ge_{40} \rightarrow Ni_{46}Ge_{30}Fe_{19}Cr_5$, $Ni_{75}Ge_{25} \rightarrow Ni_{67}Ge_{22}Fe_9Cr_2$ and for melt spun alloys: $Ni_{60}Ge_{40} \rightarrow Ni_{56}Ge_{37}Fe_6Cr_{1.5}$, $Ni_{75}Ge_{25} \rightarrow Ni_{72}Ge_{24}Fe_{3.2}Cr_{0.8}$.

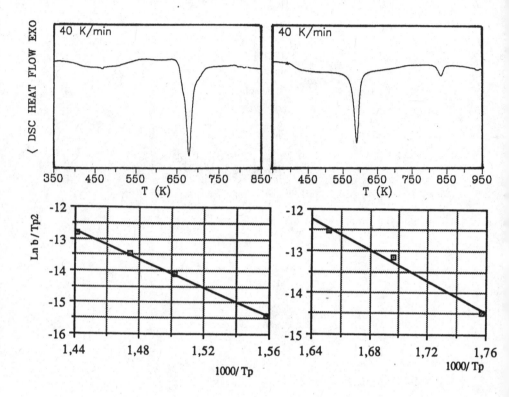

Figure 11. (left) DSC thermogram for partly amorphous, partly nanocrystalline $(NiFeCr)_{78}Ge_{22}$ obtained from the starting composition of $Ni_{75}Fe_{25}$ and Kissinger plot (see text)

Figure 12. (right) DSC thermogram for almost fully amorphous $(NiFeCr)_{70}Ge_{30}$ obtained from the starting composition of $Ni_{60}Fe_{40}$ and Kissinger plot (see text)

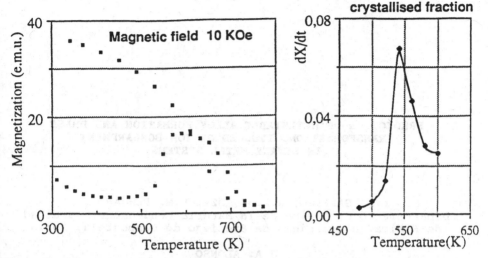

Figure 13. (left) Magnetisation versus temperature for $(NiFeCr)_{70}Ge_{30}$ amorphous alloy during heating and subsquent cooling after crystallisation.

Fig.14: (right) Plot of crystallised fraction dX/dt versus temperature for $(NiFeCr)_{70}Ge_{30}$ amorphous alloy obtained from magnetization data of figure 13.

REFERENCES

[1] Johnson , W.L , Prog. Mater. Sci. 1986, 81 30,

[2] "Amorphisation by Solid-State Reaction and Mechanical Alloying ",
    Eds : Yavari, A.R. and Desré, P.J., J. de Physique 1990-C4

[3] Yavari, A.R.and Desré, P. Phys. Rev. Lett. 1990, 65, 2571

[4] Weeber, A.W. and Bakker, H., Physica B, 1988, 93, 135

[5] Bordeaux, F. and Yavari, A.R., J. Appl. Phys. , 1990, 67, 2385

[6] Benameur, T. and Yavari, A.R., Barandarian, J.M., Mater.Sci. and Eng., 1991, A134, 1402

[7] Yavari, A.R. and Desré, P.J., Proceeding International Symposium On Mechanical Alloying, Ed: Shingu, P., Kyoto, May 1991

[8] Nash, A. and Nash, P, Binary Alloy phase Diagrams, 1991, 2 , p 1974

[9] Guinier, A., X-ray diffraction, Freeman, San Francisco ,1963, p124

[10] Dayer, A., Feschotte, P. J. of the Less. Com. Metals, 1980, 72, 51

[11] Ochioi, S., Oya, Y. and Sutuki, T., Act. Metall. 1984, 32, 2, p.289

[12] Bouanha, L., Desré, P., Hicter, P. and Yavari, A.R. in "Ordering and Disordering in Alloys" Ed Yavari, A.R., Elsevier Science in press 1991

PREDICTION OF METASTABLE ALLOY FORMATION AND PHASE TRANSFORMATIONS INDUCED BY ION BOMBARDMENT IN BINARY METAL SYSTEMS

L.J. GALLEGO, J.A. SOMOZA, H.M. FERNANDEZ
Departamento de Física de la Materia Condensada, Facultad
de Física, Universidad de Santiago de Compostela, Spain

J.A. ALONSO
Departamento de Física Teórica,
Facultad de Ciencias, Universidad de Valladolid, Spain

ABSTRACT

Thermodynamic aspects of the formation of metastable alloys by
ion bombardment, and phase transformations, are analysed using
a model which combines classical elasticity theory and
Miedema's model of heats of formation of alloys. This model has
recently been used successfully to predict the glass forming
ranges of binary and ternary transition metal alloys.

INTRODUCTION

Among the various techniques available for producing crystalline or amorphous metastable alloys, ion mixing (IM) is particularly attractive for studying phase formation processes [1]
due to the researcher's direct control over the irradiation
parameters, i.e. ion species (typically Xe^+), dose, energy
(typically several hundred keV) and temperature. Ion implantation is a related technique, but its efficiency is limited by
sputtering effects, which limit the concentration of implanted
atoms. IM, on the contrary, allows various alloy phases to be
obtained without limitation of composition by simply adjusting
the relative thickness of the metal layers.

In the past few years, considerable effort has been spent
on working towards a better understanding of IM phase formation
mechanisms (see, e.g., [1,2]). Currently, a two-step process is
postulated: the first step is a fast process in which the
bombarding ions generate atomic collision cascades, after which
the atoms come to rest; and the second is a relaxation process
in which the excited mixture falls to a state of lower energy.
The ion-mixed structure is assumed to arise during this latter

process, in which quenching rates as high as 10^{14} K/s are achieved.

As for any other amorphization technique, the glass forming ability (GFA) of a particular system under IM is the result of a favourable combination of kinetic and thermodynamic factors. The kinetic aspects of ion-induced atomic mixing in diffusion couples have been studied by several authors using an isotropic random walk model, and have recently been reviewed by Johnson [2]. A theory of radiation-induced chemical disorder has also been developed using the Bragg-Williams model to describe the crystalline-to-amorphous transformation (see [2] and refs. therein), which has also been studied by Massobrio et al. [3] by computer simulation. Complementing these studies, thermodynamic analyses of ion-induced alloy phase formation have focused on the question of what the main thermochemical factors influencing glass formation are. The first rule suggested for the amorphization of metal multilayers was based on the crystalline structures of the constituent metals (see [1] and refs. therein). More recently, Liu [4] has proposed a phenomenological model for predicting IM GFA from two characteristic parameters: the maximum possible amorphization range, which is inversely related to the terminal solid solubilities in the equilibrium phase diagram, and the heat of formation of the system. An alternative scheme proposed by Alonso and López [5] involves the construction of two-dimensional maps allowing discrimination between systems which do and do not form glasses under IM. In one of our most recent papers on this topic [6], the coordinates used to construct the maps were the heat of mixing of the (equiatomic) liquid alloy, as obtained from Miedema's model [7], and the size mismatch contribution to the heat of formation of the solid solution, which was calculated using the elasticity theory developed by Eshelby [8].

The advantage of the latter approach is that it provides a means of calculating metastable free energy diagrams for binary alloys. This combination of Miedema's and Eshelby's models has been used by Miedema and coworkers [9,10] and by us [6,11,12] to predict the possible glass-forming composition ranges of binary metal systems, and more recently we have extended the treatment to ternary systems [13], for which experimental information is still scarce. In the present paper we survey some of the main experimental IM results for binary alloys in the light of this theory, with particular emphasis on ion-induced crystalline-to-amorphous and amorphous-to-crystalline transitions. In so doing, we extend a recent preliminary study of ion-bombardment-induced transformations in the system Mo-Ni [14]. As indicated above, the thermodynamic aspects considered in this paper are only part of the factors governing alloy phase formation under IM. Kinetic aspects have also to be considered in a complete theory of ion-induced transformations. In this respect, it is worth mentioning that some of the thermochemical quantities which appear in the thermodynamic description, such as the heat of mixing, also play an important role in the kinetics of the processes involved (see ref. [2]).

The layout of this paper is as follows. In the next Section we present a two-dimensional map containing information on numerous binary alloys treated by IM. This map separates amorphous from non-amorphous alloys nearly perfectly, and can therefore be used predictively as a first approximation to the question of the IM GFA. Several systems are then studied in more detail using the model mentioned above. Because the ingredients of this model have been explained in detail elsewhere [6,9-12], we describe them only very briefly here.

GLASS-FORMING ABILITY MAPS FOR ALLOYS TREATED BY IM

Figure 1 is a thermochemical map containing information on 54 systems whose behaviour under IM has been investigated [4]. Small dots represent alloys forming amorphous phases, and open circles alloys that do not form amorphous phases. The coordinates used in the map are ΔH_M, the heat of mixing of the equi-atomic liquid alloy, as calculated from Miedema's theory, and ΔH^e, the size mismatch contribution to the heat of formation of the solid solution phase at equi-atomic composition, which was obtained by averaging the corresponding size mismatch contributions to the heats of solution of one component in the other as calculated from Eshelby's elastic model. We have not identified every system in Fig. 1 because it is unnecessary for the present discussion. What is important to appreciate is that a curve can be drawn which, with few exceptions, separates readily amorphized systems from hard-to-amorphize systems (see ref. [6] for a more detailed discussion). This suggests that the coordinates ΔH_M and ΔH^e are important parameters for the formation of amorphous alloys by IM. Different irradiation conditions (type of ion, ion dose, temperature) are not distinguished in Fig. 1, and this is the reason for some of the exceptions (small dots in the hard-to-amorphize region). The separation improves if only experiments performed under similar conditions are compared.

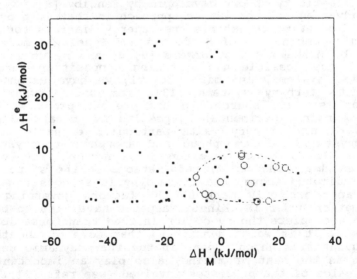

Figure 1. Glass forming ability map for alloys treated by IM.

As was mentioned in the Introduction, thermochemical maps,

though useful, are only a first approximation to the prediction of IM GFA. Proper interpretation of the experimental results requires a more quantitative treatment. The model used in this paper is briefly described in the next section, where we study in detail some illustrative examples of ion-induced crystalline-to-amorphous and amorphous-to-crystalline transitions.

TRANSFORMATIONS CONSEQUENT UPON ION IRRADIATION

We shall study the thermodynamic aspects of some transformations consequent upon ion irradiation of an illustrative sample of binary transition metal alloys for which detailed experimental information is available. To do so, we shall construct the corresponding free energy diagrams of the relevant phases using the following model. First, we assume that the entropies of mixing for both solid solution and the amorphous phase are given by the simple ideal solution model. The enthalpy of mixing of a solid solution of transition metals is considered as a sum of three contributions: a chemical contribution ΔH^c, which takes into account electron redistribution interactions generated at the surface of contact between dissimilar atomic cells; the elastic or size-mismatch contribution ΔH^e; and a structural contribution, ΔH^s, due to differences between the two metals as regards valence and crystal structure. For the amorphous phase and intermetallic compounds only the chemical contribution is present (see, e.g., [9,11]). Details of the calculation of these various contributions can be found in refs. [9-12]. It is worth pointing out that in the calculations presented in this paper the chemical contributions ΔH^c were computed assuming the same degree of short-range order in the solid solutions as in the amorphous phases [15]. In calculating the structural contribution ΔH^s in alloys including the ferromagnetic metal Ni, we have taken into account the corrected structural stability curves given by Miedema and coworkers [16] for magnetic metals.

Let us first compare a transition metal system that does undergo ion-induced crystalline-to-amorphous transformation, Au-Ti, with one that does not, Fe-W [1]. In the former case, the intermetallic compound $AuTi_3$ and the crystalline mixture $AuTi_3+AuTi$ were examined, and a transformation to an amorphous state was induced at room temperature by low irradiation doses of 5×10^{15} Xe/cm^2. The compound Fe_7W, however, remained crystalline when similarly irradiated. To understand in detail this difference in behaviour, one must examine the mechanisms by which the free energy of intermetallic compounds can be increased as a result of irradiation. Here, we shall only consider the question from the thermodynamic point of view.

In Fig. 2 we have constructed the free energy curves of the relevant amorphous phases and crystalline compounds. The arrows in Fig. 2(A) indicate the ion-induced crystalline-to-amorphous transformations observed. The amorphous state is thought to be attained by a fast process similar to that occurring in the IM of metal layers, and to be retained during the subsequent relaxation process due to kinetic hindrances. In the case of the system Fe-W, however, the amorphous phase is unstable at room temperature (Fig. 2(B)), in keeping with the observed absence of amorphization.

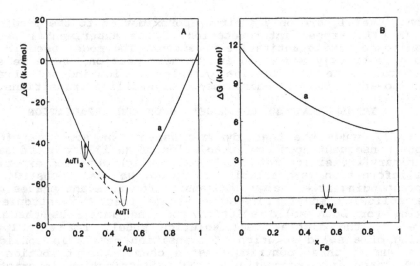

Figure 2. Calculated free energy diagrams of Au-Ti(A) and Fe-W (B) at 300 K, showing the free energies of the amorphous phases (a) and intermetallic compounds referred to in the text. In Fig. 2(A), the dashed line is the common tangent to the free energies of the intermetallic compounds, and the arrows represent the transitions discussed in the text.

Let us now consider two systems, Mo-Ni and Nb-Ni, for which the possibility of an ion-bombardment induced transformation from amorphous to crystalline states has been investigated [1]. The characteristic features of the first of these systems were discussed in detail in ref. [14], and we only consider here those aspects that are most relevant for the present discussion. In particular, we note that, as shown by Liu [1,17], a metastable crystalline (MX) phase of hcp structure can be formed by ion irradiation of alternate layers of Mo and Ni of overall composition $Mo_{35}Ni_{65}$. A similar MX phase has been observed in Nb-Ni, among other systems [17]. Although the composition of the MX phase is not exactly known because it is always mixed with the amorphous phase, it is estimated to be around $x_N = 0.75$ in both systems. The transformations observed in multilayered $Mo_{35}Ni_{65}$ and $Nb_{35}Ni_{65}$ samples treated by increasing doses of 300 keV Xe ions are [17]

$$Mo_{35}Ni_{65} \xrightarrow{5 \times 10^{15} Xe/cm^2} a+MX \xrightarrow{7 \times 10^{15} Xe/cm^2} a \xrightarrow{2 \times 10^{16} Xe/cm^2} a+MX$$

$$Nb_{35}Ni_{65} \xrightarrow{5 \times 10^{14} Xe/cm^2} a+MX \xrightarrow{5 \times 10^{15} Xe/cm^2} a \xrightarrow{1.7 \times 10^{16} Xe/cm^2} a$$

In the last step of the first of these two reactions the amorphous (a) phase dissociates and the MX phase, which was obtained in the early stages of the irradiation of the multi-layered sample, is again formed. It is thought that this ion-induced demixing effect is brought about at high irradiation levels as the result of localized regions of the amorphous phase attaining the composition of the MX phase and crystallizing as MX during the relaxation process; the rest of the material remains amorphous. This ion-induced transformation from amorphous to crystalline states appears to be much more difficult for the system Nb-Ni, which remains amorphous up to the largest dose studied, 1.7×10^{16} Xe/cm^2.

To help understand why these processes occur, we have constructed free energy diagrams for the Mo-Ni and Nb-Ni systems at room temperature (Fig. 3). In this figure we show the calculated free energies of the terminal solid solutions and the amorphous phases. The free energies of the metastable

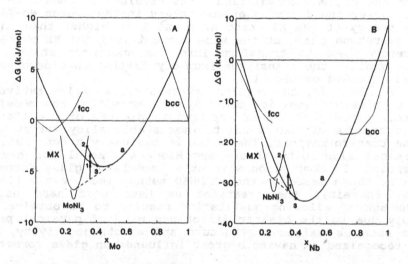

Figure 3. Calculated free energy diagrams of Mo-Ni(A) and Nb-Ni (B) at 300 K, showing the free energies of the terminal solid solutions, the amorphous phases(a), and the metastable crystalline MX phases. For comparison, we have represented the free energies of the intermetallic compounds MoNi$_3$ and NbNi$_3$ (this representation is schematic since only the minima of the compound curves are actually given by the model [7]). Arrows represent some of the transitions indicated in the text.

crystalline MX phases are more difficult to obtain from the experimental data, and perhaps further work should be devoted to fully characterizing these intriguing phases, which Liu has suggested may be considered as electron compounds [1,17]. Within the context of our model, reasonable predictions can be achieved by considering them to be compound-like (and hence with no elastic or structural contribution to the heat of formation), with a broad homogeneity range and a low chemical ordering energy similar to that of the solid solution phases. Under these assumptions, the processes occurring in the transformations presented above can be explained as follows.

We first note that in Fig. 3 the amorphous phases are more stable than the terminal solid solutions over a wide range of compositions, which explains the ready amorphization of the two systems. The formation of the two-phase MX+amorphous mixtures in the early stages of irradiation can be explained by the metastable equilibria indicated in the figure by dashed lines. Further irradiation can evidently bring about complete amorphization, leading to the states labelled 1 in Fig. 3. For Mo-Ni yet greater irradiation can lead to demixing, after which the system is a mixture of non-stoichiometric MX (indicated by point 2 in Fig. 3(A)) plus amorphous alloy (point 3 in the same figure). In the case of the system Nb-Ni, however, there needs to be a large increase in free energy for the demixing effect to occur (see Fig. 3(B)), which is in keeping with the experimental observation that this system remains amorphous under the highest irradiation doses studied. It should be noted that, after the demixing effect occurs in the system Mo-Ni, the free energy of the MX phase at point 2 is higher than that of the amorphous phase at the same composition; the MX phase will therefore tend to transform into the amorphous phase, which would explain the transition from crystalline to amorphous upon aging observed by Liu [1].

To conclude, the results obtained for an illustrative sample of binary transition metal systems show that the model used in this paper provides an appropriate picture of the thermodynamic aspects of ion-irradiated metastable alloy formation and phase transformations. The model is basically a combination of classical elasticity theory and Miedema's model for heats of formation. In comparison with other models of glass formation, such as those based on the CALPHAD method (see, e.g., [18] and refs. therein), the treatment we have used here has the advantage of allowing qualitative results to be obtained more simply due to the greater direct use made of physical properties such as valence, structure and electronegativity, which are recognized as having a great influence on glass formation.

ACKNOWLEDGEMENTS

This work was supported by the DGICYT (Project PB89-0352-C02) and the Xunta de Galicia (Project XUGA20603B90).

REFERENCES

1. Liu, B.X., Ion mixing and metallic alloy phase formation. Phys. Stat. Sol. (a), 1986, 94, 11-34.

2. Johnson, W.L., Thermodynamic and kinetic aspects of the crystal to glass transformation in metallic materials. Prog. Mat. Sci., 1986, 30, 81-134.

3. Massobrio, C., Pontikis, V. and Martin, G., Molecular-dynamics study of amorphization by introduction of chemical disorder in crystalline $NiZr_2$. Phys. Rev. B, 1990, 41, 10486-97.

4. Liu, B.X., Prediction of metallic glass formation by ion mixing. Mat. Lett., 1987, 5, 322-27.

5. Alonso, J.A. and López, J.M., Glass-forming ability in binary alloys produced by ion beam mixing and by laser quenching. Mat. Lett., 1986, 4, 316-19.

6. Alonso, J.A., Gallego, L.J. and López, J.M., Glass formation in binary alloy systems: A prediction of the composition range. Phil. Mag. A, 1988, 58, 79-92.

7. de Boer, F.R., Boom, R., Mattens, W.C.M., Miedema, A.R. and Niessen, A.K., Cohesion in Metals. Transition Metal Alloys, North-Holland, Amsterdam, 1988.

8. Eshelby, J.D., The continuum theory of lattice defects. Solid State Phys., 1956, 3, 79-144.

9. Van der Kolk, G.J., Miedema, A.R. and Niessen, A.K., On the composition range of amorphous binary transition metal alloys. J. Less-Common Met., 1988, 145, 1-17.

10. Loeff, P.I., Weeber, A.W. and Miedema, A.R., Diagrams of formation enthalpies of amorphous alloys in comparison with the crystalline solid solution. J. Less-Common Met., 1988, 140, 299-305.

11. López, J.M., Alonso, J.A. and Gallego, L.J., Determination of the glass-forming concentration range in binary alloys from a semiempirical theory: Application to Zr-based alloys. Phys. Rev. B, 1987, 36, 3716-22.

12. Gallego, L.J., Somoza, J.A., Fernández, H.M. and Alonso, J.A., Prediction of glass formation by solid state reaction in alloys. J. Physique, 1990, C4, 111-17.

13. Gallego, L.J., Somoza, J.A. and Alonso, J.A., Glass formation in ternary transition metal alloys. J. Phys.: Condens. Matter, 1990, 2, 6245-50.

14. Alonso, J.A., Gallego, L.J., Somoza, J.A. and Liu, B.X., On the formation of metastable alloys by ion mixing. In Proceedings C-MRS : Thin Films and Beam-Solid Interactions, Elsevier, Amsterdam, 1991, pp. 297-303.

15. Weeber, A.W., Application of the Miedema model to formation enthalpies and crystallisation temperatures of amorphous alloys. J. Phys. F: Met. Phys., 1987, 17, 809-13.

16. Niessen, A.K., Miedema, A.R., de Boer, F.R. and Boom, R., Enthalpies of formation of liquid and solid binary alloys based on 3d metals. Physica B, 1988, 151, 401-32.

17. Liu, B.X., A metastable electron compound formed by ion irradiation. Phys. Stat. Sol. (a), 1983, 75, K77-K81.

18. Bormann, R., Gärtner, F. and Zöltzer, K., Application of the CALPHAD method for the prediction of amorphous phase formation. J. Less-Common Met., 1988, 145, 19-29.

INVESTIGATION OF ORDER-DISORDER IN NON-CRYSTALLINE BINARIES USING DIFFRACTION METHODS

S. STEEB and P. LAMPARTER
Max-Planck-Institute for Metal Research,
Institute for Materials Science,
Seestr. 92, 7000 Stuttgart-1, Germany

ABSTRACT

A quantitative measure of the degree of order in amorphous or molten binaries is given by the chemical short range order parameters according to Warren and Cowley or according to Cargill and Spaepen. Both of them are based on partial coordination numbers which follow from partial structure factors via partial pair correlation functions. An overview on the chemical short range order parameters as obtained up to date is given. We restrict ourselves mainly on the discussion of the Warren Cowley short range order parameter and propose a refined short range order parameter which can be applied to alloys regardless to differences in the atomic diameters of the alloy components and which furthermore is convenient to calculate the contribution of the chemical short range order to the scattered intensity.

1. INTRODUCTION

A thorough investigation of the structure of non-crystalline binaries yields partial structure factors and partial pair correlation functions. Structural data are the atomic distances and the partial coordination numbers. From the latter we obtain the so called short range order parameters which yield a measure of the degree of chemical ordering and which give information on the question, whether the atoms of different kinds occupy within the different coordination spheres the topological sites in a statistical way, whether unlike pairs or whether equal pairs are preferred.

The preference within the first coordination sphere determines the classification of a non-crystalline binary as statistically distributed, compound forming or segregated alloy. The quantitative characterization of ordering effects in an alloy generally is done by comparison of experimental structural data with those of a corresponding statistically distributed reference alloy. As the latter necessarily is a hypothetical one, this characterization is not unique.

2. FUNDAMENTALS

2.1 Scattering with non-crystalline binaries.

The coherently scattered intensity of a non-crystalline binary is given by eqn. (1):

$$I_{coh}(Q) = \left[<f^2> - <f>^2 \right] + <f>^2 \left[1 + 4\pi \int R^2 \left[\rho(R) - \rho_o \right] \frac{\sin QR}{QR} \, dR \right] \quad (1)$$

with

$$<f^2> = c_1 f_1^2 + c_2 f_2^2$$

$$<f> = c_1 f_1 + c_2 f_2$$

c_1, c_2 = concentration of component 1, 2 in atomic fractions

f_1, f_2 = scattering length of component 1, 2

R = coordinate in real space

$\rho(R)$ = local atomic number density

ρ_o = mean atomic number density

Q = momentum transfer

$\quad = 4\pi \frac{\sin\Theta}{\lambda}$

2Θ = scattering angle

λ = wavelength

The first term in eqn. (1) is called Laue monotonic scattering (LMS) and it stands

$$LMS = \left[<f^2> - <f>^2 \right] = c_1 c_2 \left[f_1 - f_2 \right]^2 \quad (2)$$

2.2 Warren-Cowley short range order parameter.

In [1,2,3] it was shown how to transfer the concept of short range order in crystalline solid solutions to non-crystalline material. For this purpose eqn. (1) has to be transcribed into eqn. (3):

$$I_{coh}(Q) = c_1 c_2 \left[f_1 - f_2 \right]^2 + <f>^2 \left[1 + 4\pi \int R^2 \left[\rho^*(R) - \rho_o \right] \frac{\sin QR}{QR} \, dR \right]$$

$$+ c_1 c_2 \left[f_1 - f_2 \right]^2 \sum_j Z^j \alpha^j \frac{\sin QR^j}{QR^j} \quad (3)$$

with

$\rho^*(R) = c_1 \rho_{11}(R) + c_2 \rho_{22}(R) + 2 c_1 \rho_{12}(R)$

$\rho(R)$ = atomic number density of k-atoms
lk   around a central l-atom

j = number of coordination sphere (I, II, III, etc.)

Normally the short range order parameter of the first coordination sphere is the largest. It can be calculated from the partial coordination numbers [4,5,6] and was applied to non-crystalline substances in the following form [7]:

$$\alpha^I = 1 - \frac{Z_{12}}{Z_{12}^{stat.}} \quad (4)$$

with Z_{12} = number of atoms of kind 2 around a central atom of kind 1 in the real alloy; $Z_{12}^{stat.}$ = number of atoms of kind 2 around a central atom of kind 1 for the statistical case.

The following relationships between the different coordination numbers are valid:

$$Z_1 = Z_{11} + Z_{12} \tag{5}$$
= number of all atoms around an atom of kind 1.
$$Z_2 = Z_{21} + Z_{22} \tag{6}$$
= number of all atoms around an atom of kind 2.

Since the number of unequal pairs is independent of the kind of the central atom it stands

$$c_1 Z_{12} = c_2 Z_{21} \tag{7}$$

The Warren Cowley short range order parameter according to eqn. (4) amounts to +1 for $Z_{12} = 0$, i.e., total segregation. It amounts to zero for $Z_{12} = Z_{12}^{stat.}$, i.e., statistical distribution and it is negative for compound formation, i.e. for $Z_{12} > Z_{12}^{stat.}$. In the following we first describe its contribution to the short range order (SRO-) scattering. Then we will discuss how to obtain α^j experimentally.

2.2.1 Intensity I_{SRO} and the Warren Cowley short range order parameter.
According to eqn. (3) the contribution of short range ordering I_{SRO} to the coherently scattered intensity is given by

$$I_{SRO}(Q) = c_1 c_2 (f_1 - f_2)^2 \sum_j Z^j \alpha^j \frac{\sin QR^j}{QR^j} \tag{8}$$

Apparently the difference $(f_1 - f_2)$ must be large for large I_{SRO}.

Table 1 shows the positions and heights of maxima (+) and minima (−) of a $\frac{\sin QR}{QR}$ -function with QR = x.

$\dfrac{\sin x}{x}$	+ 1	− 0.217	+ 0.128	− 0.091	+ 0.071
x	0	4.49	7.72	10.9	14.06

Table 1. Positions x and heights of maxima (+) and minima (−) of $\frac{\sin x}{x}$ [8].

2.2.1.1 Small angle scattering. The contribution of $\alpha > 0$, i.e. segregation, to the coherently scattered intensity yields small angle scattering. This is caused by the contribution of the $\frac{\sin x}{x}$ -function at small x-values.

2.2.1.2: Premaximum. The contribution of $\alpha < 0$, i.e. compound formation, yields a maximum at $Q \cdot R = 4.49$ which normally occurs at smaller Q's than the main maximum. Such a so called premaximum first was observed for Mg–Sn–melts [9] and is a strong hint on compound formation. It corresponds to the formation of superlattice lines in crystalline superstructures. So far, the Warren Cowley short range order parameter is useful also if we know only whether it is positive or negative that means, if we know α only in a qualitative way. Let us now discuss its quantitative determination. Corresponding to the number of diffraction experiments available for one alloy with a certain concentration, there are different approaches for the determination of α. Normally these demand different definitions of $Z_{12}^{stat.}$ in eqn. (4).

2.2.2 Zero'th approach determination of α: The zero'th approach is based on Keating's assumption, which means the atomic volume of the atoms of kind 1 to be equal to that of the atoms of kind 2.

Then for the mean coordination number Z we obtain $Z = Z_1 = Z_2$ (9)

Furthermore it stands $Z_{12}^{stat.} = c_2 Z$ (10)

and thus we obtain from eqn. (4) $\alpha_{zero}^I = 1 - \dfrac{Z_{12}}{c_2 Z}$ (4')

α_{zero}^I is + 1 for total segregation; α_{zero}^I is 0 for statistical distribution; α_{zero}^I is $-\dfrac{c_1}{c_2}$ with c1 < c_2 for maximum compound formation.

The experimentally determined coordination number Z_{exp} depends in the Faber–Ziman formalism [10] on the partial coordination numbers according to eqn. (11):

$$Z_{exp} = \frac{1}{<f>^2} \left[c_1 f_1^2 Z_{11} + c_2 f_2^2 Z_{22} + 2c_1 f_2 f_1 Z_{12} \right]$$ (11)

Thus from eqns. (5,6,7,11) we obtain

$$Z_{12} = \frac{Z_{exp} <f>^2 - Z(c_1 f_1^2 + c_2 f_2^2)}{2c_1 f_1 f_2 - c_1 (f_1^2 + f_2^2)}$$ (12)

Now one further assumes that the mean coordination number Z can be calculated from the coordination numbers of the pure components:

$$Z = c_1 Z_{comp.1} + c_2 Z_{comp.2}$$ (13)

Apparently using eqns. (12,13) α in eqn. (4') can be determined in a zero'th approach from one experiment which yields Z_{exp}.
We note that beside Keating's assumption this approach requires f_1 and f_2 to be sufficiently different, because for the case $f_1 = f_2$ eqn. (12) cannot be applied.

2.2.3 First approach determination of α: If the partial coordination numbers Z_{11}, Z_{22} and Z_{12} are determined experimentally for an alloy with a certain concentration, α can be obtained without the assumption of eqn. (13) in the form

$$\alpha_{first}^{I} = 1 - \frac{z_{12}}{c_2(Z_{11}+Z_{12})} = 1 - \frac{z_{21}}{c_1(Z_{22}+Z_{21})} \qquad (4'')$$

The upper and lower limits for α_{first}^{I} are the same as for α_{zero}^{I}.
In eqn. (4") we still presume Keating's assumption.

2.2.4 Determination using the Bhatia-Thornton partials.

Depending on the binary system under consideration the Faber-Ziman partial structure factors S_{11}, S_{22}, and S_{12} follow more or less directly from the experiments. They are connected with the coherently scattered intensity via eqn. (14).

$$\frac{I_{coh}(Q)-c_1 c_2 \left[f_1-f_2\right]^2}{<f>^2} = S_{tot}^{FZ}(Q) = \frac{c_1 f_1^2}{<f>^2} S_{11}(Q) + \frac{c_2^2 f_2^2}{<f>^2} S_{22}(Q) + 2 \frac{c_1 c_2 f_1 f_2}{<f>^2} S_{12}(Q) \qquad (14)$$

with S_{tot}^{FZ} = total Faber-Ziman structure factor .

Correspondingly, the total local atomic number density can be split up in partial densities:

$$\rho_{tot}(R) = \frac{c_1 f_1^2}{<f>^2} \rho_{11}(R) + \frac{c_2 f_2^2}{<f>^2} \rho_{22}(R) + 2 \frac{c_1 f_1 f_2}{<f>^2} \rho_{12}(R) \qquad (15)$$

An alternative approach to partial functions is given by the Bhatia-Thornton procedure [11]:

$$\frac{I_{coh}(Q)}{<f^2>} = S_{tot}^{BT}(Q) = \frac{<f>^2}{<f^2>} S_{NN}(Q) + \frac{c_1 c_2 \left[f_1-f_2\right]^2}{<f^2>} S_{CC}(Q) + \frac{2<f> \left[f_1-f_2\right]}{<f^2>} S_{NC}(Q) \qquad (16)$$

with $S_{tot}^{BT}(Q)$ = Bhatia Thornton total structure factor

$\qquad S_{NN}(Q)$ = contribution to the structure factor from density-density correlations

$\qquad S_{CC}(Q)$ = contribution of the concentration-concentration correlations

$\qquad S_{NC}(Q)$ = contribution of the concentration-density correlations

Since in certain cases the BT-partial structure factors S_{NN}, S_{CC}, and S_{NC} follow more or less directly from the experiments it is convenient to express the short range order parameter α also in terms of the BT-partials. In the present case these are especially the partial densities ρ_{CC}, ρ_{NN}, and ρ_{NC} or the corresponding pair correlation functions G_{CC}, G_{NN}, and G_{NC}.

Eqn. (17) shows the linear relationship between ρ_{CC} and the Faber Ziman ρ_{1j}:

$$\rho_{CC} = c_1\left[\rho_{22}+\rho_{21}\right] + c_2\left[\rho_{11}+\rho_{12}\right] - \frac{\rho_{12}}{c_2} \qquad (17)$$

Using Keating's assumption (eqn. (9)), now expressed in terms of the ρ_{1j}, we obtain

$$\rho_{CC} = \rho_{11} + \rho_{12} - \rho_{12}/c_2 = \rho_{22} + \rho_{21} - \rho_{12}/c_2 \qquad (18)$$

For the product $\alpha \cdot Z^{exp}$ finally follows

$$\alpha \cdot Z^{exp} = \int 4\pi R^2 \rho_{CC}(R) dR \tag{19}$$

Integration along the first maximum of ρ_{NN} yields Z^{exp}:

$$Z^{exp} = \int 4\pi R^2 \rho_{NN}(R) dR \tag{20}$$

At this point we stress the fact that Z^{exp} from eqn. (19) is unequal to Z_{exp} from eqn. (11). Z_{exp} depends from the scattering lengths whereas Z^{exp} doesn't.

Thus we finally obtain a short range order parameter α^{BT} which follows from the Bhatia–Thornton functions:

$$\alpha^{BT} = \frac{\int 4\pi R^2 \rho_{CC}(R) dR}{\int 4\pi R^2 \rho_{NN}(R) dR} \tag{21}$$

or

$$\alpha^{BT} = \frac{\int R G_{CC}(R) dR}{\int \left[R G_{NN}(R) + 4\pi R^2 \rho_o \right] dR} \tag{22}$$

Eqn. (21) was derived in [12] and eqn. (22) first was derived in [13]. For the determination of α^{BT} one needs two independent diffraction experiments.

2.3 Warren Cowley short range order parameter as not based on Keating's assumption.

If Keating's assumption is not valid as in most practical cases (see, for example, Tab. 5) the mean coordination number Z in eqn. (10) has to be defined in order to obtain a better approach to $Z_{12}^{stat.}$ than given by eqns. (9,10).

2.3.1 Calculation of Z according to Vegard's law:

Often the averaged coordination number Z is calculated according to

$$Z = c_1 Z_1 + c_2 Z_2 \tag{23}$$

Z_1, Z_2 are obtained using eqns. (5,6) from the experimentally determined partial coordination numbers. From eqn. (23) we learn that for
$c_1 \rightarrow 0 \quad Z \rightarrow Z_2$ and for $c_1 \rightarrow 1 \quad Z \rightarrow Z_1$.

Eqn. (4) now becomes

$$\alpha_{Veg.}^{I} = 1 - \frac{Z_{12}}{c_2(c_1 Z_1 + c_2 Z_2)} \tag{4'''}$$

Total segregation yields $\alpha_{Veg.}^{I} = 1$, statistical distribution yields $\alpha_{Veg.}^{I} = 0$, and maximum compound formation ($Z_{11} = 0$; $Z_{12} = Z_1$) yields

$$\alpha_{Veg.max.}^{I} = \frac{Z_{12}(2c_1 - 1) + c_2 Z_{22}}{Z_{12}(2c_1) + c_2 Z_{22}} \tag{24}$$

2.3.2 Calculation of Z according to Wagner [14].

Wagner [14] proposed the so called generalized Warren-Cowley short range order parameter to be

$$\alpha^I_{gen.} = 1 - \frac{Z_{12}}{c_2(c_1 Z_2 + c_2 Z_1)} \tag{4^{IV}}$$

Total segregation yields $\alpha^I_{gen.} = 1$, statistical distribution yields $\alpha^I_{gen.} = 0$, and maximum compound formation yields

$$\alpha^I_{gen.\ max.} = \frac{Z_{12}(c_1 + c_2^2 - 1) + c_1 c_2 Z_{22}}{Z_{12}(c_1^2 + c_2^2) + c_1 c_2 Z_{22}} \tag{25}$$

Here for Z stands

$$Z = c_1 Z_2 + c_2 Z_1 \tag{26}$$

Thus for $c_1 \to 0$ $Z \to Z_1$ and for $c_1 \to 1$ $Z \to Z_2$.

Compared to eqn. (23) Z_1 and Z_2 are just interchanged.

2.3.3 Calculation of Z using maximalization of entropy.

Cargill and Spaepen [15] have obtained

$$Z = \frac{Z_1 Z_2}{c_1 Z_1 + c_2 Z_2} \tag{27}$$

by maximalization of the entropy.

Thus for $c_1 \to 0$ $Z \to Z_1$ and for $c_1 \to 1$ $Z \to Z_2$.

These are just the same limits as in eqn. (26). However, since eqn. (27) was deduced in [15] using precise physical arguments, we propose for the case, where the partial coordination numbers are known, to use the refined short range order parameter $\alpha^I_{ref.}$ as given in eqn. (28):

$$\alpha^I_{ref.} = 1 - \frac{Z_{12}(c_1 Z_1 + c_2 Z_2)}{c_2 Z_1 \cdot Z_2} \tag{28}$$

At this point we mention the Cargill-Spaepen [15] short range order parameter η_{12} to be

$$\eta_{12} = -\alpha^I_{ref.} \tag{29}$$

The refined short range order parameter $\alpha^I_{ref.}$ as presented in eqn. (28) has the following properties

i) $\alpha_{ref.}^I$ is positive for segregation with a maximum value of +1 for total segregation ($Z_{12}=Z_{21}=0$).

ii) $\alpha_{ref.}^I$ is zero for statistical distribution ($Z_1=Z_2$).

iii) $\alpha_{ref.}^I$ is negative for compound formation with a maximum value of $-c_1Z_1/c_2Z_2$ for total compound formation ($Z_{22}=0$).

iv) $\alpha_{ref.}^I$ describes according to eqn. (8) the intensity contribution I_{SRO} which is caused by short range ordering.

The normalized refined short range order parameter $\alpha_{ref.}^I$ is obtained as

$$\alpha_{norm.}^I = \alpha_{ref.}^I \cdot \frac{c_2Z_2}{c_1Z_1} \tag{30}$$

3.1 Molten binaries
Some of the short range order parameters obtained so far with compound forming molten binaries are compiled in Tab. 2.

Alloy / Method	T [°C]	α_{first}^I / $\alpha_{first}^I \cdot \frac{c_2}{c_1}$	Reference
$Ag_{28}Li_{72}$ N $<f>=0$	near T_m	−0.15 0.39	12
$Bi_{30}Mg_{70}$ X, N	near T_m	−0.44 1.0	3
$Cd_{25}Mg_{75}$ X	near T_m	−0.13 0.39	16
$Cu_{37}Sb_{63}$ N	near T_m	−0.37 0.63	17
$Cu_{66}Ti_{34}$ N[+]	950	−0.08 0.16	18,19
$Ga_{23}Li_{77}$ N $<f>=0$	475	−0.09 0.30	20
$Li_{70}Mg_{30}$ N	near T_m	−0.04 0.09	12
$Li_{80}Pb_{20}$ N $<f>=0$	995	−0.25 1.0	13
$Li_{72.5}Sn_{27.5}$ N $<f>=0$	750	−0.16 0.42	21

Table 2.
Molten binaries with compound forming tendency.
α_{first}^I = Warren–Cowley short range order parameter.
X = X-ray diffraction.
N = neutron diffraction.
$<f>$ = 0 means zero scattering alloy.
[+] S_{NN} calculation according to Percus–Yevick hard sphere model and S_{NC} = 0.

Thus the degree of chemical short range ordering reaches from 9 % to 100 %.

Up to now almost no molten binary is known with $\alpha = 0$, i.e. with statistical distribution of the atoms of both kinds. In Tab. 3 we have compiled segregated molten binaries.

Alloy Method	T [°C]	α^I_{first}	Reference
$Al_{100-x}In_x$ $(x=4.7$ to $90)$ SAXS	near T_m	0.05 to 0.21	22
$Cd_{50}Ga_{50}$	near T_m	0.23	23
$Cu_{100-x}Pb_x$ $(x=35, 65)$ N	near T_m	$x = 35$ $\alpha^I = 0.58$ $x = 0.65$ $\alpha^I = 0.46$	24
$Li_{61}Na_{39}$ SANS	317: 452:	$\alpha^I = 0.5$ $\alpha^I = 0.3$	25

Table 3. Molten segregated alloys. Warren–Cowley short range order parameter α^I_{first}. SAXS = small angle X-ray scattering. SANS = small angle neutron scattering. N = neutron scattering.

3.2 Amorphous binaries

Table 4 shows some results of chemical short range order parameters as obtained with amorphous binaries.

Table 4. Relative chemical short range order parameters of amorphous binary alloys.

Substance A_xB_{1-x}	Ref.	$\alpha^I_{norm.}$	$\alpha^I_{rel.} = \dfrac{\alpha^I_{first}}{\alpha^I_{first\ max}}$	ϕ_B/ϕ_A
$B_{20}Fe_{80}$	26	1	–	1.43
$B_{19}Ni_{81}$	27	1	–	1.40
$B_{33}Ni_{67}$	28	0.86	–	1.40
$B_{36}Ni_{64}$	29	0.81	–	1.40
$B_{43}Hf_xZr_{57-x}$	30	–	0.27	1.75
$[(Co,Fe)_{0.77}Tb_{0.23}]_{87}H_{13}$	31	– 0.2	–	1.37
$(Co,Fe)_{80}Tb_{20}$	31	0.18	–	1.37
$Co_{80}P_{20}$	32,33	1	–	0.91
$Co_{83}P_{17}$	33	–	1.0	0.91
$Cu_{75}Ti_{25}$	34	–	0.79	1.13
$Cu_{70}Ti_{30}$	35	–	0.44	1.13
$Cu_{66}Ti_{34}$	18	–	$\alpha_{am}=2\cdot\alpha_{melt}$	1.13
$Cu_{50}Ti_{50}$	36	0.07	–	1.13
$Cu_{40}Ti_{60}$	37	–	0.11	1.13
$Cu_{35}Ti_{65}$	35	–	0.11	1.13
$Cu_{33}Y_{67}$	38	0.05	–	1.38
$Cu_{60}Zr_{40}$	15	0.32	–	1.24
$Fe_{67}Tb_{33}$	39	0.09	–	1.36
$Gd_{80}Si_{20}$	40	1	–	0.69
$Ni_{31}Dy_{69}$	41	0.09	–	1.38

$Ni_{62}Nb_{38}$	42	0.14	–	1.16
$Ni_{63}Nb_{37}$	43	0.15	–	1.16
$Ni_{56}Nb_{44}$	43	0.11	–	1.16
$Ni_{50}Nb_{50}$	43	0.08	–	1.16
$Ni_{40}Nb_{60}$	43	0.11	–	1.16
$Ni_{80}P_{20}$	44	1	–	0.92
$Ni_{40}Ti_{60}$	45,46	0.33	0.3	1.15
$Ni_{40}Ti_{60}$	47	–	0.15	1.15
$Ni_{35}Ti_{65}$	46	–	0.19	1.15
$Ni_{26}Ti_{74}$	48	–	0.37	1.15
$Ni_{33}Y_{67}$	38	0.40	–	1.41
$Ni_{63.7}Zr_{36.3}$	49,50	0.07	–	1.25
$Ni_{64}Zr_{36}$	51	0.13	–	1.25
$Ni_{64}Zr_{36}$	52	0.05	–	1.25
$Ni_{50}Zr_{50}$	53	0.19	–	1.25
$Ni_{36}Zr_{64}$	54	0.07	–	1.25
$Ni_{35}Zr_{65}$	55	–	0.15	1.25
$Ni_{25}Zr_{75}$	56	0.13	–	1.25
$MS-Ni_{31}Zr_{69}$	57	–	0.18	1.25
$MA-Ni_{30}Zr_{70}$	57	–	0.32	1.25
$MS-Ni_{63.7}Zr_{36.3}$	57	–	0.44	1.25
$MA-Ni_{65}Zr_{35}$	57	–	0.80	1.25
$MS-Ni_{34}Zr_{66}$	57	–	0.05	1.25
$SP-Ni_{34}Zr_{66}$	57	–	0.07	1.25
$Ni_{35}(Zr,Hf)_{65}$	55	– 0.06	–	1.25
$Ni_{33.3}Zr_{66.7}$	58	0.06	–	1.25
$Pd_{80}Si_{20}$	59	1	–	0.87

The following abbreviations are used in Tabl. 4: MS = melt spinning. MA = mechanical alloying. SP = sputtering. ϕ_1 = atomic radius of component i.

3.3 Concentration dependence

An interesting example to study the concentration dependence of the short range order parameter is given by Ni_xNb_{100-x} alloys as studied in [43].

Alloy	Z_{11}	Z_{22}	Z_{12}	Z_{21}	Z_1	Z_2	α^I_{first} /rel. from Z_1	α^I_{first} /rel. from Z_2	$\alpha^I_{ref.}$	α^I_{norm}
$Ni_{40}Nb_{60}$	3.8	9	8.2	5.5	12	14.5	−0.21	0.08	−0.06	0.11
$Ni_{50}Nb_{50}$	5.0	7.5	7.4	7.4	12.4	14.9	−0.19	0.01	−0.1	0.08
$Ni_{56}Nb_{44}$	5.5	6.5	6.6	8.4	12.1	14.9	−0.30	0.01	−0.1	0.11
$Ni_{63}Nb_{37}$	6.6	5.6	5.9	10.0	12.5	15.6	−0.47	0.03	−0.1	0.15

Table 5.   Amorphous Ni_xNb_{100-x}. Partial coordination numbers and short range order parameters.

In Tab. 5 the experimentally obtained partial coordination numbers are compiled together with the Warren-Cowley short range order parameter in its original first approach as well as in its refined form as proposed in chapter 4.

The total coordination number around a Nickel-atom amounts to about 12 atoms, whereas the total coordination number around a Niobium-atom amounts to about 15 atoms. Thus Z_1 is unequal to Z_2 and we obtain for each central atom a different relative Warren-Cowley short range order parameter. Around a nickel atom the compound formation lies between 21 and 47 % of the maximum possible value. Around a Nb-atom α_{rel}^{I}. corresponds to statistical distribution. Thus these alloys are a good example to show that for $Z_1 \neq Z_2$ the Warren-Cowley parameter cannot be applied uniquely in its zero'th and first approximation. . It should indeed be replaced by the refined parameter α_{ref}. as introduced in the present paper.

3.4 Temperature dependence

Following the transition from molten to amorphous state by diffraction methods with the $Mg_{70}Zn_{30}$-alloy [60], the short range order paramter shows an increase in compound forming tendency. The same also stands according to [18,19] for Cu-Ti-alloys. In addition to this [18,19] showed a similar concentration dependence of α once in the molten and once in the amorphous state. The strongest compound formation, however, does not occur at the concentration of an intermetallic crystalline alloy.

References

1. Hezel, R. and Steeb, S., Z. Naturforschung, 1970, 25a, 1085.
2. Boos, A., Doctorthesis, University of Stuttgart, 1976.
3. Boos, A. and Steeb, S., Phys. Letters, 1977, 64A, 333.
4. Taylor, A., X-Ray Metallography, John Wiley & Sons, New York, 1961.
5. Cowley, J.M., J. Appl. Phys., 1950, 21, 24.
6. Warren, B.E., X-Ray Diffraction, Addison-Wesley Publ., Reading (Mass.), 1969.
7. Steeb, S. and Hezel, R., Z. Physik, 1966, 191, 398.
8. Int. Tab. Crystallography, The Kynoch Press, Birmingham, 1959, 2.
9. Steeb, S. and Entress, H., Z. Metallkunde, 1966, 57, 803.
10. Faber, T.E. and Ziman, J.M., Phil. Mag., 1965, 11, 153.
11. Bhatia, A.B. and Thornton, D.E., Phys. Rev., 1970, B2, 3004.
12. Chieux, P. and Ruppersberg, H., J. de Physique, 1980, C8, 145.
13. Ruppersberg, H. and Egger, H., J. Chem. Phys., 1975, 63, 4095.
14. Wagner, C.N.J., J. Non-Cryst.Sol., 1980, 42, 3.
15. Cargill III, G.S. and Spaepen, F., J. Non-Cryst.Sol., 1981, 43, 91.
16. Boos, A. and Steeb, S., Z. Naturforschung, 1977, 32a, 1229.
17. Knoll, W. and Steeb, S., Z. Naturforschung, 1978, 33a, 472.
18. Sakata, M., Cowlam, N., and Davies, H.A., J. Phys. F: Met. Phys., 1981, 11, L157.
19. Fenglai, H.E., Cowlam, N., Carr, G.E., and Suck, J. B., Phys. Chem. Liq., 1986, 16, 99.
20. Reijers, H.T.J., Van der Lugt, W., Van Tricht, J.B., and Vlak, W.A.H.M., J. Phys.: Cond. Matter, 1989, 1, 8609.
21. Alblas, B.P., Van der Lugt, W., Dijkstra, J., and Van Dijk, C., J. Phys. F: Met. Phys., 1984, 14, 1995.
22. Hoehler, J. and Steeb, S., Z. Naturforschung, 1975, 30a, 775.
23. Hermann, G., Rainer-Harbach, G., and Steeb, S., Z. Naturforschung, 1980, 35a, 938.
24. Lamparter, P. and Steeb, S., Z. Naturforschung, 1980, 35a, 1178.
25. Ruppersberg, H., and Knoll, W., Z. Naturforschung, 1977, 32a, 1374.

26. Nold, E., Lamparter, P., Olbrich, H., Rainer-Harbach, G., and Steeb, S., Z. Naturforschung, 1981, 36a, 1032.
27. Lamparter, P., Sperl, W., Steeb, S., and Blétry, J., Z. Naturforschung, 1982, 37a, 1223.
28. Ishmaev, S.N., Isakov, S.L., Sadikov, I.P., Sváb, E., Kőszegi, L., Lovas, A., and Mészáros, Gy., J. Non-Cryst. Sol., 1987 94, 11.
29. Cowlam, N., Guoan, Wu., Gardner, P.P., and Davies, H.A., Proc. LAM V, 1984, 1, 337.
30. Maret, M., Soper, A., Etherington, G., and Wagner, C.N.J., Proc. LAM V, 1984, 1, 313.
31. Heckele, M., Doctorthesis, University of Stuttgart, 1991.
32. Sadoc, A. and Dixmier, J., J. Mater. Sci. and Eng., 1976, 23, 187.
33. Blétry, J. and Sadoc, J.F., J. Phys., 1975, F5, L110.
34. Sakata, M., Cowlam, N., and Davies, H.A., J. Phys., 1979, F9, L235.
35. Sakata, M., Cowlam, N., and Davies, H.A., Proc. RQ IV, 1982, 327.
36. Müller, A., Bellissent, R., Bellissent, M.C., Armbroise, J.P., Bühler, E., Lamparter, P., and Steeb, S., Z. Naturforschung, 1987, 42a, 421.
37. Fukunaga, T. and Suzuki, K., RITU, 1981, 29a, 153.
38. Maret, M., Chieux, P., Hicter, P., Atzmon, M., and Johnson, W.L., J. Phys. F: Met. Phys., 1987, 17, 315.
39. O'Leary, W.P., J. Phys., 1975, F5, L175.
40. Masumoto, T., Fukunaga, T., and Suzuki, K., Bull. Am. Phys. Soc., 1978, 23, 467.
41. Wildermuth, A., Lamparter, P., and Steeb, S., Z. Naturforschung, 1985, 40a, 191.
42. Sváb, E., Mészáros, Gy., Konczos, G., Ishmaev, S.N., Isakov, S.L., Sadikov, I.P., and Chernyshov, A.A., J. Non-Cryst. Sol., 1988, 104, 291.
43. Lamparter, P., Schaal, M., and Steeb, S., Conf. on Neutron- and X-Ray Scattering, Complementary Techniques, Canterbury, U.K., 1989, Inst. of Phys. Conf. Ser., 1990, No. 101, 51, IOP Publishing Ltd.
44. Lamparter, P. and Steeb, S., Proc. RQ V, 1985, 459.
45. Fukunaga, T. Watanaba, N., and Suzuki, K., Proc. LAM V, 1984, 343.
46. Ruppersberg, H., Lee, D., and Wagner, C.N.J., J. Phys., 1980, F10, 1645.
47. Ruppersberg H. and Wagner, C.N.J., Exp. Report No. 06-08-021, ILL, 1979.
48. Fukunaga, T., Kai, K., Naka, M., Watanabe, N., and Suzuki, K., Proc. RQ4, 1982, 347.
49. Lefèbvre, S., Quivy, A., Bigot, J., Calvayrac, Y., and Bellissent, R., J. Phys. F: Met. Phys., 1985, 15, L99.
50. Lefèbvre, S., Harmelin, M., Quivy, A., Bigot, J., and Calvayrac, Y., Proc. LAM VI, 1988, 365.
51. Sadoc, A. and Calvayrac, Y., J. Non-Cryst. Sol., 1986, 88, 242.
52. de Lima, J.C., Tonnerre, J.M., and Raoux, D., J. Non-Cryst. Sol., 1988, 106, 38.
53. Fukunaga, T., Hayashi, N., Watanabe, N., and Suzuki, K., Proc. RQ V, 1985, 475.
54. Mizoguchi, T., Yoda, S., Akutsu, N., Yamada, S., Nishioka, J., Suemasa, T., and Watanaba, N., Proc. RQ V, 1985, 483.
55. Lee, A., Etherington, G., and Wagner, C.N.J., Proc. LAM V, 1984, 349.
56. Kuschke, W.M., Doctorthesis, University of Stuttgart, 1991.
57. Kuschke, W.M., Schultz, L., Lamparter, P., and Steeb, S., Z. Naturforschung, 1991, in press.
58. de Lima, J.C., Tonnerre, J.M., Raoux, D., and Morrison, T.I., 1989.
59. Suzuki, K., Misawa, M., and Fukunaga, T., Bull. Am. Phys., 1978, 23, 467.
60. Bühler, E., Lamparter, P., and Steeb, S., Z. Naturforschung, 1987, 42a, 507.

DIRECT IMAGING USING SCANNING TUNNELING MICROSCOPY OF THE ATOMIC STRUCTURE IN AMORPHOUS $Co_{1-x}P_x$

J.M. GONZALEZ, F. CEBOLLADA, M. VAZQUEZ, M. AGUILAR[*],

M. PANCORBO[*] and E. ANGUIANO[+]

Instituto de Ciencia de Materiales (sede A). CSIC.

C/Serrano 144, 28006 Madrid, Spain.

[*]Instituto de Ciencia de Materiales (sede B). CSIC.

U. Autónoma de Madrid. C-III. 28049 Cantoblanco (Madrid), Spain

[+]IBM Scientific Center. U. Autónoma de Madrid. C-XVI.

28049 Cantoblanco (Madrid), Spain

ABSTRACT

STM images of the free surface of electrodeposited amorphous CoP alloys have been obtained, in the scale of 50 Å to 1400 Å. From images showing atomic resolution the interatomic distances for first neighbours have been measured and the radial distribution function (extrapolated to three dimensions) has been calculated, showing two peaks which are in agreement with results obtained from scattering experiments for samples of similar composition. Calorimetric and magnetic characterization of the samples have also been carried out.

INTRODUCTION

Amorphous metallic alloys can be prepared by a wide variety of techniques, all of them allowing the formation of a metastable structure whose main characteristic is the absence of long range atomic periodicity as in crystalline materials. However, many experimental results show clear indications that a certain local order, that is preserved within a few interatomic distances, exists. Experimental evidence, from EXAFS and scattering experiments to the phenomenology of the induction of magnetic anysotropies, yield information about the local short range order (SRO) of amorphous alloys. The classical structural description of the amorphous alloys is obtained through the so-called Radial Distribution Function (RDF).

The RDF is related to the probability of finding an atom at a distance r from any atomic site in the structure, and can be obtained, by Fourier transform, from the reduced interference function measured in a (X-ray, electron or neutron) scattering experiment [1]. The information contained in the RDFs is essentially unidimensional and consequently no complete description of the short range order can be extracted from their evaluation.

A different approach leading to the description of the structure of amorphous alloys is that based on EXAFS measurements. Local atomic arrangements can be determined by means of EXAFS in a more direct way than in other scattering experiments [2] but, unfortunately, as in X-ray, electron or neutron experiments, EXAFS studies are in practice basically restricted to biatomic alloys.

Finally, it is worth mentioning that imaging of the microstructure of amorphous alloys in the scale of several tens of Å, where some details of the atomic structure were resolvable [3], has been carried out (in thin deposits or cuts) by means of field-ion microscopy (atomic probe) and high resolution electron microscopy [4].

Although scanning tunneling microscopy (STM) [5] could become an important tool for the study of disordered systems, few studies [6-8] have been published about STM observations of this type of system, none of them showing images with atomic resolution as far as we know. We present in this paper a STM study of a series of amorphous alloys of composition $Co_{1-x}P_x$ from several thousand Å down to the scale of tens of Å, with atomic resolution. The observed microstructural features are correlated to previous descriptions of the SRO obtained for CoP alloys, as found in the literature [9, 13]. Concerning CoP amorphous alloys, few data on their SRO structure are available. Probably, the most thorough study could be considered to be that of Sadoc and Dixmier [9] which is based on X-ray and polarized neutron scattering experiments.

In our case, the STM imaging of the surface of amorphous ferromagnetic samples has been used for studying both the short wavelength roughness of the surfaces (from the range of hundreds of Å) and the local SRO through atomic imaging.

SAMPLES AND EXPERIMENTAL TECHNIQUES

$Co_{1-x}P_x$ (0.10<x<0.21) sample , 50μm thick, were obtained by electrodeposition, using a standard bath at a temperature of 80°C, on a Cu substrate that was previously electrolytically and mechanically polished. X-ray diffractograms showed no traces of peaks, down to the resolution of the technique. EDX analysis was performed by using a 1μm x 1μm window and showed compositional homogeneity throughout most of the samples length, but a slight increase of about 1% in the Co content was observed when coming close to the edges of the samples.

Magnetic characterization of the samples was carried out at room temperature by means of an induction device, in maximum applied fields of 1T, and domain observation was performed using the Bitter technique. Calorimetric measurements were carried out with a DSC-II equipment, in Ar atmosphere and at a rate of 10 K/min.

The STM system that we have employed has commercial head from W.A. Technology, which gives a maximal resolution of the order of 0.1 Å for vertical displacements and 2 Å for horizontal displacements. The feedback circuit was developed in the Electronics Departament of our Institute and STM automation and data and image processing were made in collaboration with the IBM Scientific Centre in Madrid.

The electronics consists of a current-voltage converter and a logarithmic amplifier followed by a PID amplifier with several filters (low-pass) and a feedback system. The maximum data acquisition rate was rather high, 100 KHz, and the A/D converter resolution was 16 bits. The scanning rate we have employed is rather high (0.1KHz) in order to minimize the influence of the noise due to the air induced vibration of the tip and 1/f noise, the number of scannings per image usually being 256. The images were displayed by means of a PS2/80 and a 8514 display (768 x 1024 pixels), in which we could obtain 256 levels of grey corresponding to 8 bits resolution. Due to the high scanning rate there is a certain integration of the signal along the scanning direction but little influence of the 1/f noise which, however, has to be taken into account in the y direction if low time for image acquisition (two and a half seconds to a few tens of seconds) is employed, and thus, an asymmetric Wiener filter [10, 11] has been implemented in order to restore the images.

RESULTS AND CONCLUSIONS

All samples were characterized as X-ray amorphous as no diffraction peaks were observed down to the resolution of the technique. Further confirmation of the amorphousness of the samples was obtained through the calorimetric measurements, which showed two exothermic peaks corresponding to two different crystallization processes, each one of them yielding a crystallization enthalpy of about 60 Jg^{-1}. Hysteresis loops for fields up to 0.1T showed the existence of an easy axis perpendicular to the sample plane, which is characteristic of amorphous CoP amorphous alloys [13], this inducing a strip domain pattern which appears to be homogeneous throughout the whole sample extension.

Topographic images in the scale of a few thousand Å were obtained in order to study the typical surface roughness. Fig. 1 shows a typical image of 1400 Å x 1400 Å from the same sample, in which relatively smooth structures are evident, its wavelength and height being of the order of a few hundreds Å. The same structure was observed for samples with different Co percentages. Surface roughness may influence the approach to magnetic saturation in medium and low magnetostriction samples through the factor R^2/Ld (R and L being respectively the amplitude and wavelength of the surface irregularities and d the sample thickness). This factor in our samples is of the order of 10^{-4} [14]. This value indicates that the volume in which magnetization is affected by the poles appearing at the surface is negligible. Preliminary calculations of the fractal dimension, in this scale, are being carried out, yielding values close to 2.5. These values can eventually be used for correlation with the factor R^2/L in samples of a fixed composition and thickness, but with different medium surface roughness.

Fig. 2 shows an atomic resolution image (50 Å x 50 Å) of the surface of a $Co_{82}P_{18}$ sample. The structure of the surface is characterized by a high atomic packing factor and by the absence of a clear angular correlation between first neighbours. The presence of five-fold symmetry sites (some of them are marked by a circle in the figure) can be observed, but also sites of different coordination can be seen. From images similar to this one we evaluated the (2-D) reduced radial distribution functions corresponding to the

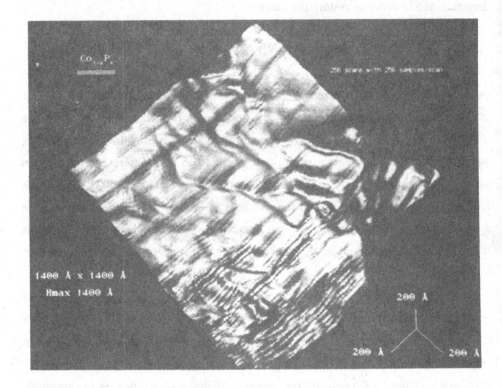

Figure 1. 1400 Å x 1400 Å image of the surface of $Co_{82}P_{18}$, in which smooth undulations are observed.

surface of the samples by measuring the dispersion of the heights in rings of width Δr, at distance r from a given reference atom and taking a mean value over different references. A direct extrapolation to 3-D is then made.

In order to test the obtained reduced RDF, we made a different approach by obtaining it through a previous calculation of the interference function (IF): new images were obtained with a single point in place of each atom, then an average over all directions of the Fourier Transform (FT) was made, and finally the reduced RDF was calculated from the IF (see fig.3). The peaks thus obtained are almost coincident.

Results obtained from scattering experiments [9] for samples of similar composition show four peaks corresponding to: Co-P (first neighbours) at 2.34 Å, Co-Co (first neighbours) at 2.54 Å, P-P at 3.34 Å and Co-Co pairs (second neighbours in a pentagonal ring) at 4.2 Å respectively. A first peak that is apparent from our RDF is centered at about 2.2 Å and a second one at about 4.1 Å. The peak at 2.2 Å can be related to the distance between first neighbours (Co-Co) and the second one at 4.1 Å to

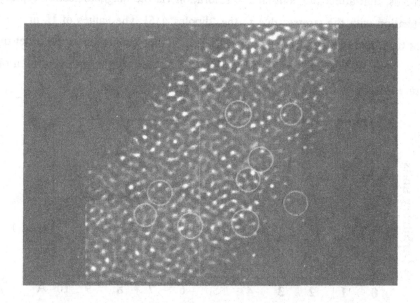

Figure 2. Atomic resolution image of the surface of a $Co_{82}P_{18}$ sample. Tunnel current: 0.4 nA; bias voltage: 50 mV; acquisition time: 40 s.

the second neighbour distance in a pentagonal ring formed by Co atoms. The peaks corresponding to Co-P and P-P distances can either be included in the structure of the two peaks at 2.2 Å and 4.1 Å or will not be apparent if P atoms are located in the holes of stable structural units, thus not being visible at the surface.

Another evaluation of the first neighbour's distribution of frequencies for interatomic distances was made by counting the atoms inside a circular region of 5 Å diameter centered in one atom and averaging over all the similar images. First neighbour interatomic distance is 2.5 ± 0.25 Å (fig.4). A second peak is observed for distances in the range of 3-4 Å, yielding the characteristic shoulder of amorphous alloys [13]. This result confirms those obtained from the previous calculations for the RDF.

Magnetic measurements in the reversible magnetization region, up to 1T, were performed in three samples ($x = 0.15$, 0.18 and 0.21 respectively) in order to study the $1/H$ and $1/H^2$ terms appearing in the law of approach to saturation. For fields under a certain value H_t a $1/H$ type law is observed, while for higher fields a $1/H^2$ type law holds. The H_t field can be related to length of the "quasidislocation dipoles" present in the samples, that affect the approach to saturation via the magnetoelastic coupling of magnetization with the stresses due to the "dipoles" [15]. The values of H_t are of the order of 0.2T, and the length of the "quasidislocation dipoles" is then of the order of 50 Å. Some kind of "defects" resembling dislocations have been observed in atomic resolution images by forcing the integration along a certain direction [14].

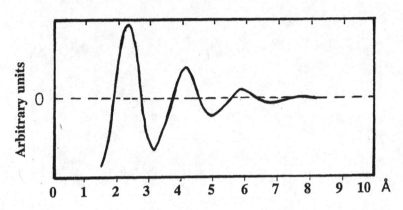

Figure 3. Reduced radial distribution function obtained from the IF calculated from a 50 Å x 200 Å image.

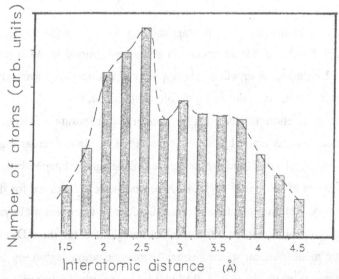

Figure 4. Number of atoms as a function of the distance from a fixed atom. Data have been taken from several photographs and averaged.

REFERENCES

1. Wagner, C.N.J. In Amorphous Metallic Alloys, ed. F.E. Luborsky, Butterworths, London, 1983, pp 58-73.

2. Wong, J. In Glassy Metals I, ed. H.J. Guntherodt and H. Beck, Springer Verlag, Berlin, 1981, pp 45-77.

3. Bolt, P.J., Hoving, W., Bronsveld, P.M. and de Hosson, J.T.M. In Rapidly Quenched Metals, ed. S. Steeb and H. Warlimont, Elsevier Science Publishers B.V., Amsterdam, 1985.

4. Zweck, J. and Hoffmann, H. In Amorphous Metallic Alloys, Ed. F.E. Luborsky Butterworths, London, 1983.

5. Hansma, P.H. and Tersoff, J., J. Appl. Phys., 1987, $61(2)$, R1-R23.

6. Zaluska, A., Zalusky, L. and Wiek, A., Mat. Sci. Eng. A, 1989, 122, 251-255.

7. Gimzewsky, J.K., Möller, R., Myhra, S., Schlittler, R.R., Stoll, E., Rivière, J.C., Calka, A. and Rdlinski, A.P., J. Non Cryst. Sol., 1990, 116, 253-261.

8. Habib, K., Eling, V. and Wu, C., J. Mat. Sci. Let., 1990, 9, 1194.

9. Sadoc, J.F. and Dixmier, J., J. Mater. Sci. Eng., 1976, 23, 187-192.

10. Pancorbo, M., Anguiano,E., Diaspro, A., Aguilar, M., Transact. Roy. Microsc. Soc., 1990, 1, 99-108.

11. Pancorbo, M., Anguiano,E., Diaspro, A., Aguilar, M., Pattern Recogn. Lett., 1990, 11, 553-556.

12. Riveiro, J.M., and Sanchez-Trujillo, M.C., IEEE Trans. on Mag., 1980, 16, 1426-1428

13. Polk, D.E., Act. Metall., 1972, 20, 485-491.

14. Cebollada, F., González, J.M., Vazquez, M., Aguilar, M., Pancorbo, M., Anguiano, E., J. Magn. Magn. Mat., to be published.

15. Grimm, H. and Kronmüller, H., Phys. Stat. Sol. (b), 1983, 117, 663-674.

COMPUTER SIMULATION OF GLASS FORMATION IN TWO DIMENSIONAL NETWORKS

F. MONTOYA and J.M. DUBOIS
Laboratoire de Science et Génie des Matériaux Métalliques
CNRS - URA159
Ecole de Mines de Nancy
54042 NANCY CEDEX - FRANCE

ABSTRACT

Although amorphous metals exhibit high degree of disorder, increasing evidence has been pointed out in recent years that they also yield a significant degree of order extending over distances as large as tens of angströms. The main features which characterize such a medium range order are most often reminiscent of both the local units observed in the parent crystal and of the topological connection scheme between units. In glass forming alloys, we relate the occurence of amorphization to an order-disorder transition that takes place together with a lattice instability mechanism which is mainly due to an atomic size mismatch effect. A numerical simulation has been designed in two dimensions in order to explore into more detail the relevant parameters calculated as a function of physical parameters.

I INTRODUCTION

Criteria for amorphisation in metallic alloys are probably almost as diverse as the preparation processes. However, the relative size of the atomic constituents is a predominant parameter to account for the tendency to amorphisation [1]. In this paper, we will focus our attention to this mechanism and particularly to the case of $M_{1-x}X_x$ binary alloys (M is a metal element and X a metalloid of smaller size).

A way of understanding the action of the atomic size mismatch is to consider the amorphisation process as a structural transformation of an intersticial solid solution. This solid solution does not exist necessarily since it is drastically unstable, due to size mismatch. It is much more to be considered as a (virtual) state of reference [2]. It will be shown that the presence of point defects (topological/chemical) in the 2D network may lead to a disordered medium having no (linear) grain boundaries, i.e. to a true amorphous state. This demonstration will be based on computer simulation.

II SIMULATION METHOD

The basic idea of our method is to consider the process of disordering as being due to degenerated structural transformations taking place in an initial and unstable atomic structure. The elementary entity in this model is composed of a point in physical space (in the initial structure) and a function labeled on an appropriate basis. This couple represents a structural unit (more precisely its core) and a local transformation describing the evolution from the initial state to the resulting one. Thus, a model in this framework is build on a set of structural unit positions and an associated discrete field of transformation.

The dynamical behaviour is governed by the interaction between the elementary entities described above. This interaction encapsulates all the microscopic details (geometrical constraints, structural units connectivity, etc.). One important consequence of this simulation method is that only elementary entities of the model are modified by the simulation and not directly the real atomic positions. More precisely, if no diffusion is considered as we shall do in this first report, only the transformation coordinates of the elementary entities evolve when building the model. From computer simulation point of view, this means that the initial list of neighbours of a given point will be maintained during the whole transition process and that only few parameters are to be modified. For the purpose of the present paper, we used a Monte-Carlo importance sampling method of simulation [3,4]. A fictive temperature is naturally introduced in the energy jump acceptance probability and allows to study the model under some kind of "heat treatment".

III A SIMPLE 2D EXAMPLE

In a previous paper [2], a simple 2D model of topological instability was presented that could lead to a transition to the amorphous state. The present paper describes this model in more detail and particularly studies the pertinent parameters that control the simulation. In this simple model, the initial structure is taken as an interstitial solid solution of a metalloid X in a square matrix of metals M. This interstitial solution is unstable because of geometrical constraints: X atoms do not fit into assigned sites of the M matrix because they are slightly too big. These constraints make X atoms move to more convenient positions so that they get more room. **Figure 1** shows the initial structure at a given composition with maximum chemical order and the resulting structure after transition to the fundamental state: each X atom has now moved so as to generate a six-neighbours larger site of M atoms. In this example, the perfect chemical order induces by collective motion a crystalline order in the resulting structure.

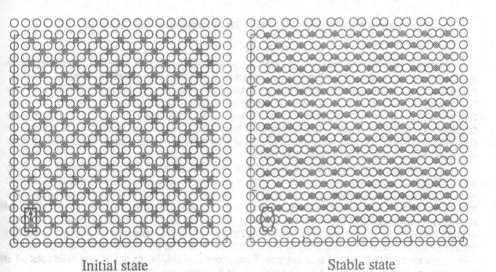

Initial state   Stable state

Figure 1. Perfect Chemical Order.

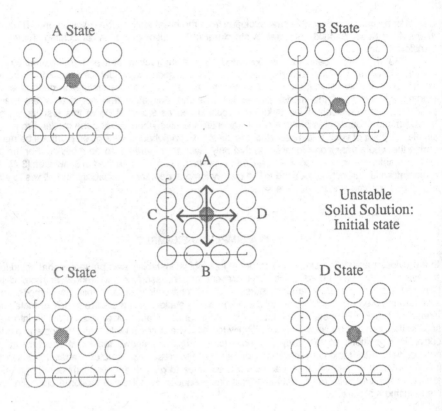

Figure 2. Initial State, Structural Transformations, Stable States.

Figure 2 presents the four possible kinds of geometrical transformations that can be applied to an X atom site to bring it to the fundamental structure. If X atoms are neighbour free, then the four transformations are energetically equivalent. An arbitrairy choice is necessary to determine the kind of transformation to apply. So, in a highly dilute system, the M-matrix would accomodate all lattice distorsions induced by X atoms and the transition towards another state would not occur. But at larger concentration and since X atoms are second shell neighbours, geometrical constraints can propagate themselves and lead to collective motion. Moreover, if the initial chemical order is not perfect, local frustration and disorder will eventually take place.

This practical explanation can be viewed in a more formal framework: Consider the symmetry group G0 (point symmetry group 4mm in Hermann-Mauguin notation) of the unstable initial solid solution and the symmetry group G (point symmetry group 2mm) of the resulting structure. The transformation considered here is thus a group to non trivial sub-group transformation by loss of symmetry. In this particular case, while an X atom moves, the symmetry is locally broken. Chemical defects induce local frustrations and thus break the symmetry independently of the material actual nature. Since the initial structure is highly unstable, transformations are irreversible: it is impossible to reversibly transform to the initial solid solution. Simulations can only be done from the solid solution at high temperature down to the transformed structure at low temperature.

IV FORMAL DESCRIPTION

The model of local transformations is quite simple in this case: Only translations along the four principal directions of the lattice are allowed (see **figure 2**). The amplitude of the translation can vary continuously (from 0, no transformation, to 1, full transformation). Thus, the transformation functions can be represented by the corresponding translation vectors. Let us define the transformation vector by:

$$\underline{M}_x = (\alpha \cos \theta, \, \alpha \sin \theta)$$

where θ labels the direction of the transformation and α is the amplitude of the transformation.

The position of the corresponding X atom at any time is given by:

$$\underline{X}' = \underline{X}_0 + \underline{M}_x$$

where $\underline{X}_0$ is the initial position in the unstable solid solution and $\underline{M}_x$ the corresponding transformation vector.

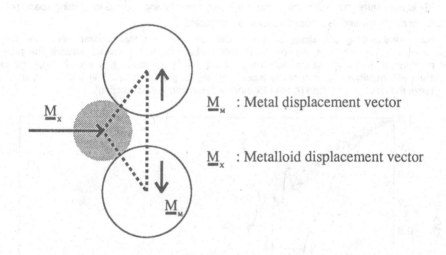

$\underline{M}_M$: Metal displacement vector

$\underline{M}_x$: Metalloid displacement vector

Figure 3. Real Image Computation.

The positions of M atoms are computed from the displacement induced by the movement of surrounding X atoms. **Figure 3** shows the specific choice of interaction (totally geometrical) between an X atom and the two nearest M atoms that we consider in the present paper. The M atom displacement is computed from the amplitude of the transformation vector applied to the X atom (the direction of motion is perpendicular to the corresponding transformation vector):

$$M_M = (\, (r_x + r_M)^2 - a^2/4 \, (1-M_x)^2)^{1/2} - a/2$$

where r_x is the X atom radius, r_M the M atom radius, a the parameter of the initial square lattice and M_x the amplitude of the corresponding transformation vector.

The contribution of all X atom neighbours are added vectorially to give the total displacement. This type of interaction is adapted to hard discs which pretend to maximum close packing. Though it does not incorporates any kind of chemical interaction, it also suffices to describe many properties of metal atoms [5].

The above description gives the static behaviour of the model which allows us to compute the real image from the elementary entities. The dynamic behaviour is based on the local interaction between transformation vector. This interaction must obey to at least two constraints: In one hand it must tend to align transformation vectors in case of perfect chemical order (i.e. for second shell neighbours), and in the other hand it must be repulsive for first shell neighbours. The simplest interaction that verify these two constraints is made of two parts:

For second shell neighbours, a ferromagnetic-like interaction:

$$H_1 = -J_1 \, \underline{M}_1 \cdot \underline{M}_2$$

where M_1 and M_2 are the interacting transformation vectors and J_1 a positive exchange coefficent.

For first shell neighbours:

$$H_2 = J_2 \, \underline{M}_1 \cdot \underline{r}$$

where $\underline{M}_1$ is the centre transformation vector, $\underline{r}$ the unit vector joining the two interacting space points (in the initial structure) and J_2 a positive exchange coefficent.

For characterizing the state of the model we can use three parameters: the mean transformation vector amplitude, the correlation function between transformation vectors, the second order momentum of the amplitude distribution of the mean transformation vector. The first two parameters will identify the phase of the model and the degree of order. The last one is used to characterize the disordered phase (if used together with the correlation function).

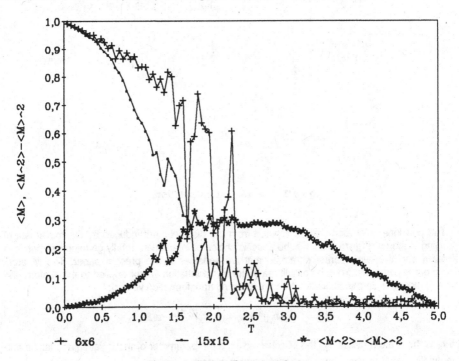

Figure 4. Perfect Chemical Order

V CASE OF PERFECT CHEMICAL ORDER

The way the perfect initial structure transforms under the effect of atom size mismatch is a way of verifying the model behaviour and adjusting the simulation control parameters for the resulting structure is known. In this case no point defect can induce further frustration, the final structure is thus the perfect crystal shown in **figure 1** provided that the cooling rate is sufficiently slow to avoid poly-crystal growth.

Two effects must be taken into account: the finite size effect and the influence of the Monte-Carlo cooling scheme. In the case of a small sample and for free boundary conditions, the surface effect is predominant. It induces strong fluctuations in the behaviour of the model. It is to be noted that in the case of periodic boundary conditions, the finite size effect tends to smooth the transition [3].

Figure 4 shows the size effect on a 6x6 and 15x15 model (crosses and dots, respectively) with the same simulated heat treatment on the mean transformation vector amplitude with free boundary conditions. A value of 1 of the mean transformation vector amplitude represents a totally ordered sample. **Figure 4** (15x15 lattice size) includes also the second order momentum of the transformation vector amplitude distribution (stars). It starts from 0 at high temperature because of the ordered aspect of the initial structure. This second order momentum reaches a maximum and then decreases to zero for the completely ordered final phase. The ordering temperature depends strongly on the cooling rate but the whole behaviour remains identical.

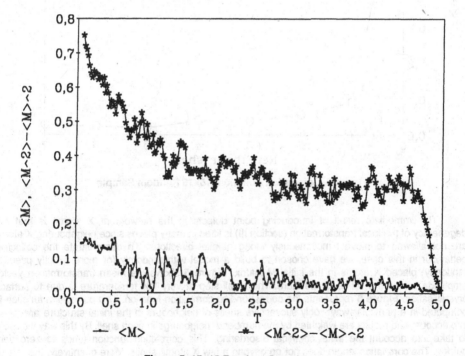

Figure 5. 15x15 Random Sample.

IV DISORDERING BY POINT DEFECTS AND TRANSITION TOWARDS THE GLASSY STATE

Our purpose in this section is to relate the glassy state to an initial state of high entropy for it corresponds obviously to the actual experimental conditions. The unstable solid solution already introduced is taken as the state of reference with the constituent atoms arranged with maximum chemical order compatible with the composition. Here, the composition is M_2X. In a 2D lattice the configurational entropy raises up either through linear defects or through point defects. In this paper, we shall consider only point defects and furthermore still assume that no atomic diffusion takes place during the transition (a more advanced simulation, incorporating the effect of diffusion, is to be published in next future).

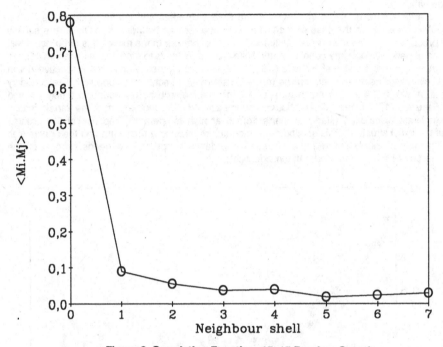

Figure 6. Correlation Function, 15x15 Random Sample.

The immediate result of introducing point defects in the network of X atoms is that the degeneracy of the local transformation (section III) is lifted in many places since neighbouring X atoms are not allowed to move simultaneously along parallel directions. To demonstrate the collective behaviour in this case, we have chosen to build a model with randomly (or more exactly pseudo-randomly) placed X atoms in the initial M-matrix. **Figure 5** shows the mean transformation vector amplitude versus the fictive temperature. The small step at very low temperarure is due to surface crystallisation. **Figure 6** represents the corresponding correlation function. This correlation function is computed in a particular way: only successive shells of neighbours in the initial structure are taken into account and results are weighted by the number of neigbourgs in each shell. By this way we avoid to take into account the initial chemical disordering. This correlation function tends to zero very quickly. The correlation length does not go beyond a few X atoms shells. More quantively, the raising part of the curve can be fitted with a function of the form:

$$<M_i . M_j> = A \, Exp(-|r_i - r_j|/l)$$

where A is a normalisation constant and l the correlation length between transformation vectors.

With this method we find l=0.6 (in unit of the initial lattice parameter) which indicates that no correlation between structural units occurs in this sample.

A more interesting parameter is the second order momentum of the distribution of the mean transformation vector amplitude (stars in **figure 5**). The simulation starts by heating the unstable structure which immediatly undergoes the symmetry breaking. The disorder is stabilized in the flat part of the curve, then the freezing starts with increasing disorder. When ($<M^2>-<M>^2$) departs from the horizontal, the sample crosses the region of glass transtion temperature.

VII DISCUSSION AND CONCLUSION

The preliminary results presented in this paper show clearly how a simple atomic size mismatch transforms a periodic lattice into an amorphous phase by means of atomic rearragement that yet obey locally to the crystallographic point symmetry.

In addition to this point, the proposed model is purely geometrical and can be adapted to more complicated situations: 3D models with a similar mechanism [2] or models involving multiple types of structural units and taking into account a possible diffusion mechanism. This work is in progress.

REFERENCES

1. Egami,T., Waseda, Y., J. Non Cryst. Sol., 64, 1984, 113
2. Dubois, J.M., J. Less Common Metals, 145, 1988, 309
3. Moritsen, O.G., Computer Studies of Phase Transitions and Critical Phenomena, Cabannes, H., Holt,M., Keller, H.B., Killeen, J., Orszag, S.A., Rusanov, V.V (Eds), Springer-Verlag, 1984
4. Kirkpatrick, S., Gelatt, C.D. Jr. and Vecchi, M.P., Optimization by Simulated Annealing, Science, 1983, 220, 4598, 671
5. Bernal, J.D., Proc. R. Soc., A280, 1964, 299

CRYSTALLIZATION AND STRUCTURAL TRANSFORMATIONS IN CoZr$_2$ STUDIED BY HYDROGEN-INDUCED INTERNAL FRICTION

HANS-RAINER SINNING and NORBERT MATZ
Institut für Werkstoffe
Technische Universität Braunschweig
Langer Kamp 8, W-3300 Braunschweig, Germany

ABSTRACT

Hydrogen-induced internal friction may be a valuable tool for crystallization studies especially in binary transition metal glasses like Co$_{33}$Zr$_{67}$.
The series of reorientation peaks observed in this alloy was used both phe-
nomenologically to follow the transformation sequence and as a local probe
to study the role of specific structural units. The latter is discussed especially for the initial transformation into a nanocrystalline state of a
metastable CoZr$_2$ phase with the cubic E9$_3$ structure. In this state, a discontinuity in the height and position of the internal friction peak is
found at H concentrations near 5 at%, which is interpreted as a transition
from the Zr$_4$ tetrahedra in the grain boundaries to the Zr$_3$Co tetrahedra
within the grains as the dominant sites for the reorientation jump. The additional influence of hydrogen on the transformation sequence is correlated
with the density of Zr$_4$ tetrahedra, which mainly favours the equilibrium
phase with the tetragonal C16 structure.

INTRODUCTION

Among the vast number of metallic glasses available to date, the group of
binary AB alloys where A is a late and B is an early transition metal is
of special interest with respect to quite different fundamental as well as
applied aspects. Many of such glasses can be produced both by melt-quenching
and by the more recent solid-state amorphization techniques; on the other
hand, the amorphous state may in turn serve as a basis for producing crystalline intermetallic compounds that are metastable or difficult to handle
by conventional techniques. Both the amorphous and crystalline AB alloys
of this type are usually able to absorb large amounts of hydrogen.

For the amorphous state, the details of hydrogen absorption as well as
the resulting internal friction processes have been particularly well described. It is known that with increasing hydrogen content, the H atoms are
occupying tetrahedral sites in the sequence B$_4$, AB$_3$ etc., with broad distributions of site energies [1], according to Fermi-Dirac statistics [2].

A classical, thermally activated internal friction peak is caused by a Snoek-type reorientation relaxation due to jumps of the H atoms between these tetrahedral sites (e.g. |3|). In contrast, the available information on H-induced internal friction in crystalline intermetallics is still rather poor.

An important model system in this context is the alloy $CoZr_2$ where several amorphous and crystalline intermetallic phases of the same composition form during a crystallization sequence with at least three exothermic transformations |4-7|, giving rise to a series of H-induced internal friction peaks |8|. The peak observed in the stable $CoZr_2$ phase with the tetragonal C16 ($CuAl_2$-type) structure has been identified - for the first time in a crystalline intermetallic compound - with a jump of the H-atom between specific interstitial sites (Zr_4 tetrahedra) in the C16 crystal lattice |8,9|. The whole series of internal friction peaks has been used phenomenologically for following the formation and annihilation of the different phases during the transformation sequence of this alloy |10|.

The first step in this sequence, the formation of a nanocrystalline state of a metastable, complex-cubic $E9_3$ ($NiTi_2$-type) phase with grain sizes below 10 nm, has recently been investigated in detail, with a thorough description of the nucleation - and - growth kinetics |6|. However, there is still lack of microscopic information on the transformation processes.

In addition to a merely phenomenological use, the main subject of the present paper is the application of hydrogen-induced internal friction as a local, structure-sensitive probe, in order to contribute to a deeper microscopic understanding of the crystallization and phase transformation processes in rapidly quenched $CoZr_2$. However, according to the preliminary nature of some of the results, we will confine ourselves to show selected examples which demonstrate some principle tendencies and possibilities of the method, rather than to give already definite answers to open questions.

EXPERIMENTAL PROCEDURE

Melt-spun ribbons of amorphous $Co_{33}Zr_{67}$ were obtained from two different laboratories at the universities of Göttingen and Münster (see below). Phases and microstructures were checked by X-ray diffraction and transmission electron microscopy. The hydrogen content was measured by vacuum extraction analysis. A special technique was developed for the handling of embrittled specimens which were wrapped in a copper foil; details will be given elsewhere |11|. The as-quenched ribbons were found to contain between 0.1 and 0.3 at% hydrogen as an impurity.

The internal friction Q^{-1} and resonance frequency f were measured on flexural vibrations ($10^2 - 10^4$ Hz) of clamped samples using the vibrating-reed technique |8|. The measurements (including also the heat treatments to crystallize the samples) were performed under vacuum, mostly with constant heating rates of 1 - 2 K/min, in a temperature range between 80 and 900 K.

In addition to the quenched-in hydrogen content, most of the samples were charged with H_2 gas either in the vibrating-reed apparatus (at low pressures) or, in most cases, in a special pressure chamber. In the latter case, the whole sample holder with the clamped sample was transferred from the vibrating-reed apparatus so that the internal friction spectrum could be measured before and after charging without any change in sample geometry.

RESULTS

Sequences of internal friction spectra and their use to follow crystallization

Figure 1 shows the sequences of internal friction (IF) spectra measured in the different crystallization stages of three CoZr$_2$ ribbons from different sources (a and b) or with different hydrogen content (b and c). Each of the peaks denoted 1 to 4 has been well characterized with respect to its frequency dependence (activation energy) as well as to its response to changes in microstructure and hydrogen content |8|. The peaks 1, 2 and 4 have all been ascribed to reorientation jumps of hydrogen between Zr4 tetrahedral sites: peak 1 in the amorphous state, peak 2 in the grain boundaries of the above-mentioned, metastable nanocrystalline state, and peak 4 in the tetragonal equilibrium phase. The relaxation mechanism of peak 3 is not yet clear, but this peak is only observed in an intermediate, multi-phase state during the crystallization sequence. It is clearly correlated with the occurrence of still unidentified X-ray diffraction lines (figure 2) and may probably be ascribed to a further metastable CoZr$_2$ phase forming partially from the cubic E93 phase prior to the final transformation into the equilibrium phase. The existence of such a phase has indeed been proved in recent up-quenching experiments |7|. The slight shift of peak 4 to higher temperatures between the last two curves in figure 1b is accompanied by an increase of the c/a ratio in the tetragonal C16 phase from 0.85 to 0.86.

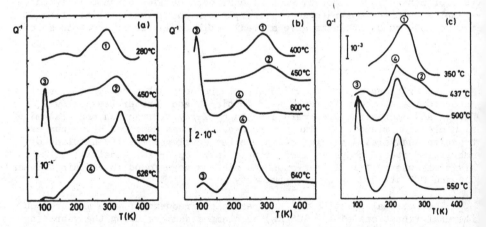

Figure 1. Hydrogen-induced internal friction spectra in the different crystallization stages of amorphous Co$_{33}$Zr$_{67}$, measured on three different ribbons after heat treatments with a heating rate of 1 or 2 K/min up to the temperatures indicated. (a) ribbon from Münster (quenched-in H content about 0.1 at%; f $\approx$ 900 Hz); (b) ribbon from Göttingen (quenched-in H content about 0.3 at%; f $\approx$ 600 Hz); (c) ribbon from Göttingen, but initially charged to a H content of a few atomic percent (f $\approx$ 450 Hz).

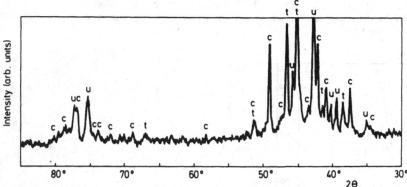

Figure 2. X-ray diffractogram (with Co K_α radiation) of a $Co_{33}Zr_{67}$ ribbon from Münster in the intermediate, multi-phase state after linear heating with 10 K/min up to 540°C, which corresponds to the third internal friction spectrum in figure 1a. The diffraction lines belonging to the cubic E9₃ (with a = 12.19 Å) and tetragonal C16 (with a = 6.40 Å, c = 5.44 Å) $CoZr_2$ phases are denoted with c and t, respectively; "u" stands for still unidentified lines (see text).

The identification of these hydrogen-induced IF peaks with the different crystallization stages gives rise to the above-mentioned, phenomenological application of the internal friction spectra to study crystallization. For instance, the development of the relative amounts of the different crystalline phases (e.g. in the different ribbons of figures 1a and b) can be followed more easily by the relative intensities of the peaks 2 to 4 in figure 1 than by X-ray diffraction (figure 2) or TEM |6|. As another example, the early presence of peak 4 in figure 1c shows that an enhanced hydrogen content in the $Co_{33}Zr_{67}$ glass favours the direct crystallization into the tetragonal equilibrium phase, in competition with the metastable cubic phase. This was also confirmed by TEM |8|.

Comparison of the peaks in the amorphous and nanocrystalline states

Low H concentrations: At low hydrogen concentrations which are still close to the quenched-in value, we have already seen that the properties of the H-induced IF peaks observed in the amorphous and nanocrystalline states (peaks 1 and 2 in figure 1) are almost identical |8,9|. This is true for the average activation energies and preexponential factors (as determined in the usual way from frequency shift measurements |12|) as well as for the qualitative response to hydrogen charging. On increasing the peak height by moderate H charging, both peaks shift to lower temperatures in the same typical way as usually observed in amorphous alloys (e.g. |13|) and can be made almost coincident by choosing appropriately different H concentrations |9|. This has been qualitatively interpreted as a structural correspondence between the grain boundaries in the nanocrystalline state and the amorphous phase |8,9|.

High H concentrations: A more thorough probing of the structure, however, requires both a variation of the hydrogen concentration over a larger range and a quantitative knowledge of the actual concentration values. Due to the scatter in charging conditions from the gas phase, the latter was possible only by analysing each individual sample after the internal friction measurement (which prevents, as a destructive method, further investigations on the same sample).

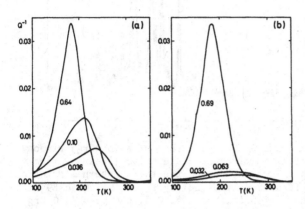

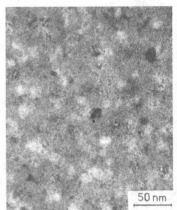

Figure 3. Response of the H-induced IF peak in amorphous (a) and nanocrystalline (b) Co$_{33}$Zr$_{67}$ (at f ≈ 300 Hz) to strong hydrogen charging treatments; indicated are the analysed H concentrations as number of H atoms per metal atom (H/M). (In contrast to figure 1c, the charging treatment in figure 3b was applied after crystallization.)

Figure 4. Transmission electron micrograph of an initially nanocrystalline CoZr$_2$ sample after a very severe H$_2$ exposure for 118 h at 37 MPa and 100°C.

As figure 3 shows, the IF peak grows differently in amorphous and nanocrystalline samples when the H concentration is raised to high values. In the amorphous state, the behaviour is still qualitatively similar to the low-concentration range, with a continuous increase in peak height and a shift to lower temperatures. In the nanocrystalline state, however, there is a gap in peak height and H concentration which is hardly accessible by varying the charging conditions. Up to a few atomic percent of hydrogen, the peak apparently becomes wider on the expense of its growth in height and thus remains more and more smaller in relation to the amorphous state. At much higher concentrations approaching H/M = 1, however, we find again a coincidence of the IF peaks in both cases.

The hydrogen charging treatments in figure 3 were performed by exposing the samples to H$_2$ gas of a pressure between 10 and 15 MPa for times between 2 and 60 hours, at a temperature of 100°C. In order to test if any structural changes take place under these conditions (e.g. a re-amorphization of the nanocrystalline samples that could account for the coincidence of the largest peaks in figure 3a and b), TEM investigations were applied. Both for the amorphous and nanocrystalline states, no changes could be detected under the charging conditions of figure 3. However, after charging a nanocrystalline sample at the same temperature for 118 h at 30 MPa, the microstructure showed more grains larger than 10 nm than before and also an internal contrast within some grains (figure 4). The latter is usually not observed for the cubic E9$_3$ phase but is typical for the beginning of the next stage in the crystallization sequence of CoZr$_2$ |6,14|.

The concentration dependence of the height and position of the H-induced internal friction peaks is summarized in figures 5 and 6 over the whole accessible range of H concentrations. Although the statistics of these results do not yet allow for recognizing finer details, the nanocrystalline state clearly shows a discontinuity at H concentrations near 5 at% which is not found for amorphous samples.

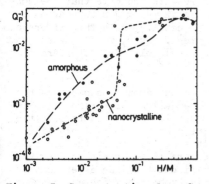

Figure 5. Concentration dependence of the height Q_p^{-1} of the H-induced internal friction peaks in amorphous and nanocrystalline CoZr$_2$.

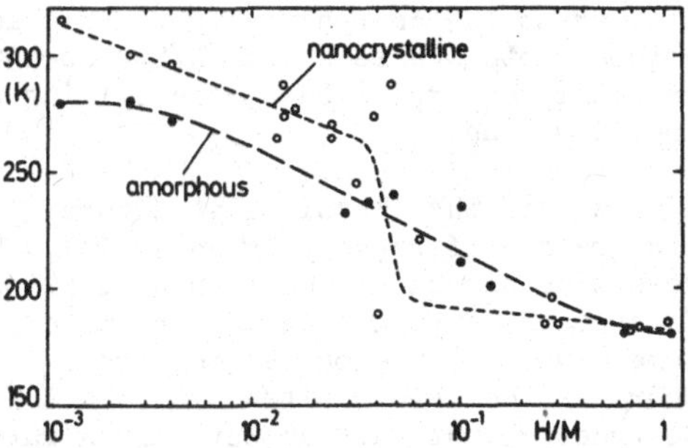

Figure 6. Concentration dependence of the peak temperature T_p, for a constant frequency of 300 Hz, of the H-induced IF peaks in amorphous and nanocrystalline CoZr$_2$.

DISCUSSION

Hydrogen-induced internal friction as a structural probe

As the H-induced IF peaks discussed in this study are due to thermally activated jumps of H atoms betwee specific structural units in the host materials (e.g. Zr4 tetrahedra), it should be possible to learn something about these structural units and their role in the different phase transformations by a comprehensive investigation of the H-induced IF spectra and their changes during the crystallization sequence. The results presented above do not yet allow for very detailed conclusions, but some principal features and tendencies can already be recognized.

Amorphous state: The decrease of the peak temperature T_p with increasing H/M ratio in figure 6 reflects the behaviour of the average activation energy Q_a (i.e. the difference between the site and saddle point energies for the most frequent reorientation jumps of the H atoms) if the preexponential factor τ_0 is assumed as a constant [12]. This general behaviour has often been discussed in the literature (e.g. [13,15]). On the other hand, a separate determination of Q_a and τ_0 over a range of H concentrations in Co33Zr67 was only carried out in one case as a function of peak height, but without quantitative values of the H concentration [8]. For a more specific structural description of amorphous Co33Zr67, these two types of results should be brought together by further experiments.

From the model of Harris et al. [1], the effective concentrations of the Zr4 and Zr3Co tetrahedral sites in amorphous Co33Zr67 are calculated as 0.376 and 0.751, respectively. The tentative shape of the concentration dependence of Q_p^{-1} drawn in figure 5 takes account of the expected additional contribution of the Zr3Co sites at high H/M values.

Nanocrystalline state: At low H concentrations, the H-induced internal friction reflects the properties of the Zr_4 tetrahedra in the grain boundaries, because in the lattice of the metastable, cubic $CoZr_2$ crystallites the only sites with a pure Zr coordination shell are regular Zr_6 octahedra which do not contribute to internal friction [8]. The results indicate that the concentration and distortion of the Zr_4 tetrahedra in the grain boundaries are similar to the amorphous state. This is not in conflict with the differences between both states at low H/M values in figures 5 and 6, which may at least partially be explained by the filling of the "invisible" Zr_6 octahedra within the grains.

Within the classical structural picture of the amorphous state as a mixture of different polyhedra [16], these observations give rise to an interesting aspect of the crystallization process: if one considers only the Zr atoms, the formation of the nanocrystalline state of $CoZr_2$ looks like some kind of "topological decomposition" with a separation of the Zr_6 octahedra in the crystallites from the Zr_4 tetrahedra in the grain boundaries. We presume that this aspect, and a further refinement of the experimental data, might help to understand the microscopic processes which are responsible for the metastable, nanocrystalline state of $CoZr_2$.

With respect to the higher H concentrations, we know that there is a large number of distorted Zr_3Co tetrahedra available as anisotropic interstitial sites in the cubic E93 structure [17]. We believe that the sudden increase of Q_p^{-1} and decrease of T_p at a H concentration near 5 at% in figures 5 and 6 is due to the beginning occupation of these Zr_3Co sites, so that above that threshold concentration the main contribution to the IF peak no longer comes from the grain boundaries but from the grain interior.

Tetragonal equilibrium phase: The reorientation jump of the H atom underlying the IF peak in the C16 equilibrium phase of $CoZr_2$ (peak 4 in figure 1) has been described elsewhere [8,9]. In the present context it is worth noting that this reorientation jump possibly may serve as a model also for the interpretation of the other peaks, insofar as in the C16 structure the interatomic distances and thus the geometry e.g. of the Zr_4 tetrahedra are known [18]. Moreover, as obviously the c/a ratio can be varied to a certain extent by thermal treatments, it should be possible to study the response of the IF peak to a known change in the geometry of the relevant interstitial sites.

Influence of hydrogen on the transformation sequence

As a hydrogen atom in the center of a Zr_4 tetrahedron is generally known as a very stable configuration, we would expect a favourable influence of hydrogen for the phases with a high density of Zr_4 tetrahedra. The total number of Zr_4 tetrahedral sites in amorphous $Co_{33}Zr_{67}$ may be estimated as about one site per metal atom (which is different from the above-mentioned effective number due to nearest-neighbour blocking) [1]. The corresponding number in the C16 equilibrium phase is as high as 1.67 (i.e. 20 sites in the unit cell with 12 atoms), whereas the E93 lattice does not offer any Zr_4 tetrahedra. The effective site densities (with nearest-neighbour blocking) show a similar ratio. Hence, the most favoured phase under the influence of hydrogen would be the equilibrium phase, followed by the amorphous phase. This is consistent both with the partially direct crystallization into the equilibrium phase in figure 1c and with the beginning transformation - without a re-amorphization - of a nanocrystalline sample after severe hydrogen charging in figure 4.

ACKNOWLEDGEMENT

We wish to thank D.Plischke (Kristallabor der physikalischen Institute, University of Göttingen) and Dr.W.Dörner (Institut für Metallforschung, University of Münster) for the production and supply of amorphous $Co_{33}Zr_{67}$ ribbons.

REFERENCES

1. Harris, J.H., Curtin, W.A. and Tenhover, M.A., Phys.Rev., 1987, B 36, 5784.

2. Kirchheim, R., Acta Metall., 1982, 30, 1069.

3. Stolz, U., Weller, M. and Kirchheim, R., Scripta Metall., 1986, 20, 1361.

4. Buschow, K.H.J., J.Less-Common Metals, 1982, 85, 221.

5. Jansson, K., Nygren, M. and Östlund, A., Mater.Res.Bull., 1984, 19, 1091.

6. Nicolaus, M.M., Sinning, H.-R. and Haeßner, F., submitted to Mater.Sci. Eng.

7. Köster, U., private communication (DFG meeting on microcrystalline materials, Bonn, 27/28 Feb. 1991).

8. Sinning, H.-R., J.Phys.: Condensed Matter, 1991, 3, 2005.

9. Sinning, H.-R., submitted to ECIFUAS-6, Cracow, 4-7 Sept., 1991, Proceedings to be published.

10. Sinning, H.-R., Mat.Sci,Eng., 1991, A 133, 371.

11. Matz, N., Diploma thesis, TU Braunschweig, in preparation.

12. Nowick, A.S. and Berry, B.S., Anelastic Relaxation in Crystalline Solids, Academic Press, New York, 1972.

13. Berry, B.S and Pritchet, W.C., in: Hydrogen in Disordered and Amorphous Solids, eds. G.Bambakidis and R.C.Bowman, Plenum, New York, 1986, p.215.

14. Nicolaus, M.M., private communication.

15. Berry, B.S. and Pritchet, W.C., Z.Phys.Chemie Neue Folge, 1989, 163, 381 and references therein.

16. Bernal, J.D., Proc.Roy.Soc., 1964, A 284, 299.

17. Yurko, G.A., Barton, J.W. and Gordon Parr, J., Acta Crystallogr., 1959, 12, 909.

18. Havinga, E.E., Damsma, H. and Hokkeling, P., J.Less-Common Met., 1972, 27, 169.

ENERGETICS AND STATISTICS OF ORDER IN ALLOYS WITH APPLICATION TO OXIDE SUPERCONDUCTORS

D. de FONTAINE[1,2], G. CEDER[3], M. ASTA[1,4] and R. McCORMACK[2]

[1]Materials Sciences Division, Lawrence Berkeley Laboratory, Berkeley, CA 94720, USA
[2]Department of Materials Science and Mineral Engineering,
University of California, Berkeley, CA 94720, USA
[3]Department of Materials Science and Engineering,
Massachusetts Institute of Technology, Cambridge, MA 02139, USA
[4]Department of Physics, University of California, Berkeley, CA 94720, USA

ABSTRACT

Now that first-principles calculations of ordering transformations are becoming increasingly accurate, the deficiencies of earlier mean field methods are becoming increasingly apparent. New techniques, based on cluster expansions, are now alleviating many of the earlier problems and are producing very satisfactory results. These ideas will be illustrated for the case of oxygen ordering in the $YBa_2Cu_3O_z$ superconducting compound, for which a very simple two-dimensional Ising model has been developed. The model features nearest-neighbor repulsive effective pair interactions and anisotropic (attractive/repulsive) next-nearest-neighbor interactions. CVM (cluster variation method) calculations based on this model have produced a phase diagram in remarkable agreement with experimentally determined phase boundaries. Monte Carlo simulations have confirmed the validity of the model and have provided a rationalization for the influence of oxygen order on the value of T_c (superconducting transition temperature) in off-stoichiometric compounds.

INTRODUCTION

The study of order-disorder phenomena in crystalline alloys has progressed considerably over the last ten years or so. No longer is the field confined to finding increasingly sophisticated methods of calculating critical temperatures and critical exponents in somewhat physically artificial but mathematically tractable systems. Today, effort is placed on investigating ordering in real systems, so that the emphasis is shifting away from the pure mathematical problem of, say, Ising model statistical mechanics, to the more physical problem of the quantum mechanics of interacting atoms. More importantly, the most fundamental task is that of combining the quantum and statistical mechanics into one coherent whole so that a free energy may be constructed from which all thermodynamic alloy properties may be deduced. It has been apparent for some time that the basic building block for a theoretical treatment, combining both statistics and energetics of ordering, is the *cluster* of lattice points.

The notion of cluster is introduced quite naturally when one attempts to describe the state of order of a crystalline solid solution. For a binary system (atoms A,B), on may attach a pseudo-spin variable σ_p at each lattice point (p), equal to +1 if the point is occupied by an A atom, −1 if by a B. Such detailed information is not available experimentally, of course, and would in any case be of no use theoretically. Instead, one seeks averages of site occupancies. The problem is, which averages?

In this article, we shall describe briefly the principle of the cluster method, then apply it to the case of oxygen ordering in the superconducting compound $YBa_2Cu_3O_z$ (YBCO). It will be shown that oxygen ordering is intimately related to the phenomenon of superconductivity itself.

CLUSTER EXPANSION

In 1982, Sanchez, Ducastelle and Gratias [1] showed that products of pseudo-spins on the sites of lattice-point clusters

$$\varphi_\alpha(\sigma) = \sigma_p \, \sigma_{p'} \dots \sigma_{p''} \tag{1}$$

form a complete set of orthonormal functions over the space of all possible 2^N configurations of a system of N lattice points. In Eq. (1), α denotes the cluster of points $\{p, p', \dots p''\}$; in principles, all possible clusters of points must be considered. In more recent publications [2,3], some of the present authors showed that other choices of cluster functions and of configuration space also produce possible sets of orthonormal functions.

Regardless of the particular choice of basis sets, it follows that any function of configuration $\{\sigma\}$ can be expanded in the set $\varphi_\alpha(\sigma)$, the expansion is unique, and the (generalized) Fourier coefficients of the expansion are obtained in the standard way by taking scalar products of the function in question with the corresponding cluster function.

The energy $E(\sigma)$, being a function of configuration, can therefore be expanded in this manner. The expectation value $<E>$ of the energy at given temperature and concentration can then be written as

$$\langle E \rangle = E_0 + \sum_{\alpha \neq 0} E_\alpha \xi_\alpha \ . \tag{2}$$

In this equation, originally derived by Sanchez [4], E_0 is a configuration-independent energy, ξ_α are expectation values $<\varphi_\alpha(\sigma)>$ of cluster variables, i.e., *multisite correlation functions*, and E_α are so-called *effective cluster interactions* (ECI), the generalized Fourier coefficients referred to above.

Expansion of the scalar product for the ECI's [2,3] produces exact formulas for "point" interactions [$\alpha = (p)$],

$$E_p = \tfrac{1}{2}(W_A - W_B) \ , \tag{3a}$$

for "pair" interactions [$\alpha = (p,q)$],

$$E_{pp} = \tfrac{1}{4}(W_{AA} + W_{BB} - W_{AB} - W_{BA}) \ , \tag{3b}$$

and so on. In these formulas, W_I is rigorously shown to be the average energy of all configurations of the system having atom of type I (A or B) at point p, and W_{II} is the average energy of all configurations having atom of type I at p and J and q. Various averaging processes have been suggested for calculating the W energies from first principles. One such method, the Direct Configurational Averaging method, is described elsewhere [3,5].

The correlation functions required in Eq. (2) are obtained at thermodynamic equilibrium by minimizing an appropriate free energy functional F expressed in terms of these variables. The formalism used is that of the Cluster Variation Method (CVM), originally proposed in 1951 by Kikuchi [6]. In this approach, the energy is that of Eq. (2) and the configurational entropy is expressed by means of logarithms of cluster probabilities, the latter being themselves expressed as linear combinations of the correlations ξ_α. The CVM has yielded excellent results, for instance for *ab initio* computation of phase diagrams [7]. Here, we shall apply the method to oxygen ordering in YBCO. Before doing so, however, let us emphasize how the cluster approach differs from earlier ones.

CRITIQUE OF MEAN FIELD METHODS

Because of their extreme simplicity, mean field models are quite popular, even today. By "mean field" we denote that class of models which includes the regular solution (or sub-regular) models (for "clustering" systems), the Bragg-Williams (BW) model (for "ordering" systems with sublattices), or the concentration wave model, which is *identical* to the BW model: merely, the mean-field energy is Fourier-transformed, which changes nothing in the nature of the approximation. Let us denote collectively these models by the acronym BW.

It is often stated that these models behave poorly because of their inadequate entropy formulation. This is true: the entropy is a "point" approximation, hence is the lowest form of the hierarchy of CVM approximations; hence, the BW entropy typically tends to overestimate the configurational entropy by a factor of two or more. The major fault of the BW approximation lies with the energy, however, which is expanded as products of average concentrations (system average or sublattice average), rather than as averages of products. As a consequence, the BW model cannot (a) account for short-range order, (b) handle frustration effects such as ordering on an fcc or triangular lattice, and (c) distinguish between various lattice geometries: BW approximations merely count the number of sublattices, but the geometrical relationship between them is ignored. Such models produce exactly the same phase diagrams in either one, two or three dimensions and, in particular, incorrectly predict phase transitions on one-dimensional systems.

It is not surprising that all of these serious deficiencies should be encountered in BW models: the mean field energy is simply not included in the exact formulation of Eq. (2) and Eqs. (3a, 3b, ...). Thus, there is no justification at all for replacing averages of products by products of averages.

The concentration wave method [8], of course, does no better: it simply replaces a product of sublattice averages by a product of Fourier transforms of the same averages, while maintaining the "point" approximation for the configurational entropy. It is sometimes stated that, since long-range interactions are included implicitly in the concentration wave (CW) formalism, and since the mean field approximation is correct in the limit of infinite range interactions, therefore the CW formalism is acceptable for ordering in metallic systems, say. That argument is triply fallacious.

(a) Long-range interactions are of course included in the CVM as well, they are merely more difficult to handle computationally. Moreover, as mentioned earlier, the CW model is identical to the Bragg-Williams, and represents not the slightest improvement on it.

(b) It is true that for "clustering" (ferromagnetic) systems with monotonically decaying infinite-range *negative* interactions, the second-order transition at *zero field* (average concentration 1/2) approaches the exact value. For ordering (antiferromagnetic) systems, of interest for this Workshop, no such limiting property exists. For example, no amount of long-range ordering interactions will ever produce a BW (or CW) phase diagram on an fcc lattice which even approaches the right overall *shape*, let alone the approximately correct numerical values for the transitions.

(c) As first shown by Ducastelle and Gautier [9], *effective* pair (cluster) interactions *do* decay rapidly in magnitude with distance. Recent numerical studies have confirmed the rapid convergence properties of the effective interactions, particularly in transition metal alloys [2,3,10]. In fact, the first pair interaction is responsible for over 90% of the *ordering energy* in typical cases, although the latter is but a small part of the total *cohesive energy*. The latter cannot be expressed in terms of "pair potentials" at all.

Nonetheless, Fourier transform methods, handled properly, are useful as a diagonalization method in perturbation analyses of ordering instabilities [11]. It is best, however, to perform such analyses, i.e., to evaluate generalized susceptibilities, on the basis of a CVM free energy [12], rather than its BW counterpart. Short-range order intensity can be treated similarly [13]. A BW-based stability analysis may give the misleading impression that the unstable ordering wave corresponds to a minimum (in k-space) of the Fourier transform of the pair interaction energy, $V(k)$. This is not always the case, as the configurational entropy must play a role in determining the ordering wave as well. There are good indications, for example, that equilibrium ordering and spinodal ordering waves belong to different "special point ordering families" in the Pd-V system [14], an effect which can be explained in the CVM, but not in the BW context [14]. *A fortiori*, k-space stability analysis cannot predict the ordering wave spectra of stable ordered structures. Hence, the oft-quoted statement that "the most stable superstructure is generated by the star whose ordering wave vectors provide the absolute minimum of $V(k)$"[8] is quite simply incorrect.

The CVM is no panacea, however, and suffers form deficiencies of its own. Mainly, the CVM recognizes only a limited range of correlations, those corresponding to the largest clusters used in the CVM approximation. It therefore follows that second-order transitions, which are characterized by infinite-range correlations, are not treated very accurately by the CVM. Still, with reasonable large-cluster approximations, one can approach transition temperatures to within a few percent of the exact (or best known) value, as opposed to, typically, 100% error for the BW.

In closing these sections on general considerations, let us mention that excellent reviews of the CVM, its advantages and limitations, are to be found in the doctoral dissertations of A. Finel [15], M. Sluiter [16], and G. Ceder [17], and also in a recent review by Inden and Pitsch [18].

YBCO OXYGEN ORDERING MODEL

The compound $YBa_2Cu_3O_z$ (YBCO) was the first one discovered whose superconducting transition temperature (T_c) exceeded the technologically important limit of 77 K [19]. At stoichiometry z=7, T_c is about 93 K, but drops progressively to very low values as oxygen is removed. Oxygen gain or loss occurs mostly on the so-called basal planes, or chain planes, those containing the Cu1 sites and O1 (normally occupied) and O5 (normally vacant) oxygen sites. Not only oxygen content but oxygen arrangements influence T_c, hence a careful study of oxygen ordering is critical to an understanding of the mechanism of superconductivity in this material.

Most oxygen-ordering phenomena in YBCO can be mapped onto a two-dimensional Ising model consisting of two interpenetrating square sublattices, the sites (α:O1, β:O5) of which may be occupied or empty [20]. A third (identical) sublattice is fully occupied by Cu. In the simplest approximation, three effective pair interactions are defined: V_1 between nearest neighbor sites, V_2 between next nearest neighbor (nnn) sites bridged by Cu (the $\mathbf{b}$ direction), and V_3 between nnn sites along the a direction of the orthorhombic cell. These three interactions are precisely those defined in Eq. (3) although, in the present case, the actual computation of W_{IJ} energies was not carried out. Instead, the values of the effective pair interactions were obtained by parametrizing the energies of a number of two-dimensional superstructures and determining V_1, V_2, V_3 by matrix inversion [21]. When α and β sublattices are equally occupied, the structure has tetragonal symmetry; α/β sublattice ordering produces orthorhombic structures. Ground state analysis [22] reveals that eight ordered superstructures are stabilized by these three interactions at stoichiometries z=7 and z=6.5. Two of these structures have been observed in YBCO: one is the orthorhombic structure (Ortho I) at z=7, consisting of parallel O-Cu-O chains, the other is the Ortho II structure consisting of alternating "full" (O-Cu-O) and "empty" (V-Cu-V; V=vacancy) chains. Another predicted ground state superstructure, of tetragonal symmetry [23], has been observed in Ca-substituted compounds [24]. The YBCO structures are stabilized when the following inequalities are verified [23]: $V_2<0<V_3<V_1$. As expected, the Cu-mediated interaction V_2 must be attractive (negative) and V_3 must be repulsive (positive) due to Coulomb interaction between O ions. First-principles calculations [21] confirm these conclusions. The Ca-substituted structure is stabilized by a similar set of inequalities but with $V_1<V_3$. Other, more complex structures with larger unit cells can be stabilized when the V_1, V_2, V_3 scheme of effective interactions is extended (see below). In all cases, the nnn interactions must have opposite signs, producing an Ising model with unique, highly anisotropic properties. Models which do not possess this characteristic anisotropy are unphysical and cannot reproduce the correct features of the system, claims to the contrary notwithstanding.

PHASE DIAGRAM

A pseudo-binary (oxygen-vacancy) phase diagram was calculated by the CVM, first with phenomenological interactions $V_2/V_1 = -1/2$, $V_3/V_1 = +1/2$ [25]. The high-temperature tetragonal-to-orthorhombic transition was found to be of second-order type, as expected. The first complete and consistent CVM phase diagrams were calculated independently by Zubkus et al. [26] and by Kikuchi and Choi [27], with parameters taken from [25]. These authors showed that only second-order transitions existed at very low temperature and that a new phase, now called anti-Ortho I [28], must be present for low oxygen concentrations and low temperatures.

The phase diagram, calculated [28] with parameters V_1, V_2 and V_3 obtained from first-principles LMTO-ASA electronic structure calculations [21], is shown in Fig. 1. The four structures mentioned above are illustrated by portions of Monte Carlo simulation outputs: small black dots in the inserts are Cu atoms, large dark ones are oxygen atoms, open circles are empty oxygen sites. In these simulated structures, Ortho I (OI) is shown with two vacancies, Ortho II (OII) contains no defects, and anti-Ortho I ($\overline{\text{OI}}$) is shown to consist of a random dispersion of a few filled stable chains in a predominantly empty background, the opposite of off-stoichiometric OI which consists of a dispersion of empty chains in a predominantly full background. The symmetry of OI or $\overline{\text{OI}}$ is orthorhombic, whereas that of the tetragonal structure (T), consisting of small segments of fluctuating chains in orthogonal orientations, is tetragonal, though locally orthorhombic. The CVM phase diagram, in this and earlier calculations [25-27] and in unpublished ones using larger cluster approximation [29-30], predicts narrow two-phase regions on the left border of the Ortho II phase region. This could be an artifact of the approximation, as Monte Carlo simulations [31] yield only higher-order transitions throughout. The existence of the $\overline{\text{OI}}$ phase and of higher-order transitions at very low temperatures is required by symmetry [28] and is not an artifact of the CVM. Other models, predicting a wide miscibility gap between T and OI structures, with possible spinodal mechanism, are in error.

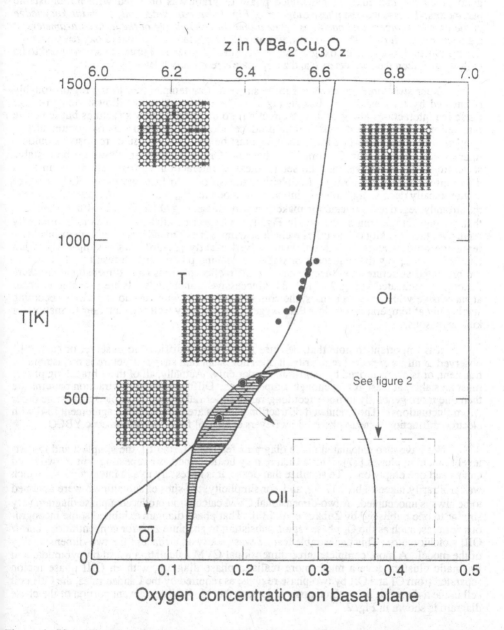

Figure 1. Phase diagram calculated by the CVM approximation, for parameters determined by Sterne and Wille [21] (full lines). Experimental points (filled circles) from Andersen et al. [32]. Inserted structure diagrams were obtained from Monte Carlo simulations (small filled circles in inserts denote copper ions, large filled circles denote oxygen ions, and open circles denote vacant sites).

The extensive experimental data of Andersen et al. [32] for the tetragonal-to-orthorhombic transitions are shown as filled circles in Fig. 1. The agreement is extraordinarily good given the fact that the calculated phase diagram was obtained with no adjustable parameters. In essence, *the phase diagram of Fig. 1 was calculated only from the knowledge of the atomic numbers and positions of the atoms and knowledge of the lattice parameters of the tetragonal unit cell.* The lower concentration (c) scale in Fig. 1 is that of oxygen content in the chain plane. If all vacant oxygen sites in the three-dimensional structure are assumed to lie in this plane, then the parameters c and z (top scale) are linearly related by $z=6+2c$.

Other structures are expected to be stable at low temperature in the region roughly delineated by the dashed-line box in Fig. 1. The reasoning is as follows: longer-range attractive interactions along chains (b direction) do not produce new structures but repulsive interactions along the a direction may produce additional ordered superstructures at low temperatures. We know that interaction V_3 must be positive because of repulsive Coulomb interactions between nnn O atoms, and because Ortho II has been shown to be a stable structure [33]. It follows that the second nearest interaction V_4 between chains in the a direction must also be positive (repulsive), and so on. In fact, we expect [34] a set of monotonically decreasing "convex" interactions along a: V_4, V_5, V_6, ... At low temperature, the strongly negative V_2 interaction makes for very stable long O-Cu-O and V-Cu-V chains, so that, roughly in the area of the box in Fig. 1, we expect parallel-like chains to be mutually repulsive, theoretically out to infinite chain spacings. The ground state problem of infinitely repulsive objects along a line has been solved exactly [35-36], and a simple algorithm determines uniquely the sequences of stable "branching phases" which result [34]. Some of the predicted structures and their respective diffraction patterns (one-dimensional structure factors) are indicated in Fig. 2 of Ref. 34. Successive branching levels are expected to occur at successively lower temperatures, the complex structures with very long periods becoming stable only at temperatures so low that oxygen mobility may well prevent their formation for kinetic reasons.

It is important to note that, because a structure is difficult to observe, or cannot be observed in all its expected regularity, it does not mean that such a structure is metastable, or transient, or is characterized by short-range order only. Actually, all of these branching phases must be stable, at least at low enough temperatures. Diffuse intensity diffraction patterns are then due more generally to imperfect long-range order rather than to stable short-range order (from fluctuations). The calculated diffraction patterns are in remarkable agreement [34] with electron diffraction patterns observed by Beyers et al. [36] for off-stoichiometric YBCO.

We have also calculated phase diagrams featuring Ortho III, the simplest and lowest-level branching phases [37]. This structure may be described as a repeating unit of two filled chains and one empty one. To stabilize that phase, it is necessary to add interaction V_4, which was arbitrarily taken to be 0.17 V_3, and, for simplicity, the sites of sublattice β were assumed to be always unoccupied. A two-dimensional CVM calculation produced a phase diagram very similar to one published by Zubkus et al. [38]. That phase diagram exhibited some incongruous features, such as the OI phase region subsisting to absolute zero for concentrations below OIII stoichiometry. This undesirable feature was clearly an artifact of the two-dimensionality of the model. A more complete three-dimensional CVM calculation based on pyramidal and prismatic clusters gave a much more realistic phase diagram, with an OIII phase region separated from OI and OII by two-phase regions, as required by the Landau rules, the OIII unit cell in the a direction being three times as long as that of OI. The relevant portion of the phase diagram is shown in Fig. 2.

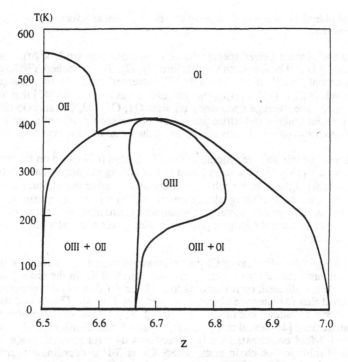

Figure 2. Portion of phase diagram corresponding roughly to boxed area of Figure 1. Ortho III phase is obtained by including a small repulsive V_4 interaction along direction **a**.

ORDERING AND SUPERCONDUCTIVITY

In February 1990 we wrote [34]:

Although neither the theoretical branching temperatures nor the temperatures at which oxygen configurations are frozen in during sample preparation are known at present, the agreement between recent experimental findings and our theoretical model is really quite striking. We are thus tempted to relate the plateau structure of the T_c versus concentration curve to the existence of a theoretically infinite set of branching phases, the T_c plateaus corresponding to states of quasi-one-dimensional order frozen in at a certain "generation" level. The number of plateaus and their range of oxygen content should depend critically on sample preparation. Recent experimental and theoretical work support these ideas. For example, Farneth et al. [39] report that the T_c plateau for low-temperature vacuum-annealed samples of $YBa_2Cu_3O_z$ is sharply defined and entirely located at temperatures higher than the monotonically decaying T_c versus oxygen-content curve for samples quenched from high temperatures. Lambin [40] has used a tight-binding model to show that the hole concentration for well ordered Ortho II and Ortho III phases should be higher than those for the corresponding disordered compounds. It is clear that chain formation minimizes the fraction of incorrectly coordinated copper ions in the basal plane. Hence, chain ordering can be regarded as a form of doping in this compound.

This conjecture received spectacular confirmation in the work of Argonne scientists on oxygen ordering at and below room temperature [41,42]. For quenched YBCO single crystals of oxygen content z=6.45, it was found that T_c increased markedly on subsequent aging, in fact by as much as 27 K [41]. During this process, it was ascertained [42] that oxygen content was not altered: the average occupancy on sites O1, O2, O3, O4 and O5 did not change, although the molar volume and lattice parameters contracted significantly, the a more than the b. Such behavior can only be interpreted as being due to *additional* oxygen ordering.

Recent "quench and age" Monte Carlo simulations [43] based on the model described above, i.e., with V_1, V_2, V_3 parameters used in calculating the phase diagram of Fig. 1, have shed considerable light on the problem. Immediately after the essentially instantaneous "computer quench," α/β ordering takes place during a very short transient, so short that it is probably unobservable experimentally. Subsequently, ordering of chain segments parallel to themselves, so as to form the longest possible chains, takes place at a rate characterized by a lower time constant.

The oxygen coordination of Cu1 atoms was monitored continuously during simulated aging at room temperature for samples quenched from 800 K. In the chain plane, Cu can be 0-, 1-, or 2-fold coordinated, corresponding to 2-, 3-, or 4-fold three-dimensional coordination if it is assumed that O4 sites are always occupied (see Fig. 3). Three- and four-fold planar coordination are forbidden because of the strong V_1 nn repulsive interaction. Electronic structure calculations [44] reveal that (spatially) 3- and 4-fold coordinated Cu is mostly present as Cu++, and 2-fold coordinated Cu is present as Cu+. As a result, chain "healing," i.e., progressive elimination of chain ends, where Cu is 3-fold coordinated, gives rise to the following "reaction" in Kröger-Vink notation:

$$2Cu_{3x} \rightarrow Cu_{4x} + Cu_{2x} + h^{\bullet}$$

thus creating holes ($h^{\bullet}$), accompanied by an excess negative charge on 2-fold coordinated Cu. The increasing hole concentration should favor a high T_c, but oxygen loss tends to destroy holes by the following mechanism:

$$O_O \rightarrow V_O + \tfrac{1}{2}O_2(g)$$
$$V_O^x \rightarrow V_O^{\bullet\bullet} + 2e'$$
$$2e' + 2h^{\bullet} \rightarrow \phi$$

The first of these equations describes oxygen loss to the gas phase (g), the second describes ionization of vacancies and creation of electrons, the third describes electron-hole recombination.

Monte Carlo simulations [43] clearly show that the fraction (f_{n-Cu}, with n=2, 3 or 4) of 2-fold or 4-fold coordinated sites increases with annealing time, that of 3-fold sites decreases. These trends (see Fig. 4) reproduce those of the rise in T_c observed experimentally by the Argonne group [41,42]. Moreover, since it is argued that oxygen ordering promotes hole formation, and that oxygen loss destroys holes, these two conflicting tendencies may well produce non-monotonic decay of T_c as oxygen content is decreased, possibly giving rise to the "plateaus" observed in the T_c vs. oxygen content curve.

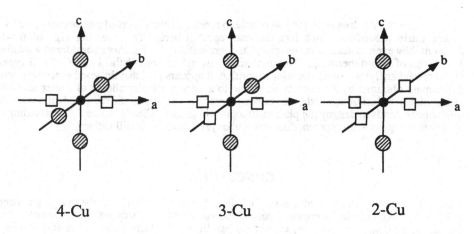

4-Cu   3-Cu   2-Cu

Figure 3. The important configurations of oxygen around a Cu1 atom. It is assumed that the axial oxygens (along the c-axis) are always in place.

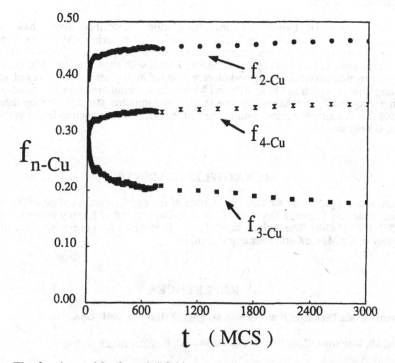

Figure 4. The fractions of 2-, 3- and 4-fold coordinated Cu atoms in the chain planes as a function of Monte Carlo time (MCS). The rapid initial rise (or decay) of n-fold coordinated functions corresponds to α/β sublattice ordering. The slower evolution is due to "chain healing," i.e., gradual elimination of chain ends (3× coordination).

Recently, Poulsen et al. [45] were able to reproduce the observed plateau structure by a Monte Carlo procedure, much like the one reported here. The counting algorithm was different, however; instead of monitoring Cu-coordination, these authors considered a relative population of two-dimensional "minimal-size clusters" of both Ortho I and Ortho II types. Upon careful analysis of their counting algorithm, it appears [46] that the good agreement with experiment obtained by Poulsen et al. is due to a very particular choice of rather artificial assumptions. Substituting another set of equally valid assumptions, differing but little from the original one, tends to destroy the plateau structure altogether. Clearly, a deep understanding of the relationship between superconductivity and oxygen ordering is still lacking.

CONCLUSION

A new "first-principles thermodynamics" of alloys is emerging thanks to the unifying concept of *clusters*: both ordering energy and configurational entropy are expressed in terms of cluster functions, multisite correlations, cluster probabilities. Clusters provide the link between quantum and statistical mechanics and, by means of the cluster variation method, with classical thermodynamics itself. Calculated phase diagrams, such as the one for oxygen ordering in YBCO, have yielded excellent agreement with experimentally determined ones. By contrast, mean field models (BW, CW) produce phase diagrams which are not only quantitatively but even qualitatively in error.

The asymmetric two-dimensional Ising model described above has explained successfully many features of the YBCO system: the experimentally observed phases (tetragonal, Ortho I, Ortho II) are ground states of the model; the calculated phase diagram (with no adjustable parameters) agrees remarkably well with experimental data; the observed electron diffraction maxima correspond closely with calculated structure factors of additional "branching phases" (such as Ortho III); and Monte Carlo simulation reproduces correctly the trends in change in T_c with annealing time in off-stoichiometric samples. It may therefore be concluded that this simple model captures most of the physics of oxygen ordering in the YBCO and related systems.

ACKNOWLEDGEMENTS

This work was supported by the Director, Office of Energy Research, Office of Basic Energy Sciences, Materials Sciences Division of the U.S. Department of Energy under Contract No. DE-AC03-76SF00098. The authors thank Dr. H. F. Poulsen for generously providing them with a copy of his Monte Carlo counting algorithm.

REFERENCES

1. Sanchez, J. M., Ducastelle, F. and Gratias, D. Physica (Utrecht), 1984, **128A**, 334.

2. Asta, M., Wolverton, C., de Fontaine, D. and Dreyssé, H., Phys. Rev. B (in press).

3. Wolverton, C., Asta, M., Dreyssé, H. and de Fontaine, D. Phys. Rev. B (in press).

4. Sanchez, J. M., private communication to D. de F.

5. Dreyssé, H., Berera, A., Wille, L. T. and de Fontaine, D. Phys. Rev. B, 1989, **39**, 2442-2452.

6. Kikuchi, R. Phys. Rev., 1951, **81**, 988-1003.

7. Sluiter, M., de Fontaine, D., Guo, X Q., Podloucky, R. and Freeman, A. J. Phys. Rev. B, 1990, 42, 10460-10476.

8. Khachaturyan, A. G., Theory of Structural Transformations in Solids, John Wiley & Sons, New York, 1983, p. 68.

9. Ducastelle, F. and Gautier, F. J. Phys. F, 1976, 6, 2039.

10. de Fontaine, D., Wolverton, C., Ceder, G. and Dreyssé, H. To be published in NATO ASI Series.

11. de Fontaine, D. Acta Metall., 1975, 23, 553-571.

12. Sanchez, J. M. and de Fontaine, D. Phys. Rev. B , 1980, 21, 216-228.

13. Sanchez, J. M. Physica, 1982, 111A, 200.

14. Solal, F., Caudron, R., Ducastelle, F., Finel, A. and Loiseau, A. Phys. Rev. Lett., 1987, 58, 2245-2248.

15. Finel, A., Contribution à l'étude des effets d'ordre dans le cadre du modèle d'Ising: Etats de base et diagrammes de phase. Doctoral dissertation, Université Pierre et Marie Curie, Paris, 1987 (unpublished).

16. Sluiter, Marcel H. F., On the first principles calculation of phase diagrams. Ph.D. dissertation, University of California, Berkeley, 1988 (unpublished).

17. Ceder, Gerbrand, Alloy theory and its applications to long period superstructure ordering in metallic alloys and high temperature superconductors. Ph.D. dissertation, University of California, Berkeley, 1991 (unpublished).

18. Inden, G. and Pitsch, W., Atomic ordering. In Materials Science and Technology, Vol. 5, ed. Peter Haasen, VCH Press, Weinheim, 1991, pp. 497-552.

19. Wu, M. K. et al. Phys. Rev. Lett., 1987, 58, 908-911.

20. de Fontaine, D., Wille, L. T. and Moss, S. C. Phys. Rev. B, 1987, 36, 5709-5712.

21. Sterne, P. A. and Wille, L. T. Physica C, 1989, 162, 223-224.

22. Wille, L. T. and de Fontaine, D. Phys Rev. B, 1988, 37, 2227-2230.

23. de Fontaine, D. J. Less-Common Metals, 1990, 160, L5-L8.

24. Fu, W. T. et al. J. Less-Common Metals, 1989, 151, 213-220.

25. Wille, L. T., Berera, A. and de Fontaine, D. Phys. Rev. Lett., 1988, 60, 1065-1068.

26. Zubkus, V. E., Lapinskas, S. and Tornau, E. E. Physica C, 1989, 159, 501-509.

27. Kikuchi, R. and Choi, J. S. Physica C, 1989, 160, 347-352.

28. Ceder, G., Asta, M. Carter, W. C., Kraitchman, M., de Fontaine, D., Mann, M. E. and Sluiter, M. Phys. Rev. B, 1990, 41, 8698-8701.

29. Sanchez, J. M., private communication to D. de Fontaine.

30. Ceder, G., unpublished work at U.C. Berkeley.

31. Bartelt, N. C., Einstein, T. L. and Wille, L. T. Phys. Rev. B, 1989, 40, 10759-10765.

32. Andersen, N. H., Lebech, B. and Poulsen, H. F. J. Less-Common Metals, 1990, 164-165, 124-131.

33. Reyes Gasga, J. et al. Physica C 1989, 159, 831-848.

34. de Fontaine, D., Ceder, G. and Asta, M. Nature, 1990, 343, 544-546.

35. Hubbard, J. Phys. Rev. B, 1978, 17, 494-505.

36. Beyers, R. et al. Nature, 1989, 340, 619-621.

37. Ceder, G., Asta, M. and de Fontaine, D. Physica C (in press).

38. Zubkus, V. E., Lapinskas, S. and Tornau, E. E. Physica C, 1990, 166, 472-475.

39. Farneth, W. E. et al. Solid St. Comm., 1988, 66, 953-959.

40. Lambin, Ph., Tight-binding investigation of the electronic properties of ordered and disordered defects in the YBaCuO system. In Oxygen Disorder Effects in High T₅ Superconductors, ed. J. L. Morán-López and Ivan K. Schuller, Plenum Press, New York, 1990, pp. 101-116.

41. Veal, B. W. et al. Phys. Rev. B, 1990, 42, 4770-4773.

42. Jorgensen, J. D. et al. Physica C, 1990, 167, 571-578.

43. Ceder, G., McCormack, R. and de Fontaine, D. Phys. Rev. B (in press).

44. Latgé, A., Anda, E. V. and Morán-López, J. L. Phys. Rev. B, 1990, 42, 4288-4297.

45. Poulsen, H. F., Andersen, N. H., Andersen, J. V., Bohr, H. and Mouritzen, O. G. Nature, 1991, 349, 594-596.

46. McCormack, R., Ceder, G. and de Fontaine, D. To be published.

Characterization of the mechanical disordering process in the Y-Ba-Cu-O System

M. SIMONEAU, F. LAVALLEE, G. L'ESPERANCE

Département de métallurgie et génie des matériaux
École Polytechnique de Montréal
Montréal, P.Q., Canada, H3C-3A7

M. TRUDEAU and R. SCHULZ

Technologie des matériaux, V.P. Recherche
Hydro-Québec, Varennes, P.Q., Canada, J3X-1S1

ABSTRACT

High energy ball milling induces order-disorder transitions on the $Y_1 Ba_2 Cu_3 O_{7-x}$ orthorhombic structure. A new metastable phase with a simple cubic perovskite structure $(Y_{0.33} Ba_{0.67}) CuO_{3-y}$ is formed. This new phase is created by a disorder on the Y and Ba cationic sites and on the oxygen sublattice. The evolution of the microstructure and superconducting properties during the milling process is presented. The thermal stability of the new metastable structure is also investigated.

INTRODUCTION

In the past few years, several processing routes, such as the melt-growth process, have been proposed for the synthesis of highly textured $YBa_2 Cu_3 O_{7-x}$ bulk superconducting oxides [1,2]. Most of these techniques involve a solidification of a liquid phase for achieving the texture. An alternative procedure to these melt-texturing processes is the controlled solid-state directional crystallisation of an amorphous compound or the directional recrystallisation of a fine grained nanocrystalline sample.

Mechanical alloying by high energy ball milling is a simple and well recognized method to synthesize metastable phases such as amorphous [3] and nanocrystalline [4] compounds. In most cases, the milling process increases the configurational disorder of the initial system. For high Tc superconductors, defects and structural disorder induced by irradiation [5-7] have been found to enhance flux pinning properties and increase the current carrying capacity (Jc) of the material. Disorder induced by mechanical deformations may therefore also improve the superconducting properties.

The present paper reports on the microstructural and phase changes occurring during the milling of YBaCuO crystalline powders. Prelimnary results on the associated superconducting properties are also presented.

EXPERIMENTAL PROCEDURE

The starting powder was a commercially available RHONE-POULENC high purity Y123 crystalline powder. Conventional thermal treatments were applied to produce a fully orthorombic polycrystalline powder. The samples were then enclosed in a sealed steel container (internal diameter: 5.7 cm, height: 7.6 cm) under an argon atmosphere with three steel balls (diameters: 1.43, 1.11 and 0.95 cm). The balls to powder weight ratio was 5:1 and milling was performed in a SPEX 8000 MIXER MILL.

X-ray measurements were carried out with a SIEMENS D-500 diffractometer using CuKα radiation. A sealed sample holder was used in order to avoid reaction of the powder with the atmosphere during X-ray scans. Transmission (TEM) and scanning (SEM) electron microscopy were performed on Hitachi H-9000 NAR and S-570 electron microscopes respectively. The H-9000 TEM was coupled to an Energy Dispersive X-ray (EDX) LINK detector for local chemical analysis. The total oxygen content was measured by carbothermal reduction on a LECO TC-136 apparatus. Thermal analysis was done using a PERKIN-ELMER differential scanning calorimeter (DSC). Magnetic measurements were taken using a Quantum Design SQUID magnetometer.

RESULTS AND DISCUSSION

Fig. 1 shows the X-ray diffraction patterns of the Y123 powder after various milling times. After less than one hour of milling, the structure can no longer be identified as orthorhombic because of peak broadening. The broadening is caused by a rapid reduction of the particle size and by strain accumulation. Complete amorphisation of the 123 compound was not observed. After 45 hours of milling a decomposition into chemically different phases (Y_2BaCuO_5 and $BaCuO_2$) is taking place.

After 20 h. of milling and before chemical segregation occurs, many of the major peaks of the 123 structure are still present, but several relatively intense orthorhombic lines are

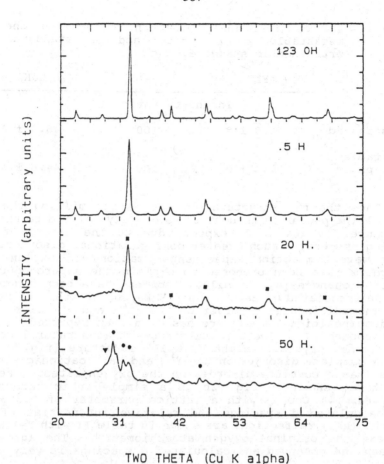

Figure 1. X-ray diffraction patterns of the 123 orthorhombic powder after various milling times. Y_2BaCuO_5 (●), $BaCuO_2$ (▼) and the disordered cubic phase (■).

missing, namely the (001), (002) and (014) reflections. No peak can be found below 22°. The absence of these reflections and the coincidency of the remaining peaks with the original orthorhombic structure suggest that the new phase may be a disordered state associated with the 123 structure. Furthermore, EDX chemical analysis shows that the cationic ratio of this new phase is similar to that of the orthorhombic standard (table I) and carbothermal reduction indicates no significant change in the total oxygen content throughout the first 20 hours of milling within the accuracy of the apparatus.

Table I. EDX chemical analysis (thin regions) of the new metastable cubic phase and a standard 123 orthorhombic standard.

	YL	BaL	CuK
	Intensity ratio (%)		
123 standard	66.0 ± 3	100	46.5 ± 3
metastable cubic phase	65.6 ± 3	100	44.6 ± 3

Thus the new structure obtained after milling the 123 crystalline powder is thought to be closely related to the 123 structure. However, we expect, due to the nature of the milling process, a much greater configurational disorder. In shock wave compaction experiments, cation and oxygen site disorders have been proposed to explain the superconducting property changes [8]. Similar disorders have been suggested for the irradiated $Gd_1 Ba_2 Cu_3 O_7$ and $Y_1 Ba_2 Cu_3 O_{7-x}$ thin films [5-7]. Thermal induced disorder on the Ba^{2+} and Y^{3+} sites near the decomposition temperature has also been reported for the 123 compound [9]. These results suggest as a natural choice for the new crystalline phase, a 123 structure (fig. 2 (a)) with a complete disorder on the Y^{3+} and Ba^{2+} cationic sites as well as a complete disorder on the oxygen sites. The new structure, shown in fig. 2b is a simple cubic perovskite $(Y_{0.33} Ba_{0.67}) CuO_{3-y}$ with a lattice parameter of 0.386 nm. The calculated d spacings and relative intensities of the various (hkl) relfections are shown in table II with Y=1 which is near the original oxygen stoechiometry. The agreement between the observed and calculated d spacings is very good.

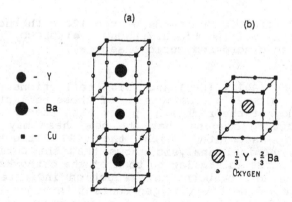

(a) (b)

● - Y
● - Ba
▪ - Cu

⬡ $\frac{1}{3} Y + \frac{2}{3} Ba$
○ OXYGEN

Figure 2.a) Atomic structure of the $Y_1 Ba_2 Cu_3 O_7$ orthorhombic phase,b) proposed atomic structure of the new disordered cubic $(Y_{0.33} Ba_{0.67}) CuO_{3-x}$ oxygen deficient perovskite.

Table II. Experimental and calculated d spacings and relative X-ray intensities of the new metastable $(Y_{0.33}Ba_{0.67})CuO_{3-y}$ cubic structure

(hkl)	d_{meas} (nm)	d_{cal} (nm)	I_{meas} (%)	I_{cal} y
100	0.387	0.386	9.2	5.7
110	0.273	0.273	100	100
111	0.224	0.223	8.0	8.4
200	0.193	0.193	25.2	29.2
210	0.173	0.173	4.0	2.8
211	0.158	0.158	34.7	41.0
220	0.137	0.137	7.7	18.7

The milling process produces small quasi spherical particles (fig. 3a) with sizes ranging from 0.1 to 5 μm in diameter. The microstructure of these particles consists of a large number of nanocrystals and some highly disordered (quasi-amorphous) regions (fig.3b).

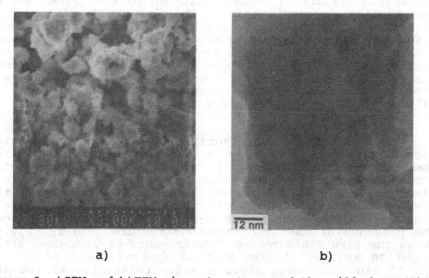

12 nm

a)   b)

Figure 3.a)SEM and b)TEM microstructures of the milled 123(8h)

The average crystal size as determined by X-ray peak broadening using the Scherrer formula [10] showed a rapid decay to a limit of about 10 nm in less than 1 hour of milling. Fig. 4 presents two high resolution electron microscopy pictures taken near Scherzer conditions. The first (fig. 4a) shows details of the atomic arrangement in the early stages (2 h.) of the milling. The crystal size has been greatly reduced and defects are widely present, but the orthorhombic c axis can still be seen in some regions.Fig. 4b) presents an 18 hours milled powder with a similar microstructure but no fringe spacing larger than 0.38 nm is observed. This is to be expected since the major phase present is the metastable cubic phase as revealed by X-ray diffraction.

The mechanism of formation of the new metastable cubic phase is similar to an order-disorder transition. The repeated high energy impacts of the milling process produce a large amount of structural defects such as grain boundaries, dislocations and vacancies on the oxygen and cationic sites. As a result, atomic diffusion proceeds until a completely random state is achieved. Based on thermodynamic considerations, the first order-disorder transition to occur should be on the oxygen sublattice. An orthorhombic to tetragonal transition should therefore be seen first during the milling process. After 30 minutes of milling the structure as seen by X-ray diffraction is still orthorhombic (fig. 1). After this period of time peak broadening rapidly becomes to important to allow any differenciation between the tetragonal and the orthorhombic structure. The second disorder appears on the Y^{3+} and Ba^{2+} cationic sites and produces the cubic phase. It takes more than 20 hours of milling to achieve a completely random state.

Fig. 5a shows the normalized maximum diamagnetic signal measured at 30K as a function of milling time. The powder sample is cooled in zero field then the magnetic moment is measured as the field increases until the maximum diamagnetic signal is reached. The figure shows that the superconductivity is pratically lost within the first half hour of milling even though x-ray diffraction still shows the presence of the orthorhombic structure (Fig.1). Fig.5b shows the zero field cooled (ZFC) moments of Y-Ba-Cu-O after 0 and 30 min. of milling with an applied field of 20 Oe. The 30 min. ZFC curve shows a major transition at 60K and a small one at 92K indicating that a structural change occured on the oxygen sublattice in the early stages of the milling process.

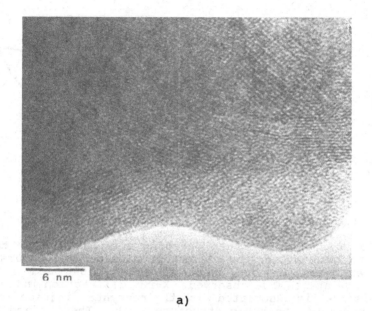

a)

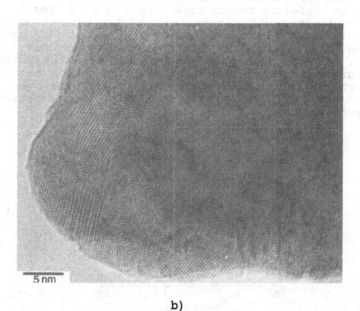

b)

Figure 4. HRTEM micrograph near Scherzer conditions of milled
123 powder particles. a) after 2 hours and b) after 18 hours

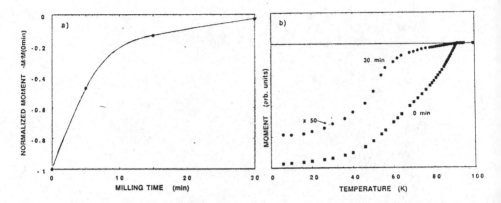

Figure 5.Magnetic properties of milled 123 powders

Fig. 6 show a DSC scan (heating rate 20° C/min) taken from 200°C to 600°C on the 123 powder milled for 20 hours. A single exotherm with an onset at about 510°C and a peak temperature at 547°C is observed. X-ray diffraction indicates that this peak is associated with the recrystallisation or the grain growth of the cubic structure [11]. The Y^{3+} and Ba^{2+} disorder is therefore temperature stable up to at least 600°C. Due to the limited temperature range of the DSC, the ordering temperature was not reached. Agostinelli and al [12] produced a similar 123 cubic phase in thin film by excimer laser ablation. Their cubic phase was formed when the substrate temperature was held near 560°C. The transition to the 123 orthorhombic structure took place at temperatures over 900°C indicating the relatively high stability of the metastable disordered state.

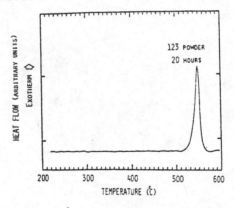

Figure 6.DSC scan (20° C/min) of a milled 123 powder
(20 h.) showing an exothermic peak at 547°C.

CONCLUSION

A new metastable cubic perovskite structure ($Y_{0.33}Ba_{0.66}$) CuO_{3-y} has been synthesized by high energy ball milling. This phase is a consequence of an order-disorder transition occuring on the Y^{3+} and Ba^{2+} sites and on the oxygen sublattice. Complete amorphization of the 123 powder is not observed during the process. The recrystallization or grain growth temperature of the metastable cubic structure is about $540°C$ but the disorder on the yttrium and barium is stable up to $600°C$. Superconductivity disappears during the first hour of milling when a structural change occurs on the oxygen sublattice.

REFERENCES

1. S. Jin., T.H. Tiefel, R.E. Sherwood, M.E. Davis, R.B. Van Dover, G.W. Kamlott, R.A., Fastnacht H.D. Keith, App. Phys. Lett., 52, 2074 (1988).

2. M. Murakami, M. Morita, K. Doi and K. Miyamoto, J. of Appl. Phys., 28, 1189 (1989).

3. C. Koch, O.B. Cavin, C.G. Mckanney and J.O. Scarbrough, Appl. Phys. Lett., 43, 1017 (1983).

4. E. Hellstern, H.J. Fecht, Z. Fu and W.L. Johnson, J. of Appl. Phys., 65, 305 (1989).

5. J.D. Jorgenson, B.W. Veal, W.K. Kwok, G.W. Carbtree, A. Mezawa, L.J. Nowicki and P. Paulikas, Phys. Rev. B36, 5731 (1987).

6. J.W. Lee, H.S. Lessure, D.E. Laughlin, M.E. McHenry, S.C. Sankar, J.O. Willis, J.R. Cost and M.P. Maley, Appl. Phys. Lett., 57, 2150 (1990).

7. S. Kohiki, S. Hatta, K. Setsune, K. Wasa, Y. Higashi, S. Fukushima and Y. Goshi, Appl. Phys. Lett 56, 298 (1990).

8. L.E. Murr, C.S. Niou, S. Jin, T.H. Tiefel, A.C.W. P. James, R.C. Sherwood and T. Sieggrist, Appl. Phys. Letter, 55, 1575 (1989).

9. T. Kajitani, K. Ohishi, M. Kikuchi, Y. Syono and M. Hirabayashi, J. of Appl. Physics, 26, L1144 (1987).

10. B.D. Cullity, Elements of X-Ray Diffraction (Addison-Wesley Massachusetts, 1978).

11. F. Lavallée, M. Simoneau, G. L'Espérance and R. Schulz, Phys. Rev. B.,(1991),(in press).

12. J.A. Agostinelli, S. Chen, G. Braunstein, Appl. Phys. Letter,(1991),(in press).

GRAIN GROWTH AND GOLD TRACER DIFFUSION

IN THIN FILMS OF NANOCRYSTALLINE Fe and $Fe_{95}Zr_5$

J. BOTTIGER, N. KARPE and K. KYLLESBECH LARSEN

Institute of Physics, University of Aarhus, DK-8000, Aarhus C, Denmark

ABSTRACT

Nanocrystalline Fe and $Fe_{95}Zr_5$ thin films with average grain sizes ranging from 5nm to 15nm made by e-beam codeposition have been investigated. The kinetics of grain growth during thermal annealing and 500keV Xe^+ ion irradiation have been studied using dark-field transmission electron microscopy. In contrast to nanostructured pure elements produced by compaction of ultrafine powders, a considerable grain growth is observed at 673K in thin films of nanostructured Fe, but not in $Fe_{95}Zr_5$. The tracer diffusion of implanted gold have been investigated in these films and, for comparison, in crystallized, initially amorphous, $Fe_{80}Zr_{20}$. The broadening on annealing of the approximately Gaussian gold concentration profile was measured by Rutherford Backscattering Spectrometry. The diffusion constant corresponding to the time evolution of the tracer concentration can be expressed as

$$D^* = 8.4 \cdot 10^{-10} \, m^2/s \quad \exp(- \, 150 \, kJ \, mole^{-1} \, / \, RT)$$

for $Fe_{95}Zr_5$ in the temperature region around 800K. The diffusion constant is close to that observed for crystallized $Fe_{80}Zr_{20}$ and amorphous $Ni_{65}Zr_{35}$ thin films. This is in contrast to the reported much higher diffusion constants for nanostructured metals made of compacted powders.

INTRODUCTION

The possibility of obtaining crystalline materials with very small grain sizes by evaporation techniques was demonstrated[1] several years ago. The feasibility of large scale synthesis of "nanocrystalline" or "nanostructured"[2] materials by compaction of powders has urged[2-3] a renewed interest in this field. These materials have an exceptionally high density of grain boundaries, which is believed to give rise to physical properties differing from those of large-grain crystalline materials. Very high ductility and hardness[3] and a high magnetic permeability combined with good thermal stability[3,4] have for example been reported. The relatively unknown structure of the grain boundaries is believed[5] to be decisive for the properties of nanostructured materials. Many factors like composition, grain size distribution, density, porosity, orientation and impurities are all well-known to influence the properties of the grain boundaries and these factors, in turn, depend critically on the production technique. From this point of view, quite diverse physical properties might be expected for nanostructured materials produced by different methods. Nanostructured pure metals produced by compaction of powders have been reported[4] to exhibit a high stability against thermally induced grain growth, which has been attributed[6] to a relatively large entropy of the disordered grain boundaries. Diffusion experiments in compacted nano-structured copper have suggested an exceptionally fast diffusion with very low activation energies[7,8].

Amorphous metals and their crystallization products, which often are nanostructured alloys, have been considered to be potentially useful as thin-layer diffusion-barriers. Since the types of grain boundaries and their properties may depend strongly on preparation methods, it is of interest to investigate if nanostructured thin films exhibit a fast diffusion as observed in the compacted materials. Therefore, we have studied thermally and heavy-ion irradiation induced grain growth and gold tracer diffusion in nanostructured thin films of Fe and $Fe_{95}Zr_5$.

EXPERIMENTAL METHODS

The thin films were prepared by codeposition of Fe and Zr from two independent electron-gun sources. The vacuum measured immediately before and after evaporation was better than $5 \cdot 10^{-7}$ Pa. The evaporations were performed with or without liquid nitrogen cooled substrates. Annealing, irradiation and gold implantation were performed in a vacuum better than $6 \cdot 10^{-5}$ Pa. The films, being 900-1000 Å thick, were prepared on sapphire substrates partially covered with a NaCl layer so that part of each sample could be floated off in deionized water for transmission electron microscopy (TEM, a Philips CM-20 microscope). Average grain diameters were estimated from dark-field image photographs by simply measuring the distribution of grain diameters assuming approximately spherical grains. Structural determinations were made using the diffraction mode.

Irradiations were performed with 500keV Xe^+ ions, doses ranging from $1 \cdot 10^{15}$ ions/cm² to $1 \cdot 10^{17}$ ions/cm², and implantations with 300keV Au^+, doses of $1-2 \cdot 10^{15}$ ions/cm². According to a TRIM[9] simulation, the projected ranges in $Fe_{95}Zr_5$ are estimated to be 930Å for Xe^+ and 420Å for Au^+. The gold implantations were found to give an approximately Gaussian concentration profile of gold with a maximal tracer concentration of approximately 0.8 at. % at a depth of about 300 Å and an overall gold concentration of less than 0.2 at % in the films.

Using Rutherford Backscattering Spectrometry[10] (RBS) with 2MeV $^4He^+$ ions, the composition profiles and their time evolution on annealing in the films were measured. The concentration profile of the implanted gold tracer, $c(x,t)$, was analyzed[11] using a best fit to a Gaussian function,

$$c(x,t) = A_o \exp (-(x-x_o)^2 / 2b^2) \qquad (1)$$

where A_o, b and x_o are fitting parameters and x is the depth. This fitting procedure was performed three times for each temperature, before annealing and after two subsequent anneals. The diffusion constant, D^*, was then deduced[11] from the changes in the width of the Gaussian concentration profile, b, using the expression

$$D^* = (b_2^2 - b_1^2) / 2t \qquad (2)$$

where t is the annealing time. For further details on this method see Ref. 11.

RESULTS

Nanocrystalline films of Fe and $Fe_{95}Zr_5$ were produced by codeposition with and without liquid nitrogen cooled substrates. Amorphous $Fe_{88}Zr_{12}$ with traces of fcc iron and fully amorphous $Fe_{85}Zr_{15}$ and $Fe_{80}Zr_{20}$ films were obtained directly

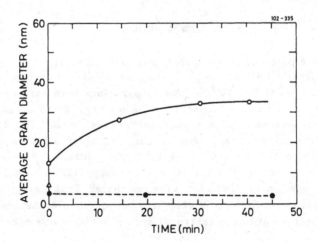

Fig.1 Average grain diameters in Fe and $Fe_{95}Zr_5$ thin films on thermal annealing at 673K. The symbols denote Fe (Δ) deposited with liquid nitrogen cooled substrates, Fe (O) and $Fe_{95}Zr_5$ (●) both deposited without cooled substrates.

from codeposition without using liquid nitrogen cooled substrates. The complete crystallization of the amorphous films at 723K-823K using the in situ hot-stage of the TEM was found to give a two-phase nanostructured material of bcc iron and traces of ZrO_2 with average grain diameters ranging from 5-10nm. Heavy-ion irradiation with 500keV Xe^+ of the same films with $2 \cdot 10^{15}$ Xe^+/cm^2 at 573K were also found to cause crystallization into the same phases with similar grain sizes.

In Fig. 1, the average grain diameters of Fe and $Fe_{95}Zr_5$ films after in situ thermal annealing at 673K in the hot-stage of the TEM are shown. In contrast to the pure Fe film, the $Fe_{95}Zr_5$ film shows no detectable grain growth, not even after annealing at 873K. In Fig. 2, the average grain diameters after 500keV Xe^+ irradiation at room temperature are shown as a function of dose for the Fe films (deposited both with and without liquid nitrogen cooled substrates) and the $Fe_{95}Zr_5$ films. All the films exhibit a rapid grain growth. No tendency of saturating grain sizes can be observed even for the highest doses. (For the highest dose, $1 \cdot 10^{17}$ Xe^+/cm^2, the average grain diameter is likely to be somewhat reduced due to sputtering.)

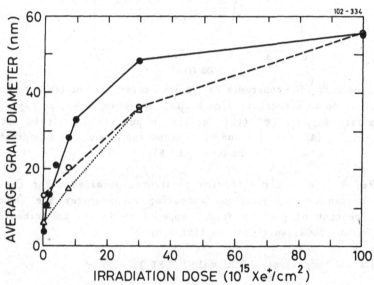

Fig 2. Average grain diameters on annealing of Fe and $Fe_{95}Zr_5$ after 500keV Xe^+ irradiation at room temperature. The symbols denote the same as in Fig. 1.

The tracer diffusion of gold in initially amorphous $Fe_{80}Zr_{20}$ and in nanostructured Fe and $Fe_{95}Zr_5$ was investigated in the temperature region around 800K. After all anneals, these films were investigated in TEM. The amorphous film was completely crystallized after 2 hours in 773K and the Fe films showed much increased grain sizes after all anneals. Only for the $Fe_{95}Zr_5$ films, which showed no signs of grain growth or structural changes, can the diffusion experiments be assumed to have taken place under close to constant conditions.

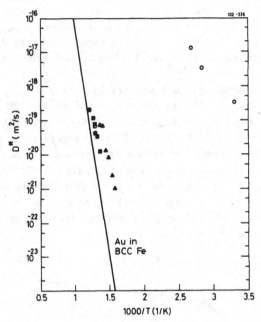

Fig. 3. Tracer diffusion constants for various materials and temperatures. The symbols denote Au in nanocrystalline $Fe_{95}Zr_5$ (■) (this work), in crystallized nanocrystalline $Fe_{80}Zr_{20}$ (▼) (this work), in pure Fe (□) (this work), in amorphous $Ni_{65}Zr_{35}$ (▲) (ref. 11) and Ag in compacted powders of nanocrystalline Cu (O) (ref. 8).

In Fig. 3, the tracer diffusion constants, obtained using the method described previously, for various annealing temperatures are shown. The diffusion constant of gold in $Fe_{95}Zr_5$ appears to follow an Arrhenius-like behaviour around 800K, which can be fitted by

$$D^* = 8.4 \ 10^{-10} \ m^2/s \cdot exp(-150 \ kJ \ mole^{-1} \ / \ RT) \qquad (3)$$

corresponding to an activation energy of approximately 1.6eV. The pure Fe films showed a wide distribution of grain diameters with grain diameters ranging from 20 to 150 nm after annealing. The data point in Fig. 3 for pure Fe does therefore correspond to an Fe film having this grain size distribution. The initially amorphous $Fe_{80}Zr_{20}$ was, as mentioned above, found to crystallize within a few minutes at 773K into bcc Fe(Zr) with traces of ZrO_2, with an average grain diameter of about 10nm. The measured tracer diffusivity for this film corresponds to a differently prepared nanocrystalline material than the $Fe_{95}Zr_5$ films, but this appear to have no significant influence on the tracer diffusivity of gold.

DISCUSSION

a) Grain growth

The reduction of the interface energy was very early recognized as the driving force for grain growth. Nevertheless, the basic understanding of grain growth has remained very empirical. Existing models[12,13] can only be applied with great success to quite idealized cases, and in many real cases the grain boundary mobility is influenced by many factors simultaneously[14], like the grain orientation, the distribution of grain sizes, impurities, and pinning of grain boundaries at precipitates and surfaces, etc. For normal grain growth empirical expressions of the type below have frequently been used[15]

$$\langle D \rangle^n - \langle D_o \rangle^n = K \cdot t \qquad (4)$$

where t is the time and K is an empirical constant which depends on the grain boundary mobility and interface grain boundary free energy and n is an empirical constant typically found to be between 2 and 3.

Thermal annealing indicates a large difference in grain growth behaviour between the pure iron and $Fe_{95}Zr_5$ films. This appears to be in good agreement with the well-established observation[14] from polycrystals that solute atoms like zirconium can exert a drag that may reduce the grain boundary mobility by orders of magnitude. Although physically very interesting effects like "entropy-stabilization"[6] have been proposed to explain the stability of nanostructured compacted powders, the "solute-drag" effect[14] appears sufficient to explain the behaviour of our nanostructured thin films. In the case of the iron based films studied here, high thermal stability could only be achieved by addition of zirconium impurities. This indicates that impurities, added intentionally or as a consequence of the fabrication conditions used, could be one reason for the thermal stability of some nanostructured materials.

Several detailed investigations[18-24] of grain growth during ion irradiation in nanostructured pure metals have previously been published. Quite generally, during ion irradiation grain growth is found at low temperatures, where thermally induced growth is not observed. This has been attributed to increased grain boundary mobility due to radiation induced defects[22] and epitaxial regrowth of the thermal spike regions[20]. Our observations are in qualitative agreement with previous reports, but our measurements differ in two ways; higher doses (up to $1 \cdot 10^{17}$ ions/cm^2) have been used and the effect of a zirconium impurity has been investigated. Even for doses as high as $1 \cdot 10^{17}$ Xe$^+$/cm^2 (corresponding to approximately 1000 displacements per atom), which is close to the maximum that can be allowed in order to avoid very severe sputtering , no saturation of the average grain size could be found. In contrast to the case with thermally induced grain growth, no drastic difference in the grain growth characteristics on heavy-ion irradiation is found to arise from adding 5 atomic percent zirconium. The Zr atoms in the grain boundaries will be spread out due to the irradiation, so no significant "solute-drag" effect should be expected.

b) Tracer diffusion

Diffusion in polycrystalline materials has been classified by Harrison[16] into three types, A, B or C, depending on the importance of grain boundaries for the diffusion. Type A and C represent the two extremes where grain boundary (or dislocation) diffusion are insignificant and dominating, respectively. Type B represents the intermediate case. Very careful diffusion measurements combined with microstructural investigations are often needed[25] to establish the type of diffusion. In nanostructured materials such investigations are difficult, because of the fine grain sizes. It is, however, often suggested that the grain-boundary-influenced diffusion (i.e. type B or C) is likely to be important in these materials, due to the large volume fraction of grain boundary regions. For example, if a typical grain boundary width is assumed to be 1.5-2 nm, an average grain diameter of 10 nm is estimated[17] to give a 20% volume fraction of grain boundaries.

From Fig. 3, the tracer diffusivity of gold in nanostructured $Fe_{95}Zr_5$ is found to be close to that observed for gold tracer diffusion in thin films of crystallized $Fe_{80}Zr_{20}$, pure Fe and amorphous Ni-Zr[11] and also close to an extrapolation to low temperatures for gold tracer data[26] in large-grain bcc Fe. The activation energy measured to be approximately 1.6eV differs, however, from that observed at higher temperatures for polycrystalline iron (2.7eV) and appears to be close to that reported[11] for amorphous $Ni_{65}Zr_{35}$. At this early stage, it can only be concluded that the diffusional behaviour is very different from that reported[7,8] previously in compacted nanostructured metals, where an extremely rapid tracer diffusivity with an activation energy close to that for surface diffusion was found. This shows that the intriguing rapid diffusion[7,8] is not likely to be an general property of nanostructured metals, but rather a characteristic feature of compacted powders. Most likely the grain boundaries of compacted nanostructured materials are different from those of thin film materials in terms of typical angular orientations, impurities, density and connections with voids.

REFERENCES

1. C G Granqvist and R A Buhrman, J Appl Phys 47, 2200 (1976)

2. R W Cahn, Nature 348, 389 (1990)

3. H Gleiter, Prog Mater Sci 33, 1 (1990)

4. E Jorra, H Franz, J Peisl, G Wallner, W Petry, R Birringer, H Gleiter and T Haubold, Philos Mag B60, 159 (1989); G Wallner, E Jorra, H Franz, J Peisl, R Birringer, H Gleiter, T Haubold and W Petry, Mater Res Soc Symp Proc 132, 149 (1989)

5. Panel report on Materials Science, J Mater Res 4, 704 (1989)

6. H J Fecht, Phys Rev Lett 65, 610 (1990)

7. J Horváth, R Birringer and H Gleiter, Solid State Comm 62, 319 (1987)

8. S Schumacher, R Birringer, R Strauss and H Gleiter, Acta Metall 37, 2485 (1989)

9. J P Biersack and L G Haggmark, Nucl Instr Methods 174, 257 (1980)

10. L C Feldman and J W Mayer, "Fundamentals of Surface and Thin Film Analysis", North Holland 1986

11. J Bøttiger, K Dyrbye, K Pampus and B Torp, Inter J Rapid Solid 2, 191 (1986)

12. M Hillert, Acta Metall 13, 227 (1965)

13. I M Lifshitz and V V Slyozov, Zh Eksp Teor Fiz 35, 479 (1958)

14. D A Porter and K E Easterling, "Phase Transformations in Metals and Alloys", Chapman and Hall, 1990

15. R W Cahn, in "Physical Metallurgy" eds R W Cahn and P Haasen, pp. 1595, North Holland 1983

16. L G Harrison, Trans Faraday Soc 57, 1191 (1961)

17. H J Fecht, E Hellstern, Z Fu and W L Johnson, Adv Powder Techn 1-3, 111 (1989)

18. J C Liu, M Nastasi, and J W Mayer, J Appl Phys 62, 423 (1987)

19. J Li, J C Liu, and J W Mayer, Nucl Inst and Meth B36, 306 (1989)

20. J C Liu, J Li and J W Mayer, J Appl Phys 67, 2354 (1990)

21. J Linnros, B Svensson and G Holmén, Phys Rev B30, 3629 (1984)

22. H A Atwater, C V Thompson and H I Smith, Phys Rev Lett 60, 112 (1988)

23. P Børgesen, D A Lillienfeld and H Msaad (preprint)

24. D E Alexander, G S Was and L E Rehn, (preprint)

25. A Atkinson and R I Taylor, Philos Mag A4, 979 (1981)

26. R J Borg and D Y F Lai, Acta Metall 11, 861 (1963)

MÖSSBAUER STUDY OF MECHANICALLY ALLOYED POWDERS

SABURO NASU
Department of Material Physics, Faculty of Engineering Science,
Osaka University, Toyonaka, Osaka 560, Japan

and

PAUL HIDEO SHINGU
Department of Metal Science and Technology, Kyoto University,
Sakyo-ku, Kyoto 606, Japan

ABSTRACT

The microscopic nature of mechanically alloyed powders has been investigated using the [57]Fe and [119]Sn Mössbauer spectroscopy. The following alloy systems have been treated by a low-energy ball mill; Al-Fe, Fe-Sn, Ag-Fe, Fe-C and Fe-B. To understand the elementary process of kneading, the repeated rolling method has been employed for the Al-Fe system and compared with the results from the ball-milling method. Pure Fe powders have also been ball-milled in order to understand the effects of heavy deformation using the same conditions and investigated by the [57]Fe Mössbauer spectroscopy. Hyperfine interaction parameters obtained show that atomic dispersion and the formation of an amorphous, non-equilibrium solid solution and intermetallic compounds occur during mechanical alloying. From the experimental findings for each powder the characteristic features of the mechanical alloying process have been discussed.

INTRODUCTION

The formation of metastable and non-equilibrium alloy phases has been achieved by various methods of "energising and quenching". One well known method is rapid quenching from liquid and gas phase. Such rapid cooling methods from high temperature can modify materials by refining the microstructure, homogenizing the composition, extending solid solubilities, creating metastable crystalline phases and often producing metallic glasses. Similar material modification can be achieved with the relatively less-studied method of mechanical alloying (MA) which is one of the so-called room temperature solid state reaction methods [1]. The aim of this investigation is to understand the elemental process of MA and the microscopic nature of the metal powders during the MA of various alloy systems using hyperfine interaction parameters obtained from Mössbauer spectroscopy. Mössbauer active atoms provide the hyperfine interaction parameters and should be one of the powders; constituents.

The following alloys have been studied: Al-Fe, Fe-Sn, Ag-Fe, Fe-C and Fe-B. For the Al-Fe system, the repeated-rolling method has been also applied and compared with the results obtained from the ball-milling method. Pure Fe powders have been also ball-milled and their microstructure investigated by ^{57}Fe Mössbauer spectroscopy. Results obtained using complementary techniques such as X-ray diffraction, transmission electron microscopy, scanning electron microscopy and differential scanning calorimetry have been prevbiously reported, along with other Mössbauer results [2-8]. We arrange this paper as follows: first, a brief description of Mössbauer spectroscopy is given followed by an explanation of the materials and methods used for MA. Experimental results obtained for each of the alloy combinations is followed by a discussion and summary of these results.

MÖSSBAUER SPECTROSCOPY

Mössbauer spectroscopy is now well established as one of the nuclear techniques used to investigate the local electronic structure of Mössbauer active atoms in solids through hyperfine interaction parameters [10]. Every phase containing Mössbauer active atoms exhibits a spectrum with characteristic hyperfine interaction parameters. In other words, the phases containing order/disorder, stoichiometric/non-stoichiometric, amorphous, para-, ferro- and antiferromagnetic ordering can be identified and, in principle, quantitatively determined by Mössbauer spectroscopy. In general, the following Mössbauer hyperfine parameters can be used to distinguish each phase (1) Isomer shift, which is derived from the center of gravity of the spectrum and is related to the charge density at nucleus. (2) Magnetic hyperfine interaction, derived from the maximum splitting in spectra which have a sextet for 57Fe and/or 119Sn nuclei, and which shows the magnitude of the magnetic field arising from the electron spin polarization at the nucleus. (3) Electric quadrupole interaction, which is derived from the splitting in spectrum having a doublet for 57Fe and/or 119Sn nuclei and is related to the electric field gradient at the nucleus. In total, Mössbauer spectroscopy can provide information of the local electronic structure of Mössbauer active atoms which is reflected by their near neighbor configurations in each phase and can also indicate the formation of alloying, ordering and amorphization in the samples. In the present study we used Fe powder as one of the elements for MA and thus 57Fe Mössbauer measurements have mainly been used to investigate the microscopic nature of Fe atoms during and after mechanical alloying. Since the starting Fe powders are bcc ferromagnetic particles showing a magnetic hyperfine field of 330 kOe at room temperature, our interests are mainly concerned with the change in the magnetic hyperfine interaction of Fe during MA. ^{119}Sn Mössbauer measurements have also been performed on the Fe-Sn system. The initial Fe powder has an average particle size of about 50 μm and is suitable for the gamma-ray absorber in transmission Mössbauer measurements. ^{57}Co in Rh and ^{119m}Sn in CaSnO$_3$ were used as gamma-ray sources. A conventional constant-acceleration spectrometer was used and the velocity scale was calibrated by measurements of bcc Fe and BaSnO$_3$ spectra at room temperature.

MATERIALS AND METHODS

As the starting materials of mechanical alloying (MA), the following elemental powders were used: Al (purity 99.9%, average particle size 50 μm), Fe (purity 99.9%, average particle size 50 μm and purity 99%, under 150 mesh), Sn (purity better than 99.9%, average particle size larger than 100 μm), Ag (purity 99.9%, under 350 mesh), B (purity 97% amorphous boron powder, average particle size on the order of several μm) and C (purity 99.9%, graphite particle, average particle size 5 μm). Combinations of these powders were put into a conventional ball-milling vial of 150 mm inner diameter. Both the vial and the balls were made of stainless steel. Sample powders to ball ratio in weight was maintained at 1/90 or 1/100. The powders were sealed in the vial under an argon atmosphere with 4 wt.% ethyl alcohol. Rotation speed of the vial was 90 or 105 rpm. In this investigation, the above conventional ball-milling method was deliberately chosen in order to insure that the low energy mechanical alloying (MA) process was achieved. The advantage of such a low energy MA process is due to the increased possibility of relating the non-equilibrium structures formed by the mechanical energy. The temperature of the vial during MA process was about 340 K. However, as will be discussed in a later section, repeated rolling method was used for MA of Al-Fe

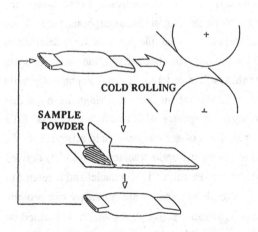

Figure 1. Schematic drawing of the repeated rolling method.

system and compared with the conventional low-energy ball milling method in order to clarify whether the MA process by ball-milling is indeed a process of repeated folding and thickness reduction of powder particles. Figure 1 shows a schematic drawing of the repeated rolling method for the mechanical alloying of Al and Fe powders. As shown in Fig. 1, Al and Fe powders are packed in a stainless steel pipe with the end of the pipe closed by press forming. The dimensions of the pipe are 18 mm inside diameter, 20 mm outside diameter and 150 mm in length. This pipe containing the specimen powders was rolled at room temperature to a thickness of about 0.5 mm by gradually reducing the roll-gap. The specimen powders were taken out and again packed in another pipe. The [57]Fe Mössbauer spectra of these powders at room temperature were obtained as a function of repeated rolling times and then compared to results obtained from the powders prepared by the conventional ball-milling method.

RESULTS AND DISCUSSION

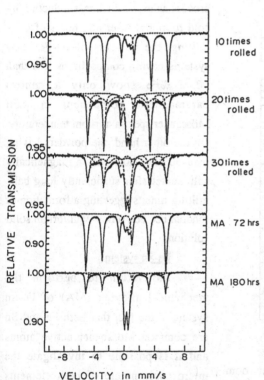

Figure 2. [57]Fe Mössbauer spectra at room temperature obtained from Al-24.4at.%Fe MA powders.

Al-Fe system: We performed mechanical alloying of Al and Fe powders with blended compositions of 4, 10, 17.4, 24.7, 33, 50, 60, 67, 80 and 90 at% Fe. Figure 2 shows typical [57]Fe Mössbauer spectra at room temperature obtained from blended powders of Al-24.4at%Fe. The upper three spectra were obtained by a repeated rolling method and the lower two show the spectra obtained by a ball-milling method. [57]Fe Mössbauer spectra obtained after MA by both methods consist of two main components; one is a ferromagnetic sextet and the other is a central paramagnetic doublet. The central paramagnetic doublets observed in the spectra have the isomer shift value of 0.19±0.03 mm/s relative to bcc Fe at room temperature and the quadrupole splitting 0.52±0.05 mm/s. These values are in fair agreement with those from a spectrum obtained from an amorphous Al-43.4at.%Fe-4.2at.%Ti powder reported previously [4]. For 4at.%Fe powder, a similar doublet has been observed and identified as an amorphous phase. Together with this amorphous phase, the formation of extended solid solution of Fe in Al was detected as a singlet absorption in the spectrum, depending on the mechanical energy of rotation speed and confirmed by X-ray diffraction experiments. Ferromagnetic components differ slightly from each other and the repeated rolling method leads to broadening of components, suggesting the existence of increased amounts of Al atoms in Fe grains , compared to the results from ball-milling method. This implies that the energy generated by the MA process and thus the degree of the mutual atomic dispersion depends on the methods depends on the method used. From the above experimental finding, we can conclude that the energy generated by kneading depends on the methods but the kneading process is principally controlled by the folding process of each grain. Folding by an impact between colliding balls and between ball and wall is the same process as in the thickness reduction by cold rolling. After 450 hours of ball-milling, the powders of Al-Fe alloy systems having compositions less than 50 at.%Fe show only a central paramagnetic component in their Mössbauer spectra at room temperature. On the other hand, the powders having over 50%Fe show only a magnetically split sextet after sufficiently long ball-milling time, suggesting a formation of random bcc Fe-Al non-equilibrium solid solutions.

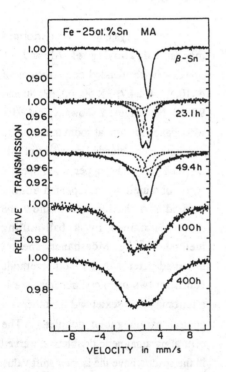

Figure 3. ^{119}Sn Mössbauer spectra at room temperature obtained from Fe-25at.%Sn MA powders.

Fe-Sn system:

Motivation to perform the mechanical alloying (MA) of Fe-Sn system is the fact that both Fe and Sn are common Mössbauer active atoms and it is possible to investigate the microscopic nature of both elements from Mössbauer measurements. Figure 3 shows the typical ^{119}Sn Mössbauer spectra at room temperature as a function of ball-milling time obtained from Fe-25at.%Sn blended powders. At the top of the figure is a spectrum of metallic Sn as a reference. After 23.1 and 49.4 hours ball-milling, the spectra consist of three components; one is a retained metallic Sn, the second is a rather sharp component having an isomer shift of 2 mm/s and the third is a widely spread component by the magnetic hyperfine interaction at room temperature. From the isomer shift value the second sharp component is determined to be due to the ^{119}Sn in FeSn$_2$ which is also identified by the ^{57}Fe Mössbauer spectra (not shown) as a central paramagnetic component. When the ball-milling time exceeds 100 hours, the ^{119}Sn Mössbauer spectra at room temperature consist only of magnetically broadened components, and their shape does not show any change by increasing milling time. This result was confirmed by the ^{57}Fe Mössbauer measurements (not shown). The results from ^{119}Sn and also ^{57}Fe Mössbauer measurements clearly show a formation of a non-equilibrium bcc solid solution as a final product of Fe-25at.%Sn MA powders, but in the early stage of the MA process the intermetallic compound FeSn$_2$ has most probably been formed at the boundaries between Fe and Sn grains. Further mechanical alloying destroys the FeSn$_2$ compound as Fe and Sn tend to dissociate with each other, and forms the non-equilibrium bcc Fe-Sn solid solution.

Ag-Fe system: Ag-Fe system is known to be an immiscible system even in the liquid state and one of the very interesting candidates for investigation by MA. When atomic dispersion has occurred in the Ag-Fe system by MA, Fe atoms associated with Ag atoms will show different hyperfine interaction parameters than Fe in a bcc Fe lattice. We performed MA of Ag-Fe powders having compositions of 10, 30, 40, 50, 56.3, 70 and 90at.%Fe. Figure 4 shows the ^{57}Fe Mössbauer spectra at room temperature obtained after 800 hours of ball-milling. The spectra in Fig. 4 consist of a magnetically split sextet attributed to the retained bcc Fe grains and broad components having hyperfine fields smaller than 330 kOe, due to the Fe atoms associated with Ag atoms. With an increase in Ag content, the broad components and the central paramagnetic components increase suggesting the occurrence of atom-mixing, most probably near boundaries between Ag and Fe grains. As reported previously [5], these MA powders were annealed after mechanical alloying and the precipitation of fcc γ-Fe is evident in the 56.3at.%Fe alloy. The above experimental results show clearly evidence of the mutual atomic dispersion for the immiscible Ag-Fe system. Direct observation of grains using high resolution electron microscopy show the formation of nano-meter sized crystalline grains in the Ag-Fe system. The above mutual atomic dispersion most probably occurred near grain boundaries. When the Ag concentration is sufficiently high the Fe atoms in this region segregate to form fcc (γ)-Fe precipitates during a post-annealing treatment.

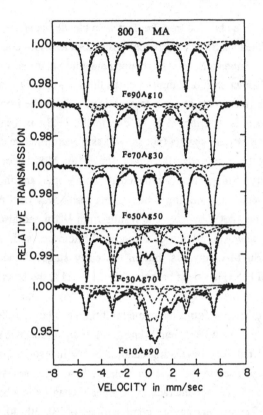

Figure 4. ^{57}Fe Mössbauer spectra at room temperature obtained from Fe-10, 30, 50, 70 and 90at.%Ag MA powders after 800 hours ball-milling. Dotted lines show the extra-component overlapped to bcc Fe component to reproduce the experimental data, to be used only as a guide. Velocity scale is relative to bcc Fe at room temperature.

Fe-C and Fe-B systems:: The formation of metastable and non-equilibrium alloy phases has been investigated intensively for Fe-metalloid systems using the rapid quenching technique. It is interesting to investigate what kinds of metastable and non-equilibrium phases will be formed by MA and whether these products are the same as those obtained by rapid quenching. We performed MA of iron and graphite powders having compositions between 17 and 90at.%C. Figure 5 shows typical ^{57}Fe Mössbauer spectra at room temperature obtained from Fe-25at.%C MA powders as a function of milling time. After 200 hours milling time, the spectrum consists of a retained bcc Fe component showing sharp sextet and intense broad contribution showing a distribution of the magnetic hyperfine fields. When the sharp bcc Fe component is subtracted from the spectrum, a broad

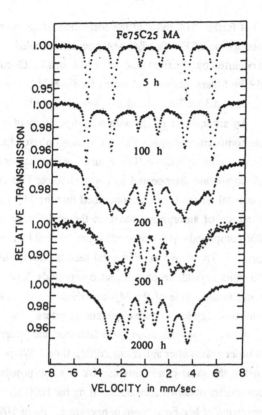

Figure 5. ^{57}Fe Mössbauer spectra at room temperature obtained from Fe-25at.%C MA powders as a function of milling time. Velocity scale is relative to bcc Fe at room temperature.

contribution having a distribution of the hyperfine magnetic fields is obtained. This subtracted spectrum was compared to the spectrum obtained from an amorphous ribbon of $Fe_{60}Si_{17}C_{33}$ prepared by splat-quenching technique. The spectra are in fairly good agreement and it was concluded that the Fe-25at.%C MA powders contain a large amount of amorphous phase after 200 hours of ball-milling. However, further ball-milling induces a reduction of averaged hyperfine field and after 2000 hours ball-milling, the spectrum shows a rather distinct sextet accompanied by a broad background contribution. Averaged hyperfine field of the sextet was determined to be 208 kOe which almost exactly agrees with that of Fe_3C (cementite) in spite of its broad pattern being due to the defective structure. The above experimental results suggest that in the early stage of MA an amorphous phase was formed in

Fe-25at.%C powders but further MA induced the formation of the more stable compound Fe₃C. A similar investigation of Fe-60at.%C MA powders showed that the formation of amorphous phase also occurred but the final product of MA was Fe_7C_3 compound. For Fe-B system, we performed [57]Fe Mössbauer measurements for Fe-B MA powders having blended compositions of 20, 30, 35 and 50at.%B. Results obtained are similar to those obtained from Fe-C MA powders. An amorphous phase with a composition of nearly 30at.%B first appeared for the whole composition range. This amorphous phase changed to a tetragonal Fe_2B compound by further milling time of 1000 hours. After further milling of 50at.%B MA powders, the Fe_2B compound disappeared and orthorhombic FeB compound appeared. From the above experimental results it can be concluded that the MA powders of Fe-C or B systems show the formation of amorphous phase in the early stage of MA but the final products are more stable compounds which depend on the original blended compositions.

Pure Fe powders : The above experimental results clearly show the formation of amorphous, non-equilibrium crystalline phases and compounds dependent on the original blended composition and milling time of the MA powders. It is important to understand what effect the mechanical energy has in pure elemental powders. Pure elemental powders, such as Fe powders in the present investigation, were flattened and folded by ball-milling and can create nano-meter order grains after sufficient milling time. We performed ball-milling of pure Fe powders using the same conditions as MA of alloy powder. Figure 6 shows typical [57]Fe Mössbauer spectra obtained after ball-milling for 1090 hours. At the top of the figure is a reference spectrum of bcc Fe at room temperature. After 1090 hours milling, the spectrum consists of a pure bcc Fe component and a satellite component having smaller hyperfine field at room temperature. At 77 K the intensity of the satellite component slightly decreased but there is no appreciable change in shape. After short time annealing at 873 K, the satellite component disappeared and an oxide component appeared. By a careful analysis of the impurity concentration in the ball-milled powders there is no appreciable contamination by gaseous impurities in the powder and the oxide formation is due to the insufficient vacuum condition during the annealing. In order to determined the grain size, the line broadening of the X-ray diffraction pattern has been analyzed. Figure 7 shows the results from X-ray line broadening analysis using Fe Kα and Mo Kα radiations. Open circles and squares show the line broadening of MA powders and the closed circles and squares show the broadening after annealing for 5 minutes at 873 K. As shown in Fig. 7, the coherent domain size for MA powders and annealed powders is about a few ten nano-meter, but the gradient of the broadening is much steeper for MA than the annealed powders. It is suggested that immediately after ball-milling the powders contain a large amount of strain located most probably near the boundary regions of each grain. Fe atoms at such regions contribute to the satellite component in Mössbauer spectrum.

After annealing for 5 minutes at 873 K, grain growth is not appreciable but the release of this strain near grain boundaries resulted in the disappearance of the satellite component in Mössbauer spectrum.

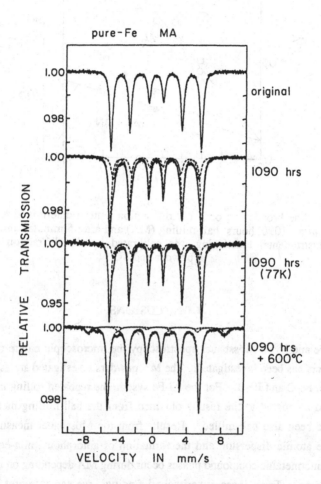

Figure 6. ^{57}Fe Mössbauer spectra obtained from pure Fe MA powders after 1090 hours ball milling. Uppermost spectrum was obtained from the original non-MA pure Fe powders. Lowest spectrum shows the result from after 5 minutes annealing of MA powders and the extra-components are due to Fe oxide. Velocity scale is relative to pure Fe at room temperature.

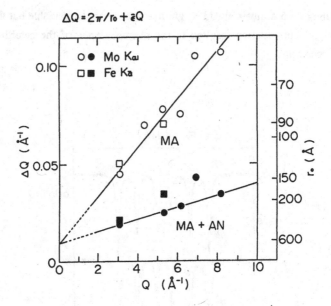

Figure 7. Extra-line broadening of X-ray diffraction pattern determined from the pure Fe MA powders after 1090 hours ball-milling (MA) and after 5 minutes annealing at 873 K (MA+AN). Instrumental broadening was calibrated by the diffraction of original Fe powder and Si.

CONCLUSIONS

Using the [57]Fe and [119]Sn Mössbauer spectroscopy the microscopic nature of mechanically alloyed powders has been investigated. The MA powders investigated are as follows: Al-Fe, Fe-Sn, Ag-Fe, Fe-C and Fe-B. For the Al-Fe system the repeated rolling method has been employed and compared to the results obtained from the ball-milling method. Pure Fe powders have been also ball-milled. Results from the Mössbauer measurements clearly show that the atomic dispersion and the formation of amorphous, non-equilibrium solid solution and intermetallic compound phases occur during MA depending on the alloy species and milling time. From these experimental findings, we can interpret the mechanical alloying, (MA) process as having the following characteristics (1) Elemental process of MA is controlled by the folding of each powder grain and kneading by an impact between colliding balls or between ball and wall, the same process that occurs in the thickness reduction by cold rolling. During such a kneading process, the formation of nano-meter order grains, atom mixing and storing of strains or defects near boundaries occurs simultaneously. (2) Atom mixing occurs near boundaries of grains and either simultaneously or subsequently the local structure of such regions transforms to more stable states depending on the local compositions. Such local lower thermodynamic potentials are gained most probably by local atomic diffusion. (3) Mechanical alloying is surely controlled by the low temperature solid state reactions and in this concept the MA is quite different from rapid cooling from liquid and gaseous phases.

ACHNOWLEDGEMENTS

The authors acknowledge Prof. emeritus F.E. Fujita of Osaka University for his encouragements for this work. We are indebted to Prof. K.N. Ishihara, Dr. B. Huang, Mr. T. Tanaka and Mr. J. Kuyama for their specimen preparations and various contributions to this work and also to Dr. H. Tanimoto, Mr. S. Imaoka, Mr. H. Nakagawa and Mr. S. Morimoto for their Mössbauer measurements and various contributions to this work.

REFERENCES

1. Koch, C.C., Material Synthesis by Mechanical Alloying., Annu. Rev. Mater. Sci., 1989, **19**, 121-43.
2. Shingu, P.H., Huang, B., Nishitani, S.R. and Nasu, S., Nano-meter Order Crystalline Structures of Al-Fe Alloys Produced by Mechanical Alloying., Suppl. to Trans. JIM. 1988, **28**, 3-10.
3. Shingu, P.H., Huang, B., Kuyama, J., Ishihara, K.N. and Nasu, S., Amorphous and Nano-meter Order Grained Structures of Al-Fe and Ag-Fe Alloys Formed by Mechanical Alloying., In New Materials by Mechanical Alloying Techniques. eds. E. Arzt and L. Schultz, Deutsche Geselleschaft für Metallkunde, Germany, 1989, 319-27.
4. Nasu, S., Shingu, P.H., Ishihara, K.N. and Nasu, S., Mössbauer Study on Mixing and Kneading of Metallic Powders., Hyperfine Interact., 1990, **55**, 1043-50.
5. Nasu, S., Morimoto, S., Tanimoto, H., Huang, B., Tanaka, T., Kuyama, J., Ishihara, K.N. and Shingu, P.H., Mössbauer Study of Mechanical Alloyed Powders., Hyperfine Interact., 1991, in press.
6. Kuyama, J., Inui, H., Imaoka, S., Nasu, S., Ishihara, K.N. and Shingu, P.H., Nanometer-Sized Crystals Formed by the Mechanical Alloying in the Ag-Fe System., Jpn. J. A. P., 1991, **30**, L854-56.
7. Okumura, H., Ishihara, K.N., Shingu, P.H., Park, H.S. and Nasu, S., Mechanical Alloying of Fe-B Alloys., J. of Mat. Sci., 1990, submitted for publication.
8. Tanaka, T., Nasu, S., Ishihara, K.N. and Shingu, P.H., Mechanical Alloying of High Carbon Fe-C System., J. of Less-Comm. Metals, 1990, submitted for publication.

MECHANICAL ALLOYING OF IMMISCIBLE ELEMENTS

A.R.YAVARI and P.J. DESRE,

LTPCM-CNRS UA29, Institut National Polytechnique de Grenoble, BP 75,

DomaineUniversitaire 38402 St Martin d'Hères, France.

ABSTRACT

We propose for the first time, a quantitative thermodynamic explanation for the formation of solid-solutions of immiscible elements by mechanical alloying. It is shown how codeformation of a hard and a soft phase results in the formation of disc or needle shape particles possessing large aspect ratios. As the particle size drops to about 10nm with continued deformation, the particle tips and edges generate nanometer size fragments upon subsequent necking. Using simple thermodynamic arguments we show how otherwise immiscible atoms of these fragments will enter in solution and that this process can lead to their full dissolution.

INTRODUCTION

The formation of metastable phases during solid-state reaction or mechanical alloying of elements with heats of mixing $\Delta H_{mix} \leq 0$ has been explained both from thermodynamic and kinetic view points [1-4] but mechanical alloying of immiscible elements is not yet understood. In the past year solid solutions (crystalline or amorphous) of several totally immiscible elements have been obtained by mechanical alloying. Examples are the Cu-Fe[5], Cu-Ag-Fe[6], Cu-W [7], Cu-Ta[8] and Cu-V[9]. While Ta,V and W are usually contaminated by oxygen and nitrogen which may remove miscibility-gaps [10], systems such as CuFe are clearly binary. The formation of such copper solid-solutions is in apparent violation of the rules of thermodynamics as these systems exhibit large, positive heats of mixing ΔH_{mix}.

We consider the co-deformation behavior of two immiscible metals. If the codeforming phases have similar mechanical properties a contineous multilayer structure is formed and further deformation results in continued decrease in thickness per layer. On the other hand, for metal couples with different crystal structures and work-hardening capabilities, one of the constituent metal layers will break up into particles of needle or disc shape with large aspect ratios. After heavy deformation, fragments, tips and edges will attain the nanometer range.

The effect of the capillary pressure due to small radii of curvature on the free energy of the atoms is considered and applied to spinodal systems with large positive heats of mixing. It is found that a critical radius exists which is the smaller, the larger the heat of mixing but that the critical radius increases sharply after the spinodal composition c_{sp} is attained. The results are supported by x-ray diffraction and microscopic observations.

EXPERIMENTAL PROCEDURE

Fcc-solid solutions of Fe and Cu were prepared by mechanical alloying of pure Cu in thin foil form and Fe in powder form in a Fritsch vibrating ball-mill (pulverisette-0) modified to allow vacuum levels of 10^{-5} torrs in the vial before introducing an operating pressure of 1.5 atmospheres of inert gas (Ar or He). Milling was done in a stainless steel vial with a 6 cm diameter steel ball. X-ray diffraction patterns were obtained using a Siemens diffractometer with Cu-K$_\alpha$ radiation. Line widths were corrected for instrumental broadening by substracting the line width of unmilled materials from that of each peak after ball-milling. Cross-sectional SEM observations were preformed on sandwiches of superimposed foils of the chosen metals cold-rolled in an electric Jolio rolling machine with indirect cold-rolling in steel sheets as described elsewhere [11]

RESULTS AND DISCUSSIONS

It is well established that when powder mixtures of metals A and B are ground together, they quickly form agglomerates of multilayers as the A and B powder particles are repeatly flattened and welded by colliding balls [12] leading to microstructures nearly the same as those obtained by cold-rolling. We have studied in detail the deformation structures obtained by the latter process [11].

These observations indicate that depending on the mechanical properties of the constituents at the deformation temperature, two distinct types of structures are obtained : increasingly thin, continuous multilayers or a fine distribution of saucer or needle shaped particles of one phase in the other. In all cases deformation proceeds by necking of one of the constituent phases in regions where the local resolved stress reaches the critical value for the onset of deformation. At this stage the surrounding matrix in the necking region undergoes heavy local deformation and work-hardening. This increases the applied force required for continued local deformation while the reduced cross-section of the layers undergoing necking reduces this force. Depending on the net balance, necking might cease or continue until rupture of the necking layer. Whether or not necking continues under the available applied stress determines which of the two deformation structures are obtained. If it ceases, layer rupture will not occur and the same layer will neck and deform at other locations which will reach the critical stress for plastic flow. This type of behaviour leads to the formation of multilayers with thin thickness per layer [13] as for example shown in the SEM micrographs of figure 1. On the other hand if the work-hardening of the layers surrounding the necking region is insufficient to raise the critical stress beyond that for the onset of necking, rupture will occur and a second deformation structure constituted of globules of the necking layer dispersed in a matrix of the second constituent will be obtained. This can be seen in the SEM micrographs of figure 2. Depending on the nature of the constituent-metals and their volume fractions, a very fine scale dispersion of say A particles in a B matrix can be obtained. Multilayers of Cu with Ta, V, W and Fe which are immiscible with Cu but do form Cu-solid solutions by mechanical alloying are all typical of composites which will produce a fine dispersion-type microstructure after heavy codeformation.

Some of the rules concerning the criteria for the occurence of this type of deformation have been discussed in [11] and it is expected that high-melting bcc metals such as Ta, V, W and Fe will break up into a fine dispersion of particles when codeformed with ductile lower-melting fcc-copper as long as the Cu volume-fraction is sufficient for the fcc phase to percolate.

Figure 3 and 4 show various SEM viewgraphs of the necking to rupture mode observed for increasingly thin slices of the hard phase in Cu-Ta and Cu-Fe composites. Ta and Fe are seen to break into particles with slice or needle shapes with large aspect ratios. While TEM observations are needed to study further microstructural refinement with continued deformation, it can be argued that with the observed large aspect ratios, tip or edge radii of curvature will attain 1 nm or less if the particle size can be reduced to 10 nm for example.

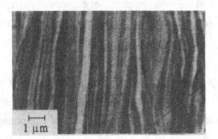

Figure 1. SEM images of Al/Cu (left) and Al/Pt (right) multilayers that avoid rupture by necking [11, 13]

Figure 2. SEM images of Al/Zr (left) and Al/Ti (right) multilayers that undergo layer fragmentation by repeated necking [11]

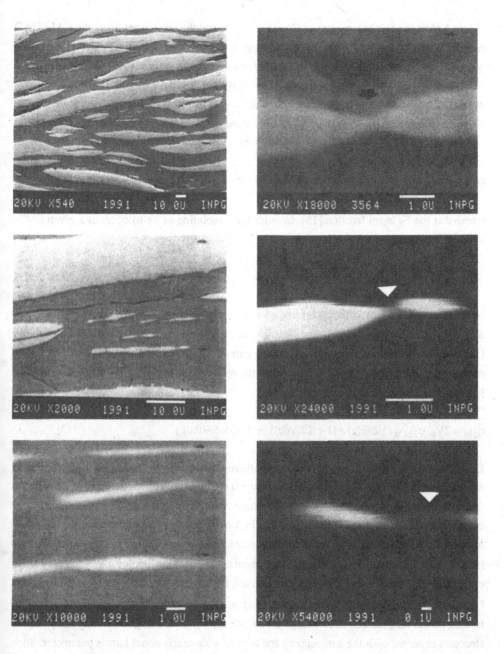

Figure 3. SEM(secondary electron) images of Cu/Ta multilayers showing Ta layer rupture by necking at various microstructural scales. Pictures with black background (right middle and bottom) are backscattered images of rupture by necking of needles as thin as 100 nm

Figure 5 plots the free energy curves G(c), for the fcc and the bcc phases using the CALPHAD calculations [14]. FeCu is a phase-separating system with spinodal points close to elemental compositions.In order for pure iron to enter a fcc solution, the tangent to the latter's G(c) curve must cut the G axis at or below the partial molar free energy ΔG_{Fe} of pure fcc iron. The insert of figure 5 shows how the tangent $\partial\Delta G(c)/dc$ turns with increasing Fe content in fcc-copper. The intersection with the G-axis reaches a maximum for the tangent at the spinodal composition c_{sp} with $\partial^2\Delta G/dc^2 = 0$ and decreases subsequently. The intersection ΔG_{Fe} is given by

$$\Delta G_{Fe} = \Delta G(c) + (1-c)\ (\partial G\ /\ \partial c)\ + \Delta G(bcc \to fcc) \tag{1}$$

where c is the Fe atom fraction [15]. In order for dissolution of Fe to occur at a given Fe/Cu interface the following relation must hold

$$\Delta G_{Fe} = 2V_m\ \sigma_{FeCu}/r\ > \ \Delta G(c) + (1-c)\ (\partial G/dc) + \Delta G_{Fe}\ (bcc \to fcc) \tag{2}$$

Here ΔG_{Fe} (bcc $\to$ fcc) is the molar free energy for Fe-bcc to fcc transformation and $2V_m\ \sigma_{FeCu}/r$ is the capillary pressure experienced by an Fe particle with a radius of curvature r [16] and the interfacial energy σ_{FeCu} which can be shown to be about 1370 mJ/m^2 for Fe particles at a temperature T $\approx$ 700 K [17] and the reference states $G^0{}_{Fe}$ and $G^0{}_{Cu}$ are pure fcc. Reordering of equation (2) yields

$$r_{min} = 2V_m\ \sigma_{FeCu}\ /\ [\ \Delta G(c) +(1-c)(\partial G/\partial c))\ + \Delta G(bcc \to fcc)\ ] \tag{3}$$

Using the calculated G(c) curve given by Kauffman [14] for the fcc-phase during deformation at a temperature T= 700 K [18], we plot in figure 6 the minimum diameter $2r_{min}$ of Fe particles that can dissolve in the surrounding fcc phase with Fe content c. The minimum radius corresponds to the spinodal composition c_{sp}. Figure 6 indicates that in order for the fcc-phase to reach Fe contents $c_{Fe} > c_{sp}$, some particles must attain $2r \approx 1$ nm but not all of the particles have to reach this size. In particular, it is sufficient for some of the particles to reach very small curvatures for Fe content of the fcc phase to reach the spinodal concentration c_{sp}. Beyond this Fe content, dissolution can occur at increasingly larger curvatures as a consequence of the turning of the tangent dG(c)/dc of figure 5. Also, the small bcc fragments may adopt an fcc structure coherent with the surrounding fcc copper with nearly equal lattice parameter. Such pseudomorphism that reduces the free energy of the mixture has been observed in Mo/Ge and Nb/Zr multilayers (19). It would correspond to the disappearance of the term ΔG (bcc $\to$ fcc) in equations (2) and (3) and increase the minimum size of soluble fragments. Like the fragmentation process and the capillary forces, such phase change prior to dissolution would be mechanically driven.

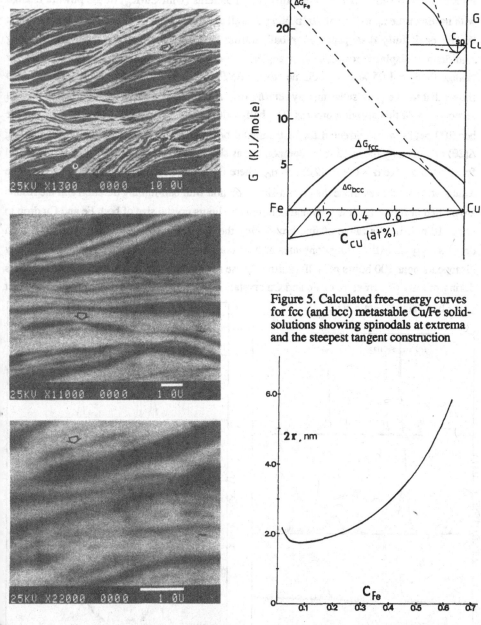

Figure 4. Backscattered SEM images of Cu/Fe multilayer showing Fe layer rupture by necking occuring simultaneously at various microstructural scales.

Figure 5. Calculated free-energy curves for fcc (and bcc) metastable Cu/Fe solid-solutions showing spinodals at extrema and the steepest tangent construction

Figure 6. Minimum calculated diameter 2r of Fe particles soluable in a copper solid solution with iron content C_{Fe}

Figures 7 and 8 show how the x-ray diffraction patterns (with Cu-K$_\alpha$) of the powder change with milling time up to 2 weeks of milling as well as after subsequent annealing. It is seen that the Fe peak fully disappears after ball-milling and only fcc-peaks remain while being significantly displaced towards lower angles. The displacement represents a lattice parameter change $\Delta a/a_0 = 1.05$ % and a volume change $\Delta V/V = 3.14\%$ for the fcc phase. Figure 7 also shows that the Fe peak subsequently returns after annealing at 500° C thus indicating the full reversibility of this alloying process. The half-width at half-max $\Delta(2\theta)$ of the fcc-111 and the bcc-200 peaks were corrected for instrumental broadening by deducting the corresponding $\Delta(2\theta)$ produced by the initial pure components as shown in figure 9. The Scherrer formula [20] for crystallite size $d = 0.9 / \Delta(2\theta) \cos \theta_0$ where θ_0 is the diffraction angle was used on the results of figure 8 to estimate the crystal-size evolution with ball-milling duration and the results are plotted in figure 10. It is seen that after only 50 hours crystal size of both Fe and Cu drop to about 10 nm. In this range of line broadening the Fe-peak intensity remains strong. It then drops slowly as ball milling continues at near constant fcc-phase crystal size and gradually disappears near 200 hours of milling time. These results indicate that alloy formation occurs during milling of a mixture of Fe and Cu crystals of size $d \approx 10$ nm and that the fcc-product phase has a similar grain size.

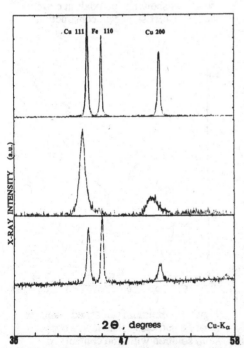

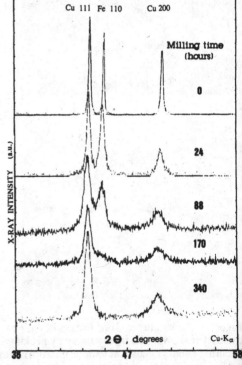

Figure 7. X-ray diffraction patterns of Cu/Fe mixture prior to milling (top), after 2 weeks of milling (middle) and subsequent annealing at 500°C (bottom)

Figure 8. X-ray diffraction patterns of Cu/Fe mixture after increasing ball-milling times showing occurence of line-broadening before the disappearance of the Fe peak

The 10 nm average crystal size together with the large aspect ratios that the Fe-particles develop as observed in figure 6, indicate that sharp tips and edges with subcritical radii $r \leq r_{min}$ of figure 7 and equation (3) exist and are continously formed by the mechanical deformation. Necking on such nano-sized tips continuously generates nanometer size particles with $r \leq r_{min}$ and thus provides the driving force needed to overcome the positive heats of mixing in these systems. It can also be shown that in these spinodal systems, the minimum wave-length of concentration fluctuations that can grow in amplitude is more than twice the dimension $2\, r_{min}$ of dissolving particles such that the small fragments cannot grow by spinodal decomposition [17]. A similar mechanism can develop to dissolve Cu atoms from Cu particles into a surrounding Fe matrix but this will only occur when the Cu phase breaks down into particles which cease to percolate. Because of the much higher ductibility of Cu, this occurs only when the Cu volume fraction is below about 30 at % such that the fcc solution extends over 70 % and the bcc over 30 % of the entire composition range as observed by Shingu et al [5].

Finally, it should be pointed out that when large atom size differences exist as between Cu and Ta, V or W the dissolution of the larger atoms can amorphise the Cu-fcc solid solution while Fe which has nearly the same atomic size as Cu does not destabilise the fcc structure. This is consistent with criteria of amorphisation based on atomic level stresses [21]. More recent results will be presented elsewhere [17].

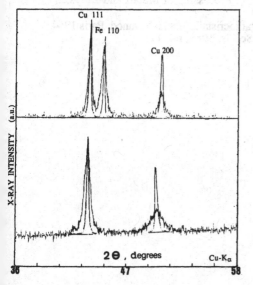

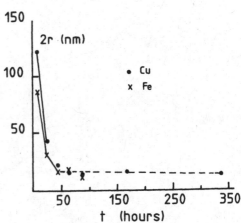

Figure 9. Superposition after displacement, of elemental Bragg peaks showing instrumental line-broadening and the Cu-fcc and Fe-bcc peaks of the mixture showing broadened due to ball-milling

Figure 10. Evolution of crystal size with milling time for Cu/Fe mixture

REFERENCES

1) Meng W. J., Nieh C.W.and Johnson W. L. , J. Appl. Phys.1987, **51** , 1963
2) Highmore R.J. , Greer A.L, Leake J.A. and .Evetts J.E , Mater.Lett. ,1988, **6**, 401
3) Desré P.J. and Yavari A.R. , Phys. Rev. Lett.,1990, **64**, 1533.
4) Yavari A.R. and Desré P.J, Phys. Rev. Lett.,1990 , **65**, 2571
5) Shingu P.H., Ishihara K.N., Uenishi K. , Kuyama J Huang,B.and Nasu,S. , Solid State Powder Processing, Eds : Clauer A.H. and de Barbadillo J.J , The Minerals, Metals and Materials Society 1990, pp. 21-34
6) Uenishi K. ,Kobayashi K.F. , Ishihara K.N. and Shingu,P.H. proceedings of the International symposium on Mechanical Alloyiong, May 1991, Kyoto
7) Gaffet E., Louison C. ,Harmelin M. and Faudet F., Mater. Sci and Eng., 1991 **A134**, 1380
8) Veltl G., Scholz B. and Kunze H.D, Mater. Sci and Eng.,1991, **A134**, 1410
9) Fukunaga T., Mori M., Inou K. and Mitzutani U., Mater. Sci and Eng, 1991, **A134**, 863
10) Yavari A.R and Desré P.J, Proceedings of the International Symposium on Mechanical Alloying, May 1991, Kyoto
11) Bordeaux F. and Yavari A.R. , Z. Metallkd.,1990, **81**, 130
12) Schwarz R.B., Petrich R.R. and Saw C.K, J. Non cryst. Solids,1985, **76**, 281
13) Bordeaux F. and Yavari A.R., J. Appl. Phys.,1990,**67**, 2385
14) Kaufman L. , CALPHAD, 1978, **2**, 117
15) Gaskell D.R., "Introduction to Metallurgical thermodynamics" McGraw Hill,1973, p.305,
16) Shewmon P.G., "Phase Transformation In Solids" Mc Graw Hill series on Materials Science and Engineering, p.151
17) Yavari A.R. and Desré P.J , to appear in the Journal of Materials Research
18) Benjamin J.S., Proceedings of the International Symposium on Mechanical Alloying, May 1991, Kyoto
19) Greer A.L. and Somekh R.E., in "Materials Saimu and Technology", Edited by Cahn R.W. Haasen P. and Kramer E.J. (Vol. 15 : Processing of Metals and alloys,ch.8) VCH, Weihiem, 1991 in press
20) Guinier A., "Théorie et Technique de la Radiocristallographie" Dunod, Paris 1964
21) Egami T. and Waseda Y., J. Non-Cryst. Solids, 1984, **64**, 113

AMORPHIZATION INDUCED BY BALL - MILLING IN SOME PURE ELEMENTS : SI, GE

E. GAFFET, F. FAUDOT, M. HARMELIN
Centre d'Etudes de Chimie Métallurgique - C.N.R.S.
15 Rue G. Urbain - F94407 - Vitry / Seine Cedex - France

ABSTRACT

We report on the crystal to amorphous phase transition induced by ball - milling in pure Si and pure Ge powders. Based on X - ray diffraction patterns, scanning and transmission electron micrographs, EDX microanalyses, differential thermal analyses and differential scanning calorimetry, it seems that amorphization occurs during the continuous decrease of the particle grain size and the crystalline lattice parameter expansion which both lead to the destabilization of the diamond cubic structure. The analyses of the X - ray diffraction patterns reveal the coexistence of microcrystallites and of an amorphous phase. The transmission electron microscopy investigations reveal that some particles of about 100 to 200 nm exhibit an amorphous structure (bright field image and selected area diffraction patterns). Only some slight effect of the ball - milling conditions are noted on the critical values of the grain size and the lattice expansion of such a destabilization.

INTRODUCTION

The crystal to amorphous phase transition induced by mechanical alloying (MA) was first reported for the Co - Y and Ni - Nb binary systems [1, 2]. Such a phase transition was assumed to require some prerequisite conditions : a negative heat of mixing and the presence of a fast diffuser [3]. Such a phase transition was claimed to be analogous to amorphization by solid state diffusion in the multilayers systems [4]. Nevertheless, amorphization induced by MA was reported for the V - Zr system for which no fast diffuser is known [5]. Recently, the crystal to amorphous phase transition was reported for binary systems exhibiting a positive heat of mixing : Si - Sn, Si - Zn, [6, 7], Cu - W [8, 9], Cu - V [10], Cu - Ta [11 - 13].

The mechanical alloying of some brittle components such as a germanium and silicon powder mixture was reported by R.M. Davis and C.C. Koch [14] : the Spex Mill 8000 end product structure was found to correspond to a so called true crystalline solid solution but no amorphous phase was detected.

In our previous works, amorphization induced by ball - milling (BM) was reported in pure systems : Si [6, 7, 15] and Ge [16] powders.. Such a crystal to amorphous phase transition was attributed to a crystalline lattice instability in relation with a Si or Ge grain size decrease. Based on scanning electron microscope observations and energy dispersive chemical analyses, X - ray diffraction and transmission electron microscopy investigations (bright field images and selected area diffraction patterns) and differential thermal analysis (DTA), it was the first time that such a crystal to amorphous phase transition induced by planetary ball milling in pure systems was observed. The former results are summerized in this paper.

EXPERIMENTAL DETAILS

BALL - MILLING CONDITIONS

10 g of pure germanium (Ge 99.999+%, CERAC Inc.) and pure silicon (Hyperpure Polycrystalline Silicon, Wacker Chemitronic, GmbH ; 300 Ω cm n and 3000 Ω cm p) pieces - the investigated compositions (25, 50, 75, in Si Wt%) - were introduced into a cylindrical tempered steel container of capacity 45 ml. This procedure was performed in a glove box filled with purified argon. Each container was loaded with five balls (diameter, 1.5 cm ; mass, 14 g). The containers were sealed in the glove box with a Teflon O - ring and the milling proceeded in stationary argon atmosphere. Mechanical alloying (MA) was carried out using two classical Fritsch planetary high energy ball - milling machines (Pulverisette P7/2 and P5/2). In the case of the latter machine, two intensity settings were chosen which are referred later to as P5/2(10) and P5/2(5). Ω is the rotation speed per minute (r.p.m.) of the disc on which the vial holders are fixed. The latters turn at a rotation speed of ω. The Ω rotation radius is equal to R_Ω. Therefore the shock kinetic energy is proportional to $\Omega^2 R_\Omega$. The Ω value has been checked by a tachometer. The specific parameters as a function of the various MA conditions are listed in Table 1 :

TABLE 1
Specific milling parameters as a function of the milling machines
(Ω, R_Ω, ω, see the text)

MA conditions	Ω (r.p.m.)	R_Ω (10^{-2}m)	ω (r.p.m.)	$\Omega^2 R_\Omega$ (m/s^2)
P7/2	650 - 700	6.75	1300 - 1400	311.5
P5/2(10)	340	12	765	153.8
P5/2(5)	200	12	450	51.3

SCANNING ELECTRON MICROSCOPY AND CHEMICAL MICROANALYSES (SEM/ EDX)

Some BM particles were taken from the vial for further electron microscopy investigations : the particle morphology was characterized using a digital scanning electron microscope (Zeiss DSM 950) in the secondary electron image mode. In order to evaluate possible container contamination which may have occurred by friction of the particles on the balls and the walls of the container, EDX analyses were performed using the Si - Li detector and the TRACOR EDX analyzer in conjunction with the scanning electron microscope. A semi - quantitative program with internal references (SQ from TRACOR) was used to analyse the EDX spectra. The (Fe,Cr) container contamination has been found to be less than 0.5 At% and 0.2 At % for the ball - milled Ge and Si powders respectively.

Figure 1 . Typical scanning electron microscopic image corresponding the P5(10) ball - milled Si powders (secondary electron image mode)

Figure 2 . Typical scanning electron microscopic image corresponding the P5(5) ball - milled Ge powders (secondary electron image mode)

X - RAY INVESTIGATIONS (XRD)

After continuous milling, a small amount of MA powder was extracted from the container and glued onto a silica plate for X - ray investigations. The XRD patterns were obtained using a (θ - 2θ) Philips diffractometer with the CoK$_\alpha$ radiation (λ = 0.17889 nm). A numerical method (the ABFfit program [17]) was used to analyse the XRD patterns and to obtain the position and the full - width at half height (FWHH) of the various peaks. As shown in [15, 16], the various deconvolutions based on the different peak shapes (i.e. gaussian, Lorentz or Cauchy) lead to the same results. Therefore, a deconvolution based on the gaussian peak shapes was chosen.

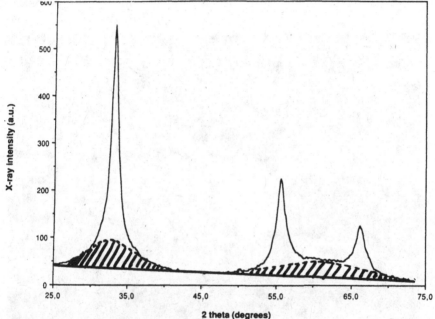

Figure 3. Typical X - ray diffraction pattern corresponding to the P5(5) ball - milled Si powders. The dashed areas correspond to the first and second diffuse halos corresponding to the amorphous phase. These areas are schematically given. For further details, see [6, 7, 15, 16].

Bragg expression has been applied to determine the d parameter corresponding to the diffraction peak position (θ) :

$$\lambda = 2 d \sin\theta. \tag{1}$$

The effective diameter of the particles (hereafter referred to as Φ) has been calculated from the Scherrer expression :

$$\Phi = 0.91 \lambda / (B. \cos \theta) \tag{2}$$

where λ is the X - ray wavelength, B the linewidth (in 2θ) and θ the diffraction angle.

TRANSMISSION ELECTRON MICROSCOPY INVESTIGATIONS (TEM)

Some MA (Ge - Si) particles were spread on copper TEM grids in order to study the microstructural state using a JEOL 2000FX transmission electron microscope. The structural state was investigated by selected area diffraction patterns on the particles which were thin enough to be observed and which remained on the edges of the copper grid. No electrochemical polishing nor ionic pulverisation were used to obtain thin foils. Therefore, no structural artefact could have been introduced by such classical foil preparation techniques.

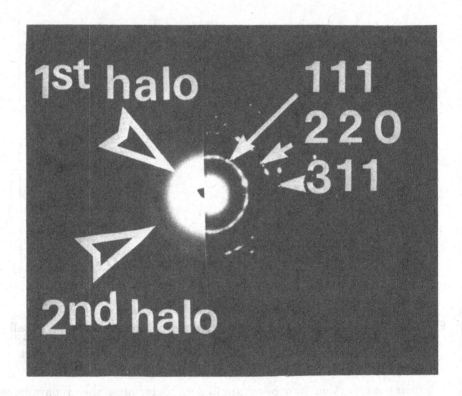

Figure 4 . Typical selected area diffraction pattern (SAD pattern) corresponding to the various states of the ball - milled diamond cubic material (Ge powders).The SAD pattern corresponding to the ball - milled crystalline state is given on the right side of Figure 4. The SAD pattern corresponding to the amorphous state is given on the left side of Figure 4.

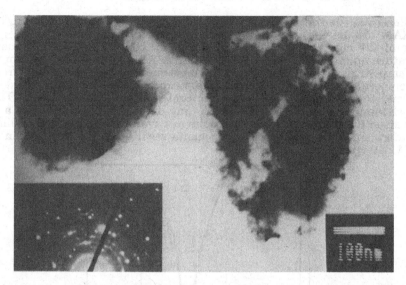

Figure 5 . Typical transmission electron microscopic image (bright field image) corresponding to the ball - milled powders which remain crystalline during the BM process.

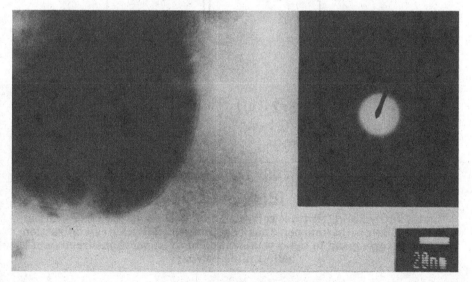

Figure 6 . Typical transmission electron microscopic image (bright field image) corresponding to the ball - milled powders which become amorphous during the process.

DIFFERENTIAL THERMAL ANALYSIS (DTA)

The differential thermal analysis (DTA) was carried out using a micro DTA SETARAM (Model M4) apparatus with a heating rate of 40°C min^{-1}. A sample of 20 mg was put in an alumina crucible, heated from room temperature up to 1050°C under flowing pure argon and then cooled down to room temperature. Some specific DTA experiments were performed from room temperature up to an intermediate temperature which was found to correspond to the end of the exothermic contribution. Therefore, the DTA annealed samples were then taken for X - ray diffraction investigations and further structural analysis. No DTA results are available for the ball - milled Ge powders since the performed experiments resulted in the destruction of the DTA detecting head.

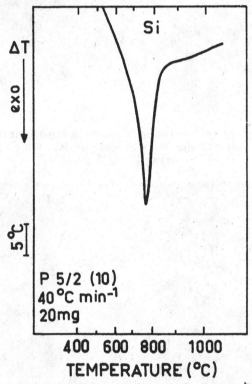

Figure. 7 . Typical DTA trace corresponding to the P5(10) Si ball - milled powder. The exothermic peak has been confirmed (by X - ray diffraction pattern) to correspond to the crystallization of the amorphous fraction of the ball - milled powder.

DISCUSSION AND CONCLUSION

According to our previous works [6, 7, 15, 16], the crystal to amorphous phase transition may be explained as following : the ball - milling method leads to a continuous refinement of the powder grain size. For a grain size lower than a first critical grain size value (hereafter referred to as Φ_{crit1} in the Table 2), such a process leads to an expansion of the crystalline lattice parameter. If the grain size is reduced to a value lower than a second critical value (hereafter referred to as Φ_{crit2}) - to which corresponds the largest relative increase of the lattice parameter (hereafter noted Δ) – , then the crystalline lattice becomes unstable. Such an effect will lead to the amorphization process.

Table 2
Critical grain size value and the corresponding lattice parameter. Δa is the largest lattice expansion of the crystalline fraction of the ball - milled powders as a function of the milling conditions. The position of the first halo (d) and the related full width at half height ($\Delta(2\theta)$) corresponding to the amorphous phase are also given.

Si	Milling Condition	Grain size nm	Lattice parameter nm 10^{-1}	Δa %	Amorphous phase d value, $nm * 10^{-1}$	$\Delta(2\theta)$ °
Ge*			5.658			
Ge	P7/2	$\Phi \geq \Phi_{crit1} = 7.6$	5.672		3.259	7.2
		$\Phi_{crit2} = 3.5 \pm 0.5$	5.681	+ 0.2		
	P5/2(10)	$\Phi \geq \Phi_{crit1} = 6.5$	5.672		3.253	6.8
		$\Phi_{crit2} = 3.2 \pm 0.5$	5.687	+ 0.4		
	P5/2(5)	$\Phi \geq \Phi_{crit1} = 5.5$	5.667		3.223	8.5
		$\Phi_{crit2} = 2.5 \pm 0.5$	5.682	+ 0.3		
Si	P7/2	$\Phi \geq \Phi_{crit1} = 10$	5.425 ± 0.015		3.148	7.5
		$\Phi_{crit2} = 4.0$	5.435 ± 0.015	+ 0.2		
Si	P5/2(10)	$\Phi \geq \Phi_{crit1} = 10$	5.430 ± 0.015		3.165	7.1
		$\Phi_{crit2} = 4.0$	5.435 ± 0.015	+ 0.1		
Si	P5/2(5)	$\Phi \geq \Phi_{crit1} = 8$	5.430 ± 0.025	0.0	3.153	4.7
Si*			5.430			

* given as the reference value for pure Si and pure Ge powders.

Based on experimental investigations (SEM, EDX, XRD patterns and TEM), the crystal to amorphous phase transition induced by ball - milling of pure Ge and pure Si has been observed. Such a phase transition seems to be the result of the continuous refinement of the diamond cubic particle grain size leading to a lattice expansion which destabilizes the crystalline lattice structure. No significant effect of the planetary ball - milling conditions was noted in the critical values of such a destabilization.

References

1. A. Y. Yermakov, Y.Y. Yurchikov and V.A. Barinov
   Phys. Met. Metall., 52(1981) 50
2. C.C.Koch, O.B. Cavin, C.G. McKamey and J.O. Scarbrough
   Appl. Phys. Lett., 43 (1983) 1017
3. For a review :
   a) A.W.Weeber and H. Bakker
   Physica B, 153 (1988) 93
   For recent conferences :
   b) Proc. "DGM Conference, New Materials by Mechanical Alloying
   Techniques", 3 - 5 October, 1988, Hirsau, F.R.G.,
   E. Arzt and L. Schultz (Eds)
   c) Proc. " International Conference on Amorphization by Solid - State
   Reaction", 21 - 23 February, 1990, Grenoble, France,
   Les Editions de Physique, Colloque de Physique, C4, suppl. N14, T51
4. J. Eckert, L. Schultz, K. Urban
   Proc. " 7th International Conference on Rapidly Quenched Metals",
   Stockholm - 1990 - Mater. Sci. Eng. A, A134 (1991) 1389 - 1393
5. A.W.Weeber, H. Bakker
   Z. Chem. Phys. NF, 157 (1988) 221
6. E. Gaffet and M. Harmelin
   Proc. " International Conference on Amorphization by Solid - State
   Reaction", 21 - 23 February, 1990, Grenoble, France, Les Editions de
   Physique, Colloque de Physique, C4, suppl. N14, T51, C4 - 139
7. E. Gaffet, M. Harmelin
   Proc. ASM Conference on "Structural Applications of Mechanical
   Alloying", Myrtle Beach SC, USA 1990
   Eds F.H. Froes, J.J. deBarbadillo, 257
8. E. Gaffet, C. Louison
   "International Conference on Amorphization by Solid - State
   Reaction", 21 - 23 February, 1990, Grenoble, France,
   Oral communication
9. E. Gaffet, C. Louison, M. Harmelin, F. Faudot
   Proc. " 7th International Conference on Rapidly Quenched Metals",
   Stockholm - 1990 - Mater. Sci. Eng. A, A134 (1991) 1380 - 1384
10. T. Fukunaga, M. Mori, K. Inou, U. Mizutani
    Proc. " 7th International Conference on Rapidly Quenched Metals",
    Stockholm - 1990 - Mater. Sci. Eng.A, A134 (1991) 863 - 866
11. C. H. Lee, T. Fukunaga, U. Mizutani
    Proc. " 7th International Conference on Rapidly Quenched Metals",
    Stockholm - 1990 - Mater. Sci. Eng. A, A134 (1991) 1334 - 1337
12. G. Veltl, B. Scholz, H.-D. Kunze
    Proc. " 7th International Conference on Rapidly Quenched Metals",
    Stockholm - 1990 - Mater. Sci. Eng. A, A134 (1991) 1410 - 1413
13. K. Sakurai, Y. Yamada, C.H. Lee, T. Fukunaga, U. Mizutani
    Proc. " 7th International Conference on Rapidly Quenched Metals",
    Stockholm - 1990 - Mater. Sci. Eng. A, A134 (1991) 1414 - 1417
14. R.M. Davis, C.C. Koch
    Scr. Metall., 21 (1987) 305

433

15. E. Gaffet and M. Harmelin
    J. Less - Common Met., 157 (1990) 201
16. E. Gaffet
    Mater. Sci.Eng. A, A136 (1991) 161 - 169
17. A. Antoniadis, J. Berruyer, and A. Filhol, Lauë Langevin Institute,
    Grenoble, France, Internal Report, 1988, 87AN22T

THE EFFECT OF CONCENTRATION GRADIENTS ON THE STABILITY OF AMORPHOUS LAYERS OF TERNARY ALLOYS

L. BOUANHA, P.J. DESRE, P. HICTER and A.R. YAVARI

LTPCM-CNRS UA.29, Institut National Polytechnique de Grenoble,

BP 75, Domaine Universitaire, 38402 Saint-Martin-d'Hères, France

ABSTRACT

We have extended the previously developed expression for the effect of sharp concentration gradients on the driving force for crystallisation of binary amorphous interlayer to the case of ternary systems. We find that appropriately chosen thirs elements should further reduce this driving force. It was previously shown that the driving force for crystallisation is reestablished as the amorphous layer grows and the concentration gradient drops to a critical value. Appropriately chosen ternary elements should increase the critical thickness beyond which crystallisation may occur. These concepts were applied to Ni/Zr and (Ni-Cu)/Zr multilayers and the experimental results were found to support the theory.

INTRODUCTION

Schwarz and Johnson [1] first reported an amorphization of isothermally annealed A-B mulilayers obtained by vapor deposition. Since then , experiment showed that of amorphous phases grows at the A-B interfaces up to a critical thickness at which a crystalline compound nucleates. Several authors tried to explain this observation by both thermodynamics and kinetics considerations. U. Gösele and K.N. Tu, for instance, using their more general "interfacial reaction barrier" approach [2], give an expression of the critical thickness which takes into account a competition between diffusion controlled and interfacial reaction controlled growth of various phases. Meng et al's model [3] is based on a nucleation-time argument and assume that a critical nucleus of the crystalline phase in the amorphous phase can grow only if its length in the direction of moving interface exceeds a typical value Highmore et al [4] developped a transcient nucleation model which considers that beyond a critical thickness of the amorphous layer, the velocity of the interface is too weak to consume subcritical crystalline nuclii.

Based on radical different assumptions the model of Desré and Yavari [5] shows that sharp concentration gradients in the amorphous interfacial layer can suppress the thermodynamic driving force for nucleation of the crystalline phase. While the thickness of the layer increases, these gradients flatten out and the driving force for crystallization is restored.

Let us recall that this latter model considers a cubic nucleus (volume V) of the crystalline phase in the amorphous layer placed in a concentration gradient ∇c. Figure 1 shows that each section of the cube contributes to the Gibbs free energy of transformation into a crystalline cube of concentration c^* according to :

$$\Delta G_{ac}(c) = \Delta G_{pc}(c^*) + \Delta G_{fl}(c,c^*)$$ (1)

since $\Delta G_{pc}(c^*)$ represents the molar Gibbs energy for polymorphous crystallysation of a cube of concentration c^* ΔG_{fl} the molar Gibbs energy to make the concentration of the slice fluctuate from c to c^* using the thermodynamic approach of Cahn and Hilliard for non uniform systems [6], the Gibbs energy of a volume V of the amorphous layer is given by :

$$\Delta G_a = \rho \int_V [G_0(c) + N_a \chi (\nabla c)^2] \, dV$$ (2)

where ρ is the number of moles of atoms per unit volume, N_a the Avogadro number and χ a constant value and $G_0(c)$ is the Gibbs energy per atom of the amorphous phase of composition C.. Moreover, the Gibbs free energy for fluctuation, ΔG_{fl}, is given by :

$$\Delta G_{fl} = \frac{1}{2} (c-c^*) \left(\frac{\partial^2 G}{\partial c^2}\right)$$ (3)

Since G is the Gibbs energy of the amorphous phase. These results lead, by integration to an expression of the Gibbs energy of crystallisation of the cube :

$$\Delta G_{ac} = 8\rho \left[\Delta G_{pc}(c^*) - N_a \chi(\nabla c)^2 \right] r^3 + \frac{4}{3}\rho \left(\frac{\partial^2 G}{\partial c^2}\right)(\nabla c)^2 r^5$$ (4)

It was also said [5] that the term $N_a \chi(\nabla c)^2$ could be generally neglected by virtue of a crude evaluation which showed that it represents in most cases less than 1% of ΔG_{pc}.
Thus, in taking into account the interfacial energy between the amorphous and the crystalline phases of concentration c^* and setting $\alpha = \left(\frac{\partial^2 G}{\partial c^2}\right)$ we obtain :

$$\Delta G_N = 24\sigma_{pc} r^2 + 8 \rho \left[\Delta G_{pc}(c^*) \right] r^3 + \frac{4}{3}\rho \alpha (\nabla c)^2 r^5$$ (5)

Figure 2 representing ΔG_N versus the radius r of the crystalline embryo for different values of ∇c implies the existence of a critical gradient ∇c_c from which there is no longer a driving force for crystallization. Its expression is given by :

$$\nabla c_C = \frac{\rho \left(2\left|\Delta G_{pc}\right|\right)^{3/2}}{9 \, \sigma_{pc} \, \alpha^{1/2}} \qquad (6)$$

Thus, the critical thickness can be evaluated by $\varepsilon_C = \dfrac{1}{\nabla c_C}$.

A numerical application of this formula in the case of the NiZr system taking: ΔG_{pc}= -4.5 kJ/mol, σ_{pc}= 300mJ/m2 ,and α= 468kJ/mol [5] leads to ε_C= 200Å for ∇c_C=5.105 /cm.

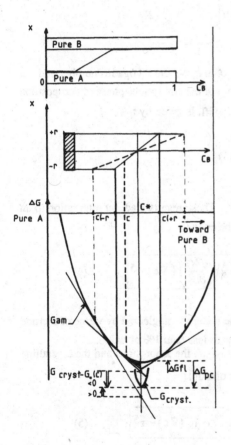

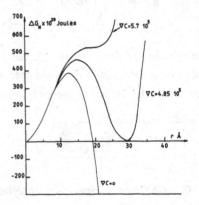

Figure 2.
Gibbs free-energy $\Delta G_N(r)$ of formation of crystalline embryo of radius r in amorphous phase with different concentration gradients ∇C. For ∇C above a critical value, $\Delta G_N > 0$ and nucleation is disallowed

Figure 1. Gibbs free-energy tangent construction for compositions at the extrema of critical nucleus of diameter 2r of intermetallic phase in an amorphous layer subject to concentration gradient ∇C

EXTENSION TO TERNARY SYSTEMS

We consider the case of obtaining ternary amorphous composite multilayers AB-C since AB is an alloy of two metals which have separately the required conditions for amorphisation by Solid State Reaction (SSR)with C, i.e very different diffusivities $D_A \gg D_B$, $D_A \gg D_C$ and very negative mixing enthalpy $\Delta H_{AC} \ll 0$ and $\Delta H_{BC} \ll 0$.

With this aim we consider that the diffusion at the AB-C interface follows the development of Kirkaldy [7] for multicomponent diffusion, applied here to a ternary system.

It is showed that the three concentrations C_A, C_B and C_C are not independant, because $\Sigma\, c_i = 1$, and the diffusion equations lead to :

$$
\begin{aligned}
c_A &= a\ \mathrm{erf}\left(\frac{\lambda}{2\sqrt{u}}\right) + b\ \mathrm{erf}\left(\frac{\lambda}{2\sqrt{v}}\right) + c \\[2mm]
c_B &= d\ \mathrm{erf}\left(\frac{\lambda}{2\sqrt{u}}\right) + e\ \mathrm{erf}\left(\frac{\lambda}{2\sqrt{v}}\right) + f
\end{aligned}
\tag{7}
$$

Where a,b,c,d,e,f,u,v depend on the ternary diffusivities and initial concentrations on each side of the interface and $\lambda = \dfrac{x}{\sqrt{t}}$ (x : distance in the direction of diffusion; t : time).

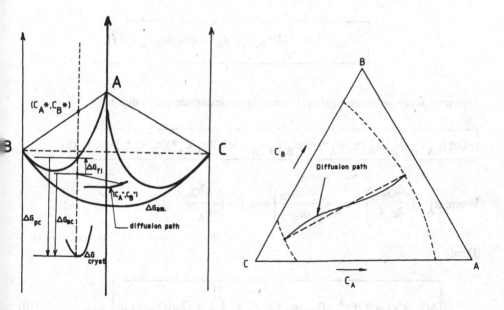

Figure 3. (left) Gibbs free-energy curves showing free-energy for polymorphous crystallisation ΔG_{ac} and concentration fluctuation ΔG_{fl} as in figure 1 but for ternary system also showing diffusion path for transformation

Figure 4. (right) Diffusion path for transformation as defined by Kirkaldy [8], see text

These results imply a correlation between C_A and C_B which, in a simplifying view, can be considered as a function $f(C_A) = C_B$ called diffusion path in a ternary diagram [8].

An equivalent diagram of Fig.1 applied to a three components system is thus drawn in Fig. 3 including the diffusion path which represents in this case the only permitted values of Gibbs energy available for the amorphous phase at the interface. We make the crude assumption that the diffusion path represented in terms of concentrations in a ternary diagram (see Fig.4) follows a straight line (i.e a quasi-binary section). Thus we can write that :

$$X = \frac{dC_A}{dC_B} = \frac{\frac{dC_B}{dx}}{\frac{dC_A}{dx}} = \frac{\nabla C_A}{\nabla C_B}$$

is a constant value along the diffusion path.

Besides this hypothesis, we keep neglecting the "Cahn term" equivalent of $Na\,\chi(\nabla c)^2$ for the binary system.

By the same argument as in the binary system we consider the ability of the amorphous cube of concentration $(C_A{}^\circ, C_B{}^\circ)$ to crystallize into a compound of concentration $(C_A{}^*, C_B{}^*)$.

$(C_A{}^\circ, C_B{}^\circ)$ is thus a point of the diffusion path (see Fig. 4).

Therefore, the driving force for crystallization of the cube, at constant ρ, is :

$$\Delta G_{ac}(V) = \int_V [\Delta G_{pc}(c_A{}^*,c_B{}^*) + \Delta G_{fl}(c_A,c_B)]\, dV$$

(8)

Moreover, it can be shown by simple thermodynamic considerations that :

$$\Delta G_{fl}(c_A,c_B) = \frac{1}{2}[\alpha_A(c_A-c_A{}^*)^2 + \alpha_B(c_B-c_B{}^*)^2 + 2\beta(c_A-c_A{}^*)(c_B-c_B{}^*)] + \dots \quad (9)$$

Where $\alpha_A = \left(\dfrac{\partial^2 G_{am}}{\partial c_A{}^2}\right)$, $\alpha_B = \left(\dfrac{\partial^2 G_{am}}{\partial c_B{}^2}\right)$ and $\beta = \left(\dfrac{\partial^2 G_{am}}{\partial c_A\, \partial c_B}\right)$

Then:

$$\Delta G_{ac}(V) = 8\,\rho\, r^3\, \Delta G_{pc}(c_A{}^*,c_B{}^*) + \int_{-r}^{+r} 4\,\rho\, r^2\, \Delta G_{fl}(r)\, dr \qquad (10)$$

One of the two concentrations, say c_A can be taken as a parameter for integration since it exists a relation $f(c_A)=c_A$. Consequently, $dr = dc_A/\nabla c_A$, $c(-r) = c_A{}^\circ - r\nabla c_A$ and $c(+r) = c_A{}^\circ + r\nabla c_A$ so that:

$$\Delta G_{ac}(V) = 8\rho r^3 \Delta G_{pc}(c_A{}^*,c_B{}^*) + \int_{c_A{}^\circ - r\nabla c_A}^{c_A{}^\circ + r\nabla c_A} \frac{4\rho r^2}{\nabla c_A} \phi(c_A,c_B)\, dc_A \qquad (11)$$

Where : $\phi(c_A,c_B) = \frac{1}{2}[\alpha_A (c_A-c_A{}^*)^2 + \alpha_v(c_B-c_B{}^*)^2 + 2\beta(c_A-c_A{}^*)(c_B-c_B{}^*)]$ with $c_B = f(c_A)$

Since $X = \dfrac{\nabla c_B}{\nabla c_A} = \dfrac{c_B - c_B{}^*}{c_A - c_A{}^*}$ is a constant value in the range $c_A{}^\circ - r\nabla c_A < c < c_A{}^\circ + r\nabla c_A$, it

follows that :

$$\Delta G_{ac}=[8\rho r^3 \Delta G_{pc}(c_A{}^*,c_B{}^*)+4\rho(\alpha_A+\alpha_B X^2+2\beta X)(c_A{}^\circ - c_A{}^*)^2]r^3 + 4/3\rho(\alpha_A+\alpha_B X^2+2\beta X)\, r^2$$
$$(12)$$

Taking into account the classical interfacial term, the Gibbs energy to crystallize the cube can be written :

$$\boxed{\Delta G_N = 24\sigma_{pc}\, r^2 + [8\rho r^3 \Delta G_{pc} + 4\rho\, Y(X)\, (c_A{}^\circ - c_A{}^*)^2]\, r^3 + \frac{4}{3}\rho\, Y(X)\, r^5}$$
$$(13)$$

where $Y(X) = (\alpha_A + \alpha_B X^2 + 2\beta X)$. This formula is analogous to (5) and thus leads to a critical concentration gradient by setting both ΔG_N and $\partial \Delta G_N/\partial r$ to zero:

$$\boxed{\nabla c_{Ac} = \frac{\rho}{9\sigma_{pc}} \frac{\left[2\left|\Delta G_{pc}\right| - Y(X)\, (c_A{}^\circ - c_A{}^*)^2\right]^{3/2}}{[Y(X)]^{1/2}}} \qquad (14)$$

In practice, one of the problems in SSR should be to increase the critical thickness and therefore to reduce the critical gradient. The expression (14) for a ternary system compared to (6) for a binary system shows that, even in the most unfavourable case where $c_A° = c_A*$, the ternary critical gradient could be less than the binary one under certain conditions concerning $Y(X)$. Fig.5 represents $Y(X)$ for both cases $\beta<0$ and $\beta>0$. α_A and α_B are supposed to be positive because they represent the concavities of the curves A-C and B-C for amorphous binary systems in which no unmixing is expected (in the ternary system this stability condition becomes $\alpha_A\alpha_B-\beta^2>0$). One can notice that in each case, the ternary denominator $Y(X)$ may be more than the binary $\alpha_A=\alpha$ in a certain X-range which means $\nabla c_{AC}(\text{tern.})<\nabla c_C(\text{bin.})$ and therefore $\varepsilon_C(\text{tern.})>\varepsilon_C(\text{bin.})$. For instance, in a system with $\beta>0$ and $X>0$ (or $X< -2\beta^2/\alpha_B$), the third element theoretically improves the ability to amorphize the multilayer by increasing its critical thickness).

To have an idea of this theoretical improvement, a numerical application of the "ternary effect" has been carried out on the system NiCu-Zr setting A : Ni, B : Cu and C : Zr, with the composition $(Ni0.5Cu0.5)60Zr40$ under the assumptions of a regular solution. The Gibbs enthalpy of such a system is given by:

$$\Delta G/RT = c_A\ln c_A + c_B\ln c_B + (1-c_A-c_B)\ln(1-c_A-c_B) + lc_Ac_B + mc_A(1-c_A-c_B) + nc_B(1-c_A-c_B)$$

where l, m and n are the classical regular solutions parameter.

This leads to:

$\alpha_A/RT=1/c_A+1/c_C-2m$, $\alpha_A/RT=1/c_B+1/c_C-2m$ and $\beta/RT=1/c_C+l-m-n$. By the Miedema's model [9] we found : $l(NiCu)=+2.4$, $m(NiZr)= -38.46$ and $n(CuZr) = -17.63$ with the chosen composition $X=c_B/c_A=0.25$. For the most unfavourable case where the composition of the crystal compound would be on the diffusion path (i.e : $c_A*=c_A°$ and $c_B*=c_B°$). $\varepsilon_C(\text{tern.})/\varepsilon_C(\text{bin.}) = 1.42$ so that the effect should be observable.

Figure 5. Y[X] function of the second concentration derivatives α_A, α_B and β, of the amorphous phase free-energy curve versus the concentration gradient ratio $X = \nabla C_A/\nabla C_B$ (see equations 13 and 14)

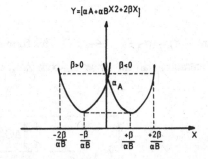

EXPERIMENTAL PROCEDURE

Ribbons of pure Zr and Cu alloyed nickel (50 at %) were obtained by rolling ingots to an average 130 μm thickness. The ribbons were then annealed to restore full ductility, co-wound in a spiral and indirectly rolled between two steel sheets according to the method described by Bordeaux and Yavari [10]. The evolution of thickness per layer of the resulting multilayer composites were monitored by cross-sectional SEM observations.X-ray diffraction spectra using Cu-K_α radiations were used to monitor the formation of the amorphous alloy phase after intensive cold-rolling and during subsequent annealing up to the beginning of crystallisation . The heat-treatments were all conducted in sealed tubes under secondary vacuum.

 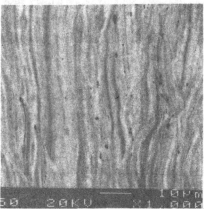

Figure 6. SEM secondary electron images of binary NiZr (left) and ternary (NiCu) Zr (right) multilayers after 50 rolling passes

RESULTS AND DISCUSSION

Figure 6 shows SEM micrographs of binary NiZr and ternary CuNiZr multilayers after 50 rolling passes following the previously described procedure [10]. The thickness per layer is seen to be of the same order. Figures 7 and 8 show x-ray diffraction patterns for the two composites in their as-prepared states corresponding to the micrographs and after 24 hours and 12 days of annealing at 285 ° C. The appearance and the relative intensity (with respect to the pure component Bragg peaks) of the amorphous background existing in the spectra of the two as-prepared samples also indicates that the two multilayers have equivalent starting structures prior to annealing.

The small framed peaks that appear at low angles in the as-prepared state are not definitely identified but become insignificant during annealing and amorphisation.

After 24 hours at 285 °C, the amorphous halo of the ternary alloy has become significantly more important than in the spectrum of the binary alloy which presents proportionally more important Bragg peaks of the remaining pure Ni ans Zr.

The faster progress of amorphisation in the ternary alloy is proof of faster diffusion kinetics that control amorphous layer growth. This is generally considered to depend on the diffusivity of the faster-diffusing species (Ni or Cu) across the amorphous layer into the remaining Zr layer [11] but such kinetics are very sensitive to presence of residual gases [11, 12]. The faster amorphisation of the ternary multilayer is consistent with a faster diffusivity in the ternary amorphous interlayers as the latter usually increases exponentially with decreasing melting temperatures and Cu addition lowers the liquidus of NiZr alloys. However these results do not prove an increased stability of the ternary alloy in a concentration gradient.

Figure 7 and 8 also show x-ray diffraction patterns for the two multilayers after 12 days of annealing at 285°C. This annealing duration was selected because it corresponds to the onset of crystallisation at 285° C. Both spectra show growth of new Bragg peaks corresponding to the formation of intermetallic compound crystals but at this point nearly all of the initially unmixed layers have disappeared in the ternary alloy as indicated by the disappearance of the Zr and CuNi peaks while significant amounts persists in the binary alloy.

Since the initial thickness per layer of the unmixed layers in the two multilayers were similar, the occurence of crystallisation in the binary multilayer while unmixed components persists indicates that the critical thickness as defined by the model is thinner for the binary amorphous layer than in the ternary alloy. In the latter the amorphous phase apparently succeeds in consuming all of the thickness of the initial un-mixed layers and thus attains a higher thickness before the onset of crystallisation.

While further evidence preferably from cross-sectional TEM observations on vapor-deposited multilayers with more regular morphologies is desirable, the present results seem to indicate that the onset of crystallisation occurs under higher concentration gradients in the binary amorphous interlayers. It can therefore be concluded that consistent with the developped thermodynamic model, addition of appropriately chosen third elements reduce further the thermodynamic driving force for compound formation in an amorphous phase under sharp concentration gradients.

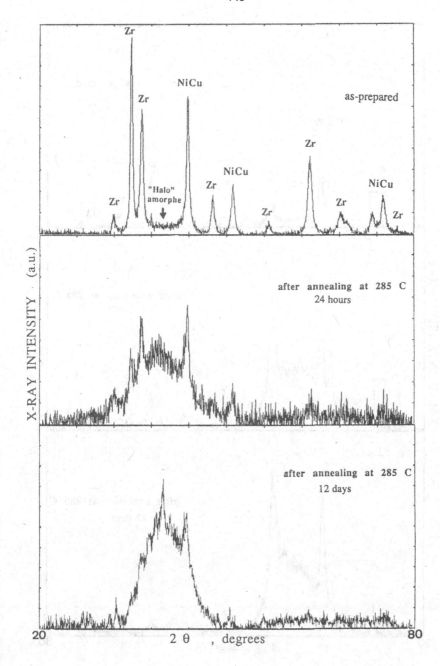

Figure.7 X-ray diffraction patterns of ternary (NiCu) Zr multilayer in the as-prepared state and after annealing for 24 hours and 12 days at 285 C

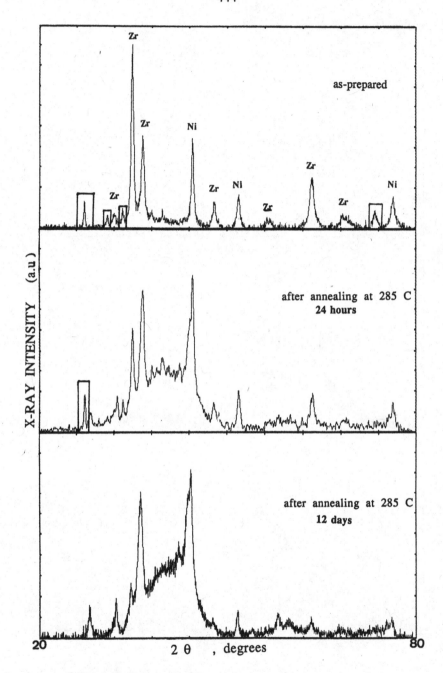

Figure.8 X-ray diffraction patterns of binary NiZr multilayer in the as-prepared state and after annealing for 24 hours and 12 days at 285 C

REFERENCES

[1] Schwartz R.B and Johnson W.L., Phys. Rev. Lett., **51**, 415 (1983)

[2] Gösele U.and Tu K.N., J. Appl. Phys., **66**, 2619 (1989)

[3] Meng W.C., Nieh C.W. and Johnson W.L., Appl. Phys. Lett., **51**, 1693 (1987)

[4] Highmore R.J, Greer A.L., Leake J.A. and Evetts J.E., Mat. Lett., **6**, 401 (1988)

[5] Desré P.J. and Yavari A.R., Phys. Rev. Lett., **64**, 1533 (1990)

[6] Cahn J.W.and Hilliard J.E., J. Chem. Phys., **28**, 258 (1958)

[7] Kirkaldy J.S., J. Can. Phys., **36**, 899 (1958)

[8] Kirkaldy J.S, lane J.E. and Mason G.R., J. Can. Phys., **41**, 2174 (1963)

[9] Miedema A.R, de Boer F.R. and Boom R., Physica B, **103**, 67 (1981)

[10] Bordeaux F. and Yavari A.R., Z. Metallkde, **81**, 130 (1990)

[11] Yavari A.R., Desré P.J. and Bordeaux F., J. Physique, C-4, 23 (1990)

[12] Bordeaux F. and Yavari A.R., Bochu B. and Givord D., in "Basic Features of the Glassy State" eds : Colmenero J. and Alegria A., World Science, pp 554-559, (1990)

THERMODYNAMIC ASPECTS OF AMORPHIZATION AND DEFORMATION IN Cu-Ti-TiH$_2$ MIXTURES

MARCELLO BARICCO, LIVIO BATTEZZATI
Dipartimento di Chimica Inorganica, Chimica Fisica
e Chimica dei Materiali - Università di Torino
INFM - Unità di ricerca di Torino
Via P.Giuria, 9   I-10125 Torino (Italy)

GIORGIO COCCO, ISABELLA SOLETTA
Dipartimento di Chimica - Università di Sassari
Via Vienna, 2 I-07100 Sassari (Italy)

STEFANO ENZO
Dipartimento di Chimica Fisica - Università di Venezia
D.D. 2137 I-30123 Venezia (Italy)

ABSTRACT

Mechanichal alloying of Cu, Ti and TiH$_2$ and grinding of γ-CuTi in H$_2$ atmosphere lead to a mixture of a hydrogenated amorphous phase and TiH$_2$. From the heat of transformation of metastable phases, their enthalpy of formation is determined along a section of the Cu-Ti-H phase diagram. The free energy of formation, estimated at 300 K, shows a driving force for amorphization at low hydrogen concentration. A phase separation is predicted for higher hydrogen contents. The effect of deformation of pure elements on amorphization is discussed.

INTRODUCTION

The Cu-Ti system is known to form amorphous phase by rapid solidification [1] and by ball milling [2] in a wide composition range. In addition, many hydrogenated amorphous and metastable crystalline phases were reported to form in this system by means of conventional electrolytic and pressure loading [3,4]. Aim of this paper is to describe, mainly from a thermodynamic point of view, the role of hydrogen in solid state amorphization reactions during ball milling of Cu-Ti. As H_2 absorption is known to be one of the routes to solid state amorphization, it is of interest to combine it with mechanical alloying. The study of the effect of hydrogen may be important also for application purposes of mechanical alloying. In fact, Ti from industrial producers contains serious amounts of it as impurity. Moreover, it evolves from the alloy on heating, so its use as alloying element in intermetallics has been suggested as a method for grain refinement [3].

From the fundamental point of view, the Cu-Ti system appears suitable for this study in two respects. It has been recently assessed and several data on thermodynamic properties are available [5]. Also, the free energy of mixing of the liquid phase and the free energy of formation of the interme- tallic compounds are only moderately negative, so even small changes due to the influence of a third element could be important in varying the stability of the phases.

EXPERIMENTAL

Mechanical alloying of various mixtures of Cu, Ti and TiH_2 were performed in a Spex Mixer-Mill 8000 in purified argon atmosphere. Grinding of γ-CuTi was performed in H_2 atmosphere. The Ti used was either high purity or commercial purity grade. Hardened steel or copper-berillium vials were used. The struc- ture of the powders at various stages of milling was checked by X-ray diffraction, using Cu $K\alpha$ radiation. The thermal stability of the powders was analyzed with a Perkin Elmer DSC7. Open gold pans were used for analysis up to 725°C and crimped aluminum pans up to 600°C in order to reduce gas evolution on heating.

RESULTS AND DISCUSSION

Hydrogen was loaded through various routes in Cu-Ti alloys by ball milling. In a first experiment, crystalline .-CuTi was ground in H_2 atmosphere for 16 hours. An amorphous hydrogenat- ed phase and TiH_2 with traces of unreacted .-CuTi are detected in fig.1, curve b from XRD pattern. Furthermore, the milling of an amorphous $Cu_{60}Ti_{40}$ powder with TiH_2 leads to the same mixture (fig. 1, curve c). Hydrogenated samples were prepared by milling for 16 hours Cu, Ti and TiH_2 according to the composition series $Cu_{45}[(1-x)Ti+xTiH_2]_{55}$ ($0 \leq x \leq 1$). A single amorphous phase was obtained for x < 0.05 the XRD pattern of which is shown in fig. 1, curve a, where the most intense reflection due to TiH_2 is barely perceivable. At higher hydro- gen content the amorphous phase coexists with unreacted TiH_2 [6]. Similar results occurred for some $Cu_x[0.86Ti+0.14TiH_2]_{1-x}$ compositions, which resulted from the use of commercial con- taminated Ti [6].

Fig. 2 shows the DSC traces of selected samples for which

an exothermal peak associated to the crystallization of the amorphous fraction in the powders is observed. At higher temperature (usually above 400°C) H_2 evolves from the alloy causing a strong endothermal peak [7], the beginning of which is evident in fig. 2. Curve a in fig. 2, related to a fully amorphous sample containing about 7 % hydrogen, shows a well defined crystallization peak at ≈410°C. In order to delay the hydrogen evolution and prevent the overlapping with crystallization, crimped pans were used to obtain the DSC trace shown in fig. 2 (curve b) for $Cu_{45}[0.95Ti + 0.05TiH_2]_{55}$. The crystallization occurs in two temperature ranges. XRD were collected after each transformation for phase identification.

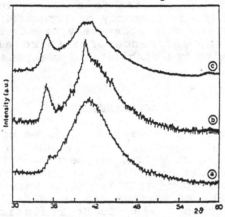

Figure 1. XRD patterns of amorphous and TiH_2 mixtures obtained from: curve a, $Cu_{45}[0.95Ti + 0.05TiH_2]_{55}$ milled for 16 hours; curve b, .-CuTi milled for 16 hours in hydrogen atmosphere; curve c, amorphous $Cu_{60}Ti_{40}$ milled with TiH_2 for 25 hours. Curve b shows a peak of .-CuTi left unreacted.

The first broad signal (≈200 °C) is typical of heavily hydrogenated alloys and gives Cu and TiH_2. The second one (≈450 °C) gives hydrogenated ɣ-CuTi and is located at a temperature close to the H_2 evolution signal so that measurements of the total heat of crystallization is prevented. The crystallization of the amorphous phase that was produced by H_2 absorption in ɣ-CuTi is shown in fig. 2, curve c. A lower heating rate was employed with respect to curve a and b to shift the crystallization peak to a lower temperature range. Again hydrogenated crystalline ɣ-CuTi is obtained coexisting with TiH_2.
As a summary, the combination of X-ray and DSC analyses shows that different solid state reactions at selected composition in Cu-Ti-H, all produce the same phase mixture which transform similarly on heating. Therefore, milling appears as a method to load hydrogen into intermetallics and amorphous phases.
Considering the loading of hydrogen in Cu-Ti alloys by means of other techniques, different results were obtained. $Cu_{50}Ti_{50}$ amorphous ribbons were electolytically charged with hydrogen to give amorphous hydrides with composi-

tions $TiCuH_{1.68}$ and $TiCuH_{0.86}$, where a phase separation in amorphous Cu and TiH_x rich regions was observed [8]. Amorphous $Cu_{50}Ti_{50}$ ribbons have also been charged by exposure to a pure hydrogen atmosphere at room temperature. Fully amorphous $TiCuH_{0.89}$ [4] and $TiCuH_{1.41}$ [9] were obtained. Also crystalline Ti_2Cu and χ-CuTi can absorb hydrogen with formation of metastable crystalline hydrides [9].

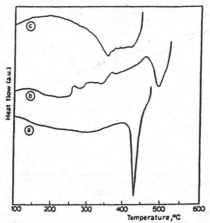

Figure 2. DSC traces of: curve a, amorphous $Cu_{60}Ti_{40}$ containig about 7 % at of hydrogen; curve b, $Cu_{45}[0.95Ti + 0.05TiH_2]_{55}$ milled for 16 hours; curve a, χ-CuTi milled for 16 hours in hydrogen atmosphere. Heating rate of 30 K/min for curve a and b and of 5 K/min for curve c.

Comparing the hydrogen charging of Cu-Ti alloys by mechanical alloying to electrochemical and gas phase absorption methods, the main difference is the formation of a mixture of hydrogenated amorphous and TiH_2 instead of a separation in two amorphous phases. In order to understand these reactions and establish a phase hierarchy, a thermodynamic analysis of a portion of the ternary diagram is performed in the following.

Owing to the composition of most of the ternary amorphous and crystalline phases reported in the literature and prepared in this work, the thermodynamic behaviour of the ternary system will be discussed around the section from $Cu_{66}Ti_{34}$ to TiH_2. The reference state was taken as the pure elements in their stable state at room temperature, i.e. hcp Ti, fcc Cu and gaseous H_2.

The enthalpies of formation of the various crystalline and amorphous phases are listed in Table I and shown in Fig. 3. The heat of formation of intermetallic Cu-Ti compounds were calculated from the assessment of the phase diagram [10].

As an example, the reactions used to calculate the thermodynamic data for the crystalline CuTiH are given:

$$TiCu + \tfrac{1}{2} H_2 = TiCuH \qquad (1)$$

$$Ti + Cu = TiCu \qquad (2)$$

$$Ti + Cu + \tfrac{1}{2} H_2 = TiCuH \qquad (3)$$

For reaction (1), the enthalpy change is estimated to be -37.5 kJ/mol [4]. A value of -18.8 kJ/mol is given for reaction (2) [11]. The enthalpy change of the overall reaction (3) turns out to be -56.3 kJ/mol (-18.8 kJ/g·at).

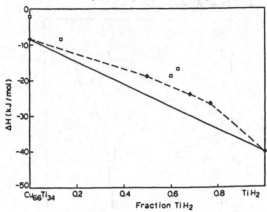

Figure 3. Enthalpy of formation as a function of TiH$_2$ fraction along a section of the ternary system Cu-Ti-H. Black squares: crystalline compounds; open squares: amorphous alloys.

The heat of formation of the Cu$_{66}$Ti$_{34}$ amorphous phase was calculated from the difference between the heat of formation of the crystalline phases and the heat of crystallization (Table I). Analogously, the heat of formation of amorphous and crystalline hydrides was derived from the heat of transformation to the stable phases. For instance, the amorphous phase obtained from the mechanical alloying of pure Cu with Ti containing 14% of TiH$_2$ (Ti$_{36}$Cu$_{57}$H$_7$) gives a heat of crystallization of -3.2 kJ/mol. Considering that the main crystallization products are TiH$_2$ and a mixture of intermetallics, its heat of formation turns out to be -8.5 kJ/g·at.

In order to get an estimate of the free energy of formation of the different phases, the entropy of formation is necessary. For crystalline Cu$_{66}$Ti$_{44}$ a value of -1.40 J/g·at K is obtained from the phase diagram calculation [10]. For metastable crystalline hydrides, an entropy change of -18.8 J/g·at is reported for reaction (3) [4]. Considering analogous entropy changes due to absorption of gaseous hydrogen in crystalline intermetallics, the entropy of formation for TiCuH$_{2.71}$ and TiCiH$_{2.10}$ is obtained as -20 J/g·at K and -22 J/g·at K respectively. The entropy of formation of TiH$_2$ is tabulated as -44 J/mol K [13]. As concerns the amorphous phases, an entropy of crystallization, ΔS_x, can be evaluated from the equation:

$$\Delta s_x = \Delta s_f - \Delta \overline{c}_p \ln(T_f/T_x) \qquad (5)$$

TABLE I
Enthalpies of formation and transformation
of various phases in the Cu-Ti-H system.

Composition	x_{Cu}	x_{Ti}	x_H	ΔH_{for} (kJ/g·at)	ΔH_T (kJ/g·at)	Ref
c-Cu$_{66}$Ti$_{34}$	.66	.34	0	-8.3		10
a-Cu$_{66}$Ti$_{34}$	.66	.34	0	-2.1	-6.2	This work
a-Cu$_{60}$Ti$_{40}$(H)	.57	.36	.07	-8.5	-3.2	This work
c-CuTiH$_{0.96}$	.34	.34	.32	-18.8	-2.7	4,12
a-CuTiH$_{1.40}$	.29	.29	.42	-18.9	-5.5	9
a-CuTiH$_{1.68}$	.27	.27	.46	-17.0	-8.1	8
c-Ti$_2$CuH$_{2.10}$	.20	.39	.41	-24.0	-2.6	9
c-Ti$_2$CuH$_{2.71}$	.17	.35	.48	-26.5	-3.4	9
c-TiH$_2$	0	.34	.66	-40.0		13

ΔH_{for} = Enthalpy of formation
ΔH_T = Enthalpy of transformation from metastable to stable phases

where ΔS_f is the entropy of fusion, $\Delta \overline{C_p}$ is the average specific heat difference between liquid and solid phases and T_f and T_x are the temperature of fusion and crystallization respectively. Data are available for amorphous Cu$_{66}$Ti$_{34}$ [14] for which ΔS_x is calculated as 3.9 J/g·at K. Such a value implies that the glassy phase exhibit a degree of short range order which is fully restored in the crystals. Owing to the impossibility of reaching the melting point, the same value of ΔS_x was used for hydrogenated amorphous phases.

A free energy picture at 300 K is shown in fig. 4. All the hydrogenated phases are metastable with respect to Cu-Ti intermetallics and TiH$_2$. The free energy for the glass is consistent with an amorphous in amorphous phase separation as reported for samples loaded electrolytically and from the gas phase [8]. Considering our starting mixture for milling experiments (Cu, Ti and TiH$_2$) a driving force for amorphization exists at low hydrogen content which disappears in the center of the diagram due to the free enthalpy rise, so that TiH$_2$ remains unreacted. A different entropy scenario can be envisaged for glassy alloys containing hydrogen, where a variety of interstitial-like sites may be available to accomodate its small atoms. In this event, the entropy of amorphous phase would approach the ideal entropy of mixing but the resulting free energy would be more negative than those of the crystals of the same composition. Therefore, it is concluded that hydrogen in amorphous Cu-Ti must be largely confined to the same sites it occupies in crystalline hydrides, i.e. Ti$_4$ tetraedra.

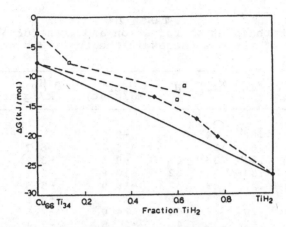

Figure 4. Free energy of formation at 300 K as a function of TiH_2 fraction along a section of the ternary system Cu-Ti-H. Black squares: crystalline compounds; open squares: amorphous alloys.

The reference state for figs. 3 and 4 was taken as the stable pure elements. In recent years there have been various reports that the elements may store a considerable amount of energy, more than ten percent of the heat of fusion, during milling [15]. On accounting for these data, the thermodynamic picture, just proposed, would be significantly modified. For Cu, the amount of heat released after extensive milling was reported as -2.8 kJ/g·at [16] and -5.8 kJ/g·at [17]. Copious data, previously obtained for pure copper rolled to various thickness, showed that the heat released during recrystallization gives a well defined DSC peak, around 250°C, increases as a function of deformation and levels off to 50 J/g·at for thickness reductions larger than 65 % [18]. The difference between this and the above figures for heat release, although scattered, is striking and must be related to a difference in the deformation mechanism by sheet rolling and powder milling. We milled pure copper for time intervals up to 12 hours. A Warren-Averbach analysis of the (111) and (222) reflections shows that the effective crystallite size is about 30 nm after 1 hour and 20 nm after 12 hours milling, whereas the effective crystallite size in the heavily rolled sheets is of the order of 50 nm. DSC traces for milled powders display broad exotherms, but, as the powders contained traces of Cu oxides, the heat evolved cannot be correlated with the data obtained from X-ray analysis.
    Considering these differences the problem of deformation of pure elements remains open. However, it appears that there is very little effect of deformation of the pure elements on the alloying mechanism. In fact, alloy formation has been achieved both by milling elemental mixtures and by rolling alternate elemental sheets in the same systems being driven by chemical effects expressed by the heat of mixing and by

interdiffusion as previously pointed out for the Ni-Ti system [18].

ACKNOWLEDGMENTS

Work performed for Progetto Finalizzato C.N.R. "Materiali Speciali per Tecnologie Avanzate" under contracts n. 89.00623.68 (Univ. Sassari) and n. 89.00605.68 (Univ. Torino).

REFERENCES

1. K.H.J.Buschow, Acta Met., 1983, **31**, 155
2. C.Politis, W.L.Johnson, J. Appl. Phys., 1986, **60**, 1147
3. D.Manzel, A.Niklas, U.Koster, Mat. Sci. and Eng., 1990, **133A**, 312
4. A.J.Maeland in Hydrides for Energy Storage eds. A.F.Andersen and A.J.Maeland, Pergamont Press, Oxford, 1978, p. 447
5. J.L.Murray, Bull. of Alloy Phase Diag., 1983, **4**, 81
6. G.Cocco, Proc. Intern. Symp. on Mechanical Alloying, Kyoto, 1991, in press
7. M.Baricco, L.Battezzati, I.Soletta, L.Schiffini, N.Cowlam, Mat. Sci. and Eng., 1991, **134**, 1398; L. Battezzati, M. Baricco, S. Enzo, G. Cocco, I. Soletta and L. Schiffini, Proc. Intern. Symp. on Mechanical Alloying, Kyoto, 1991, in press
8. B.Rodmacq, A.Chamberod, Phys. Rev. B, 1988, **38**, 1116
9. J.S.Cantrell, R.C.Bowman, G.Bambakidis, in Hydrogen in Disordered and Amorphous Solids, eds. G.Bambakidis and R.C.Bowman, Plenum Press, New York, 1986, p. 185
10. N.Saunders, CALPHAD, 1985, **4**, 297
11. M.Arita, R.Kinaka, M.Someno, Metall. Trans. A, 1979, **10A**, 529
12. R.C.Bowman, R.J.Furlan, J.S.Cantrell, A.J.Maeland, J. Appl. Phys., 1984, **56**, 3362
13. R.C.Weast, Handbook of Chemistry and Physics, CRC Press, Boca Raton, Florida (USA), 1983
14. L.Battezzati, M.Baricco, G.Riontino, I.Soletta, J. de Phys., 1990, Tome 51, Coll. C4, Suppl. 14, C4-79
15. H.J.Fecht, E.Hellstern, Z.Fu, W.L.Johnson, Metall. Trans. A, 1990, **21A**, 2333
16. P.H.Shingu, K.N.Ishihara, J.Kuyama, Proc. od 34th Jap. Conf. on Materials Research, The Jap. Soc. of Mat. Sci., Kyoto, 1991, p. 19
17. J.Eckert, J.C.Holzer, W.L.Johnson, Proc. Intern. Symp. on Mechanical alloying, Kyoto, 1991, in press
18. A.Lucci, M.Tamanini, L.Battezzati, G.Venturello, J. Therm. Anal., 1978, **14**, 93
19. G.Cocco, S.Enzo, L.Schiffini, L.Battezzati, Mat. Sci. and Eng., 1988, **97**, 43

ON THE ENERGY TRANSFER MECHANISM IN THE BALL MILLING POWDER PROCESSING

E.Paradiso, A.Iasonna, M.Magini, S.Martelli and F.Padella
Amorphous Materials Project, PCM-INN-ENEA, CRE-Casaccia, Via
Anguillarese, Rome, Italy.

ABSTRACT

Milling conditions are of fundamental importance in defining the energy transfer process and hence the nature of the final product when milling a given system starting from powders. In a planetary mill, the energy transfer is essentially governed by the hit of the ball against the vial wall. In order to analyse the hit mechanism, the velocities of the balls inside the mill have been estimated and reproduced in "free fall" experiments, in which the yield of the impact has been measured. From free fall experiments with bare balls and with powder coated balls taken out from the mill at different milling times, it follows that , at least in the early stages of milling, the hit is totally anelastic and that the energy of the ball is totally dissipated onto the powder. The outcome of the free fall experiments can be used as a guide in understanding the energy transfer per hit realized in a planetary mill.

INTRODUCTION

In the ball milling process, an energy transfer from the milling tools to the milled powder is realized. It is expected that the milling conditions should strongly affect the energy transfer and, consequently, the end products of the process. The milling conditions can vary by changing the milling device. The most common devices quoted in literature are attritors, first used to prepare ODS alloys; vibrating and planetary mills, more recently employed in investigating the Mechanical Alloying (MA) process. The milling conditions can also vary, for the same apparatus, by varying the parameters affecting the energy transfer like the mass of the ball and the rotation (or vibration) speed of the mill.
The mechanism of energy transfer from the milling tools to the milled powder is certainly different for different

apparatuses. In recent years, some attempts of modelisation have been done. Experiments and computer model simulation have been carried out by Davis et al. [1] in order to extimate the temperature rise. Maurice and Courtney [2] attempted to define the MA basic geometry from wich impact time, temperature increase and strain rate could be approximately inferred. Hashimoto and Watanabe [3] simulated the ball movement in an one-dimensional vibratory mill and evaluated the energy consumption during the ball impact. A model of energy transfer in a vibrating frame has been proposed by Chen et al. [4].

In a planetary mill, as we have described in a previous paper [5], the energy transfer is essentially governed by the collision of the balls charged with powder against the vial walls as depicted in figure 1. It would be, therefore, of extreme importance to analyse the hit mechanism in order to gain information on the energy transfer process. It is hard to conceive this kind of analysis directly in a vial in motion. It is possible, however, to estimate the velocities of the balls launched inside the vial [5] and reproduce them in free fall experiments in which the yield of the impacts can be carefully analysed.

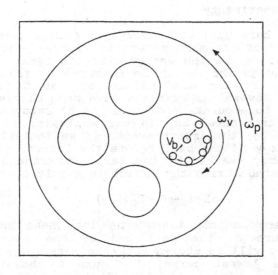

Figure 1. Scheme of a planetary ball mill W_p and W_v are non independent angular velocities of the ball mill and of the vial respectively.

EXPERIMENTAL

The ball milling processes have been carried out by a planetary ball mill (Fritsch "Pulverisette 5") provided with four cylindrical 250 cc tempered steel vials, of radius R_v of 37 mm, mounted on a supporting plate of radius R_p of 122.2 mm. Balls of the same material as the vials with different diameters have been used. The rotation speed W_p of the planetary mill's plate can be varied by a potentiometer and measured by stroboscopic device. The dependent W_v (angular speed of the vial) is determined by the constructor's relation $W_v = -1,25 \, W_p$.

For the "free fall" experiments (see below), a mixture of Ti:Al=1:1 (atomic ratio) was ball milled in order to allow the balls to be covered with a thin powder layer as usually occurring in the milling process. Powder coated balls were then taken out at increasing milling times and properly classified.

Mechanical alloying experiments on both $Pd_{80}Si_{20}$ and $Ti_{40}Al_{60}$ systems were carried out as previously described [5].

RESULTS AND DISCUSSION

"Free fall" experiments

If we let one bare ball to freely fall from a given height on a flat surface of the same material, we can define the yield of the rebound, η, as the energy fraction remaining with the ball after the impact. η can be measured by the ratio h'/h, where h and h' are the height of launch and rebound respectively. Several experiments have been carried out using a videocamera to accurately measure the h' rebounds heights.

What we are interested in, in our analysis, is the energy dissipated during the impact event. If we indicate by E_o the potential energy of the ball before the launch, ηE_o gives the potential energy reached by the ball after the jump and the energy dissipated during the impact is simply given by

$$\Delta E = E_o - \eta E_o = E_o(1-\eta) \qquad (1)$$

This energy mainly transforms into heat and increases the temperature of the ball and of the plate. A minor fraction of it will be stored onto the material as structural desorder. The lowest curve of figure 2 shows the energy dissipation per hit realized with bare balls. Free fall experiments have been then performed with balls charged with powder taken out at different times during the milling process. It is well known that in the early milling stages (ranging between about ten minutes and about one hour depending on the system under investigation and on the milling conditions) the balls, as well as the vial walls, become covered by a composite thin powder layer. At one hour of milling practically all the powder is spread out over the balls. The rebounds of the covered balls are much more irregular than those of bare balls. Notwithstanding, an

average rebound yield has been obtained for each of the balls
used in the experiments. The corresponding ΔE/hit dissipated
during the impact are given in the same figure 2. It can be
seen from figure 2 that in the early stages of milling (i.e.
the points at 1 hour) the energy dissipation per hit
approaches the boundary limit of the total dissipation energy
which is realised when the ball does not rebound at all and
remains "glued" after the impact. The kinetic energy is then
totally dissipated in the impact and the upper straight line
represents the total dissipation boundary condition. On the
contrary at 30 hours of milling the energy dissipation
approaches that of the bare balls. We have to consider that
the results shown on figure 2, for charged balls, have been
obtained using a flat surface plate not covered with a thin
powder layer. In the real milling process the vial walls are
also covered with a thin powder layer with the consequence
that the real η value should be even lower. It follows, then,
that it is reasonable to assume that, at least in the early
stage of milling, the hit is only anelastic and that the
conditions of total dissipation of the kinetic energy are
realised.

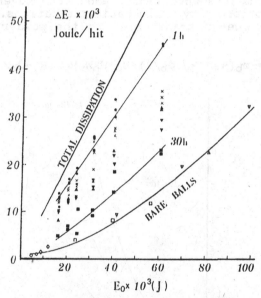

Figure 2. Energy dissipation during impact, ΔE, plotted
versus E_O (see text). The open symbols of the lowest curve
refer to bare balls of different diameters:○ = 5 mm,◇ = 6
mm,□ = 10 mm, ▽ = 12 mm,△ = 15 mm. The full symbols refer to
experiments with balls charged with powder at different
milling times: ●, x ,▼,▲, ■ = 1,2,5,15,30 hours of milling.
At each milling time one symbol has been used for each of the
balls employed in the free fall experiments. The upper
straight line represents a total dissipation energy curve
(see text). The lowest curve, as well as those at 1 and 30
hours of milling, have been drawn through the experimental
points for clarity.

458

Just comparing the two limiting curves of figure 2 (bare balls and total dissipation) one argues that, for any given E_O, most of the kinetic energy dissipated per hit goes to the powder. Moreover, it is obvious that when the ball is coated the energy dissipated into the ball, for any given E_O, is much lower than that shown in figure 2 for bare balls. The conclusion is that it is a good approximation to consider that, at least in the early milling stages, all the energy of the ball is totally dissipated onto the powder and the ball does not rebound any more.

Hit mechanism in a planetary mill

The condition of "total dissipation", in a planetary ball mill, begins from the moment at which the balls and the vial's wall are well covered with a thin powder layer. Starting from this condition, the ball, after each impact, will practically remain glued to the wall, accelerated by the vial and launched again against the opposite wall at a given moment due to the composition of inertial forces acting on it. In other words from a given moment on, the trajectory of the ball becomes identifiable and its movement can be properly described by kinematic equations. In these conditions the energy transferred to the powder in a single hit is [5]:

$$\Delta E = -m_b[W_v^3(R_v-d_b/2)/W_p+W_pW_vR_p](R_v-d_b/2) \qquad (2)$$

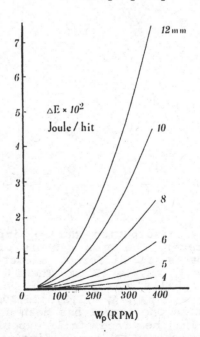

Figure 3. Energy dissipated per hit plotted versus the rotation speed of the planetary mill expressed in rounds per minute. The number near each curve is the ball diameter in millimeters.

where m_b and d_b are the mass and the diameter of the ball, R_v, W_v, R_p and W_p are the radius and the rotation speed of the vial and of the plate respectively. Figure 3 shows the energy released to the powder in a single hit in conditions of total dissipation, calculated from the previous equation.

Test of the hit model

Equation (1) can be simplified to the relation:

$$\Delta E \simeq f(d_b^3, W_p^2) \tag{3}$$

that, keeping constant the rotation speed of the milling apparatus, becomes

$$\Delta E \simeq K d_b^3 \tag{4}$$

It is, therefore, possible to change the energy transferred to the powder in a single hit, by simply varying the ball diameters.

Fig. 4-a shows the x-ray patterns of the $Ti_{40}Al_{60}$ system at various milling times, milled at $W_p = 250$ RPM and with balls of 12 mm diameter. The formation of the TiAl intermetallic phase is evident from the patterns' sequence (the little peak at $2\theta \approx 16.3$ at 25 hours is mainly due to a very little quantity of TiH_2, that is present as impurity in the starting titanium). Figure 4-b shows the same system milled at the same angular velocity, with balls of 5 mm of diameter. In this case a reaction path towards a full

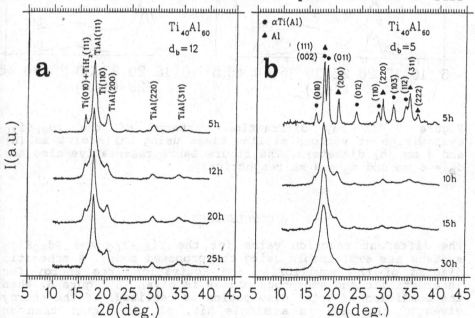

Figure 4. X ray diffraction patterns of the $Ti_{40}Al_{60}$ composition at various milling times milled at $W_p = 250$ RPM and $d_b = 12$ mm (a) and 5 mm (b).

amorphous phase is clearly recognizable.

Again different reaction paths have been verified by equation (4) by milling the $Pd_{80}Si_{20}$ system. In this case the milling velocity has been kept constant at 280 RPM and balls of 4,5,6 and 8 mm have been used. A full amorphous path is obtained by both d_b = 4 mm and d_b = 5 mm while experiments with d_b = 6 mm and 8 mm clearly show the formation of the Pd_3Si intermetallic phase. Figure 5 shows the x-ray patterns of the $Pd_{80}Si_{20}$ system milled under the two different conditions.

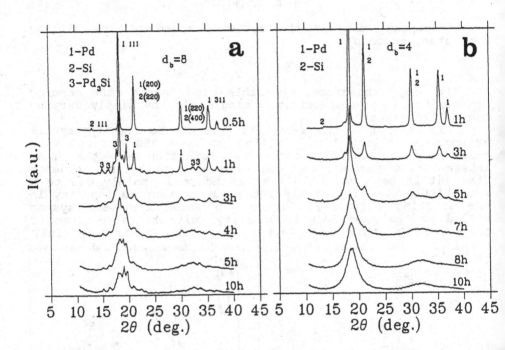

Figure 5. X ray diffraction patterns of the $Pd_{80}Si_{20}$ composition at various milling times using balls of 8 mm (a) and 4 mm (b) diameters. The figure is representative also for d_b = 6 mm and d_b = 4 mm respectively.

CONCLUSIONS

The different reaction paths for the $Ti_{40}Al_{60}$ and $Pd_{80}Si_{20}$ systems are explainable using the proposed model. A schematic diagram of the equilibrium and activation free energy for intermetallic and amorphous phases is given in figure 6. When the condition of total dissipation is realized, if the energy given by the ball in a single hit, ΔE, is higher than an intermetallic phase's activation free energy, the system evolves towards the intermetallic compound, while when ΔE is lower than the previous threshold but higher than the amorphous phase activation free energy, the system goes towards the amorphous phase.

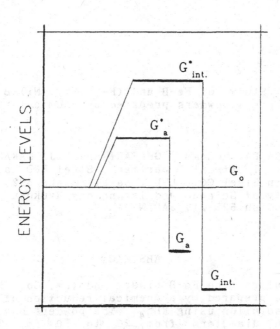

Figure 6. Schematic free energy diagram. G_0, G_a and G_{int} are the free energies of the initial, the amorphous and the intermetallic states. G_{int}^* and G_a^* are the activation energies to form the intermetallic compound and the amorphous phase respectively.

REFERENCES

1. R.M.Davis, B.Mc Dermott and C.C.Koch, Metall. Trans. 19A(1988),2867.
2. D.R.Maurice and T.H.Courtney, Metall. Trans. 21A(1990),289.
3. H.Hashimoto and R.Watanabe, Mater. Trans. 31(1990),219.
4. Y.Chen, R.Le Hazif and G.Martin, Colloq. Fis. C4(1990),273
5. N.Burgio, A.Iasonna, M.Magini, S.Martelli and F.Padella Nuovo Cimento, section D, 13(1991),459.

Mössbauer Study of Fe-B and (Fe, Co, Ni)-B ultra-fine amorphous powders prepared by chemical reduction

JUNJI SAIDA, MOHAMMAD GHAFARI and YOJI NAKAMURA*
Surface Treatment Department, Steel R&D Laboratories,
Nisshin Steel CO., LTD., Sakai, Osaka 592, JAPAN
*Faculty of Science and Technology Ryukoku University,
Seta, Otsu 520-21, JAPAN

ABSTRACT

Amorphous ultra-fine Fe-B binary and (Fe, Co, Ni)-B ternary powders were prepared by a chemical reduction of metal ions in an aqueous solution using KBH_4. The powders have a spherical shape with diameters from 20 to 50 nm. The chemical composition of ultra-fine powders depends on the mixed molar ratio of KBH_4 to metal ions in the Fe-B binary system. The Mössbauer results show that depending on the molar ratio, two kinds of distribution of internal magnetic field $P(B_{hf})$ exist. From Mössbauer, X-ray and inductively coupled plasma (ICP) spectroscopy measurements we propose the following reaction process for the formation of ultra-fine Fe-B powders;

(1) $4Fe^{2+} + 2BH_4^- + 6OH^- \rightarrow 2Fe_2B + 6H_2O + H_2$

(2) $4Fe^{2+} + 2BH_4^- + 7OH^- \rightarrow Fe_3B + Fe + BO_2^- + 5H_2O + 5/2H_2$

INTRODUCTION

Amorphous materials are usually prepared by (1) chemical deposition (electroless plating), (2) rapid quenching (melt-spinning), (3) vapor-deposition technique (sputtering), (4) irradiation and (5) solid-state reaction. Recently, amorphous materials have been prepared by chemical reaction processes [1-4], where amorphous materials are formed directly from atomic or ionic states. It is of interest to investigate the physical properties of these new materials. These materials are different from well-known amorphous alloys prepared by other techniques. In present paper, we report on results of Mössbauer, ICP and X-ray measurements. The Mössbauer results show an anomalous increase in the average magnetic hyperfine field, $\bar{B}_{hf}$, with decrease of the mixed molar ratio, M_R, below 10.0. Here, M_R is given by $M_R=M_{KBH4}/M_{Metal}$; M_{KBH4}: mol concentration of KBH_4; M_{Metal} : mol concentration of metal ions. The deduced $\bar{B}_{hf}$ are in agreement with those of the boron content of about 25 at.% for M_R below 9.0 and 33 at.% in the M_R range from 9.0 to 20.0. Similar results were observed by chemical analysis. Our observation can be explained in the context of two different chemical reaction processes in amorphous Fe-B particles;

(1) $4Fe^{2+} + 2BH_4^- + 6OH^- \longrightarrow 2Fe_2B + 6H_2O + H_2$

(2) $4Fe^{2+} + 2BH_4^- + 7OH^- \longrightarrow Fe_3B + Fe + BO_2^- + 5H_2O + 5/2H_2$

The reaction (1) is the dominant process for the M_R range from 10.0 to 20.0, while the reaction (2) takes places for $M_R<10.0$.

EXPERIMENTAL METHODS

Potassium borohydride (KBH_4) is well-known as a reduction agent in electroless plating [5]. This agent has an important property as a strong agent to metal ions: the decomposed B element reacts on metal elements during the reduction process. We expect that the materials are in a metastable state because the reaction takes place rapidly under the glass transition temperatures. A typical preparation process is shown in figure 1. A 1 mol/l KBH_4 solution was mixed with metal ion-containing solution in air at room temperature. The mixed molar ratio of the KBH_4 concentration to metal ion concentration, M_R, is an important factor for the formation of powders. We varied M_R from 1.0 to 20.0 in the Fe-B system in order to investigate the influence of the reaction condition on the formation of powders. After 600 s stirring, the block products were immediately collected on a filter, washed with distilled water followed by acetone rinse and dried in air.
The composition of powders was determined by inductively coupled plasma spectrometry. The structure was examined by X-ray diffractometry with Cu Kα radiation. Crystallization temperature was measured with a heating rate of 20 K/min by differential scanning calorimetry (DSC). Mössbauer measurements were performed using a conventional constant acceleration-type spectrometer with a moving source of Co^{57} in a Pt matrix. The isomer shift is given relative to α-Fe at room temperature.

RESULTS AND DISCUSSION

(1) Structure, composition and morphology

Experimental Procedure

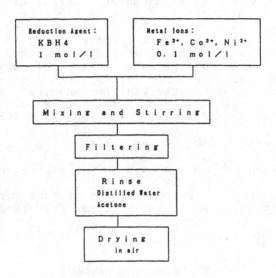

Fig. 1 Experimental procedure for the production of amorphous
ultra-fine powders.

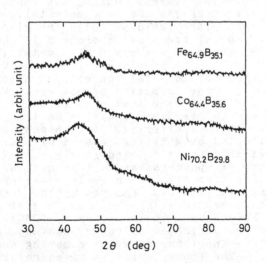

Fig. 2 X-ray diffraction patterns of (Fe, Co, Ni)-B binary
alloy powders prepared by the chemical reduction method

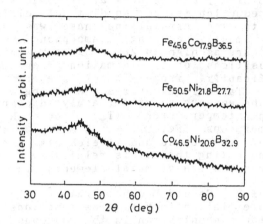

Fig. 3 X-ray diffraction patterns of (Fe, Co, Ni)-B ternary alloy powders prepared by the chemical reduction method.

Table 1 Mixing composition of metal ions in an aqueous solution, analyzed compositions of the chemically synthesized (Fe, Co, Ni)-B binary and ternary alloy powders, and their crystallization temperatures (T_x) measured by DSC at a scanning rate of 20 K/min.

Alloy	Mixing Comp. (at%)	Chemical Comp. (at%)	Crystallization Temp. T_x (K)
Fe-B	Fe_{100}	$Fe_{64.9} B_{35.1}$	737
Co-B	Co_{100}	$Co_{64.4} B_{35.6}$	755
Ni-B	Ni_{100}	$Ni_{70.2} B_{29.8}$	593
Fe-Co-B	$Fe_{70} Co_{30}$	$Fe_{45.6} Co_{17.9} B_{36.5}$	719
Fe-Co-B	$Fe_{30} Co_{70}$	$Fe_{18.4} Co_{41.7} B_{39.9}$	735
Fe-Ni-B	$Fe_{70} Ni_{30}$	$Fe_{50.5} Ni_{21.8} B_{27.7}$	709
Fe-Ni-B	$Fe_{30} Ni_{70}$	$Fe_{22.7} Ni_{52.8} B_{24.5}$	621
Co-Ni-B	$Co_{70} Ni_{30}$	$Co_{46.5} Ni_{20.6} B_{32.9}$	627
Co-Ni-B	$Co_{30} Ni_{70}$	$Co_{20.1} Ni_{46.9} B_{33.0}$	625

Figures 2 and 3 show the X-ray diffraction patterns of the
(Fe, Co, Ni)-B binary and ternary alloy powders prepared by
the chemical reduction method. No distinct diffraction peak
corresponding to the crystalline phase was observed. The
powders form mostly a single amorphous phase. The B
concentrations are as high as 28 to 37 at.%, indicating that
the alloy compositions for the formation of an amorphous phase
differ significantly from those prepared by rapid
solidification. Table 1 summarizes the mixing compositions of
metal ions in the aqueous solution, analyzed compositions and
crystallization temperatures (T_x) for the chemically
synthesized amorphous (Fe, Co, Ni)-B alloy systems. The
compositions of the Fe, Co and Ni elements agree with the
mixed ratio, indicating that it is relatively easy to achieve
the desired composition of metal elements in the prepared
powders.
The changes of Fe, Co and B concentrations in the Fe-Co-B
alloy powders are plotted in Fig. 4 as functions of the ratio
of metallic ion concentration in the aqueous solutions of
$FeSO_4$ plus $CoCl_2$. The metallic concentrations change almost
linearly with the concentration of metallic ions, while the B
concentration remains almost constant of about 37 at.%.
Figure 5 shows a bright-field electron micrograph of the
$Fe_{45.6}Co_{17.9}B_{36.5}$ alloy. The particles have a spherical shape
with diameters from 20 to 50 nm and no distinct contrast
revealing the existence of a crystalline phase can be seen.
Furthermore, the electron diffraction pattern taken with an
aperture with a diameter of 20 μm consists only of a halo
peak. This result clearly indicates the formation of ultra-
fine amorphous alloy powders.

(2) Mössbauer measurements
Figure 6 shows the representative Mössbauer spectra at room
temperature and corresponding magnetic hyperfine distributions
$P(B_{hf})$ of amorphous Fe-B powders in the M_R range from 2.0 to
20.0. In all samples we have found a quadrupole splitting
QS=0.9 mm/s in the center of Mössbauer spectra. The isomer
shift of the quadrupole splitting is 0.42 mm/s at room
temperature. These indicate the existence of the Fe^{3+} oxide.
Six broad lines correspond to ferromagnetic amorphous Fe-B
alloys. The sharp six lines with an internal magnetic field
of 33.1 T appear below M_R=9.0 indicating the crystalline α -
Fe. The relative intensities of the α -Fe increase with
decreasing the M_R value as shown in Fig. 7. At M_R=2.0 about 48
% area of the spectrum consists of the crystalline α -Fe.
The average magnetic hyperfine field ,$\bar{B}_{hf}$, is shown in Fig. 8
as a function of M_R value. The $\bar{B}_{hf}$ clearly shows a
discontinuous change at about M_R=8.5. These results show that
amorphous powders for M_R≤8.5 and M_R>8.5 are different. On
the other hand, the $\bar{B}_{hf}$ of amorphous powders decreases
linearly with increasing the boron content for 25 to 33 at.%
[6]. Applying these results, we found that the Fe-B powders
prepared in the M_R range from 8.5 to 20.0 have the boron
content of about 33 at.%, while M_R<8.5, the $\bar{B}_{hf}$ of amorphous
particles is in agreement with the $\bar{B}_{hf}$ of the amorphous alloy

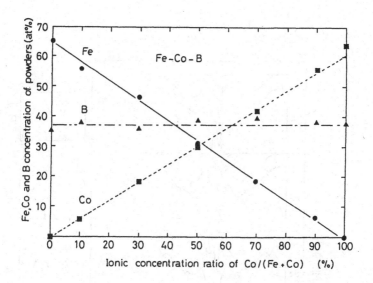

Fig. 4 Fe, Co and B concentrations in amorphous Fe-Co-B powders as functions of the ionic concentration ratio Co/(Fe+Co) in an aqueous solution.

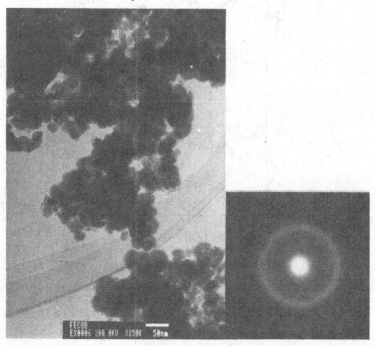

Fig. 5 Electron bright-field images and diffraction pattern of amorphous $Fe_{45.6}Co_{17.9}B_{36.5}$ powders prepared by the chemical reduction method.

468

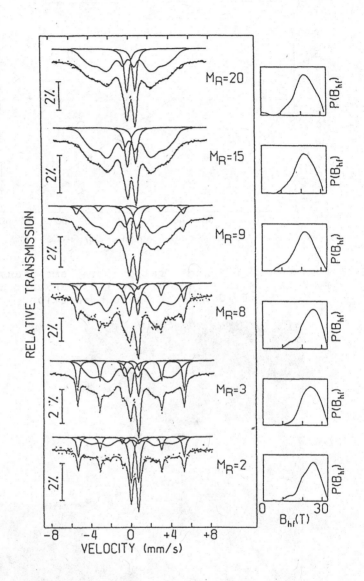

Fig. 6 Mössbauer spectra measured at room temperature in Fe-B powders prepared at different M_R values ranging from 2.0 to 20.0 and $P(B_{hf})$ of amorphous powders.

with the composition of about 25 at.% boron.
A similar tendency was observed by the ICP measurements as
shown in Fig. 9. The boron concentration is nearly constant at
about 33 at.% in the M_R range from 10.0 to 20.0. Further
decrease of M_R leads to a decrease of boron content.
Furthermore, the X-ray diffraction patterns of Fe-B binary
powders show an amorphous phase in the M_R range from 10.0 to
20.0 and below 8.0, crystalline α -Fe are observed as shown in
Fig. 10.
These results are in good agreement with Mössbauer
measurements.
From these results we suggest two reaction processes for the
formation of amorphous Fe-B powders.

(1) $4Fe^{2+} + 2BH_4^- + 6OH^- \rightarrow 2Fe_2B + 6H_2O + H_2$

(2) $4Fe^{2+} + 2BH_4^- + 7OH^- \rightarrow Fe_3B + Fe + BO_2^- + 5H_2O + 5/2H_2$

The reaction (1) is the dominant reaction showing the
formation of Fe-B amorphous alloy with about 33 at.% boron in
the M_R range from 10.0 to 20.0. The reaction (2) is the
dominant process for M_R values below 10.0. In this reaction,
the amorphous phase has a composition of about 25 at.% boron.
In order to investigate the existence of BO_2^-, we measured the
powders by using ESCA method. The measurements show the peaks
corresponding to boron oxide. In conclusion, we have shown
that by the chemical reduction method two reaction processes
take place. We are presently working on it to find details
of these two reaction processes.

CONCLUSIONS

Amorphous powders can be prepared by the chemical reduction
method using KBH_4. We investigated the structure,
composition, morphology and reaction process of these powders.
The results obtained are summarized as follows:

1. Amorphous ultra-fine Fe-B and (Fe,Co,Ni)-B powders with
   diameter range from 20 to 50 nm can be prepared by a
   chemical reduction method.
2. Mössbauer results indicate two kinds of average internal
   magnetic fields in amorphous Fe-B powders. This is due to
   the difference of the boron concentration in Fe-B
   amorphous phase.
3. Two reaction processes in the Fe-B system are as follows:

(1) $4Fe^{2+} + 2BH_4^- + 6OH^- \rightarrow 2Fe_2B + 6H_2O + H_2$

(2) $4Fe^{2+} + 2BH_4^- + 7OH^- \rightarrow Fe_3B + Fe + BO_2^- + 5H_2O + 5/2H_2$

4. The molar ratio of KBH_4 to metal ions ,M_R, plays an
   important role in determining the structure and
   composition. With decreasing M_R value, the structure of
   Fe-B powders changes from amorphous to a crystalline

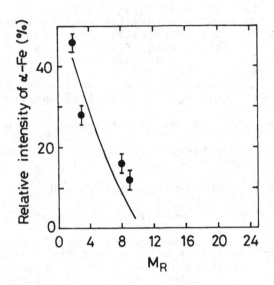

Fig. 7 Relative intensity of crystalline α-Fe as a function
of M_R obtained from Mössbauer spectra.

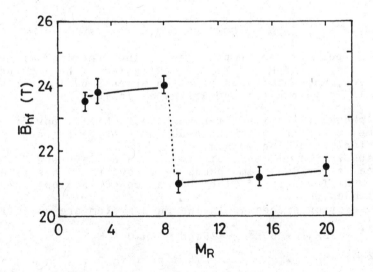

Fig. 8 Average magnetic hyperfine field as a function of M_R in
the Fe-B system.

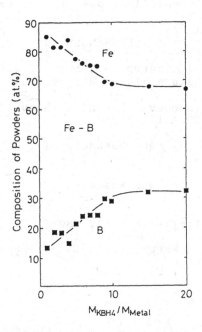

Fig. 9 Chemical composition of Fe and B in Fe-B powders as function of M_R value.

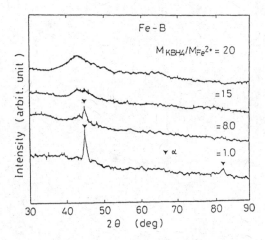

Fig. 10 X-ray diffraction patterns of Fe-B binary alloy powders prepared at M_R=20.0, 15.0, 8.0 and 1.0.

phase and the B content of powders decreases.

REFERENCES

1. van Wonterghem, J., Mørup, S., Koch, C.J.W., Charles, S.W. and Wells, S., Nature, 1986, 322, 622-3.

2. Dragieva, I., Slavcheva, M. and Buchkov, D., J. Less-Common Met., 1986, 117, 311-7.

3. Inoue, A., Saida, J. and Masumoto, T., Met. Trans., 1988, 19A, 2315-8.

4. Saida, J., Inoue, A. and Masumoto, T., Mater. Sci. Eng., 1991, A133, 771-4.

5. Narcus, H., Plating, 1967, 54, 380-1.

6. Linderoth, S. and Mørup, S., J. Appl. Phys., 1991, 69, 5256-61.

TEMPERATURE DEPENDENCE OF THE LAW OF APPROACH TO MAGNETIC SATURATION IN NANOCRYSTALLINE FERROMAGNETS

J. GONZALEZ[a], M. VAZQUEZ[b], E. T. de LACHEISSERIE[c] and G. HERZER[d]

a) Departamento de Física de Materiales, Facultad de Químicas, 20009 San Sebastián, SPAIN
b) U.E.I. de Materiales Magnéticos (C.S.I.C.), Serrano 144, 28006 Madrid, SPAIN
c) Laboratoire Louis Nèel, C.N.R.S., 38042 Grenoble, FRANCE
d) Vacuumschmelze GmbH, D-6450 Hanau, GERMANY

ABSTRACT

Coefficients of the $1/H$ and $1/H^2$ terms of the law of approach to magnetic saturation in $Fe_{73.5}Cu_1Nb_3Si_{13.5}B_9$ alloy either for amorphous or for nanocrystallized and fully crystallized samples have been determined as functions of the temperature in the range up to 300 K. The change of such coefficients with annealing as well as their temperature dependence is analyzed considering the different microstructures of the sample. Moreover, the temperature dependence of the saturation magnetic polarization in this range of temperature is also reported for the as-quenched and annealed samples.

1. INTRODUCTION

Amorphous alloys obtained by rapid quenching from the melt exhibit outstanding properties as soft magnetic materials that make them suitable for technological applications [1]. This behavior is determined by the lack of long range atomic periodicity giving rise to vanishing macroscopic magnetocrystalline anisotropy. In fact, short range ordered units are assumed to be present extended up to around 10 - 20 Å. The main macroscopic magnetic features of these samples are then governed by local magnetoelastic anisotropies arising from the coupling between the frozen-in internal stresses and the magnetostriction constant.

Fe-rich amorphous alloys show relatively high magnetostriction constant, λ_s, ($\lambda_s \sim 3 \times 10^{-5}$) so that to obtain an even softer magnetic behavior both internal stresses and magnetostriction should be reduced. Internal stress can be partially relieved by conventional treatments below the crystallization temperature but this is not the case for λ_s.

Recently, Yoshizawa et al.[2] have reported quite interesting results concerning the magnetic behavior of certain Fe-rich alloys thermally treated above the crystallization temperature. An example is the $Fe_{73.5}Cu_1Nb_3Si_{13.5}B_9$ alloy ribbon which in its as-cast state behaves as conventionally established for Fe-rich amorphous alloys. After treatments above the crystallization temperature, a stable and homogeneous α-FeSi fine grain microstructure has been observed with typical grain size of about 10 nm. The formation of this nanocrystalline structure is ascribed to the combined addition of Nb and Cu and their low solubility in α-Fe: Cu, hereby, promotes the nucleation of the bcc grains while Nb hinders the growth [3,4]. Related to the as-cast state, coercivity is decreased while susceptibility increases both magnetic parameters being dependent on the grain size [4-6]. Moreover, saturation magnetostriction is reduced at least one order of magnitude [2]. This soft magnetic behavior can be interpreted as a consequence of the magnetostriction decrease and also of the still vanishing macroscopic magnetocrystalline anisotropy for grain size smaller than the ferromagnetic exchange length according to the random anisotropy model [7,4].

The temperature dependence of saturation magnetization of as-cast and heated (1 h. at 520 °C) $Fe_{73.5}Cu_1Nb_3Si_{13.5}B_9$ alloy, above room temperature, has been previously reported above the room temperature [8]. The aim of the present work has been to get deeper information on the magnetic behavior from its temperature dependence within the range up to 300 °K. In particular, magnetization is studied for both kinds of samples under a magnetic field large enough for magnetization to approache saturation. Coefficients of the law of approach to saturation can then be obtained. Moreover, the evolution of the saturation magnetization with temperature is also obtained for the above commented range.

2. EXPERIMENTAL TECHNIQUES AND RESULTS

Amorphous alloy ribbons of nominal composition $Fe_{73.5}Cu_1Nb_3Si_{13.5}B_9$ were kindly supplied by Vacuummschmelze, Hanau. The regular material dimensions are: cross section about 20 μm thick and 0.50 cm wide and the length of the samples used for measurements was 1.50 cm.

In order to investigate the effect of the grain size on the high field behavior, the samples were submitted to two thermal treatments: 520 °C during 1 hour, and 800 °C 1 hour. While treatment at 520 °C gives rise to nucleation of α-FeSi crystallites with grain size around 10 nm, annealing at 800 °C produces fully crystallization with much larger grain size.

For analysing the magnetic approach to saturation of the as-cast, nanocrystallized and crystallized samples, the magnetic polarization, J, of these samples has been measured from 4.2 °K up to 300 °K in a magnetic field up to 60 KOe using an extraction method. Corrections of the macroscopic demagnetizing field have been considered in these measurements.

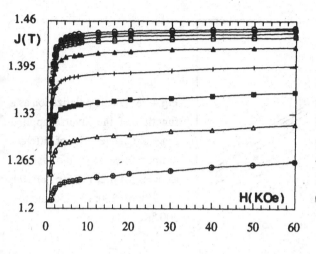

Fig.1.- Magnetization curves of as-cast sample for a series of different temperatures (°K): (o) 4.2; (●) 20; (x) 40; (◻) 60; (Δ) 100; (+) 150; (■) 200; (Δ) 250 and (⊕) 300.

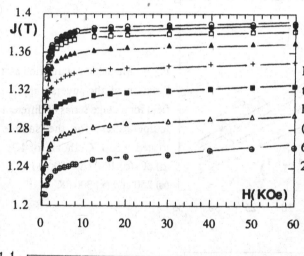

Fig.2.- Magnetization curves of the sample treated at 520 °C, 1 hour at different temperatures (°K): (o) 4.2; (●) 20; (x) 40; (◻) 60; (▲) 100; (+) 150; (■) 200; (Δ) 250 and (⊕) 300.

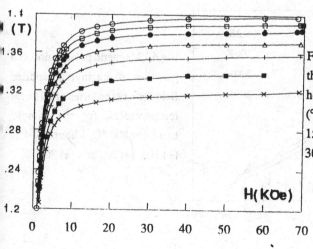

Fig.3.- Magnetization curves of the sample treated at 800 °C, 1 hour at different temperatures (°K): (o) 4.2; (◻) 60; (●) 100; (Δ) 150; (+) 200; (■) 250 and (x) 300.

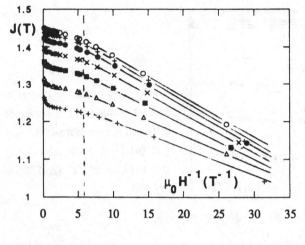

Fig.4.- Magnetic polarization as a function of the inverse applied field for a range series of different temperatures, for the as-cast sample. (o) 4.2; (+) 60; (●) 100; (x) 150; (■) 200; (Δ) 250 and (+) 300 °K.

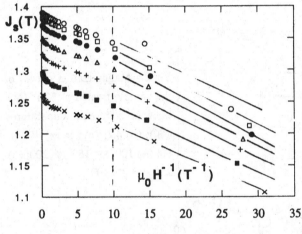

Fig.5.- Magnetic polarization as a function of the inverse applied field for a range series of different temperatures, for the sample treated at 520 °C, 1hour. (o) 4.2; (□) 60; (o) 100; (Δ) 150; (+) 200; (■) 250 and (x) 300 °K.

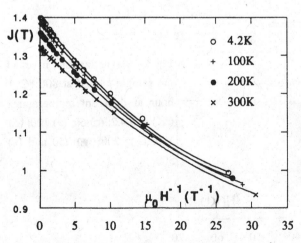

Fig.6.- Magnetic polarization as a function of the inverse applied field for a range series of different temperatures, for the sample treated at 800 °C, 1 hour. (o) 4.2; (+) 100; (●) 200 and (x) 300 °K.

Figures 1, 2 and 3 show the large field magnetization curves at different temperatures corresponding respectively to the as-cast sample and after annealing at 520 °C and 800 °C. A rough observation of the results indicates that, as expected, the sample treated at 800 °C reach saturation at much higher fields than as-cast and nanocrystallized samples. Clearly, this must be connected to the large grain size obtained as a results of such a heat treatment.

3. ANALYSIS OF THE RESULTS

It is known that magnetization of amorphous ferromagnetic alloys begins to approach its saturation value in relatively low magnetic fields typically around 100 Oe or evenless. At large applied fields, change of magnetization is only a few percentage of the total value. Under these fields, deviations of magnetization from complete saturation has been adscribed to local magnetoelastic effects arising from the coupling between the magnetostriction constant and the internal stress fields. Other mechanisms such as stray fields at the sample edges or surface roughness can also play an important role.

It has been experimentally found that for most amorphous alloys, when magnetic polarization, J, approaches the saturation value, J_s, its applied field dependence follows the law [9,10]:

$$J = J_s \left[1 - (a_1/H) - (a_2/H^2) \right] + b\, H^{1/2} \hspace{3cm} [1]$$

where the $1/H$ dependence prevails in the region of relatively low field, while for larger applied fields the term $1/H^2$ becomes the more important. The $H^{1/2}$ contribution arises from the so called "para-effect" and is relevant in the region of the highest fields region.

Figures 4, 5 and 6 show respectively the dependence of the magnetic polarization on the inverse applied field for as-cast, nanocrystallized and crystallized samples measured in a range of temperatures. Deviation from linear behavior is observed for fields larger than about 1000 and 1700 Oe respectively for nanocrystallized and as-cast samples. On the other hand, no clear behavior can be established for the sample treated at 800 °C. Figures 7 and 8 show the evolution of the magnetic polarization with the squared reciprocal applied field for as-cast and nanocrystallized samples respectively in the range of applied fields where the $1/H$ law is no longer obeyed. In its term, figure 9 shows the dependence of the magnetic polarization on the square root of the largest applied fields.

Figure 10 shows the temperature dependence of the coefficient a_1 for both as-cast and nanocrystallized samples. While for the as-cast sample a_1 clearly decreases with increasing temperature, only sligth changes are detected for the nanocrystallized sample. Similar behavior are observed for the coefficient a_2 in figure 11.

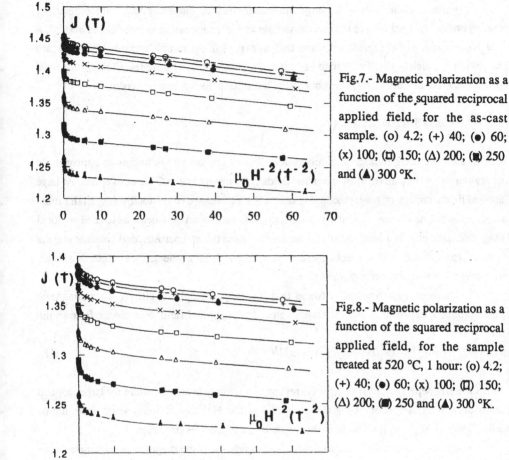

Fig.7.- Magnetic polarization as a function of the squared reciprocal applied field, for the as-cast sample. (o) 4.2; (+) 40; (●) 60; (x) 100; (▫) 150; (Δ) 200; (■) 250 and (▲) 300 °K.

Fig.8.- Magnetic polarization as a function of the squared reciprocal applied field, for the sample treated at 520 °C, 1 hour: (o) 4.2; (+) 40; (●) 60; (x) 100; (▫) 150; (Δ) 200; (■) 250 and (▲) 300 °K.

Fig.9.- Dependence of magnetic polarization on the square root of the applied field for as-cast samples at different temperatures (°K): (o) 4.2; (●) 100; (x) 150; (▫) 200; (+) 250 and (Δ) 300.

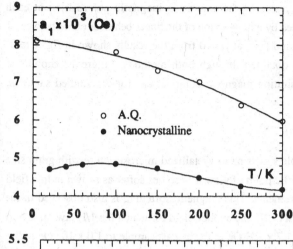

Fig.10.- Temperature dependence of the a_1 coefficient of the law of approach to saturation for as-cast and nanocrystallized samples.

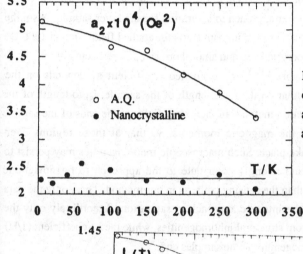

Fig.11.- Temperature dependence of the a_2 coefficient of the law of approach to saturation for as-cast and nanocrystallized samples.

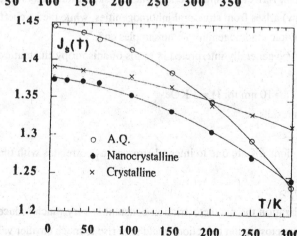

Fig.12.- Temperature dependence of saturation magnetic polarization for as-cast and treated samples at 520 and 800 °C

According to equation 1, the value of the saturation magnetic polarization at each measuring temperature can be obtained by extrapolation of the linear behavior either between J and $1/H^2$ at $1/H^2 = 0$ or between J and $H^{1/2}$ at $H = 0$ from the results shown in fig. 7 to 9. Very similar saturation values are obtained through both methods. Figure 12 shows the temperature dependence of the saturation magnetic polarization for the studied samples.

4. DISCUSSION

After annealing at 520 °C, the sample shows nanocrystallized microstructure with grain size around 10 nm, while from the magnetic point of view it becomes softer as shown in low field susceptibility and coercivity measurements. This magnetic softening is also observed in the region of approach to saturation. First of all, the applied field for which the $1/H$ law turns into the $1/H^2$ one is reduced from about 1.7×10^3 Oe in the as-cast sample to 1.0×10^3 Oe in the nanocrystallized. It indicates a quicker approach to saturation for the latter sample. Also, the region for which the para-effect becomes predominant starts for applied fields around 1500 Oe and 3000 Oe corresponding to nanocrystallized and amorphous samples respectively.

As reported in a previous work [11], the evaluated coefficient a_1 depends on the measuring technique and in particular on the ratio length of the sample, l_s, to length of the sensing coil, l_c, when using extraction method. In fact, stray fields at the ends of the sample give rise to inhomogeneities of the magnetic moments, so that at those regions even macroscopic rotation processes take place. Such macroscopic inhomogenities may persist to fields of about 1 KOe or more and strongly contribute to the approach to ferromagnetic saturation if the sample is shorter than the sensing coils, as in our method. Thus, only the data for fields $H \geq 1...2$ KOe, reflect the intrinsic approach to saturation. Accordingly only the a_2-coefficient ($1/H^2$-law) arises from structural inhomogenities, while the a_1-coefficient ($1/H$) originates from macroscopic inhomogenities at the samples end.

The $1/H^2$-law can be generally interpreted in terms of anisotropy fluctuations on a scale large than

$$l_H = \sqrt{(2A)/(JsH)} \qquad \approx 10 \text{ nm for } H \approx 1 \text{ KOe} \qquad \text{I2I}$$

The a_2-coefficicient then results as

$$a_2 = 4<\delta K^2>/Js^2 \qquad \text{I3I}$$

if the anisotropy fluctuations δK are due to internal stresses, a_2 correlates with the square of λs, i.e.

$$a_2 \propto (\lambda s^2/Js^2) <\sigma_i^2> \qquad \text{I4I}$$

After nanocrystallization the magnetostriction constant is very much reduced. But the only consideration of magnetostriction reduction would give rise to much smaller values for the coefficients of nanocrystallized samples. This circunstance may reflect the contribution of the randomly oriented magneto-crystalline anisotropy of the bcc-grains. Another explanation would be that the intrinsic a_2 becomes too small and the measured a_2-coefficient is determined again by residual inhomogenities at the samples ends.

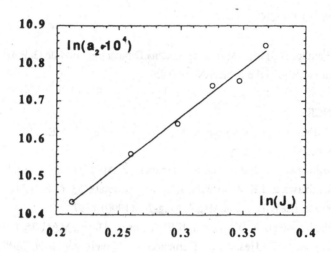

Fig.13.- $\ln[10^4 a_2(T)] \propto \ln[J_s(T)]$ for as-cast sample.

On the other hand, a_2 coefficient of as-cast sample decreases with increasing temperature and this can be related to the temperature dependence of magnetostriction. Analysing the temperature dependence of coefficient a_2 in the as-cast state and the one of the saturation magnetic polarization has led us to consider the following relationship:

$$a_2 (T) = A_2 J_s^n(T) \qquad \qquad |5|$$

where the coefficient n takes the value 2.78 according to the fitting of experimental results (see figure 13). Considering the equation 2 yields a temperature dependence of magnetostriction according to:

$$\lambda_s (T) \propto J_s^m(T) \qquad \qquad |6|$$

with m ≈ 2.4 which is smaller than the expected value m∼3 for Fe-rich amorphous alloys [12].

As regards the nanocrystallized sample, only sligth temperature dependence of a_1 and a_2 is observed. To understand this behavior, a further analysis about the influence of the peculiar microstructure of the sample is neeeded.

Regarding the temperature dependence of the saturation magnetic polarization shown in fig.12, it can be observed that for a temperature sligthly below room temperature both as-cast and nanocrystallized samples exhibit the same magnetic polarization. This result is consistent with previous measurements above room temperature[8] and the small difference with reported values at room temperature can arise from the way of extrapolating the value of saturation magnetic polarization. Such crossing temperature is a consequence of the different values of saturation polarization at zero Kelvin and of the Curie points of the as-cast sample and the nanocrystallized phase present in the treated sample.

ACKNOWLEDGMENT

One of the author (JG) is grateful to the Excma Diputación Foral de Guipúzkoa by the finantial support with charge to the Project No.660.85.

REFERENCES

(1) F.E. Luborsky in "Amorphous Metallic Alloys", Ed. F.E. Luborsky, (Butterworth, London, 1983).

(2) Y. Yoshizawa, S. Oguma and K. Yamauchi, J. Appl.Phys., 64, (1988) 6044.

(3) Y. Yoshizawa and K. Yamauchi, J.Jap.Instit..Metals, 53, (1989) 241.

(4) G. Herzer, IEEE Trans.on Magn., Mag-26, (1990) 1397.

(5) Y. Yoshizawa and K. Yamauchi, IEEE Trans.on Magn., MAG-25, (1989) 3324.

(6) R. Grössinger, R. Heszke, A. Hernando, O. Mayerhofer, K.H. Müller, Ch. Pollak, E. Pulido, G. Rivero and J. Schneider, IEEE Trans.on Magn., MAG-26, (1990) 1403.

(7) R. Alben, J.J. Becker and M.C. Chi, J.Appl.Phys., 49, (1978) 1658.

(8) G. Herzer, IEEE Trans.on Magn., MAG-25, (1989) 3327.

(9) H. Kronmüller, Atomic Energy Rev.Suppl., 1, (1981) 215.

(10) H. Grimm and H. Kronmüller, Phys.Stat.Solidi (b), 117, (1983) 663.

(11) G. Herzer, Anales de Física B, 86, (1990) 64.

(12) R.C. O'Handley in "Amorphous Metallic Alloys", Ed. F.E. Luborsky, (Butterworths, London, 1983)

ELECTRONIC ORIGIN OF STABILITY IN QUASICRYSTALS

P.M. OSSI

Dipartimento di Ingegneria Nucleare, Politecnico di Milano
Via Ponzio 34/3, 20133 Milano, Italy
Unità cINFM di Trento, 38050 Povo (Tn), Italy

ABSTRACT

The nearly free electron model has been applied to indagate the stability of the quasicrystalline state through the analysis of X-ray or neutron diffraction data. Several alloys, representative of the families of Aluminium-Transition Metal (Al-TM) and of PSD-type systems have been considered. The validity of the picture is discussed in the framework of recent measurements of electronic properties of the icosahedral state. Model predictions and experimental data agree for both classes of icosahedral systems. On such grounds a specific electron phase, characteristic of the quasicrystalline state of matter is defined.

INTRODUCTION

Probably the best known example of band structure influence on the stability of a crystal is constituted by the so called electron compounds (or electron phases). For such systems Hume-Rothery (H-R) postulated [1] the existence of limiting values of the average electron per atom concentration $\langle Z \rangle$, which were soon shown by Jones [2] to correspond to the condition of geometrical contact between Fermi Surface (FS) and Brillouin Zone (BZ). Within a nearly free electron (NFE) scheme, the condition is

$$|\bar{Q}| = 2 \bar{k}_F \tag{1}$$

with $\bar{Q}$ reciprocal lattice vector.

The proposed instability mechanism, i.e. the opening of a gap in the dispersion relation $E(\bar{k})$ which forces electrons to energetically unfavourable states, stresses the influence of FS-BZ interaction on the density of states (DOS) at the Fermi level E_F. However, even quite wide gaps in $E(\bar{k})$ are able to induce only small van Hove singularities in the DOS; in fact the origin of the structure stabilising energy-contribution is the rapid variation of the polarizability

$\chi(\overline{q})$ of the uniform electron gas, when $\overline{q}$ is near $2\overline{k}_F$. In direct space, if the wavelength λ_F of the long range Friedel oscillations of the interatomic potential matches the interatomic distance d $(d=2\pi/|\overline{Q}|)$, i.e. when

$$d = \frac{\pi}{\overline{k}_F} = \lambda_F \qquad (2)$$

a stabilising energy contribution arises.

The NFE picture was extended to interpret the stability of amorphous structures [3] under the hypotheses of spherical FS and spherically symmetric structure factor $S(\overline{q})$. When for the first peak of $S(\overline{q})$, $\overline{q}_p$, whose position is assimilated to that of a smeared out reciprocal lattice vector, holds

$$\overline{q}_p = 2\overline{k}_F \qquad (3)$$

a minimum in the DOS is expected at $\overline{q}_p$ thus providing a stabilising mechanism for a disordered structure, analogous to the one operating for crystals.

Although translational periodicity is lacking in glassy materials, the structure of $S(\overline{q})$, in particular the existence of the peak at $\overline{q}_p$, guarantees the existence of sufficiently long ranged correlations to produce a minimum in the DOS [4]. The symmetry of $S(\overline{q})$ implies that all states for which $|\overline{k}| = \overline{q}_p/2$ are affected; the minimum thus results more marked for increasing amplitude of $S(\overline{q})$.

In direct space, provided an average interatomic distance $\overline{d}$ is defined as

$$\overline{d} = 2\pi/\overline{q}_p \qquad (4)$$

the stability condition corresponds to coincidence between the pair distribution maxima and the pair potential minima.

The predicted minimum in the electronic DOS at E_F has been indagated, particularly in the case of glassy alloys. Extensive work on noble metal-polyvalent metal [5], and simple metal alloys [6;7] shows the presence of a structure induced minimum in the DOS, in agreement with calculations of the electronic structure of glassy systems which predict the existence of the minimum [8]. According to such evidences, these systems may be defined amorphous H-R alloys.

Midway between crystalline and glassy states are the quasicrystals. Soon after their discovery it was proposed that they could be considered H-R alloys [9]. Till now however no systematic investigation concerning the validity of the proposed picture was undertaken.

In the present work the idea is pursued that the quasicrystalline state can be regarded as an intermediate electron phase. As a consequence such a state should be treated in the framework of the NFE approximation; the presence of transition metals in a wide class of quasicrystalline alloys makes the hypothesis problematic. Here it is discussed, starting from recent experimental results.

RESULTS

In the case of icosahedral phase formation electronegativity (i.e. chemical effects) plays no specific role, while the size ratio is often in the immediate neighbourhood of the value pertinent to ideal icosahedral atomic packing [10]. Such kind of packing is found in the long period Frank-Kasper phases, some of which constitute crystalline approximants of realistic quasicrystals, as well as in metal glasses and in liquids. It is the simplest result of close packing of structural units, whose wide occurrence is a consequence of the spherical symmetry of the pair potentials in metals. The closest arrangement of spherical atoms is obtained when their centres coincide with the vertices of a tetrahedron. The first coordination shell resulting from a regular arrangement of tetrahedra is an icosahedron, which is known since long to be more stable than the most common close packed crystalline structures [11]. In Frank-Kasper intermetallics, to completely fill the three dimensional space, tetrahedra must be slightly distorted through the introduction of an ordered array of -72° disclination lines; polyhedra with coordination numbers z=14,15,16 alternate with icosahedra with z=12. On the opposite side, structural models of glassy alloys require to alternate icosahedra and Bernal polyhedra with z=9,10,11.

To indagate band structure effects, we assume the NFE model to hold in the case of quasicrystals; the Hume-Rothery-Jones geometric conditions are:

$$\frac{4}{3} \pi \, \bar{k}_F^3 = \frac{(2\pi)^3/2 \, <Z>}{\frac{4}{3} \pi \, <\bar{r}>^3 / \eta} \tag{5}$$

The complete spherical symmetry of the model permits to obtain the FS volume as the ratio between the volumes of the occupied electron states and the Wigner-Seitz cell. $<r>$ is the average atomic radius of the alloy and η is the packing efficiency, which for icosahedral packing, is taken to be $\eta = 0.688$ [12]. In the case of icosahedral coordination we can approximate the pseudo-Wigner-Seitz cell by a sphere due to the high multiplicity m of the most intense experimentally observed icosahedral peaks. We consider the (100000) (m=12) and the (110000) (m=30), in the case of Al-TM alloys, and the (111101) (m=30) for PSD compounds, formed by a p element (Al, Ga, Ge, Sn) an s element (Li, Na, Mg) a near noble d element (Ni, Cu, Zn, Au). Our choice reflects the experimentally found meaningful differences between diffraction patterns of Al-TM and SPD alloys. The latter present intense (111101) peaks, which are nearly absent in the former family; by example in $Al_{56.1}Li_{33.7}Cu_{10.2}$ [29] the (111101) peak shows the highest relative intensity I, while the (100000) displays I=19.5. To estimate $\bar{k}_F$ values from eqn. (5) for e.g. a binary $A_{(1-x)}B_x$ alloy, valencies Z_A and Z_B must be attributed to alloy components in the evaluation of $<Z>$,

$$\langle Z \rangle = (1-x) \, Z_A \pm x Z_B \qquad (6)$$

The sign -, in alternative to the sign + in eqn. (6), means that when TMs are involved a charge transfer to their d bands is assumed; the effect is indeed observed [13] and it is thought to allow for compensation of the unpaired spins of d electrons. While TMs are assigned 0 valency when $\langle Z \rangle$ is calculated for H-R alloys, we refer to a phenomenological model of TMs [14] and to X-ray studies on Al-TM alloys [15] to take Mn and Cr valencies while the values for Fe, Co and Ni were adopted in a NFE scheme to indagate the structural properties of glassy TM-polyvalent element alloys [16]. In Table 1 are listed atomic radii and valencies chosen for the present work.

We present a selection of data for Al-Tm concentrated binary and ternary alloys, with Al as the majority component, and for PSD alloys. Only systems for which experimental X-ray, or neutron diffraction data have been published either as diffractograms, or in tabulated form are considered here. We deliberately avoid TM-TM quasicrystalline alloys, and systems including TMs, or rare earths with uncontrolled valencies. Note that $\bar{k}_F$ values extracted directly from diffraction patterns, could present some minor imprecision.

Table 2 lists alloy compositions, experimental $\bar{k}_F$ values derived from the most significant X-ray or neutron peak and calculated values of average electron per atom ratios $\langle Z \rangle$.

TABLE 1

Elemental valencies and atomic radii used in the calculations. Negative valency implies charge transfer to the d band of the involved TM

Element	Valency	Atomic Radius (nm)
Al	+3	0.143
Mn	-3.66	0.125
Cr	-4.66	0.129
Fe	-1.11	0.126
Co	-1.03	0.125
Ni	-1.14	0.124
Cu	+1	0.128
Au	+1	0.144
Zn	+2	0.133
Si	+4	0.117
Ga	+3	0.125
Ge	+4	0.122
Li	+1	0.152
Mg	+2	0.160

TABLE 2

Alloy composition, $\bar{k}_F$ values deduced from X-ray, or neutron diffraction data and calculated average electron per atom concentrations <Z>. The reported $\bar{k}_F$ values refer to the most significant reflection, whose index is given in parentheses; for data from neutron diffraction, the superscript N is used. ■: decagonal phase. The index reported in the second column is used to number the alloys in Fig. 1.

Alloy	n°	$\bar{k}_{F_{exp}}(nm^{-1})$ $(\bar{k}_{F_{exp}}=\bar{Q}_{exp}/2)$	Ref.	<Z>
$Al_{86}Mn_{14}$	1	15.25 (110000)	17	2.068
$Al_{80}Mn_{20}$	2	14.61 (100000)	18	1.668
$Al_{77.5}Mn_{22.5}$	3	14.52 (100000)	19	1.502
$Al_{86}Cr_{14}$	4	14.99^N(110000)	20	1.928
$Al_{82}Cr_{18}$	5	14.30 (100000)	21	1.621
$Al_{80}Cr_{20}$	6	14.35 (100000)	21	1.468
$Al_{80}Fe_{20}$	7	15.42 (110000)	22	2.178
$Al_{85}Mn_{14}Si$	8	15.19^N(110000)	20	2.078
$Al_{74.5}Mn_{21}Si_{4.5}$	9	14.26 (100000)	9	1.646
$Al_{71.5}Mn_{21}Si_{7.5}$	10	14.02 (100000)	9	1.676
$Al_{74}Mn_{20}Si_6$	11	15.21 (110000)	23	1.728
$Al_{74}Cr_{20}Si_6$	12	14.51 (100000)	23	1.528
$Al_{60}Cr_{20}Si_{20}$	13	15.21 (110000)	24	1.668
$Al_{85}Mn_7Cr_8$	14	15.07^N(110000)	20	1.921
$Al_{65}Cu_{20}Mn_{15}$	15	14.70 (100000)	25	1.601
$Al_{65}Cu_{20}Fe_{15}$	16	15.71 (110000)	25	1.983
$Al_{75}Ni_{10}Co_{15}$	17	15.34 ■ (100001)	26	1.982
$Al_{75}Ni_{10}Fe_{15}$	18	15.31 ■ (100001)	26	1.970
$Al_{65}Cu_{15}Co_{20}$	19	15.54 ■ (000001)	27	2.044
$Al_{60}Ge_{20}Mn_{20}$	20	15.38 (110000)	28	1.868
$Al_{56.1}Li_{33.7}Cu_{10.2}$	21	15.74 ■ (111101)	29	2.122
$Al_{59}Li_{30}Cu_{11}$	22	15.85 ■ (111101)	30	2.180
$Al_{60}Li_{30}Au_{10}$	23	15.72 (111101)	30	2.200
$Al_{51}Li_{32}Zn_{17}$	24	15.57 (111101)	30	2.190
$Al_{55}Mg_{36}Cu_9$	25	15.30 (111101)	31	2.400
$Al_{50}Mg_{48}Cu_2$	26	15.10 (111101)	32	2.480
$Ga_{24}Mg_{36}Zn_{40}$	27	15.52 (111101)	33	2.240
$Ga_{16}Mg_{32}Zn_{52}$	28	15.59 (111101)	24	2.160

In Fig. (1) $\bar{k}_F$ values derived from eqn.(5) are plotted versus experimental $\bar{k}_F$ values. The solid line corresponds to exact coincidence between the NFE prediction and the experiment; the dashed line is the result of a least squares fit on the data points. For the consistent group of representative alloys considered, the agreement appears quite good.

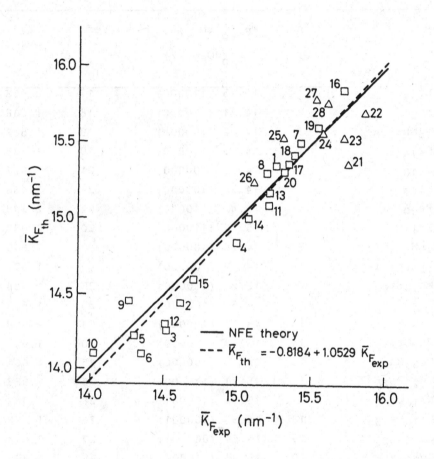

Figure 1. $\bar{k}$ values for the alloys in Table 2 as obtained by a NFE model, versus $\bar{k}_F$ values derived from experimental $\bar{Q}$ values: $\bar{k}_{Fexp} = \bar{Q}_{exp}/2$. Indexes of the data points correspond to the alloy indexes in Table 2.
Al-TM alloys: □
SPD alloys: △
The full line corresponds to perfect agreement between the behaviour of quasicrystalline system and the model.
Dashed line: least squares fit on the experimental points.

DISCUSSION

Fig. (1) deserves some comments. First, there is no apparent influence of the preparation technique used to produce quasicrystalline samples on their behaviour. The degree of disorder, strongly dependent on sample history does not appear to play a specific role; alternatively, the NFE picture is intrinsically too coarse to be sensitive to finer details of the alloy scattering behaviour. Indeed, among those alloys prepared by conventional techniques and fully annealed, such as $Al_{65}Cu_{15}Co_{20}$ (index 19) and $Al_{60}Li_{30}Cu_{10}$ (index 23) which are nearly free from phason disorder as well as from microstructural imperfections, the data point of the latter appears comparatively distant from the full line. Besides, diffraction data are available for rapidly quenched as well as for conventionally solidified $Al_{65}Cu_{20}Fe_{15}$ [25;27] and $Al_{60}Li_{30}Cu_{10}$ [29;34]: comparing the corresponding diffraction profiles no significant differences are found.

It is common to correlate the variation with alloy composition of the intensity ratio between the most intense icosahedral and impurity (e.g. fcc Al) diffraction peaks with the hypothetical stoichiometry of maximum stability of the icosahedral phase, assumed to be that at which impurity peaks disappear [26]. In the present analysis, among e.g. Al Cr alloys (indexes 4,5,6) the reported "most stable" $Al_{80}Cr_{20}$ (index 6) lies farther from the theoretical curve, taken as the locus of maximum stability for this class of alloys.

The model works equally well for icosahedral and decagonal phases, as expected: indeed, in alloys showing coexistence of both types of incommensurate structures, contributions to the diffraction pattern by the different phases result indistinguishable within the experimental uncertainty [23].

Various parameters have been used to interpret icosahedral phase occurrence and stability; the average conduction electron concentration $<Z>$ seems to be prominent; the initially promising atomic size factor was recently found to fail [35].

Several Authors indicate, starting from the results of investigations on a rather exiguous number of Al-TM alloys, the interval $1.7<<Z><1.8$ as characteristic of the icosahedral phase [35;36]. Isomer shifts and quadrupole splittings in some AlMn alloys show maxima at the composition $Al_{80}Mn_{11}Fe_9$, coincident with $<Z>$ = 1.76. As Fe substitution for Mn is not a near neighbour effect, the quite impressive changes observed in magnitude and trend of isomer shift can be reasonably attributed to alterations in the distribution of conducting electrons; the same argument may be extended to all alloys studied here.

The meaningful lowering of electronic DOS at E_F indicated by Mössbauer results, was observed also in spectroscopic measurements and electronic specific heat coefficient γ of AlMn [37] and Al Li Cu [38]. While the origin of the minimum in Al Li Cu is definitely different from that in Al Mn [39], the coincidence of the Fermi level with the minimum, revealed by the small measured γ value, was proposed to assist stabilisation of the icosahedral structure. Also Fujiwara calculation of the electronic DOS for crystalline $Al_{114}Mn_{24}$ approximant of quasicrystalline $Al_{84}Mn_{16}$ [40] indicates that Mn 3d states form sharp resonance and antiresonance peaks, with E_F falling in the resulting DOS minimum.

Starting from our results, we tentatively assign to the icosahedral phase the <Z> interval between 1.5 and 2.5; we believe that the behaviour of TM-TM icosahedral systems should not appreciably change the proposed phase extension. In Table 3 is plotted a list of Hume-Rothery phases. It includes crystalline phases, the amorphous phase discussed by Häussler and the icosahedral phase, lying between the highest <Z> values for a crystalline phase, and the lower limit for the glassy phase.

TABLE 3

Crystal structure, average electron concentration per atom <Z> and <Z> boundary values for the different electron phases

Electron phase	Crystal structure	<Z>	Hume-Rothery boundary <Z>
α	fcc	1.00-1.41	1.362
β	bcc	1.35-1.60	1.480
ζ	hcp (c/a: 1.633)	1.22-1.83	-
γ	complex cubic	1.54-1.70	1.620
δ	complex cubic	1.55-2.00	-
μ	β-Mn	1.40-1.54	-
ε	hcp (c/a: 1.570)	1.65-1.89	1.700
η	hcp (c/a: 1.750)	1.92-2.00	-
q.c.	icos.	1.50-2.50	-
glassy	disordered	>1.80	-

In conclusion it appears justified to consider the quasicrystalline state as an electron phase whose stability is due to electronic band effects, able to stabilize an otherwise energetically unfavoured packing of tetrahedral structural units.

ACKNOWLEDGEMENTS

Financial support by Ministero Università e Ricerca Scientifica e Tecnologica is acknowledged.

REFERENCES

1. Hume-Rothery, W., J. Inst. Metals, 1926, 35, 295.
2. Jones, H., Proc. Roy. Soc., 1934, A144, 225.
3. Nagel, S.R. and Tauc, J., Phys. Rev. Lett., 1975, 35, 380.
4. Hafner, J., From Hamiltonians to Phase Diagrams, Springer, Berlin, 1987, pp. 298-306.
5. Häussler, P., Baumann, F., Krieg, J., Indlekofer, G., Oelhafen, P. and Güntherodt, H.J., Phys. Rev. Lett., 1983, 51, 714.
6. Häussler, P., Baumann, F., Gubler, V., Oelhafen, P. and Güntherodt, H.-J., Zeits, Phys. Chem. Neue Folge, 1988, 157, 471.
7. Hafner, J., Jaswal, S.S., Tegze, M., Pflugi, A., Krieg, J., Oelhafen, P. and Güntherodt, H.-J. J. Phys. F: Met. Phys., 1988, 18, 2583.
8. Nicholson, D. and Schwartz, L., Phys. Rev. lett., 1982, 49, 1050.
9. Bancel, P.A. and Heiney, P.A., Phys. Rev., 1986, B33, 7917.
10. Ossi, P.M., Nuovo Cim., 1989, D11, 1123.
11. Frank, C., Proc. Roy. Soc. London, 1952, 215, 43.
12. Mackay, A.L., Acta Cryst., 1962, 15, 916.
13. Wenzer, A. and Steinemann, S., Helv. Phys. Acta, 1974, 47, 321.
14. Pauling, L., Phys. Rev., 1938, 54, 899.
15. Raynor, G.V., Phil. Mag., 1945, 36, 770.
16. Häussler, P. and Kay, E., Zeits. Phys. Chem. Neue Folge, 1988, 157, 377.
17. Bancel, P.A., Heiney, P.A., Stephens, P.W., Goldman, and Horn, P.M., Phys. Rev. Lett., 1985, 54, 2422.
18. Kimura, K., Hashimoto, T., Suzuki, K., Nagayama, K., Ino, H. and Takeuchi, S., J. Phys. Soc. Jpn., 1985, 54, 3217.
19. Inoue, A., Arnberg, L., Lehtinen, B., Oguchi, M. and Masumoto, T., Metall. Trans., 1986, 17A, 1657.
20. Bellisset, R., Bouree-Vigneron, F. and Sainfort, P., J. de Phys., 1986, 47, C-3.
21. Lawther, D.W., Lloyd, D.J. and Dunlap, R.A., Mater. Sci. Engin., 1990, A123, 33.
22. Liu, B.X., Cheng, G.A. and Shang, C.H., Phil. Mag. Lett., 1987, 55, 265.
23. McHenry, M.E., Srinivas, V., Bahadur, D., O'Handley, R.C., Lloyd, D.J. and Dunlap, R.A., Phys. Rev., 1989, B39, 3611.
24. Chen, H.S., Inoue, A., Scripta Metall., 1987, 21, 527.

25. Tsai, A.P., Inoue, A. and Masumoto, T., J. Mater. Sci. Lett., 1988, 7, 322.
26. Tasai, A.P., Inoue, A. and Masumoto, T., Mater. Trans. JIM, 1989, 30, 150.
27. Inoue, A., Tsai, A.P. and Masumoto, T., J. Non Cryst. Solids, 1990, 117/118, 824.
28. Inoue, A., Tsai, A.P., Kimura, H.M., Bizen, Y. and Masumoto, T., J. Mater. Sci. Lett., 1987, 6, 771.
29. Shen, Y., Poon, S.J., Dmowski, W., Egami, T. and Shiflet, G.J., Phys. Rev. Lett., 1987, 58, 1440.
30. Chen, H.S., Phillips, J.C., Villars, P., Kortan, A.R., Inoue, A., Phys. Rev., 1987, B35, 9326.
31. Sakurai, Y., Kokubu, C., Tanaka, Y., Watanabe, Y., Masuda, M. and Nanao, S., Mater. Sci. Engin., 1988, 99, 423.
32. Inoue, A., Nakano, K., Masumoto, T. and Chen, H.S., Mater. Trans. JIM, 1989, 30, 200.
33. Ohashi, W., Ph. D. Thesis, Univ. of harvard, 1989, unpublished.
34. Sainfort, P. and Dubost, B., J. de Phys., 1986, 47, C-3.
35. Tsai, A.P., Inoue, A., Yokohama, Y. and Masumoto, T., Mater. Trans. JIM, 1990, 31, 98.
36. Werkman, R.D., Schurer, P.J., Vincze, I. and van der Woude, F. Hyperf. Inter., 1989, 45, 409.
37. Traverse, A.L., Dumoulin, L., Beling, E. and Senemaud, C., in Quasicrystalline Materials, eds. Ch. Janot and Dubois, J.B., World Sci., Singapore, 1988, p. 399.
38. Kimura, K., Iwahashi, H., Hashimoto, T., Takeuchi, S., Mizutani, U., Ohashi, S. and Itoh, G., J. Phys. Soc. Japan, 1989, 58, 2472.
39. Wang, K., Garoche, P. and Calvayrac, Y., in Quasicrystalline Materials, eds. Ch. Janot and Dubois, J.M., World Sci., Singapore, 1988, p. 372.
40. Fujiwara, T., Phys. Rev., 1989, B40, 942.

QUASICRYSTALS AND THE CONCEPT OF INTERPENETRATION IN m$\bar{3}\bar{5}$ - APPROXIMANT CRYSTALS WITH LONG - RANGE ICOSAHEDRAL ATOMIC CLUSTERING

JEAN - LOUIS VERGER - GAUGRY
LTPCM / INPG (CNRS UA 29), BP 75, Domaine Universitaire,
38402 - Saint Martin d'Hères Cédex, France.

ABSTRACT

Interpenetrating Pseudo - Icosahedral atomic Clusters (IPIC) are found in various m$\bar{3}\bar{5}$ - approximant crystals such as α - AlMnSi, R - AlLiCu, presenting one orientational family of IPIC, c - Ti_2Ni, β - AlMnSi, μ - Al_4Mn, exhibiting several ones, and in many models of icosahedral or decagonal phases. We have developped a theory of G - approximant crystals based on IPIC, where G is a non - crystallographic point group (G = m$\bar{3}\bar{5}$, C_n, D_{2n}, with n = 5, > 6) allowing to describe what happens in reciprocal space for the intense Bragg peak positions and their intensities, which is easily generalizable to quasicrystals and provides a continuity of the perception of the symmetry group G in reciprocal space when small atomic rearrangements occur in real space during the phase transition quasicrystal <-> approximant crystal. This theory is new and completes the edifice of the theory of crystallographic groups by considering non - crystallographic point groups.

INTRODUCTION

The existence of the quasicrystalline state of matter (quasicrystals were discovered in 1984 by Shechtmann et al., [1]) has been favouring a lot of studies about forbidden symmetries in condensed

matter physics, mathematics, crystallography, physical metallurgy, ...
for a few years, and therefore a lot of cross - fertilization between
these domains. The controversy about quasicrystals is entertained by
Pauling [2] who is claiming that such objects do not exist and that
their diffraction properties can be explained by means of suitable 3D
approximant cubic crystals, constructed by packings (with tangential
connections and twinning) of similarly oriented icosahedral clusters,
rejecting all the initial work made by Duneau and Katz [3] and Gratias
[4] on the (6D) hyperspace approach, though both attitudes are not
fundamentally opposite. Our attitude does not consist in finding a
compromise between both approaches but in showing that the
concept of IPIC model gives a general framework, covering
simultaneously the constructions of approximant cubic crystals of
Pauling, the already existing approximant crystals and the N -
dimensional crystals used in the cut method which lie above
quasicrystals in a N dimensional Euclidean space (N is equal to 6 very
often).

In order to justify the present theory and to fill a certain
absence of theoretical support for handling G - approximant crystals,
with G a non - crystallographic point group, let us start by
considering two identically oriented distorted atomic G - clusters in
matter, as in Fig. 1.a (matter is not assumed periodically crystallized
and the G - invariance of the atomic positions is almost respected in
the clusters). By the principle of superposition, the response to an

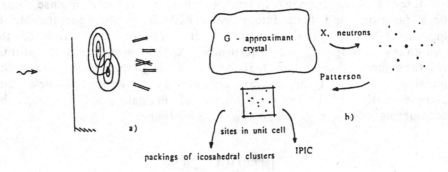

Figure 1. a) Scattering process from partially interpenetrating
clusters; b) Structural determination of an approximant crystal with
final reasearch of the spatial extent of IPIC and site multiplicities.

incident plane wave by the atomic clusters will be the addition of both contributions, whatever the way clusters interpenetrate, provided that the atoms in the intersected region are counted with their multiplicities and that a phase due to the relative positioning of the clusters centers is introduced in the structure factor. A consequence of this is the following, when multiple interpenetration occurs: we have schematically represented in Fig. 1.b the structural determination scheme of a G - approximant (single)crystal by X - ray or neutron scattering experiments. A Patterson analysis allows the characterization of the atom positions inside the unit cell. This knowledge of the unit cell is sufficient for investigating local order with respect to permitted point symmetry groups, by periodicity; however, this is not the case for non crystallographic point symmetries since we have to recover the real spatial extent of the G - clusters present in the structure, by hand (or on a computer); this last step often leads to IPIC and provides a correct conception of the approximant crystal which is necessary to know the real object responsible for the G - invariance in the distribution of the intense peaks in reciprocal space.

The notion of interpenetration is not classical in crystallography since the easiest way to build a crystal consists in reproducing a motif at each element of a lattice. Here, our attitude aims at investigating the consequences of complex overlappings of motives when some parts of them can be coherently defined (propagation of icosahedral symmetry for instance) over many motives around the central one, and therefore everywhere by periodicity. In the present scheme, an approximant crystal is not only a "crystallographically defined" crystal but more. We shall propose precise definitions in the following [5 - 8].

Since IPIC occurs in approximant crystals, it is reasonable to conceive that it should also happen in quasicrystals, but for families of clusters centers now distributed on Z - modules of rank (say m) strictly greater than 3, while in approximant crystals, the rank is equal to the dimension of space , = 3. It is a fact that all the existing models and constructions of quasicrystals and approximant crystals take into account packings of icosahedral clusters instead of interpenetration of icosahedral clusters since it is easier to conceive and to treat on a computer than when clusters overlapp [2, 9] (see the icosahedral glass model for instance, [10]).

CASE OF α - AlMnSi

We have analyzed icosahedral local order in this well - known cubic phase, in real space and in reciprocal space [8]. Its composition (with %at (Al+Si)/%at Mn ≈ 5.4) is close to that of the i - phase AlMnSi and its structure is given by $Pm\bar{3}$, a = 1.268 nm with the Wyckoff positions occupied as in Table 1 [11]. Its density is 3.55 g cm^{-3} and a correct formula is $Al_9Mn_2Si_{1.8}$. Two empty sites, at (0, 0, 0) and (1/2, 1/2, 1/2), are centers of pseudo - icosahedral clusters for which the first three layers constitute a Double Mackay Icosahedra (DMI). The structure is almost body - centered, up to small deformations. This structure was used to model the i - AlMnSi phase in 1985 by several researchers [9].

The succession of icosahedral layers around (0, 0, 0) is reported in Table 2, up to a radius equal to the lattice parameter 1.28 nm, but icosahedral polyhedra can be defined beyond this value. There is a common orientation to all these polyhedra which is, with respect to the cubic axes, for pseudo - 5 fold axis : (1, 0, τ), (-1, 0, -τ), (-1, 0, τ), (1, 0, -τ), (0, τ, 1), (0, -τ, -1), (τ, 1, 0), (-τ, -1, 0), (τ, -1, 0), (-τ, 1, 0), (0, -τ, 1), (0, τ, -1), with τ = (1+√5)/2. We observe an integral arithmetic sequence of 17 homothetic icosahedra, identically oriented

Atoms		Positions	x	y	z
12	Mn(1)	12(j)	0.	0.3271	0.2006
12	Mn(2)	12(k)	0.5	0.1797	0.3085
6	Al(1)	6(e)	0.3638	0.	0.
6	Al(2)	6(h)	0.1216	0.5	0.5
6	Al(3)	6(f)	0.2897	0.	0.5
12	Al(4)	12(j)	0.	0.1636	0.0997
12	Al(5)	12(k)	0.5	0.3342	0.3990
12	Al(6)	12(j)	0.	0.3319	0.4037
12	Al(7)	12(k)	0.5	0.1205	0.1175
24	Al(8)	24(l)	0.1185	0.1892	0.2980
24	Al(9)	24(l)	0.3897	0.3127	0.1955

Table 1. Crystallographic orbits and Wyckoff positions for α - AlMnSi.

Type (n°)	Atoms	Distance (nm)	Mean radius (nm)	Occupation rate
ico(1)	12 Al(4)	0.24293	0.24293	1.
icosi(1)	6 Al(1)	0.46130	0.46997	1.
	24 Al(8)	0.47214		
ico(2)	12 Mn(1)	0.48655	0.48655	1.
hexe(1)	12 Al(6)	0.66268		
	24 Al(7)	0.66895	0.67224	1.
	24 Al(9)	0.68032		
ico(3)	12 Al(3)	0.73273	0.73273	1.
icosi(2)	24 Mn(2)	0.77904	0.78457	1.
	6 Al(1)	0.80670		
penta(1)	12 Mn(1)	0.89034		
	24 Al(9)	0.90417	0.90364	1.
	24 Al(2)	0.90977		
trisico(1)	24 Al(5)	0.91515	0.91515	0.4
hexe(2)	12 Al(6)	0.86534	0.91115	0.6
	24 Al(8)	0.93406		
ico(4)	12 Al(6)	0.98980	0.98980	1.
penta(2)	24 Al(9)	1.03205	1.04405	0.6
	12 Al(4)	1.06806		
hexe(3)	24 Al(5)	1.07809	1.08587	0.6
	12 Al(3)	1.10143		
dodeca(1)	12 Mn(2)	1.09521	1.09637	0.6
	8 vs 1/2	1.09812		
hexe(4)	24 Al(8)	1.10559		
	24 Mn(2)	1.10575	1.11164	1.
	12 Al(6)	1.13550		
trisico(2)	12 Al(4)	1.16028	1.16726	0.6
	24 Al(5)	1.17075		
penta(3)	24 Al(9)	1.19156	1.19620	0.8
	24 Al(9)	1.20084		
hexe(5)	24 Al(8)	1.20403	1.20403	0.4
icosi(4)	6 vs 0	1.26800	1.27712	0.8
	24 Mn(2)	1.27940		

Table 2. Sequences of similarly oriented concentric icosahedral polyhedra around the empty site (0, 0, 0). (ico = icosahedron, dodeca = dodecahedron, icosi = icosidodecahedron, hexe = hexecontahedron, trisico = trisicosahedron, penta = pentakisdodecahedron, vs = empty site).

(the nth icosahedron is almost n times larger than the first one). A succession of 10 concentric icosidodecahedra centered at (0, 0, 0), forming a geometric sequence (the nth one is τ^n larger than the first one), is also put into evidence. Some layers are not complete, with occupation rates equal to 0.4 to 0.8, and some "occupied" by the empty sites of the structure (centers of clusters). By truncating the icosahedral cluster centered at (0, 0, 0) at a radius = 0.68 nm, and the other one centered at (1/2, 1/2, 1/2) at 0.47 nm, we obtain a packing of icosahedral clusters, with tangential connections, which cover entirely all the atomic sites [Fig. 2]. In this sense, we join Pauling's constructions with only two types of clusters. However, it is clear that these clusters which were arbitrarily truncated at definite radii, interpenetrate actually over larger distances (several nm, see [8]). Atomic sites are then counted several times according to the centers from which their are counted. Such clusters possess thousands of atoms, much more than the number of atoms in the unit cell.

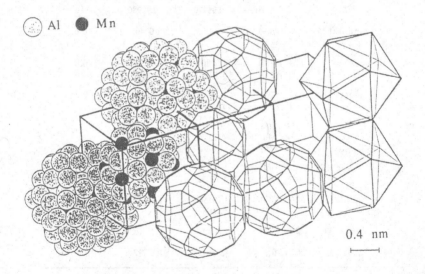

Figure 2. Packing of truncated icosahedral clusters (with tangential connections) for the reconstitution of the whole α - AlMnSi phase structure (CsCl type structure) without "glue" atoms (glue atoms are sometimes added at random in quasicrystalline models between clusters to fit the actual density; here, we use the same terminology).

INTERPENETRATION IN m$\bar{3}\bar{5}$ - APPROXIMANT CRYSTALS

We have developed [6] a logicial on the computer, on line with the (EMS)Electron Microscopy System logicial of P. Stadelmann [12], by which it is possible to obtain, from any m$\bar{3}\bar{5}$ - approximant crystal, the positioning of perfect icosahedral clusters fitting optimally the deformed ones observed in the m$\bar{3}\bar{5}$ - approximant crystal whatever their spatial extent (namely, deformations of clusters (in nm), angular deviations of clusters (in degrees), exact icosahedral positions q^{ex} in reciprocal space and 2D representations of the scattering function $S(g)^2$ of perfect icosahedral clusters,...). This program has required a lot of numerical analysis [13]. It is used to find the icosahedral polyhedra which form the successive layers in clusters.

Unique Orientational Family
It is the case of the α - AlMnSi phase and also of the R - Al$_5$LiCu$_3$ phase [9, 14, 15] for which only one orientational family of pseudo - icosahedral clusters can be defined coherently.

Multiple Orientational Family
There exist many crystals in which CN12 sites can be found, that is which contain more or less regular icosahedra. However, the number of crystals for which this icosahedral order spreads over some distance or where it is possible to define globally only one action of the group of the icosahedron is reduced. It is more usual to find in approximant crystals several types of pseudo - icosahedral clusters, oriented in non - equivalent ways, which overlapp partially. It is the case of c - Ti$_2$Ni [7], β - AlMnSi or μ - Al$_4$Mn [8], which, curiously, are not considered as "true" approximant crystals in literature. We have already refused this terminology "not true" for such phases presenting several non - equivalent orientational families of icosahedral clusters. For instance, TEM experiments have shown, in the TiNiV system [14, 16], that the intense peaks in the reciprocal space of c - Ti$_2$Ni were almost following the icosahedral symmetry and the phase transition i/c is accompanied by (continuous) alignments of the spots according to the icosahedral symmetry. We have also shown that such orientation relationships between i and c

are those of local order, and exactly those given by the pseudo - icosahedral clusters centered at the Ni(e) atoms in c - Ti_2Ni, whereas Ti(c) - pseudo - icosahedral clusters do not participate in the phase transition "visualized by diffraction".

In approximant crystals, in general, there are several ways to define the action of the (abstract) icosahedral group $m\bar{3}5$, in a Z - reduced basis of the structure, in fact as many times as there exist ($m\bar{3}5$ -) non - equivalent pseudo - icosahedral clusters (they are not necessarily equivalent under the action of the space group of the crystal). We have :

c - Ti_2Ni : 8 orientational families of pseudo - icosahedral clusters, 4 around Ni(e) atoms, 4 around Ti(c) atoms; only those around Ni(e) atoms give rise to diffracting contributions in reciprocal space which do not destroy [7, 14]. Concentric pseudo - icosahedral polyhedra of intense peaks appear in reciprocal space in the reciprocal lattice, by duality.

β - AlMnSi : the structure is hexagonal, space group = P63/mmc, a = 0.75 nm, c = 0.7722 nm, 26 atoms per unit cell [8]; 2 orientational families in real space and 2 in reciprocal space are visible. Cluster centers are Si atoms at (0, 0, 0) and (0, 0, 1/2). The first icosahedra are formed by 6 Al and 6 Mn. Si is substituted at random to Al atoms. Icosahedral local order does not propagate beyond third neighbours and this phase is considered as a rather bad approximant of i - AlMnSi.

μ - Al_4Mn : the structure is hexagonal, space group = P63/mmc, a = 1.998 nm, c = 2.4673 nm, 563 atoms per unit cell in average [8]; 5 orientational families of pseudo - icosahedral clusters can be defined in real space whereas only 2 of them appear in reciprocal space. Three are with three twofold axis respectively parallel to the a, b, c axis, the two remaining ones having a threefold axis parallel to the c axis and two twofold axis respectively parallel to the a and b axis. These two last families of IPIC are those which contribute to the pseudo - icosahedral character in reciprocal space. Three extinctions occur in this case.

After investigating a lot of approximant crystals, we believe that α- or R- are really exceptions in the familly of approximant crystals, and that it occurs much more frequently several orientational families than only one, with extinctions in reciprocal space of the diffracting contributions arising from some orientational families and IPIC for each orientational family.

THEORY OF G - APPROXIMANT CRYSTALS

Let us start by some preliminaries. We define [5] a $m\bar{3}\bar{5}$ - cluster by a finite set of points in $\mathbf{R}^3$ which is $m\bar{3}\bar{5}$- invariant around a center, center which belongs to this set or not. Denote it by X(n) if n is the number of $m\bar{3}\bar{5}$ - invariant layers in the cluster. When $m\bar{3}\bar{5}$ acts transitively on one layer, the number of sites on this layer is 12 (icosahedron), 20 (dodecahedron), 30 (icosidodecahedron), 60 (3 possibilities: hexecontahedron, pentakisdodecahedron, tris- icosahedron) or 120 (general case) (Table 2). The color of a site is the type of atom placed at one of the points of the cluster and we consider an atomic site as a couple (point, color) as usual. Let us denote by $\delta(X(n))$ a slight deformation of the colored cluster. A $m\bar{3}\bar{5}$ - approximant crystal is merely a state of (periodically) crystallized matter for which a (3D) crystallographic compatibility between such deformed colored $m\bar{3}\bar{5}$ - clusters exists.

Now, consider a crystal Ξ, for which T, F, R are respectively its translation lattice, space group and point group and such that the coloration of the crystallographic orbits inside some Wyckoff positions is known. We say, in a general way, that the crystal Ξ is $m\bar{3}\bar{5}$ - approximant if the following local conditions are satisfied : (i) in the unit cell of T, there is a finite number of centers of clusters, called CC = $\{s_1, s_2, ..., s_t\}$, (ii) CC+T is invariant under the action of F, (iii) (orientational compatibility) there exists a common $N \geqslant 1$, such that $\cap_{i=1 \text{ to } t}$ [$X_i(n_i) - s_i$] is a non empty $m\bar{3}\bar{5}$ - cluster X(N), (iv) each atom in the crystal belongs at least to one deformed colored cluster $\delta_i(X_i(n_i))$, centered at s_i, for a certain i. For instance, it is the case of the $m\bar{3}\bar{5}$ - approximant crystals α - AlMnSi, with clusters around the (0, 0, 0) and (1/2, 1/2, 1/2) sites (here t = 2), β - AlMnSi, R - AlLiCu, c - Ti_2Ni (here t = 4 for one orientation), and so on.

In this framework it is possible to show that a first - order development of the structure factor of the crystal Ξ expressed by an "incident wave - m $\overline{3}5$ - cluster" interaction term can be made. For doing this, we define and study the following scattering function :

for any q $\in$ T* (reciprocal lattice of T),

$$H(q, \gamma_1, \gamma_2, \gamma_3) = \Sigma_k \Sigma_j \ (K + \gamma_1 z_j) \ (\Lambda + \gamma_2 \lambda_j)^{-1} \ \exp[\ 2i\pi \ q \cdot (x_j + \gamma_3[\ \delta_k(x_j)$$
$$- x_j \ ]) \ ] \tag{1}$$

with $\gamma_1, \gamma_2, \gamma_3$ in the interval [0 ; 1], where the first sum is taken over the t m $\overline{3}5$ - clusters, where j runs over all the atomic sites of the k-th distorted m $\overline{3}5$ - cluster $\delta_k(X_k(n_k))$; where K is the average atomic scattering factor (average over the atomic species in the crystal), z_j the atomic scattering factor excess for the jth site, Λ is the average site multiplicity (average over all the atomic sites in a unit cell of the Bravais lattice), λ_j the site multiplicity excess for the jth site, q^{ex} is the exact icosahedral position associated with q in reciprocal space, $S(q^{ex})$ is the structure factor at q^{ex} of the perfect icosahedral unit (optimally positioned m $\overline{3}5$ - invariant cluster endowed with an average atom) which is the perfect icosahedral unit the closest to the distribution of slightly deformed m $\overline{3}5$ - atomic clusters (in the sense of the least squares method). We obtain, at the first order, the following m $\overline{3}5$ - invariant expression for the intensities, at a node q of the reciprocal lattice [5]:

$$I(q) = (\ K\Lambda^{-1}t\ S(q^{ex})\ )^2 \tag{2}$$

QUASICRYSTALS (IPIC MODEL) AND m - DIMENSIONAL APPROACH

For approximant crystals, the Z - module which is the support of all the clusters centers (CC) has a rank invariably equal to 3 = dim R^3. In quasicrystals, the atomic displacements which take place during the phase transition c/qc make such rank m greater than 3 and force us to consider quasicrystals as arising from crystals in higher dimensional Euclidean spaces. Some values of m are incompatible with a periodization in $\mathbf{R}^m$.

The above formalism is used by taking an aperiodic (infinite) distribution of IPIC, instead of a periodic one, and developing (1) with infinite sums at the first order [6, 17]. It is the analogue of the Fig. 1.b scheme for quasicrystals. In this context, $m\bar{3}\bar{5}$ - atomic clusters are slightly deformed with respect to the perfect icosahedral symmetry, but deformations are different from those existing in the crystal; they interpenetrate strongly, but are arranged aperiodically. This leads to the following expression of the intensities, at the first order :

$$I(q^{ex}) = (K\Lambda^{-1} \ T \ S(q^{ex}))^2 \qquad (3)$$

where T is now a "transparency" function associated with the characteristics of the aperiodic distribution of the centers of clusters CC. T cannot be defined here due to the lack of space and we report to [17] for its construction and properties; however, we can say that the existence of such a function arises from the fact that the intensities should converge at all points in reciprocal space. T arises from an integrability condition of a function F defined on $\mathbf{R}^{m-3}$ linked to the convergence of the intensities I(q), where q runs over reciprocal space. We analyze this function F, "complementary to T", in Fourier series to investigate in which extent it provides a band algorithm on the hyperspace. Extensive formal developments are in [17].

Having changed our look at crystals by considering IPIC instead of only classical crystallography leads us to very similar expressions in the diffraction process, that is (2) and (3), where the modulation term $S(q^{ex})$ remains the prominent term. This term allows us to envisage pseudo - inflation and inflation rules rather independently of the $m\bar{3}\bar{5}$ - invariance and its study gives the key of what happens in reciprocal space, on the reciprocal lattice of the $m\bar{3}\bar{5}$ - aproximant crystal and everywhere in reciprocal space for the aperiodic crystal.

Dubois et al's work (in MOSPOQ'91 for instance [20]) shows that some structural asymmetry of peaks occurs for some quasicrystals, that is anisotropic diffuse scattering in reciprocal space. Up to him, these effects are not coming from the mosaicity of the samples, where mosaicity means collections of microdomains of 3D crystalline approximant phase, but from the structure itself of the quasicrystal. These structural transformations in the qc phase evolute with temperature, and it is difficult to say when "a defected qc phase" appears and then disappears when these features appear and disappear. Actually, no definition of "defected qc phase" is now rigorous, and the transition point between both states can not be defined accurately. This experimental work shows, on the contrary, that there exists a continuous evolution between splitted Bragg peaks and non splitted Bragg peaks with an asymmetry of peaks resembling an apparent splitting.

On the other hand, formula (3) in the text emphasizes the rôle played by the average cluster structure factor S(qex) and we have made numerical simulations on various sizes of icosahedral clusters to understand better how intrinsic is this peak asymmetry to the spatial extent of icosahedral clusters and to the "icosahedralness". Therefore, it is believed that some atomic reorganization of matter, induced by heating, modify the size of the icosahedral clusters as far as a coherent icosahedral scattering process is concerned. We discuss in [17] this aspect in more general terms, considering T and S(qex) independantly and then grouped into the same expression.

In this framework, via the decomposition of the term $S(q^{ex})$ into subsums relative to subsequences in clusters, this *"apparent" splitting* of intense peaks observed for quasicrystals [20] has two origins: either it is a real splitting coming from diffracting subsequences of concentric icosahedral polyhedra in real space which give contributions at almost the same scattering vector with, for each of these subsequences, an important spatial extent (peaks are narrow), or it is just a spreading of the intensities of the intense peaks arising from a decrease in the spatial extent of the clusters in the IPIC. *Pseudo - inflation rules* are related to the spatial extent of the IPIC and transform into real inflation rules with the golden mean or with algebraic functions of the golden mean [8] when the icosahedral clustering extent becomes larger and larger.

Interpenetration, as IPIC, occurs naturally in some recent models (Burkov [18], or Torres et al.[19]).

REFERENCES

1. Shechtmann, D., Blech, I., Gratias, D. and Cahn, J., Phys. Rev. Lett., 1984, **53**, pp. 1951.
2. Pauling, L. in Extended Icosahedral Structures, Vol. **3**, eds. Jaric, M.V. and Gratias, D., Academic Press, London,1989, pp. 137.
3. Duneau, M. and Katz, A., Phys. Rev. Lett., 1985, **54**, pp. 2688; Katz, A. and Duneau, M., J. Physique France, 1986, **47**, pp. 181.
4. Gratias, D., in From Crystalline to Amorphous, ed. Godrèche, C., 1988, pp. 83; Duneau, M., in From Crystalline to Amorphous, ed. Godrèche, C., 1988, pp. 157.
5. Verger - Gaugry, J.-L., J. Phys. I, 1991, **1**, in press.
6. Verger - Gaugry, J.-L., Tamura, N. and Barbier, J.-N., in Methods of Structural Analysis of Modulated Structures and Quasicrystals, eds. Perez - Mato, J.M., Zuniga, F.J., Madariaga, G. and Lopez - Echarri, A., World Scientific, Singapore, 1991, in press.
7. Tamura, N., Francois, A., Loiseau, A. and Verger - Gaugry, J.-L., submitted to Phil. Mag. A, 1991.
8. Tamura, N. and Verger - Gaugry, J.L., submitted to Phil. Mag. B, 1991.
9. Audier, M. and Guyot, P., in Extended Icosahedral Structures, Vol. **3**, eds. Jaric, M.V. and Gratias, D., Academic Press, London, 1989, pp. 1.
10. Stephens, P.W., in Extended Icosahedral Structures, Vol. **3**, eds. Jaric , M.V. and Gratias, D., Academic Press, London, 1989, pp. 37.
11. Cooper, M. and Robinson, K., Acta Cryst., 1966, **20**, pp. 614.
12. Stadelmann, P.A., Ultramicroscopy , 1987, **21**, pp. 131.
13. Verger - Gaugry, J.-L., Tamura, N. and Barbier, J.-N., II., to be submitted to J. Phys. I, 1991.
14. Verger - Gaugry, J.-L. and Tamura, N., Phase Transitions, 1991, **32**, pp. 89 - 101.
15. Cherkashin, E.E., Kripyakevich, P.I. and Oleksiv, G.I., Soviet Physics - Crystallography, 1964, **8**, 681.
16. Zhang, Z., Ye, H.Q. and Kuo, K.H., Phil. Mag. Lett. B, 1985, **52**, pp. L 49; Zhang, Z. and Kuo, K.H., Phil. Mag. Lett. B, 1986, **54**, pp. L 83.
17. Verger - Gaugry, J.-L., III., to be submitted to J. Phys. I ,1991.
18. Burkov, S.E. , in Methods of Structural Analysis of Modulated Structures and Quasicrystals, eds. Perez - Mato, J.M., Zuniga, F.J., Madariaga, G. and Lopez - Echarri, A., World Scientific, Singapore, 1991, in press.
19. Aragon, J.L., Romeu, D. and Gomez, A., Phys. Rev. B, 1991, in press; Romeu, D., Acta Metall. Mater. 1990, 38, pp. 113.
20. Dubois, J.-M., MOSPOQ'91, Phase Transitions, 1991, in press.

APPLICATION OF MAGIC STRAINS TO PREDICT NEW ORDERED PHASES: A FIVE-FOLD COORDINATED CRYSTAL STRUCTURE FOR SILICON

L. L. BOYER, EFTHIMIOS KAXIRAS, M. J. MEHL, J. L. FELDMAN AND J. Q. BROUGHTON
Complex Systems Theory Branch
Naval Research Laboratory
Washington, D.C. 20375-5000 U.S.A.

ABSTRACT

The concept of magic strain is reviewed and illustrated using a simple two dimensional lattice. Practical use of the concept for making theoretical predictions of new ordered structures is demonstrated; first, using the two dimensional example, and second, using realistic calculations which predict that cubic silicon can be transformed to a stable body-centered tetragonal structure with five-fold coordination.

INTRODUCTION

A magic-strain tensor is defined to be any symmetric matrix which transforms a Bravais lattice into its identity.[1,2] Because they are symmetric tensors they have no rotational component, and thus, in principle, are experimentally accessible. They also have been referred to as physically-allowed lattice-invariant (PALI) strains.[3] In this paper we will keep the original name, because we believe it is appropriate: magic strains are unquestionably difficult to visualize, but they are easily understood once the "trick" to their derivation is revealed. Since the concept of magic strains is quite new, their possible relevance to real physical processes is just beginning to be explored.

The minimum energy path between the unstrained and magically strained lattice (magic-strain path) defines a symmetric "double well" in strain space. As we shall see, there are many equivalent magic-strain paths for a given type of magic strain, and many different types as well. Thus, the complete energy surface is a complex multi-welled surface. In the systems studied thus far, magic-strain paths have barriers that are on the order of the thermal energy available at the melting temperature.

The qualitative comparison between melting temperatures and magic-strain barriers, plus an analogy with other first-order structural transitions of known double-well origin, lead to the speculation of a possible fundamental connection between the solid-liquid transition and magic strains. In any case, models used to simulate properties of liquids could benefit by assuring accurate results for structures along magic-strain paths. The same would evidently apply for models used to simulate other materials properties involving large shear strain. In this regard, we note that Clapp and Rifkin[5] identified a magic-strain path based on results of simulations of a martensitic transformation.

MAGIC STRAINS

General

The derivation of a magic-strain transformation proceeds as follows[2]: Let A be a matrix composed of primitive lattice vectors, and let B be a matrix composed of a different set of primitive lattice vectors. The transformation $S = B A^{-1}$ gives the lattice strain which takes A to B; but S is generally not symmetric. S can be symmetrized by multiplying it by its transpose (S^T) and taking the square root of the product -- but since the latter can only be done for diagonal matrices, we must first diagonalize $S^T S$. Let U be the matrix of eigenvectors. Taking the square root of the diagonal matrix, $U^T S^T S U$, and transforming back to the original coordinates we get the magic-strain transformation, $S_m = U[U^T S^T S U]^{1/2} U^T$. The transformed lattice $S_m A$ is the same as $B = S A$, apart from a rotation (R); thus, $R S A = S_m A$, or $R = S_m S^{-1}$.

Square Lattice

A simple two dimensional example serves to illustrate the magic-strain formulation. Let A be formed by the usual primitive vectors of a square lattice, (1,0) and (0,1), and let (1,0) and (1,1) be the primitive vectors forming B. In this case A is the identity matrix, so $S = B$. We have

$$S^T S = \begin{pmatrix} 1 & 1 \\ 1 & 2 \end{pmatrix} \quad \text{and} \quad S_m = 5^{-1/2} \begin{pmatrix} 2 & 1 \\ 1 & 3 \end{pmatrix}$$

The square root of the eigenvalues of $S^T S$, $[(3 \pm 5^{1/2})/2]^{1/2}$, are the eigenvalues of S_m. They give the fractional change in distance produced by

S_m along the eigenvector directions, $(\alpha, \beta) \sim (0.526, 0.851)$ (+ value) and $(\beta, -\alpha) \sim (0.851, -0.526)$ (- value), where $\alpha = [(5 + 5^{1/2})/2]^{-1/2}$ and $\beta = [(5 - 5^{1/2})/2]^{-1/2}$. The transformation is illustrated in fig. 1: where a) shows the unstrained lattice with faces cut perpendicular to the eigenvector directions; b) shows the lattice half way through the magic transformation, i.e. $(I + S_m)/2$; At this point the lattice is approximately hexagonal; and c) shows the magically strained lattice. The matrix $R = S_m S^{-1}$ describes a rotation of $\sin^{-1}(5^{-1/2}) \sim 27°$, which brings the lattice in fig 1a into coincidence with that in 1c.

If the lattice has a basis, then at least two modes of transformation are possible, depending upon the energy as a function of basis translation. 1) Basis vector translation may accompany the transformation to preserve the crystal structure, in which case, interesting structures can arise along the magic-strain path. This mode is illustrated below with results for silicon. 2) It may not be energetically favorable for basis translation to occur in a way which produces an equivalent crystal structure. This would likely be the case for intermetallic compounds that alloy. For example, if two types of atoms occupy the square lattice, as in fig. 2a, then the transformation pictured in fig. 1 produces a new ordered structure with a rectangular lattice (fig. 2c).

The fcc Lattice

We restrict the choice of primitive unit cells (A and B matrices) for the face-centered cubic (fcc) lattice by confining the primitive lattice vectors to the nearest-neighbor shell. With this restriction there are eight different types of magic strain tensors. Representative transformations for each type are listed in table 1 in the order of increasing magnitude of distortion.

The transformations are described by the eigenvalues (λ_i) and eigenvectors (e_i) of $S^T S$, with the square root of the eigenvalues giving the fractional change in distance along the directions e_i. The maximum expansion (compression) directions are given by the e_1 (e_2) eigenvectors. A measure of the distortion is given by the ratio of the surface area of a magically strained crystal to that of an unstrained cube-shaped crystal with axes along the directions e_i, given by $A_m/A_o = \left[(\lambda_1\lambda_2)^{1/2} + (\lambda_1\lambda_3)^{1/2} + (\lambda_2\lambda_3)^{1/2}\right]/3$. N is the number of equivalent transformations for a given type. For example, in the first type, the maximum compression direction (e_2) can be along $(1,0,0)$, $(0,1,0)$ or $(0,0,1)$ crystallographic axes. Given one of these directions, say $(0,0,1)$, the maximum expansion direction can be either of two

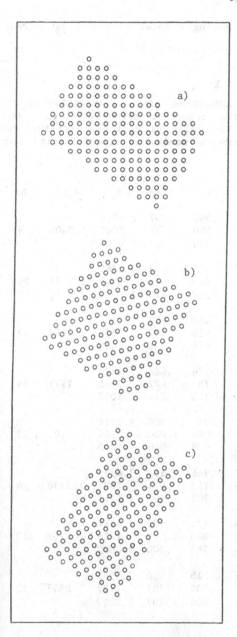

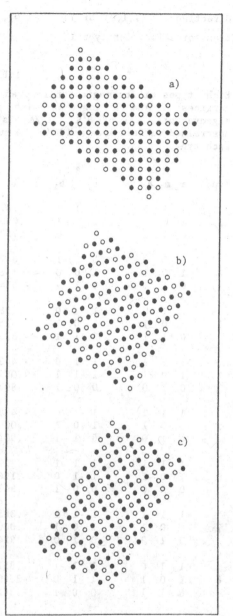

Figure 1. a) Square lattice with faces cut perpendicular to the magic-strain eigenvectors; b) the lattice transformed by half the magic strain $(S_m+I)/2$; c) the lattice transformed by S_m.

Figure 2. Same illustrations as in Figure 1, but with alternately shaded circles, to illustrate the exchange of neighbors leading to a new ordering.

directions, (1,1,0) or (1, -1,0), making a total of N = 6 equivalent transformations for type 1.

TABLE 1

Eight types of fcc magic-strain transformations, given by primitive vector matrices A and B together with the resultant eigenvalues (λ_i) and eigenvectors (e_i) of $S^t S$. A_m/A_o is the ratio of magically strained to unstrained surface area and N is the number of equivalent transformations for each type.

Type	A a_1 a_2 a_3	B b_1 b_2 b_3	e_1	e_2	e_3	λ_i	A_m/A_o	N
1	1 1 0	1 1 1	.707	.000	.707	2.0000		
	1 0 1	1 0 -1	.000	1.000	.000	.5000	1.0404	6
	0 1 1	0 1 0	-.707	.000	.707	1.0000		
2	1 1 0	1 1 -1	.849	.335	.408	3.1861		
	1 0 1	1 0 0	-.158	.899	-.408	.3139	1.1151	24
	0 1 1	0 1 1	-.504	.282	.816	1.0000		
3	1 1 0	1 1 0	.888	.460	.000	3.7321		
	1 0 1	1 0 -1	-.325	.628	.707	.2679	1.1498	12
	0 1 1	0 1 -1	-.325	.628	-.707	1.0000		
4	1 1 1	1 1 0	-.442	-.576	.688	5.5398		
	1 0 -1	1 -1 1	.089	.735	.672	.2565	1.1971	24
	0 1 0	0 0 1	.893	-.358	.274	.7037		
5	1 1 1	1 -1 0	-.433	.901	.000	6.7016		
	1 0 -1	1 0 1	.000	.000	1.000	.2984	1.2103	12
	0 1 0	0 1 -1	.901	.433	.000	.5000		
6	1 1 1	1 1 -1	-.277	.946	.167	3.8985		
	1 0 -1	1 -1 0	.196	-.115	.974	.1805	1.2330	24
	0 1 0	0 0 1	.941	.302	-.154	1.4210		
7	1 1 1	1 -1 1	-.383	.924	.000	5.8284		
	1 0 -1	1 0 -1	.000	.000	1.000	.1716	1.2761	12
	0 1 0	0 1 0	.924	.383	.000	1.0000		
8	1 1 0	1 1 -1	.857	.515	.000	3.3508		
	1 0 1	1 -1 0	-3.65	.606	.707	.1492	1.2807	12
	0 1 1	0 0 1	-.365	.606	-.707	2.0000		

RESULTS FOR SILICON

To facilitate parameterization of the magic-strain path, we write the diagonal elements of a **general** strain tensor, in the coordinates which diagonalize S_m, as $s_{11} = (1+f)^{-1/3}(1+g)^{2/3}$, $s_{22} = (1+g)^{-1/3}(1+f)^{2/3}$ and $s_{33} = (1+f)^{-1/3}(1+g)^{-1/3}$. For convenience, let f (g) correspond to the direction of maximum compression (expansion). Note that when f and g have the magic values f_0 and g_0 (-0.4398 and 0.7850 respectively for the type 2 strains in table 1), then s_{ii} are the eigenvalues of S_m. A minimum energy path between the unstrained and magic strained structures can be found by stepping f or g, whichever produces the smallest increase in energy, by some fraction of their magic values, while allowing the remaining structure parameters to relax with each step. The remaining structure parameters are the off diagonal elements of s, the volume (V) and basis vector translation (t).

We have used the Stillinger-Weber model[6] to calculate the energy along the type 2 magic-strain path. Results for the energy and structure along this path, with off-diagonal elements of s excluded, are shown in fig. 3. The small discontinuities in the strain parameters (fig. 3b) result from the finite step size, chosen to be 0.02. The components of the translation vector are with respect to the usual axes of the undistorted cubic structure. The minimum in energy at the midway point indicates the formation of a new stable crystal structure. Starting from the structure at this minimum, the energy is further lowered (to ~ 7 mRy) when the lattice is allowed to relax with respect to all possible strains, i.e. by including non-zero off-diagonal elements of s. The structure at the minimum, with a suitable coordinate transformation, is seen to be body centered tetragonal, space group I4mmm, with atoms occupying the e sites (Wyckoff notation)[7].

The lattice parameters for the new structure, based on the Stillinger-Weber model, are found to be a = 6.329 Bohr and c = 12.292 Bohr, and the internal structure parameter z = 0.1810. In this structure, each atom has four near neighbors at a distance of 4.786 Bohr, with the fifth at a distance of 4.450 Bohr. Each atom is at the approximate center of a pyramid with corners occupied by the five near neighbors.

We have carried out lattice dynamics calculations for this structure, again, using the Stillinger-Weber model, which show the structure is locally stable against small distortions of any type. More elaborate first-principles calculations[8] of elastic constants and optic modes show the structure is stable against long-wave distortions. (A short-wave instability would serve to lower the energy by distorting to increase the size of the unit cell. Presumably, five-fold coordination would be maintained by such a distortion.) Molecular dynamics calculations using the Stillinger-Weber model yield a melting temperature near that of cubic silicon.[8]

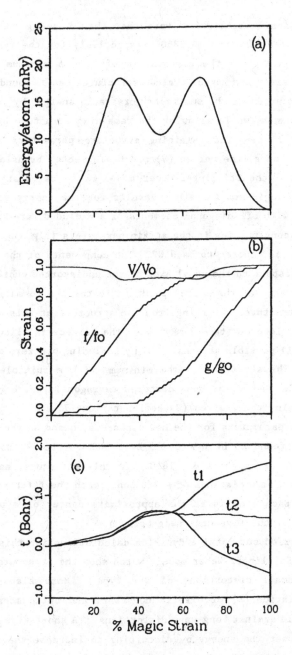

Figure 3.   a) Energy along the type 2 magic-strain path derived from the Stillinger-Weber model for silicon; b) Lattice strain parameters along the magic-strain path; c) Components of the basis vector translation, with respect to the unstrained cubic axes.

SUMMARY

The magic-strain concept has been presented in general and illustrated using a simple two dimensional lattice and realistic calculations for silicon. If the lattice has a basis, two modes of transformation are possible: 1) For systems with two types of atoms that are easily interchanged, e.g. ordered intermetallics that easily alloy, then new ordered structures can be generated by applying magic strains to the lattice which results by assuming the atoms are of one type. 2) Even if the basis shifts to preserve the crystal structure, interesting unforseen structures can arise along the magic strain path, e.g. the five-fold coordinated structure for silicon.

Our results demonstrate how the idea magic-strains can guide theoretical explorations for new ordered materials. Models used for calculating properties of disordered materials could benefit by assuring they do well for all ordered structures of similar (and lower) energy. A systematic application of the magic-strain concept can determine these competing ordered structures.

REFERENCES

1. Boyer, L.L., Magic strains in face-centered and body-centered cubic lattices. Acta. Cryst., 1989, A45, AFC29.

2. Van de Waal, B.W., A general procedure to derive magic tensors. Acta. Cryst., 1990, A46, FC17.

3. Mehl, M.J. and Boyer, L.L., Calculation of energy barriers for physically allowed lattice invariant strains in aluminum and iridium. Phys. Rev. B, 1981, 43, 9498.

4. Boyer, L.L., Kaxiras, E. and Mehl, M.J., Energy for magic strain transformations in crystals with fcc lattices. (Proceedings of the Materials Research Soc. 1990 Fall Meeting, Nov. 26 - Dec. 1, Boston Mass.).

5. Clapp, P.C. and Rifkin, J., Simulated martensitic transformations. Mat. Res. Soc. Symp. Proc., 1984, 21, 643.

6. Stillinger, F.H. and Weber, T.A., Computer simulation of local order in condensed phases of silicon. Phys. Rev. B, 1985, 31, 5262.

7. "International Tables for X-ray Crystallography", N. F. M. Henry and K. Lonsdale, eds., Kynoc Press, Birmingham England, 1954.

8. Boyer, L.L., Kaxiras E., Feldman, J.L., Broughton, J.Q. and Mehl, M.J., A new low energy crystal structure for silicon. Phys. Rev. Lett., Aug. 5, 1991, to be published.

INDEX OF CONTRIBUTORS

Printed in the United States
By Bookmasters